THE FRANCISCANA DOLPHIN

On the Edge of Survival

THE FRANCISCANA DOLPHIN
On the Edge of Survival

Edited by

PAULO CÉSAR SIMÕES-LOPES
Department of Ecology and Zoology, Federal University of Santa Catarina, Brazil

MARTA JUSSARA CREMER
Laboratory of Ecology and Conservation of Marine and Coastal Tetrapods, University of the Region of Joinville, Brazil

ELSEVIER

ACADEMIC PRESS
An imprint of Elsevier
elsevier.com/books-and-journals

Academic Press is an imprint of Elsevier
125 London Wall, London EC2Y 5AS, United Kingdom
525 B Street, Suite 1650, San Diego, CA 92101, United States
50 Hampshire Street, 5th Floor, Cambridge, MA 02139, United States
The Boulevard, Langford Lane, Kidlington, Oxford OX5 1GB, United Kingdom

ISBN: 978-0-323-90974-7

For Information on all Academic Press publications visit our website at https://www.elsevier.com/books-and-journals

Publisher: Nikki P. Levy
Acquisitions Editor: Anna Valutkevich
Editorial Project Manager: Emerald Li
Production Project Manager: Selvaraj Raviraj
Cover Designer: Vicky Esser Pearson

Typeset by Aptara, New Delhi, India

We dedicate this book to the memory of Pablo Bordino (April 18, 1963 to June 11, 2018†). He was tireless in the fight for the conservation of Franciscanas and marine life. Like few others, he filled the gap between science and public policy, conducting research and discussing alternatives with politicians, researchers, administrators, fisher, and the community to reduce impacts on marine species. Much of his work was devoted to researching the life history of the Franciscanas and developing strategies to mitigate franciscanas' bycatch. In addition, he worked in partnership with many authors of this book. His contribution is invaluable to everyone and will always be an inspiration.

We dedicate this book to the memory of Pablo Bordino (April 18, 1963 to June 11, 2018†). He was tireless in the fight for the conservation of Franciscanas and marine life. Like few others, he filled the gap between science and public policy, conducting research and discussing alternatives with politicians, researchers, administrators, fisher, and the community to reduce impacts on marine species. Much of his work was devoted to researching the life history of the Franciscanas and developing strategies to mitigate franciscanas' bycatch. In addition, he worked in partnership with many authors of this book. His contribution is invaluable to everyone and will always be an inspiration.

Contents

Contributors

Artur Andriolo
Laboratório de Ecologia Comportamental e Bioacústica, Universidade Federal de Juiz de Fora, Juiz de Fora, MG, Brazil

Carla Beatriz Barbosa
Instituto Argonauta para Conservação Costeira e Marinha, Ubatuba, São Paulo, Brasil

André Silva Barreto
Laboratório de Informática da Biodiversidade e Geomática, Universidade do Vale do Itajaí, Itajaí, SC, Brazil

Manuela Bassoi
Laboratório de Bioacústica (LaB), Centro de Biociências, Universidade Federal do Rio Grande do Norte-UFRN, Natal, RN, Brazil

Ana Paula Madeira Di Beneditto
Laboratório de Ciências Ambientais, Universidade Estadual do Norte Fluminense Darcy Ribeiro, Campos dos Goytacazes, Rio de Janeiro, Brazil

Leonardo G. Berninsone
Fundación AquaMarina, Centro de Estudios en Ciencias Marinas, Pinamar, Buenos Aires, Argentina

Carolina Pacheco Bertozzi
Instituto de Biociências (Câmpus do Litoral Paulista), Departamento de Ciências Biológicas e Ambientais, Universidade Estadual Paulista - UNESP, São Vicente, São Paulo, Brasil

Miguel Iñíguez Bessega
Fundación Cethus, Olivos, Buenos Aires, Argentina

Tatiana Lemos Bisi
Laboratório de Mamíferos Aquáticos e Bioindicadores, Faculdade de Oceanografia, Universidade do Estado do Rio de Janeiro, Rio de Janeiro, Brazil

Silvina Botta
Laboratório de Ecologia e Conservação da Megafauna Marinha, Instituto de Oceanografia, Universidade Federal do Rio Grande, Rio Grande, Rio Grande do Sul, Brazil

Ana Paula Frederico Rodrigues Loureiro Bracarense
Laboratório de Patologia Animal, Universidade Estadual de Londrina, Londrina, Paraná, Brazil

Matt K. Broadhurst
NSW Department of Primary Industries, Fisheries Conservation Technology Unit, National Marine Science Centre, Southern Cross University, Coffs Harbour, NSW, Australia

Camila Carvalho de Carvalho
Laboratório de Ecologia e Conservação da Megafauna Marinha, Universidade Federal do Rio Grande, Rio Grande, Rio Grande do Sul, Brazil

Pedro Volkmer de Castilho
Laboratório de Zoologia, Universidade do Estado de Santa Catarina (UDESC), Laguna, Santa Catarina, Brazil

Henrique Chupil
Instituto de Pesquisas Cananéia, Cananéia, São Paulo, Brazil

Rodrigo Claudino
Programa de Pós-graduação em Oceanografia Biológica, Instituto de Oceanografia, Universidade Federal do Rio Grande, Rio Grande, Rio Grande do Sul, Brazil

Adriana Castaldo Colosio
Instituto Baleia Jubarte, Caravelas, Bahia, Brazil

Marta Jussara Cremer
Laboratório de Ecologia e Conservação de Tetrápodes Marinhos e Costeiros, Universidade da Região de Joinville, São Francisco do Sul, Santa Catarina, Brazil

Enrique A. Crespo
Centro Nacional Patagonico, Consejo Nacional de Investigaciones Científicas y Técnicas, Puerto Madryn, Argentina

Haydée Andrade Cunha
Laboratório de Mamíferos Aquáticos e Bioindicadores, Faculdade de Oceanografia, Universidade do Estado do Rio de Janeiro, Rio de Janeiro, Brazil

Daniel Danilewicz
Grupo de Estudos de Mamíferos Aquáticos do Rio Grande do Sul, Torres, Rio Grande do Sul, Brazil

Pablo Denuncio
Instituto de Investigaciones Marinas y Costeras, Mar del Plata, Buenos Aires, Argentina

Liane Amaral Dias
Laboratório de Ecologia e Conservação da Megafauna Marinha, Instituto de Oceanografia, Universidade Federal do Rio Grande, Rio Grande, Rio Grande do Sul, Brazil

Isabela Guarnier Domiciano
Laboratório de Patologia Animal, Universidade Estadual de Londrina, Londrina, Paraná, Brazil

Camila Domit
Laboratório de Ecologia e Conservação, Universidade Federal do Paraná, Pontal do Sul, Paraná, Brazil

Sergio Estima
Núcleo de Educação e Monitoramento Ambiental, Rio Grande, Rio Grande do Sul, Brazil

Mauricio Failla
Fundación Cethus, Olivos, Buenos Aires, Argentina

Alejandro Fallabrino
Karumbé NGO, Montevideo, Uruguay

Ana Paula Cazerta Farro
Laboratório de Genética e Conservação Animal, Universidade Federal do Espírito Santo, São Mateus, Espírito Santo, Brazil

Sue Fisher
Animal Welfare Institute, Washington, DC, United States

Gabriel Fraga da Fonseca
Laboratório de Ecologia e Conservação, Centro de Estudos do Mar, Universidade Federal do Paraná, Pontal do Paraná, PR, Brazil

Rodrigo Forselledo
Laboratorio de Recursos Pelágicos, Dirección Nacional de Recursos Acuáticos, Constituyente, Montevideo, Uruguay

Isabel Gonçalves
KAOSA, Rio Grande, Rio Grande do Sul, Brazil

Carolina S. Gutstein
Red Paleontológica, Lab. Ontogenia y Filogenia, Dep. Biología, Fac. Ciencias, Universidad de Chile, Consultora Paleosuchus Ltda, Santiago, Chile

Danniele Firme Herbst
Institute of Environmental Science and Technology and Department of Prehistory, Universitat Autònoma de Barcelona, Bellaterra, Spain

Camila Márquez Iturriaga
Red Paleontológica, Lab. Ontogenia y Filogenia, Dep. Biología, Fac. Ciencias, Universidad de Chile, Consultora Paleosuchus Ltda, Santiago, Chile

Miguel Iñíguez Bessega
Fundación Cethus, Olivos, Buenos Aires, Argentina

Sebástian Jiménez
Laboratorio de Recursos Pelágicos, Dirección Nacional de Recursos Acuáticos, Constituyente, Montevideo, Uruguay

José Lailson-Brito Júnior
Laboratório de Mamíferos Aquáticos e Bioindicadores, Faculdade de Oceanografia, Universidade do Estado do Rio de Janeiro, Rio de Janeiro, Brazil

Cristiane Kiyomi Miyagi Kolesnikovas
Associação R3 Animal, Florianópolis, Santa Catarina, Brazil

Paula Laporta
Yaqu Pacha Uruguay, Punta del Diablo, Rocha, Uruguay

Martín Laporta
Unidad de Gestión Pesquera Atlántica, Dirección Nacional de Recursos Acuáticos, Rocha, Uruguay

Andrea Maranho
Instituto Gremar Pesquisa, Educação e Gestão de Fauna, Guarujá, SP, Brazil

Daiane Santana Marcondes
Laboratório de Ecologia e Conservação, Centro de Estudos do Mar, Universidade Federal do Paraná, Pontal do Paraná, PR, Brazil

Juliana Marigo
Laboratório de Patologia Comparada de Animais Selvagens, Departamento de Patologia, Faculdade de Medicina Veterinária e Zootecnia, Universidade de São Paulo, São Paulo, Brazil

Mariana Melcón
Fundación Cethus, Olivos, Buenos Aires, Argentina

Danielle Monteiro
Laboratório de Ecologia e Conservação da Megafauna Marinha, Instituto de Oceanografia, Universidade Federal do Rio Grande, Rio Grande, Rio Grande do Sul, Brazil

Bárbara Moura Reis Manhães
Laboratório de Mamíferos Aquáticos e Bioindicadores, Faculdade de Oceanografia, Universidade do Estado do Rio de Janeiro, Rio de Janeiro, Brazil

Nara Oliveira-Ferreira
Laboratório de Mamíferos Aquáticos e Bioindicadores, Faculdade de Oceanografia, Universidade do Estado do Rio de Janeiro, Rio de Janeiro, Brazil

Paulo Henrique Ott
Universidade Estadual do Rio Grande do Sul (Uergs), Osório, Rio Grande do Sul, Brazil

Renan Lopes Paitach
Laboratório de Ecologia e Conservação de Tetrápodes Marinhos e Costeiros, Universidade da Região de Joinville, São Francisco do Sul, Santa Catarina, Brazil

Cecilia Passadore
Vida Silvestre Uruguay, Montevideo, Uruguay; Dirección Nacional de Recursos Acuáticos (DINARA), Ministerio de Ganadería Agricultura y Pesca (MGAP), Montevideo, Uruguay

Jonatas Henrique Prado
Cooperativa para Conservação e Proteção dos Recursos Naturais, Florianópolis, Santa Catarina, Brazil

Alessandra Rocha
Universidade Federal do Rio Grande, Rio Grande, Rio Grande do Sul, Brazil

Sergio Rodriguez-Heredia
Fundación Mundo Marino, San Clemente del Tuyú, Argentina

Diego Rodríguez
Instituto de Investigaciones Marinas y Costeras, Facultad de Ciencias Exactas y Naturales, Universidad Nacional de Mar del Plata, Mar del Plata, Argentina

Alan Rosenthal
Asociación de Naturalistas Geselinos, Villa Gesell, Argentina

Elitieri Santos-Neto
Laboratório de Mamíferos Aquáticos e Bioindicadores, Faculdade de Oceanografia, Universidade do Estado do Rio de Janeiro, Rio de Janeiro, Brazil

Eduardo R. Secchi
Laboratório de Ecologia e Conservação da Megafauna Marinha, Instituto de Oceanografia, Universidade Federal do Rio Grande, Rio Grande, Rio Grande do Sul, Brazil

Angela Zaccaron Silva
Laboratório de Ecologia e Conservação, Centro de Estudos do Mar, Universidade Federal do Paraná, Pontal do Paraná, PR, Brazil; Associação MarBrasil, Pontal do Paraná, PR, Brazil

Paulo César Simões-Lopes
Laboratório de Mamíferos Aquáticos, Departamento de Ecologia e Zoologia, Universidade Federal de Santa Catarina, Florianópolis, Santa Catarina, Brazil

Federico Sucunza
Grupo de Estudos de Mamíferos Aquáticos do Rio Grande do Sul, Torres, Rio Grande do Sul, Brazil

Gabriela Tanios
Programa de Pós-graduação em Zoologia, Universidade Estadual de Santa Cruz, Ilhéus, Brazil

Rodrigo Hipólito Tardin
Laboratório de Bioacústica e Ecologia de Cetáceos, Universidade Federal do Rio de Janeiro, Rio de Janeiro, Brazil

Maurício Tavares
Centro de Estudos Costeiros, Limnológicos e Marinhos, Campus Litoral Norte, Universidade Federal do Rio Grande do Sul, Imbé, Rio Grande do Sul, Brazil

Juan Pablo Torres-Florez
Centro Nacional de Pesquisa e Conservação de Mamíferos Aquáticos, Instituto Chico Mendes de Conservação da Biodiversidade, Santos, São Paulo, Brazil

Vanessa Tossenberger
Whale and Dolphin Conservation (WDCS), Chippenham, United Kingdom

Tailise Hoppe Trevizani
Instituto Oceanográfico, Universidade de São Paulo, São Paulo, Brazil

Genyffer Cibele Troina
Laboratório de Ecologia e Conservação da Megafauna Marinha, Instituto de Oceanografia, Universidade Federal do Rio Grande, Rio Grande, Rio Grande do Sul, Brazil

Juliana Di Tullio
KAOSA, Rio Grande, Rio Grande do Sul, Brazil

Meca Valdivia
Sección Mamíferos, Museo Nacional de Historia Natural, Montevideo, Uruguay

Ana Luisa Valente
Departamento de Morfologia, Instituto de Biologia, Universidade Federal de Pelotas, Pelotas, Rio Grande do Sul, Brazil

Rodrigo del Rio do Valle
Instituto Biopesca, Praia Grande, São Paulo, Brazil

Tara Van Belleghem
Laboratório de Ecologia e Conservação, Centro de Estudos do Mar, Universidade Federal do Paraná, Pontal do Paraná, PR, Brazil

Lara Gama Vidal
Laboratório de Ecologia e Conservação, Centro de Estudos do Mar, Universidade Federal do Paraná, Pontal do Paraná, PR, Brazil

Gabriela Vélez-Rubio
Sección de Oceanografía y Ecología Marina, Instituto de Ecología y Ciencias Ambientales, Facultad de Ciencias, Universidad de la República, Montevideo, Uruguay

Randall S. Wells
Chicago Zoological Society's Sarasota Dolphin Research Program, ℅ Mote Marine Laboratory, Sarasota, Florida, United States

Timothy B. Werner
Ocean Associates Inc., Arlington, VA, United States

Maria Fernanda Zapata
Fundación AquaMarina, Centro de Estudios en Ciencias Marinas, Buenos Aires, Argentina

Camilah Antunes Zappes
Departamento de Oceanografia, Centro de Ciências Humanas e Naturais, Universidade Federal do Espírito Santo, Vitória, Espírito Santo, Brazil

Alexandre Novais Zerbini
Cooperative Institute for Climate, Ocean and Ecosystem Research (CICOES), University of Washington and Marine Mammal Laboratory, Alaska Fisheries Science Center, Seattle, WA, United States; Marine Ecology and Telemetry Research, Seabeck, WA, United States

Preface

A vulnerable species—presentation

This book provides updated and comprehensive knowledge about the most endangered species of marine mammals in the Western South Atlantic Ocean. Known by the common name of "Franciscana" in Uruguay and Argentina, and by "Toninha" in Brazil, since 2006, this species is listed on the IUCN Red List as "Vulnerable."

The Franciscana dolphin *Pontoporia blainvillei* is a coastal species and sometimes occupies estuarine environments. The conservation of the species and the environment in which it lives are worrisome, at a time when environmental policies are giving way to the unrestrained exploitation of natural resources with minimal control by environmental agencies.

In many ways, we could say that this dolphin is at its limit of survival. The next few decades (perhaps years) give us a small amount of time to interfere in this nefarious fate. The point is that we cannot sit idly by waiting for this fate to be fulfilled. We need to produce reliable scientific information, disseminate it massively, and advise environmental agencies in the best possible way. We need to discuss possible solutions with governments, communities, and the fishing industry and propose solutions, such as the creation of fishing exclusion areas, or changes in the way we exploit fisheries resources.

A few years ago when the Baiji or Yangtze River dolphin, the Chinese dolphin, was declared extinct in the wild, there was an international commotion. We are seeing the vaquita on the path of extinction, with a reduction registered year by year, and reaching the point where there is no longer any expectation of reversing the scenario. In both cases, it was too late; what a shame! But in the case of Franciscanas, there is still a lot of expectation that we can reverse the situation and we still have the problem at hand … and we have to do our part in it.

This book brings together and synthesizes, in a single compendium, the information on taxonomy, morphology, and evolution, feeding ecology, parasites and diseases, reproduction and growth, genetic diversity and stock definition, distribution and abundance estimation, behavior and movement patterns, acoustical repertoire, fisheries interactions and bycatch, contamination levels, coastal development, and habitat loss, stranding, and mortality, rehabilitation on stranded specimens, alternative fishing methods and the potential use of acoustic repellents, the characteristics of artisanal and industrial fishing communities, ethnoecological approaches and fishermen perception, marine protected areas (their potential for the conservation of the species), and finally the national and international agreements aiming at a joint conservation effort.

The book contains 19 chapters written by an internationally renowned expert team and many dedicated their lives to the study of this small dolphin from an ancient lineage. It fills substantial gaps in the knowledge and explores both aspects of the species' natural history and urgent problems. We need to look for ways to properly manage marine resources, considering their impact on threatened species, and reduce bycatch. We need to reduce impacts on coastal environments, so necessary for the survival of biodiversity and the maintenance of healthy marine ecosystems. We need to improve our ability to care for debilitated, live-stranded dolphins. We need proposals and political strength for the creation of marine protected areas, as well as international public policies and agreements that lead us to reduce the threats.

The Franciscana Dolphin book is an important resource for a broad spectrum of biologists and ecologists, oceanographers and veterinarians, decision-makers, NGO communities, and conservationists who want to contribute to the conservation of the species or who face similar threats with other dolphin species in other parts of the globe. The "Franciscana" cannot face these challenges in the coming years alone. Scientific knowledge is not the only step to be taken but a necessary and urgent one.

By Paulo César Simões-Lopes & Marta Jussara Cremer editors

CHAPTER 1

Taxonomy, skeletal morphology, and evolutionary history

Paulo César Simões-Lopes[a], Carolina S. Gutstein[b], Camila Márquez Iturriaga[b]
[a]Laboratório de Mamíferos Aquáticos, Departamento de Ecologia e Zoologia, Universidade Federal de Santa Catarina, Florianópolis, Santa Catarina, Brazil
[b]Red Paleontológica, Lab. Ontogenia y Filogenia, Dep. Biología, Fac. Ciencias, Universidad de Chile, Consultora Paleosuchus Ltda, Santiago, Chile

Taxonomy

Type: Skull only, collected by M. de Freminville and deposited in the Muséum National d'Histoire Naturelle, Paris. **Type locality**: Mouth of 'Rio da Prata' near Montevideo, Uruguay. Genus *Pontoporia Gray*, 1846, Zoology of the voyage of H.M.S. Erebus and Terror, 1 (Mammalia): 46. **Valid name**: *Pontoporia blainvillei* (Gervais and d'Orbigny) was originally described as *Delphinus blainvillei* Gervais and d'Orbigny, 1844, Bull. Soc. Philomatique, 5, 38–40. Synonyms considered by Herskovitz (1966): *Stenodelphis blainvillei* d'Orbigny and Gervais, 1847 and *Pontoporia tenuirostris* Malm, 1871.

Local names: Toninha in Brazil and franciscana in Uruguay and Argentina. La Plata dolphin was also used as a common name in English; currently, franciscana prevails as an internationally commonly accepted name.

Skeleton

It was described in general terms by Burmeister (1869) and Carvalho (1961) with details of the tympano-periotic bones in Kasuya (1973) and the hyoid apparatus in Pretto et al. (2009). Higa et al. (2002) confirmed sexual dimorphism by analyzing the size and shape components of the skull using geometric morphometric techniques, and Gutstein et al. (2009) established comparisons with the fossil genus *Brachydelphis* using the same technique.

Skull, mandible, and hyoid apparatus

In adults, the braincase is almost square, and the rostrum is extremely long, reaching 60%–73% of the condylobasal length of the skull (Fig. 1.1A). In

The Franciscana Dolphin
DOI: https://doi.org/10.1016/B978-0-323-90974-7.00002-1

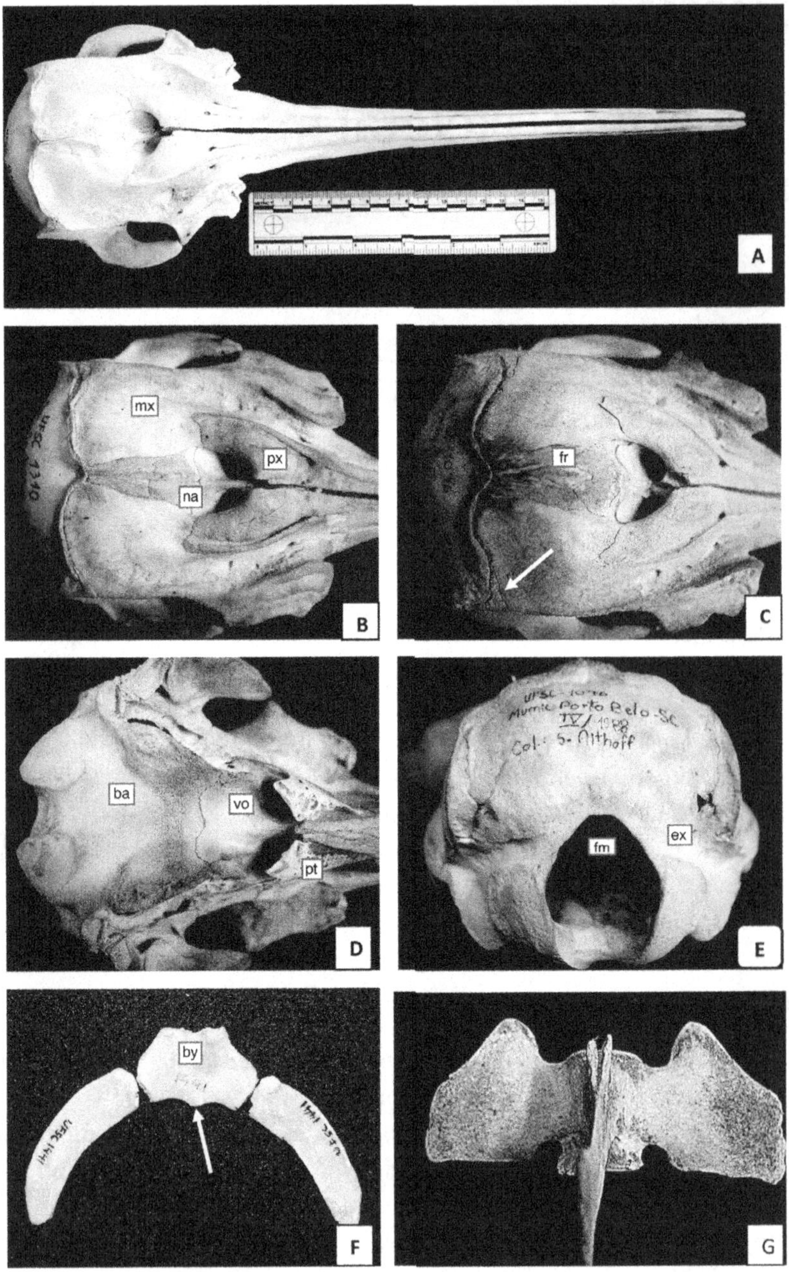

Fig. 1.1 (A) *Pontoporia blainvillei* skull in dorsal view; (B) skull detail of an adult animal in dorsal view and (C) of a juvenile animal (the arrow indicates the interposition of the parietal); (D) ventral view of the cranial floor of an adult animal; (E) fenestrations of the occipital complex; (F) basihyal and thyrohyoid bone (the arrow indicates the characteristic medial projection); and (G) lumbar vertebra with flat transverse processes. Abreviations: ba, basioccipital; by, basihyal; ex, exoccipital; fm, *foramen magnum*; fr, frontal; mx, maxilla; na, nasal; pt, pterygoid; px, premaxilla; vo, vomer.

neonates, the braincase is more rounded, and the rostrum is much shorter, reaching 54%–62% of the condylobasal length. The rostrum is narrow, with the toothrows nearly parallel, except for the last eight–nine teeth that begin to diverge. There are 53–58 pointed teeth in the upper rows and 51–56 in the lower ones. The mandibular symphysis is 54–61% of the length of each dentary bone (*dentale*), but in neonates, this proportion drops to just over 42%. It is common for the rostrum and *processus alveolaris* of the mandible to suffer some lateral or ventral curvatures.

The cranial roof is almost symmetrical, including nasal openings, except for a minimal difference in favor of the length of the right premaxillary bone (premaxilla or *os premaxillare*) (Figs. 1.1B and 1.2). This slight asymmetry is more visible in adults than in juveniles. The ascending process of the premaxillary bone has a convex portion that corresponds to the area of attachment of the premaxillary sac (Fig. 1.3A), a premaxillary plate, as in most other Inioidea and Phocoenidae, instead of a premaxillary fossa, as in most other odontocetes. The maxillary bones (*maxilla*) are elongated posteriorly and reach the supraoccipital crest (*nuchal crest*) in adults, completely covering the frontals (*os frontale*), except medially (Figs. 1.1B and 1.2A,C). In very young animals, the frontal bones appear in a narrow line behind the maxillary bones, indicating that the telescoping process of the skull is completed early. The frontal bones form the entire cranial roof but are visible only in a narrow band on the occiput. The nasals (*os nasale*) are rectangular and almost symmetrical (Fig. 1.1B). The interparietal bone (*os interparietale*) cannot be seen in the dorsal view of the skull, but inside the braincase, there is evidence of its presence strongly ankylosed to the frontals. The mesethmoid (*mesethmoideum*) protrudes anteriorly in the *lamina perpendicularis*, dividing the nasal openings. The vomer (*vomer*) plays an important role in the region, separating the nasal openings, and involving the pre-sphenoid (*praesphenoideum*) and the base of the ethmoid complex; it also supports the mesorostral cartilage, playing a critical role in the transmission of sounds via the rostrum.

The lower edge of the orbit is limited by the jugal bone (*os jugale, malare*). This is a very thin bone that loosely attaches itself to the squamosal zygomatic process. This bone is often lost during carcass preparation and is mostly absent in specimens from scientific collections. Squamosal bones (*os squamosale, os temporale*) do not participate in the delimitation of the orbit in *Pontoporia blainvillei*.

The zygomatic process of the squamosal bones is well-developed, flattened, and laterally defines a small temporal fossa. The temporal region includes the sides of the braincase and is composed of parietals (*os parietale*)

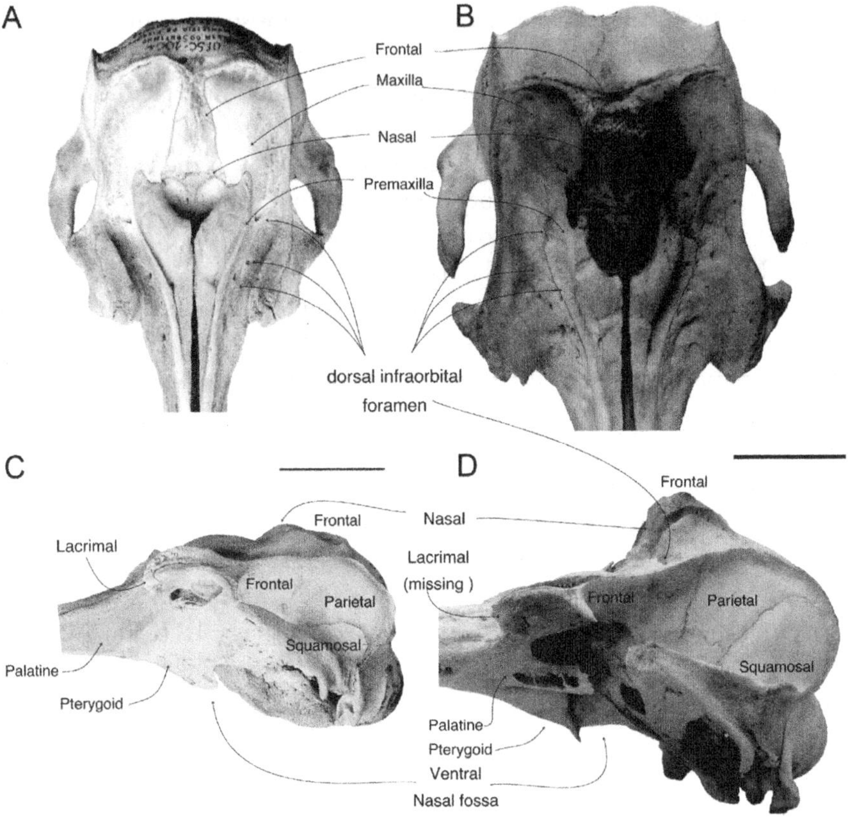

Fig. 1.2 Dorsal and lateral views (A,C); *Pontoporia blainvillei; Inia* (B,D).

and squamosals. In neonates and juveniles, the temporal region is notably convex or inflated, bounded by a smooth temporal ridge. In these specimens, the parietals are interposed in a wedge between the frontal and supraoccipital regions (Fig. 1.1C).

The occipital bone complex results from the coalescence of four endochondral bones surrounding the *foramen magnum*. Each exoccipital (*os exoccipitale*) contains the occipital condyle and remains individualized in neonate specimens. In some cases, fenestrations were observed at the junction between the exoccipital, parietal, and supraoccipital bones (Fig. 1.1E). The supraoccipital bone forms the upper part of the *foramen magnum,* but the basioccipital bone (*os basioccipitalis*) does not participate in this foramen. The cranial floor is composed of the basioccipital, basisphenoid (*basisphenoideum*), and presphenoid; the last two are then covered by the posterior

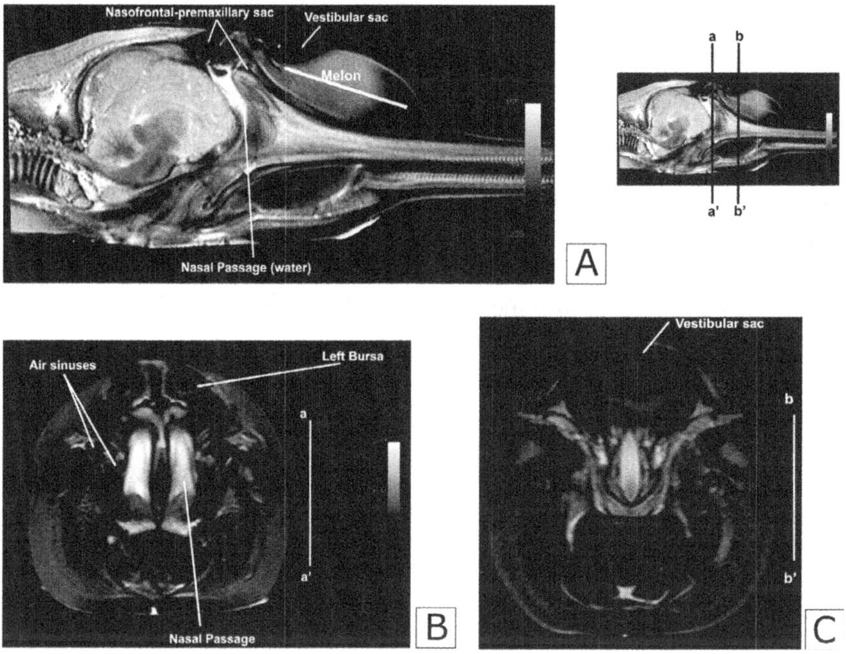

Fig. 1.3 ***Magnetic resonance images (MRI) of the head and neck of a neonate*** **Pontoporia blainvillei.** (A) Sagittal plane view of the head where the melon can be appreciated in its mesial extension, including the associated aerial sacs (vestibular, premaxillary, and nasofrontal). (B,C) Transversal plane view of the head in two sections of the anteroposterior axis, as shown in the a-a' section (B) corresponding to the nasal passage level, where the air sinuses in the ventral portion and the posteriormost portion of the melon connecting with the bursae can be observed; b-b' section (C) corresponding to the orbit level, where the vestibular sac anterior portion, can be appreciated and expanded.

laminar process of the vomer in ventral view. The suture between the basioccipital and basisphenoid (*sphenoccipital synchondrosis*) is visible only in neonate specimens.

As in the other odontocetes, the periotic-tympanic bone complex is separated from the skull and housed below the cranial hiatus. The periotic (*petrosum*) is a replacement bone, while the tympanic (*tympanicum*) is a lining bone. Both are linked by synostosis, which can be easily separated. These bones are the smallest among the living odontocetes. The periotic is very rounded with an extreme reduction of the anterior and posterior process, a delicate structure that barely attaches to the squamosal bone through ligaments. The pars cochlearis is more rounded and globose than in Inia (Gutstein et al., 2014).

The infraorbital dorsal foramina (*foramen infraorbitale*) of the maxillary bones are reduced in size and number compared to the Delphinidae and in size compared to Iniidae. There is a premaxillary foramen on each premaxillary bone. In the hard palate, there are discrete palatine foramina, followed by elongated grooves (*palatine sulcus*). These foramina are located next to the suture of the palatal bones (*os palatinum*), along with the maxillary bones (*minor* and *major palatine foramen*, respectively). The lacerated foramen, between the alisphenoid and orbitosphenoid bones (*alae orbitalies*), is covered by the pterygoid bones (*os pterygoideus*). These bones are extremely fragile and have a lacy appearance (Fig. 1.1D).

Each dentary bone has a deep ventral groove in the ventral portion of the *processus alveolaris*, where the trigeminal nerve passes. This nerve emerges from the mandibular canal through the mental foramina (*foramina mentalia*). Contrary to what occurs in Delphinidae, the mental foramina of *P. blainvillei* emerge in the proximal third of the dentary bone, where the mandibular symphysis begins.

The basihyal (*corpus, basihyoidum*) is rectangular and flat and has a characteristic medial projection on the caudal border (Fig. 1.1F). The thyrohyoid (*cornus majus, thyrohyoideum*) are curved and flat; the stylohyais (*styloideum*) are also flattened and also straight.

Vertebral column, ribs, and sternum

Seven cervical vertebrae were unfused. The vertebral formula was defined by Pinedo et al. (1989): C7 + T10 + L5-6 + Ca18-19 = 41–43. The four anterior ribs have a *capitulum* and a *tuberculum*, and the latter is articulated in the transverse process via a cartilaginous projection. The transverse process of each lumbar vertebra is very characteristic in this species, being quite flattened dorsoventrally and broad in the craniocaudal direction (Fig. 1.1G). The sternum consists of two segments and comprises four sternal ribs.

Scapula, forelimbs, and pelvic bones

The supraspinous fossa of the scapula is well developed, and the coronoid process is relatively short. The phalangeal formula was described by Pilleri and Gihr (1976): $I_0 + II_{4\text{-}5} + III_{3\text{-}4} + IV_{2\text{-}3} + V_{2\text{-}3}$. Pelvic bones are triangular, somewhat variable, and their anterior extremity reached the 2nd caudal vertebra (Ca). It was not possible to establish homology between the pelvic bones and sacral region of the vertebral column (Simões-Lopes and Gutstein, 2004).

Morphology associated with echolocation

As described above, *Pontoporia* has a mostly symmetric skull with only a slight size difference in the premaxillary ascending process. In contrast to Delphinoidea taxa and even more basal taxa such as Ziphiidae and Platanistidae, the cranial vertex of *Pontoporia* is very low, almost parallel to the antorbital notches and the rostrum. Nevertheless, the soft tissue anatomy does not reflect such a difference, presenting all the same structures of other odontocetes with well-developed and high vertex, which has been used as an argument to its reversal to a smooth vertex, as it has been noted that the facial soft tissue anatomy (Cranford et al., 1996). The structures associated with echolocation are indeed asymmetric (Huggenberger et al., 2010) as the posterior melon extension and the premaxillary and vestibular sacs. The premaxillary plates are involved with the sacs through all convex portions of the premaxillary ascending process (see Fig. 1.3).

It has also been reported that some of these structures vary during the development of the skull and head during the growth stage, a feature that has been related to differences in the bioacoustics, which might be an important factor in determining higher rates of bycatch in infants (Frainer et al., 2015).

Fossils and evolutionary history

An interesting aspect of this species is that it represents an older and relatively diverse group, indicating different species and genera, and was somewhat widespread during the Neogene (period of the geologic scale that extended from 23–22 million years ago—m.a.) and Pleistocene (22–18 thousand years). The oldest record of the family Pontoporiidae is the genus *Brachydelphis* with two species, *B. mazeasi* and *B. jahuayensis*, from the middle to late Miocene of Pisco Formation in Peru and Chile.

In addition, from the late Miocene, there are two other genera in South America, *Pontistes* and *Pliopontos*. *Pontistes rectifrons* was first described in the Late Miocene of Argentina, in the Paranean Sea record of Entre Rios, Argentina; however, the genus has also been reported informally in the Bahia Inglesa Formation, northern Chile (Gutstein et al., 2015) and possibly in Europe (Pyenson and Hoch, 2007). *Pliopontos littoralis* was first described as a Pliocene species, but its locality age has been updated to the late Miocene, consistent with informal reports from Chile (Gutstein et al., 2015). In Brazil, there are some dubious records that might be attributed to the family Pontoporiidae. From the late Miocene of the Amazon (Solimões Formation, Acre), a probable Pontoporiidae was tentatively referred to and published as *cf. Pontistes*, albeit not formally described (Negri and Bocquentin, 2000), being more conservative to consider at the family level. The cranial remains from the Pleistocene of the coast of Rio Grande do Sul, Brazil, are the oldest recognized species of the genus *Pontoporia* and possibly the species *P. blainvillei*, as there were no significant differences in the specimens analyzed (Ribeiro et al., 1998). Older records of isolated periotics from Argentina have been attributed to this genus (Cozzuol, 1985, 1996).

Nevertheless, the initial thought that Inioidea and the families within Iniidae and Pontoporiidae were complete or mostly endemic to South America (Cozzuol, 1996, 2010) has been diluted with the new findings throughout the globe in more recent years (Fig. 1.4). In actuality, the group has proven to be a widespread marine group with relatively small odontocetes that diverged during the middle Miocene. Some known South American Neogene genera, such as *Pontistes* and *Brachydelphis*,

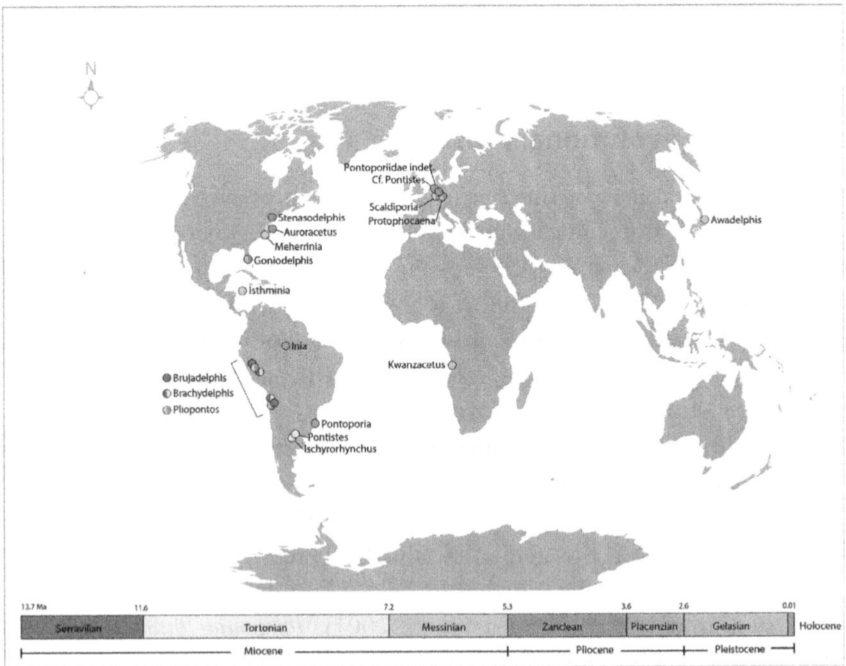

Fig. 1.4 ***World map showing the widespread and diverse fossil records of Inioidea during the Neogene.*** Genus that are generally attributed to the Pontoporiidae family: *Brachydelphis, Pliopontos, Pontistes, Scaldiporia, Auroracetus, Pontoporia, and Protophocoena* (modified from Post et al 2017).

have been tentatively reported in the North Atlantic (Pyenson and Hoch, 2007). Several other genera recovered as phylogenetically related to the Inioids and mainly pertaining to the Iniidae family have been reported in different continents and seas (see map Fig. 1.4), currently only presented by one genus that inhabits the different rivers of the Amazon and Orinoco basin but that was a diverse and abundant group in the late Neogene. It has been hypothesized that the evolution of South American extant river dolphins could be a result of a heterochronic event, with the crown genus representing a peramorphic (Inia)/paedomorphic pair (*Pontoporia*; Cozzuol, 2010) (see Fig. 1.2 for the comparisons between juvenile and adult *Inia* with *Pontoporia* and *Ischyrorhynchus*), a very robust form from the late Miocene riverine beds of Entre Rios Argentina (Ituzaingó Formation). Despite this preliminary hypothesis, most of the group records were marine and rather widespread during the entire Neogene (Pyenson et al., 2015) and have become extinct in most places, surviving only in the riverine and coastal estuarine habitats since the Pleistocene. This feature was observed in all the river dolphin groups, which might have presented a differential survival into the riverine systems, as proposed by Cassens et al. (2000). Their decline corresponds to the rise of delphinoids during the Pliocene and Pleistocene.

Phylogenetically, the Inioidea clade was recovered as monophyletic in the most recent analysis, including the extant genera *Inia* and *Pontoporia* and several fossil genera from the Neogene; depending on the analysis and its taxonomic sample, the clade may include a different assortment of the following genera: *Isthminia*, *Brujadelphis*, *Meherrinia*, *Kwanzacetus*, *Ischyrorhynchus*, *Scaldiporia*, *Pontoporia*, *Brachydelphis*, and *Pliopontos*. These taxa are widely distributed within the timespan corresponding to the Neogene (see Figs. 1.4 and 1.5). Several other genera have not been included in the phylogenetic analysis mainly due to fragmentary state or lack of revisions, but are described as inioids due to morphological similarities and diagnostic characters, as is the case for *Goniodelphis*, *Auroracetus*, *Protophocoena minima*, *Stenasodelphis russelae*, and *Pontistes rectifrons*.

In terms of morphological characteristics, compared to other lineages, the group presents a symmetric skull and a very rounded par cochlearis, with very small anterior and posterior processes (in most cases; Cozzuol, 2010; Gutstein et al., 2014; Muizon, 1988a), as for example Delphinoidea and even earlier divergent as the Ziphiids. Nevertheless, to date, the characteristics defining the synapomorphies of the clade are a very elongated rostrum with fused mandibular symphysis, a lacrimal that wraps around

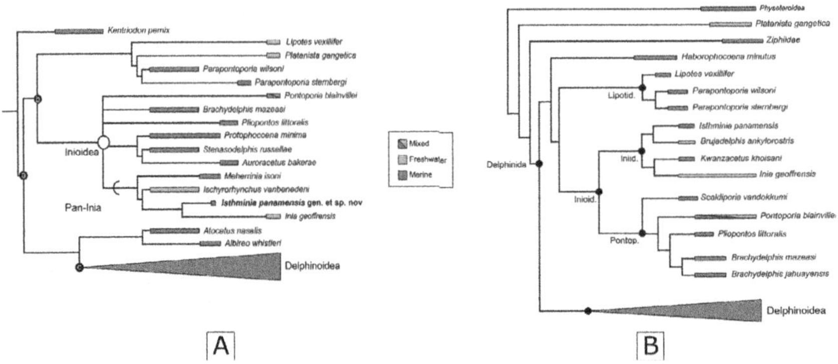

Fig. 1.5 ***Phylogenetic hypothesis regarding the Inioidea group and the Pontoporiidae family.*** (A) Hypothesis published in Pyenson et al. (2015), with no support for Pontoporiidae monophyly, and in (B) hypothesis published in Lambert et al. (2018) with support for the family. Note also the position of *Ischyrorhynchus* in (A).

the anterior edge of the supraorbital process of the frontal and slightly overlies its anterior end, and the maxilla that forms the dorsolateral edge of the ventral infraorbital foramen (Geisler et al., 2011). The analysis of Geisler et al. (2011), a super-matrix with a partition of 311 morphological characters and several extra taxa of all the Odontoceti groups, has been widely used (Geisler et al., 2012; Lambert et al., 2018) to continue coding the Inioidea-related taxa that have been described in the last decades, as shown in Fig. 1.5 from Pyenson et al. (2015) in the description of *Isthmnia panamensis* and from Lambert et al. (2018) of *Kwanzacetus khoisani*. Still this sequential coding and inclusion of taxa not always agree in coding and results as observed by Godfrey et al. (2021), in their description of a new iniid species *Isoninia borealis*. In Fig. 1.6, we show a comparison of a very disputed and previously described species, *Ischyrorhynchus vanbenedeni*. This taxon has also been classified as a Platanistidea (represented nowadays only by the river dolphins of the Ganges (Geisler et al., 2011, 2012; Pilleri and Gihr, 1979); however, a careful revision of the character coding for the species was able to recover it as an Inioidea (Pyenson et al., 2015; see Fig. 1.5A).

Alternatively, another phylogenetic matrix has been used with a few Inioidea related taxa in the new description of an Inioidea from Japan, *Awadelphis hirayamai* (Murakami et al., 2016), expanding the record of the clade to the northwest of the Pacific Rim.

In the particular case of Pontoporiidae, as defined by the last phylogenetic hypothesis that has the most comprehensive taxonomic sampling (Fig. 1.5B), the family would include *Scaldiporia* (Late Miocene of Europe),

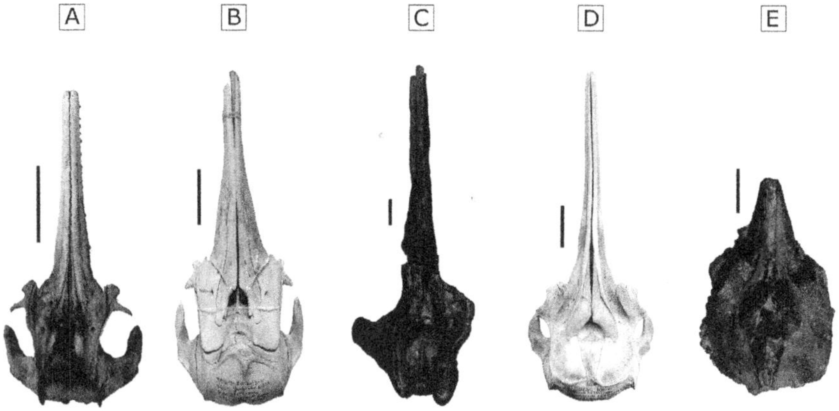

Fig. 1.6 ***Skull of known extinct and extant Inioidea from Sotuh America.*** (A) *Inia geoffrensis* (adult), (B) *Inia geoffrensis* (neonate), (C) *Ischyrorhynchus vanbenedeni*, (D) *Pontoporia blainvillei*, and (E) *Brachydelphis mazeasi*.

Pliopontos and *Brachydelphis* (from the late Miocene of Chile and Peru), and *Pontoporia*. Nevertheless, the following genera have also been recognized as belonging to the family, even if not included in the phylogenetic analysis: the extinct *Auroracetus*, *Pontistes*, *Protophocaena*, and *Stenasodelphis* (see Fig. 1.4, for the distribution in the present-day world map). In general, all these taxa have a low vertex of the skull and premaxillary plates rather than the fossa for the premaxillary sacs. In addition, the right and left premaxilla are almost completely symmetrical, and their contacts and posterior shape are diagnostic characters for the different genera; the posterior end of the ascending process of the premaxilla has different configurations that have been used as specific characters to diagnose the species among these genera and their contact with the nasal that is present in *Brachydelphis* and *Pontistes* and absent in *Pliopontos* and *Pontoporia* (Gutstein et al., 2009; Muizon, 1984, 1988a,b; Post et al., 2017).

References

Burmeister, H., 1869. Descripcion de cuatro especies de delfinides de la costa Argentina del Oceano Atlantico. An. Mus. Publ. Hist. Nat. Buenos Aires. 1, 367–442.

Carvalho, C., 1961. *Stenodelphis blainvillei* na costa meridional do Brasil, com notas osteológicas (Cetacea, Platanistida). Rev. Bras. Biol. 21, 443–454.

Cassens, I., Vicario, S., Waddell, V.G., Balchowsky, H., Belle, D.V., Ding, W., Fan, C., Mohan, R.S.L., Simões-Lopes, P.C., Bastida, R., Meyer, A., Stanhope, M.J., Milinkovitch, M., 2000. Independent adaptation to riverine habitats allowed survival of ancient cetacean lineages. Proc. Natl Acad. Sci. 97 (21), 11343–11347.

Cozzuol, M.A., 1985. The Odontoceti of the Mesopotamiense of the Parana River ravines. Systematic review. Invest. Cet. 17, 39–54.

Cozzuol, M.A., 1996. The records of the aquatic mammals in southern. South America. Münch. Geo. Abh. 30, 321–342.

Cozzuol, M.A., 2010. Fossil record and the evolutionary history of Inioidea. In: Ruiz-Garcia, M., Shostell, J.M. (Eds.), Biology, Evolution and Conservation of Riverine Dolphins Within South American and Asia. Nova Science Publishers, Hauppage, pp. 193–221.

Cranford, T.W., Amundin, M., Norris, K.S., 1996. Functional morphology and homology in the odontocete nasal complex: implication for sound generation. J. Morphol. 228, 223–285.

Frainer, G., Huggenberger, S., Moreno, I.B., 2015. Postnatal development of Franciscana's (*Pontoporia blainvillei*) biosonar relevant structures with potential implications for function, life history, and bycatch. Mar. Mamm. Sci. 31, 1193–1212.

Geisler, J.H., McGowen, M.R., Yang, G., Gatesy, J., 2011. A supermatrix analysis of genomic, morphological, and paleontological data from crown Cetacea. BMC Evol. Biol. 11, 112.

Geisler, J.H., Godfrey, S.J., Lambert, O.A., 2012. New genus and species of late Miocene inioid (Cetacea, Odontoceti) from the Meherrin River, North Carolina, USA. J. Vertebr. Paleontol. 32, 198–211. https://doi.org/10.1080/02724634.2012.629016.

Godfrey, S.J., Gutstein, C.S., Morgan, III,D.J., 2021. A new odontocete (Inioidea, Odontoceti) from the late Neogene of North Carolina. USA. Foss. Rec. 24, 275–285. https://doi.org/10.5194/fr-24-275-2021.

Gutstein, S., Cozzuol, M.A., Vargas, A.O., Suarez, M.E., Schultz, C.L., Rubilar-Rogerns, D., 2009. Patterns of skull variation of *Brachydephis* (Cetacea, Odontoceti) from the Neogene of the Southeastern Pacific. J. Mamm. 90 (2), 504–519.

Gutstein, C.S., Figueroa-Bravo, C.P., Pyenson, N.D., Yury-Yañez, R.E., Cozzuol, M.A., Canals, M., 2014. High frequency echolocation, ear morphology, and the marine-freshwater transaction: a comparative study of extant and extinct toothed whales. Palaeogeogr. Palaeoclimatol. Palaeoecol. 400, 62–74.

Gutstein, C.S., Horwitz, F.E., Valenzuela-Toro, A.M., Figueroa-Bravo, C.P., 2015. Cetáceos fosiles de Chile: contest evolutivo y paleogeobiográfico. Publicación Ocasional del Museo Nacional de Historia Natural, Chile 63, 339–387.

Hershkovitz, P., 1966. Catalog of Living Whales. Smithsonian Institution, US National Museum Bulletin. Government Printing Office, Washington, DC.

Higa, A., Hingst-Zaher, E., de Vivo, M., 2002. Size and shape variability in the skull of *Pontoporia blainvillei* (Cetacea: Pontoporiidae) from the Brazilian Coast. LAJAM 1 (1), 145–152. Special Issue 1. https://doi.org/10.5597/lajam00018.

Huggenberger, S., Vogl, T.J., Oelschläger, H.H.A., 2010. Epicranial complex of the La Plata dolphin (*Pontoporia blainvillei*): topographical and functional implications. Mar. Mamm. Sci. 26 (2), 471–481. https://doi.org/10.1111/j.1748-7692.2009.00349.x.

Kasuya, T., 1973. Systematic consideration of recent toothed whales based on the morphology of tympano-periotic bone. Rep. Whales Res. Inst. 25, 1–103.

Lambert, O., Auclair, C., Cauxeiro, C., Lopez, M., Adnet, S., 2018. A close relative of the Amazon river dolphin in marine deposits: a new Iniidae from the late Miocene of Angola. PeerJ 6 (e5556), 1–33.

Muizon, C., 1984. Les Vertébrés de la Formacion Pisco (Peru). Deuxième parties: less Odontoètes (Cetacea, Mammalia) du Pliocène inférieur de Sud-Sacaco. Travaux de Bull. Inst. Fr. Etudes Andin. 27, 1–188.

Muizon, C., 1988a. Les relation phylogénétiques des Delphinidae. Ann. Paléontol. 74, 159–227.

Muizon, C., 1988b. Les Vertébrés fossiles de la Formacion Pisco (Peru). Triosième parties: less Odontoètes (Cetacea, Mammalia) du Miocène. Travaux de Bull. Inst. Fr. Etudes Andin. 42, 1–244.

Murakami, M., 2016. A New Extinct Inioid (Cetacea, Odontoceti) from the Upper Miocene Senhata Formation, Chiba, Central Japan: The First Record of Inioidea from the North Pacific Ocean. Paleontological Res. 20 (3), 207–225. https://doi.org/10.2517/2015PR031. S2CID132209924.

Negri, F., Bocquentin, J., 2000. cf. *Pontistes rectifrons* Burmeister, 1891 (Cetacea, Odontoceti, Pontoporiidae) no Mioceno superior-Plioceno, Estado do Acre, Brasil. XVI Jornadas Argentinas de Paleontologia de Vertebrados. San Luis, p. 43.

Pilleri, G., Gihr, M., 1976. On the manus of the La Plata dolphin. *Pontoporia blainvillei*. Invest. Cetacea. 7, 119–128.

Pinedo, M.C., Praderi, R., Brownell Jr. R.L., 1989. Review of the biology and status of the franciscana, Pontoporia blainvillei. In: Perrin, W.F., Brownell Jr, R.L., Zhou, K, Jiangkang, L. [eds]. Biology and Conservation of the river Dolphins. Occas. pap. IUCN SSC, v. 3, pp. 46-51.

Post, K., Louwye, S., Lambert, O., 2017. *Scaldiporia vandokkumi*, a new pontoporiid (Mammalia, Cetacea, Odontoceti) from the Late Miocene to earliest Pliocene of the Westerschelde estuary (The Netherlands). PeerJ 5, e3991.

Pretto, D.J., Oliveira, G., Graipel, M., Simões-Lopes, P.C., 2009. Morfologia do aparato hióide de algumas espécies de odontocetos do sul do Brasil. Biotemas (UFSC) 22, 111–120.

Pyenson, N.D., Hoch, E., 2007. Tortonian pontoporiid odontocetes from the eastern North Sea. J. Vertebr. Paleontol. 27, 757–762.

Pyenson, N.D, Vélez-Juarbe, J., Gutstein, C.S., Little, H., Vigil, D., O'Dea, A., 2015. Isthminia panamensis, a new fossil inioid (Mammalia, Cetacea) from the Chagres Formation of Panama and the evolution of 'river dolphins' in the Americas. PeerJ 3, e1227. https://doi.org/10.7717/peerj.1227.

Ribeiro, A.M., Drehmer, C.J., Buchmann, F.S.C., Simões-Lopes, P.C., 1998. Pleistocene skull remains of *Pontoporia blainvillei* (Cetacea, Pontoporiidae) from the coastal plain of Rio Grande do Sul State, Brazil, and the relationships of pontoporids. Rev. Universidade Guarulhos 3 (6), 71–77.

Simões-Lopes, P.C., Gutstein, C.S., 2004. Notes on the anatomy, positioning and homology of the pelvic bones in small cetaceans (Cetacea, Delphinidae, Pontoporiidae). LAJAM 3 (2), 157–162.

CHAPTER 2

Overview of franciscana diet

Silvina Botta[a], Manuela Bassoi[b], Genyffer Cibele Troina[a]
[a]Laboratório de Ecologia e Conservação da Megafauna Marinha, Instituto de Oceanografia, Universidade Federal do Rio Grande, Rio Grande, Rio Grande do Sul, Brazil
[b]Laboratório de Bioacústica (LaB), Centro de Biociências, Universidade Federal do Rio Grande do Norte-UFRN, Natal, RN, Brazil

Introduction

The franciscana dolphin (*Pontoporia blainvillei*) is a high trophic level predator, mainly feeding on fish species, followed by crustaceans and mollusks. Previous dietary studies indicate that this dolphin preys predominantly upon bottom-dwelling juvenile teleosts, squid and shrimp. Tellechea et al. (2017) found that franciscana prey mainly on sound-producing fish within the low visibility estuarine and coastal waters in Uruguay. This behavior suggests that the species is a generalist and opportunistic predator (Crespo, 2018).

Young franciscana dolphins start their predation upon small fish, squid and shrimps during their first year of age, also mixing these solid items with milk (Bassoi et al., 2021; Denuncio et al., 2017; Troina et al., 2016). Weaned young dolphins differ from adult specimens by a higher shrimp consumption in the former (Bassoi et al., 2021; Denuncio et al., 2017; Troina et al., 2016), while some prey preference for cephalopods and bigger fish specimens is shown by adults franciscanas (Bassoi et al., 2021; Cremer et al., 2012; Henning et al., 2017). Despite their reversed sexual dimorphism, no remarkable sex differences were found in diet composition (Bassoi et al., 2021; Troina et al., 2016). Furthermore, seasonal prey fluctuations were also reported, mainly for the southern areas (e.g., Bassoi et al., 2021).

Geographical dietary variation at small spatial scales was reported for the species within Franciscana Management Areas (FMAs, Cunha et al., 2014; Secchi et al., 2003). Henning et al. (2017) reported some differences in the importance of prey species along the coast of São Paulo state (SP - FMA II). Bassoi et al. (2020) showed distinct geographical dietary compositions within Rio Grande do Sul (RS - FMA III) coast. Franciscanas

The Franciscana Dolphin
DOI: https://doi.org/10.1016/B978-0-323-90974-7.00003-3

from estuarine and coastal regions from Uruguay (FMA III, Artecona et al., 2019) and Buenos Aires (BA - FMA IV, Rodriguez et al., 2002) also showed variation in their diet, reflecting the abundance of prey types in those particular environments. However, no comprehensive comparative analysis of the diet of the franciscanas including its whole distribution was conducted to date.

Estimating the feeding habits (e.g., diet composition, prey selection) of a predator is important to understand how individuals and populations respond to ecological and environmental variability, it provides the foundation for dietary niche characterization, trophic relationships, and their functional roles in marine ecosystems.

Marine mammals' feeding habits have been reported from analyses of scat, stomach contents, direct observations, or inferred by indirect methods such as stable isotope ratios, fatty acids, and molecular identification. The most traditional methods, stomach, and scat contents, rely on the finding and identification of tissue and structure remains representing a typical meal; for example, fish bones and otoliths (ear stones) and the jaws of cephalopods ("beaks"). In particular, fish otoliths and cephalopod beaks are diagnostic structures for the identification of prey because their shape varies considerably from species to species, and the dimensions of these structures correlate well with the length and weight of the species from which they originate (Clarke, 1986; Pierce and Boyle, 1991). The advantages of these methods are: (1) knowledge of prey composition and size classes allows us to infer the spatial and temporal distribution of predators to be investigated; (2) they enable studies on predator-prey dynamics; (3) predator diets can give considerable information about poorly known prey species; (4) changes in diet can be monitored; and (5) samples can be collected from carcasses in an advanced stage of decomposition, which in several cases are the main source of data of marine mammals, especially cetaceans. The disadvantages of the methods are (1) fish otoliths can last for only a few days in the gastrointestinal tracts of marine mammals, whereas cephalopod beaks may accumulate for several days or months, leading to an overrepresentation of the latter; (2) prey lacking hard parts (e.g., invertebrates) will be underrepresented; and (3) a comprehensive reference collection of fish otoliths and cephalopod beaks from a particular area along with specific regression equations are needed for precise identification of prey, and accurate estimation of size and weight, respectively. Moreover, some studies also identify prey items through their DNA present in scats and stomach contents, but also a reference database for the genetic signature of prey is needed (Deagle and Tollit, 2007).

Another method to study diet is the analysis of stable isotopes, in which the composition of heavier and lighter isotopes of particular elements (e.g., carbon, nitrogen) assimilated through the diet analyzed in tissues of predators can be traced to those of their prey (Barros and Clarke, 2002; Latja and Michener, 1994). The advantages of this method, when compared to scat or stomach content methodologies, are that by analyzing tissues with different turnover times it is possible to investigate the feeding history relative to the last days, months, or the entire life of the dolphin. Additionally, depending on the tissue analyzed, it is less invasive and does not necessarily need to obtain samples from dead animals. However, this method also requires previous information of potential prey species (usually acquired from traditional dietary study methods) and a reference database of the isotopic signature of prey.

In this work, we conducted a meta-analysis of franciscanas' dietary studies based on stomach content and stable isotope analysis. We provide an assessment of the variation and temporal trends in the importance of prey species for franciscana dolphins throughout the species distributional range. Specifically, we wanted to assess (1) how populations differ in terms of the dominance and diversity of prey species consumed; (2) the magnitude of spatial differences in franciscanas' carbon and nitrogen stable isotope values throughout the species latitudinal range; and (3) if prey items have changed over time within each population.

Materials and methods

Data collection for meta-analysis

This compilation included all available information of franciscana feeding habits published in scientific papers or reported in scientific meetings, thesis, and dissertations (Table 2.1). The main two methods that have been used to gain insight into what franciscanas eat are the analysis of food that remains present in their stomach and, more recently, the analysis of stable isotopes of carbon and nitrogen. We describe the trend in the report of dietary studies (stomach contents and stable isotopes) and the number of stomachs analyzed across years and geographic regions.

Stomach content analysis

The prey composition data contemplated in our analysis were obtained from stomach contents studies described in Table 2.1 (References 1 to 37). The prey species frequency of occurrence (% FO) was spatially analyzed

Table 2.1 General information and references of franciscana diet studies. Argentina: Buenos Aires Province (BA); Brazil: Rio Grande do Sul State (RS), Santa Catarina State (SC), Paraná State (PR), São Paulo State (SP), Rio de Janeiro (RJ), Espírito Santo State (ES); Stomach contents analysis (ST), stable isotopes analysis (SI).

	FMA	Geographic region	Diet method	N stomachs or tissue	References	Publ. year	Sampling period
1	IV	South BA	ST	1	Burmeister, Proc. Zool. Soc. Lond. 484-489	1867	1867
2	IV	North BA	ST	1	Burmeister, Anal. Mus. Publ. Bs. As. 1(5): 389-422	1869	1868
3	IV	Argentine	ST	30	Lahille, Rev. Mus. La Plata, 9: 389–392	1899	–
4	IV	Argentine	ST	–	[a]Lahille, Rev. Zool. Buenos Aires, 11, 1: 28–82	1905	–
5	IV	Argentine	ST	–	[a]Cabrera and Yepes, Hist. Natural Ediar. Buenos Aires	1940	–
6	II	SP	ST	2	Carvalho, Revi. Bras. Biol., 21(4): 443–454	1961	1960's
7	III	Uruguay	ST	–	[a]Van Erp, Pacific Discovery, 22(2): 18–24	1969	–
8	III	Uruguay	ST	11	Brownell and Ness,Proc. Sixth Annu. Conf. on Biological Sonar and Diving Mammals, Stanford Res. Inst., Palo Alto, California, 23–28	1970	1969
9	III	Uruguay	ST	11	[b]Fitch and Brownell, Bull. Mar. Sci., 21(2): 626–636	1971	1969
10	III	Uruguay	ST	433	Brownell, J. Fish. Res. Board Can., 32(7): 1073–1078	1975	1969–73
11	III	Uruguay	ST	200	Brownell and Praderi, Advisory Committee on Marine Resources Research. Scientific Consultation on Marine Mammals, Norway	1976	1973–74
12	III	South RS	ST	277	Pinedo, Master Thesis, Universidade do Rio Grande, Brazil	1982	1976–80
13	III	Uruguay	ST	6	Praderi, I Reunión RT, Buenos Aires, 206–214	1986	1959–79
14	IV	South BA	ST	30	Perez-Macri, II Reunión RT, Rio de Janeiro, 107	1987	–
15	III	Uruguay/ RS	ST	–	[c]Brownell. Handbook of Marine Mammals: River Dolphins and the Larger Toothed Whales, 45–67	1989	1960–70
16	II	SP/PR	ST	21	Schmiegelow, Master Thesis, Universidade de São Paulo, Brazil	1990	1986–88
17	IV	South BA	ST	2	Bastida et al., III Reunión RT, Buenos Aires, 25–30	1992	1984–88
18	III	North RS	ST	36	Ott, Dissertation Thesis, Universidade do Rio Grande do Sul, Brazil	1994	1992–94

	FMA	Geographic region	Diet method	N stomachs or tissue	References	Publ. year	Sampling period
19	IV	South BA	ST	62	Perez et al., IV Reunión RT, Viña del Mar, 91	1996	1989–92
20	III	South RS	ST	36	Bassoi, Dissertation Thesis, Universidade do Rio Grande, Brazil	1997	1994–96
21	II	PR	ST	12	Oliveira et al., VIII Reunión RT, Recife, 25–29	1998	1997–98
22	Ib	North RJ	ST	89	Di Beneditto and Ramos, J. Cet. Res. Manag., 3(2): 185–192	2001	1986–99
23	IV	North BA	ST	60	Rodriguez et al., LAJAM, 1(1): 77–94	2002	1992–2000
24	III	RS	ST	270	Bassoi, PhD Thesis, University of Southampton, UK	2005	1994–2001
25	Ib	North RJ	ST	99	Bittar and Di Beneditto. Zoologia (Curitiba), 26(2): 374–378	2009	1989–2005
26	II	North SC	ST	8	Cremer et al. Iheringia. Série Zoologia, 102(2): 131–137	2012	2000–06
27	II	SP/PR	ST	86	Lopes, Master Thesis, Universidade Estadual Paulista, Brazil	2012	2005–07
28	IV	South BA	ST	66	Paso-Viola et al., Rev. Biol. Mar. Oceanogr., 49(2): 393–400	2014	2003–11
29	II	North SC	ST	28	Paitach, Master Thesis, Universidade de Santa Catarina, Brazil	2015	2000–14
30	III	Uruguay	ST	41	Tellechea et al., Aquat. Mamm., 43(4): 430–439	2017	2013–16
31	II	SP	ST	58	Henning et al., Biotropica, 50(1): 157–168	2017	1999–2010
32	IV	North BA	ST	119	Denuncio et al., J. Sea Res., 129: 70–79	2017	2000–10
33	III	Uruguay	ST	305	Drago et al., Sci. Rep., 8(1): 1–12	2018	1980–2009
34	Ia	ES	ST	18	Rupil et al., Revista Biotemas, 32: 2	2019	2012–15
35	II	SP	ST	145	Campos et al., J. Mar. Biol. Assoc. U.K., 100(2): 301	2020	2005–15
36	II	North RS	ST	76	Machado et al., MEPS, 639: 215–232	2020	1993–2016

(Continued)

Table 2.1 Cont'd

	FMA	Geographic region	Diet method	N stomachs or tissue	References	Publ. year	Sampling period
37	Ib	North RJ	ST	4	Di Beneditto and Siciliano, R. Ib-Am. Ciênc. Amb., 12(1): 745–749	2021	2010–11 Ib North RJ SI muscle Di Beneditto et al., Mar. Biol., (158): 2209-2217 2011 1987-98
38	Ib	North RJ	SI	muscle	Di Beneditto et al., Mar. Biol., (158): 2209–2217	2011	1987–98
39	Ib	North RJ	SI	muscle	Di Beneditto et al., Biota Neotropica, 13(3): 29–33	2013	1987–98
40	II	North SC	SI	muscle	Hardt et al., Biota Neotropica, 13(4): 90–98	2013	2000–06
41	Ib	North RJ	SI	muscle	Baptista et al., Environ. Pollut., (218): 1298–1307	2016	2001–13
42	Ib	North RJ	SI	muscle	Di Beneditto and Monteiro, J. Mar. Biol. Assoc. U.K., 96(4): 853	2016	2001–05
43	III	RS	SI	tooth	Troina et al., Mar. Mamm. Sci., 32(3): 1115–1137	2016	1994–2010
44	III	Uruguay	SI	skin	Franco Trecu et al., J. Exp. Mar. Biol. Ecol., (486): 230–236	2017	2009
45	IV	South BA	SI	muscle	Paso Viola et al., Mamm. Biol., (85):47–54	2017	2000–10
46	III	Uruguay	SI	bone	Drago et al., Sci. Rep., 8(1): 1–12	2018	1953–2015
47	III	Uruguay	SI	bone	Artecona et al., Wildl, 46(2): 136–144	2019	2014
48	II	SP	SI	muscle	Trevizani et al. AQUAT CONSERV, 29(12): 2045–2056	2019	2014–15
49	II/III	South SC	SI	bone	Teixeira et al., Mar. Mamm. Sci., 37(1): 111–126	2020	1985–2017
50	II	SP	SI	skin	Pascoalini et al., J. Mar. Biol. Assoc. U.K., 1–10	2021	2011–16

[a] This information is cited in Fitch and Brownell (1971).
[b] This information is cited in Drago et al. (2018).
[c] Brownell (1989): the author compiles the results obtained by Fitch and Brownell (1971), Brownell (1975), Pinedo (1982).

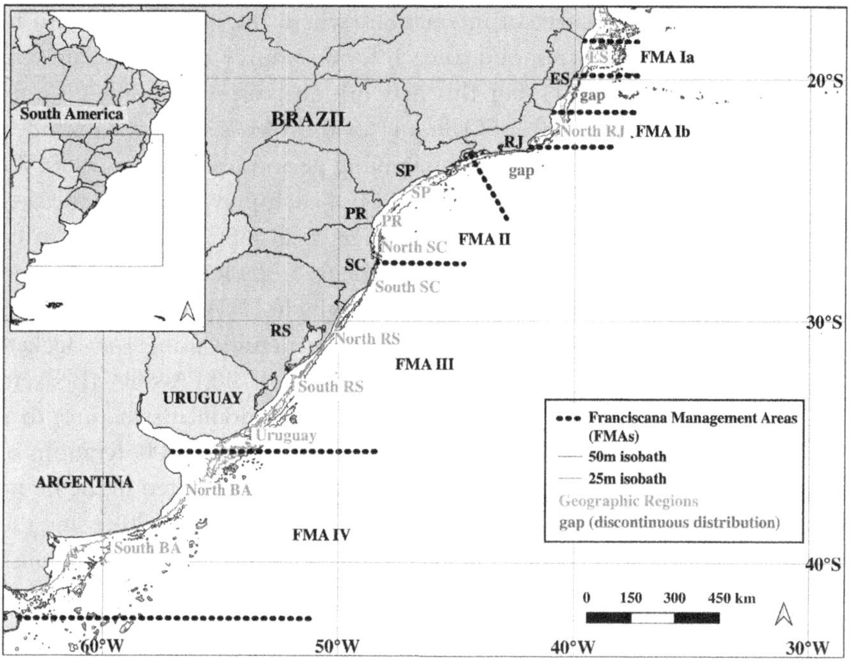

Fig. 2.1 ***Study area of franciscana dolphin distribution showing the geographic ranges and distributional gaps of the Franciscanas Management Areas (FMAs), proposed by Amaral et al. (2018), highlighting some geographic regions.*** *ES*, Espírito Santo; *PR*, Paraná; *RJ*, Rio de Janeiro; *RS*, Rio Grande do Sul; *SP*, São Paulo; *SC*, Santa Catarina.

according to FMAs and within their geographic regions (Fig. 2.1). Only prey species with a minimum of 5% FO in the stomach contents in at least one of the group categories (FMA, geographic region, or sampling period) were taken into account. Rényi diversity profiles were used to rank prey species diversity comparing geographic regions within FMAs (Kindt and Coe, 2005). This index was calculated using the vegan package (Oksanen et al., 2020). The temporal variation in franciscana diet was analyzed for those geographic regions with representative sample data throughout decadal time scales: (1) SP (FMA II), marine area, and; (2) Uruguay (FMA III), marine and estuarine areas. Furthermore, we discuss some interesting findings from the North Rio de Janeiro (RJ) region, although we could not conduct the same analysis due to the nature of the stomach content data reported for this area. These analyses were conducted by grouping sampling periods into decades from 1960 to 2020 (six groups). The former years were grouped into a single period (1860–1940). All analyses were run in R (R Core Team, 2020).

Additionally, we described the trophic interactions of franciscanas from FMAs with their feeding items as a bipartite network. We defined a weighted incidence matrix of interactions where the rows correspond to FMAs, and columns correspond to each food resource reported as prey in previous work (Table 2.1). For this network analysis, we considered only those prey with at least 10% FO in any of the FMAs. Also, we excluded from our analysis those studies with sampling periods earlier than 1980. We weighed the network by the average %FO of each prey species in the diet of franciscanas from each FMA (i.e., average of all %FO reported in studies from each FMA). We determine the degree of modularity (weighted modularity, Qw), which is the extent to which FMAs form subgroups with higher internal than external interaction density, using the Beckett weighted modularity maximizing algorithm DIRTLPAwb+ (Beckett, 2016). The modularity index ranges from zero (no modular structure) to 1 (high density of heavier links within each module). We ran 100 iterations of the algorithm and reported the modular structure that resulted in the highest value of Qw. All analyses were run in R (R Core Team, 2020), using the package bipartite (Dormann et al., 2008) for network metrics and graphs.

Stable isotope data

Isotopic data available from published literature (Table 2.1) were used to develop isotopic maps characterizing the spatial patterns in franciscana carbon and nitrogen stable isotopes throughout the species distributional range. Isoscapes were generated separately for each tissue type (tooth, bone, skin, and muscle) using the freely available software ocean data view version 5.1.5 (Schlitzer, 2002) with data-interpolating variational analysis gridding method for data contouring.

Results and discussion

A brief history of franciscana diet

The first insights about franciscana prey ingestion were obtained near 150 years ago (References 1 and 2, Table 2.1), when cephalopod beaks (identified as Loliginidae) and remains of the fish *Odonthestes argentinensis* found in the stomach contents of two animals from the Argentinean coast were reported. Since then, 37 research studies (References 1 to 37, Table 2.1) analyzed nearly 2700 franciscana stomachs, mainly sampled during the 1970s (Fig. 2.2A and B) throughout its distribution. Despite the increasing number of published studies applying the stomach contents methodology (Fig. 2.2A), the number of stomachs analyzed has decreased (Fig. 2.2B). This could be explained by the increase in the effort toward franciscana research

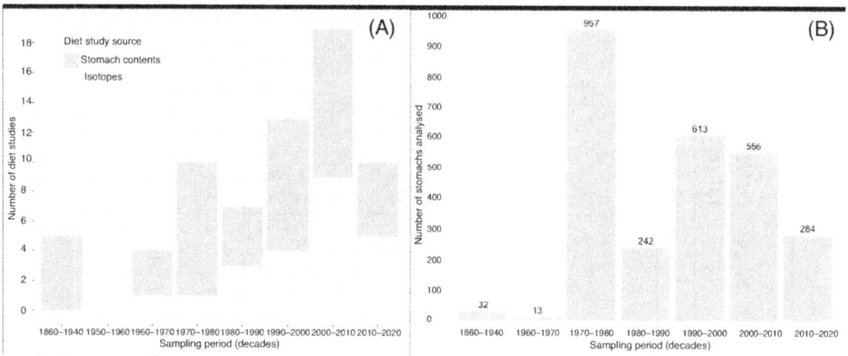

Fig. 2.2 ***Franciscana dietary studies from stomach content and stable isotopes (bone, muscle, skin and teeth) methodologies published from 1867 up to 2021.*** The sampling period refers to that of the franciscana sampling, instead of the year of publication of the study. (A) number of studies per decade. (B) Number of stomachs analyzed per decade.

in the northern areas through the years. These northern areas are geographically smaller and studies count on fewer franciscana samples, in comparison with the southern areas, where the main source of dietary information was from higher numbers of bycaught dolphins in gillnet fisheries.

It was only a decade ago that the first study of the franciscana feeding habits using stable isotope analysis was published (Reference 38, Table 2.1), with a total of 13 research studies (References 38 to 50, Table 2.1) analyzing samples of muscle, bone, skin, and tooth. However, as stable isotopes can be analyzed in a variety of preserved samples, many of these studies investigated the diet of animals collected since the 1950s (Fig. 2.2A). The feeding habits of franciscanas have been well recorded in the southern areas of its distribution, with the majority of feeding studies based on the analysis of stomach contents. On the other hand, in the northern regions, the analysis of stable isotopes has played an equally important role as a source of dietary information.

The increase in stable isotope studies in the last few years reflects the strengths of this methodology, as samples obtained from different tissues represent different temporal scales, and data collected and preserved in the past can be used nowadays for analysis.

Prey ecological overview

Franciscana feeding studies have reported at least 100 prey species from 41 families. A total of 88 prey items have been identified to the species level, 72 of which are fish, 8 cephalopods, and 8 crustacean species (Table 2.2).

Table 2.2 Ecological information of franciscana prey species from diet studies (n=37) of stomach contents analysis (n = 2697) throughout its distribution, from 1867 to 2021.

Familiy	Code	Species	Common name	Habitat	Relevant information bottom and pelagic environment	Depth ranges (m)[a]	Fisheries target[b]	Bycatch
Fish								
Sciaenidae	Cg	*Cynoscion guatucupa* (sin. *C. striatus*)	Striped weakfish	Dem-pel	Muddy and sandy bottoms	0–190 (<50)	HE	
	Cj	*Cynoscion jamaicensis*	JAMAICA weakfish	Demersal	Muddy and sandy bottoms	0–70 (20–55)	OE	Yes
	Cl	*Cynoscion leiarchus*	Smooth weakfish	Demersal	Muddy and sandy bottoms	10–50 (<25)	Yes	Yes
	Cm	*Cynoscion microlepidotus*	Small-scale weakfish	Demersal	Muddy and sandy bottoms	0–30	Yes	
	Cv	*Cynoscion virescens*	Green weakfish	Demersal	Muddy and sandy bottoms	0–70	Yes	
	Ctg	*Ctenosciena gracilicirrhus*	Barbel drum	Demersal	Muddy and sandy bottoms	10–130	Yes	
	Ip	*Isopisthus parvipinnis*	Bigtooth corvina	Demersal	Muddy and sandy bottoms	0–45	Yes +	Yes
	Lb	*Larimus breviceps*	Shorthead drum	Demersal	Muddy and sandy bottoms	0–60	Yes +	Yes
	Ma	*Macrodon ancylodon*	King weakfish	Demersal	Muddy and sandy bottoms	0–120 (<40)	HE	
	Mat	*Macrodon atricauda*	Southern king weakfish	Demersal	Soft bottoms	0–100 (<30)	HE	
	Me	*Menticirrhus americanus*	Southern kingcroacker	Demersal	Sandy mud to hard sand bottoms	0–60 (<20)	OE	Yes

Familiy	Code	Species	Common name	Habitat	Relevant information bottom and pelagic environment	Depth ranges (m)[a]	Fisheries target[b]	Bycatch
	Ml	*Menticirrhus littoralis*	Gulf kingcroaker	Demersal	Muddy and sandy bottoms (enters estuaries)	0–50	OE	Yes
	Mf	*Micropogonias furnieri*	White croaker	Demersal	Muddy and sandy bottoms	0–180 (<80)	OE	
	Mu	*Micropogonias undulatus*	Atlantic croaker	Demersal	Muddy and sandy bottoms	0–100	Yes	Yes
	Pbr	*Paralonchurus brasiliensis*	Banded croacker	Demersal	Muddy and sandy bottoms	0–80 (<60)	Yes	Yes
	Pc	*Pogonias cromis*	Black drum	Demersal	Muddy and sandy bottoms	0–50	Yes	
	Sb	*Stellifer brasiliensis*	Cangoá	Demersal	Soft muddy bottoms	0–50 (<20)	Yes	Yes
	Sr	*Stellifer rastrifer*	Rake stardrum	Demersal	Soft muddy bottoms	0–40 (<20)	Yes	Yes
	Uc	*Umbrina canosai*	Argentine croaker	Demersal	Muddy and sandy bottoms	0–180 (>40)	HE	
Engraulidae	Af	*Anchoa filifera*	Longfinger anchovy	Pelagic	Schooling species	0–35	Yes	
	Al	*Anchoa lyolepis*	Shortfinger anchovy	Pelagic	Schooling species	0–50	Yes	
	Am	*Anchoa marinii*	Anchovy	Pelagic	Schooling species	0–200 (<40)	Yes	
	As	*Anchoa spinifer*	Spicule anchovy	Pelagic	Schooling species	0–55 (5–30)	Yes	Yes
	At	*Anchoa tricolor*	Piquitinga anchovy	Pelagic	Schooling species (enters brackish water)	0–50	Yes +	
	Aal	*Anchoviella lepidentostole*	Broadband anchovy	Pelagic	Schooling species found inshore	0–50	OE	

(Continued)

Table 2.2 Cont'd

Familiy	Code	Species	Common name	Habitat	Relevant information bottom and pelagic environment	Depth ranges (m)[a]	Fisheries target[b]	Bycatch
	Ce	*Cetengraulis edentulus*	Atlantic anchoveta	Pelagic	Inshore and forms quite large schools	0–25	Yes +	
	Ea	*Engraulis anchoita*	Argentine anchovy	Pelagic	Forming dense schools	0–200 (>30)	Yes +	
	Lg	*Lycengraulis grossidens*	Atlantic sabre-tooth anchovy	Pelagic	Enters brackish waters and freshwater	0–60 (<20)	Yes	Yes
Clupeidae	Ba	*Brevoortia aurea*	Brazilian menhaden	Pelagic	Schooling species	0–20	Yes +	
	Bp	*Brevoortia pectinata*	Argentine menhaden	Pelagic	Schooling species (enters brackish water)	0–20	Yes	
	Oo	*Opisthonema oglinum*	Atlantic thread herring	Pelagic	Solitary and schooling species	0–50	Yes +	
	Ra	*Ramnogaster arcuata*	Jenyns's sprat	Pelagic	Schooling species	0–50	Yes	Yes
	Rm	*Ramnogaster melanostoma*	Uruguay river sprat	Pelagic	Confined at mouths of rivers or closer	0–10		
	Sab	*Sardinella brasiliensis*	Brazilian sardinella	Pelagic	Forming compact schools (enters estuaruies)	0–60	OE	
Atherinopsidae	Oa	*Odontesthes argentinensis*	Pejerrey	Pelagic	Enters brackish water and freshwater	0–25	Yes +	
	Ob	*Odontesthes bonariensis*	Argentinian silverside	Pelagic	Enters brackish water and freshwater	0–25	Yes +	
	Oi	*Odontesthes incisa*	Silverside	Demersal	Enters brackish water and freshwater		Yes	

Familiy	Code	Species	Common name	Habitat	Relevant information bottom and pelagic environment	Depth ranges (m)[a]	Fisheries target[b]	Bycatch
Haemulidae	Hp	*Haemulon plumierii*	White grunt	Demersal	Dense aggregations on sandy bottoms	0–40	Yes +	
	Hc	*Haemulopsis corvinaeformis*	Roughneck grunt	Demersal	Sandy or rocky bottoms (enters estuaries)	0–50	Yes +	
	Pr	*Orthopristis ruber*	Corocoro grunt	Demersal	Sand, rock, or mud bottoms	0–70	Yes +	
Pristigasteridae	Cb	*Chirocentrodon bleekerianus*	Dogtooth herring	Pelagic	Enters brackish water	0–60	Yes	
	Om	*Odontognathus mucronatus*	Guiana longfin herring	Pelagic	Inshore (enters brackish water)	0–40	Yes	Yes
	Ph	*Pellona harroweri*	American coastal pellona	Pelagic	Muddy bottoms	0–35	Yes	Yes
Carangidae	Cc	*Chloroscombrus chrysurus*	Atlantic bumper	Pelagic	Soft bottoms (enters brackish water)	0–110	Yes	
	Ps	*Parona signata*	Parona leatherjacket	Dem-pel	Enters brackish water	0–80 (<40)	HE	Yes
	Tla	*Trachurus lathami*	Rough scad	Dem-pel	Schooling species	0–350 (20–180)	Yes +	
Stromateidae	Ppa	*Peprilus paru*	American harvestfish	Dem-pel	Forming large schools (near brackish water)	0–200 (25–70)	Yes	
	Sbr	*Stromateus brasiliensis*	Butterfish	Dem-pel	Schooling species	0-200 (>20)	Yes	
Paralichthyidae	Pi	*Paralichthys isosceles*	Flounder	Demersal	Hard or sandy substrates, rarely on mud	20–250 (60–250)	HE	
	El	*Etropus longimanus*	Flatfish	Demersal	Soft bottoms	0–190		Yes

(Continued)

Table 2.2 Cont'd

Familiy	Code	Species	Common name	Habitat	Relevant information bottom and pelagic environment	Depth ranges (m)[a]	Fisheries target[b]	Bycatch
Mugilidae	Mli	*Mugil liza* (sin. *M. platanus*)	Mullet	Dem-pel	Form sizeable schools (near estuaries)	0–200 (<20)	OE	
	Cd	*Chelon dumerili*	Grooved mullet	Dem-pel	Coastal estuarine, brackish, or marine waters	0–300 (<20)	Yes	
Congridae	Co	*Conger orbignianus*	Argentine conger	Demersal	Sandy bottoms	10–380	Yes	Yes
	Asp	*Ariosoma* sp.	Conger	Demersal	Sandy bottoms			
Serranidae	Da	*Dules auriga* (sin. *Serranus auriga*)	Cochero/ mariquita	Demersal	Sandy bottoms	15–140	Yes	Yes
Batrachoididae	Ppo	*Porichthys porosissimus*	Lantern mid-shipman	Demersal	Sandy bottoms	0–250 (20–120)	Yes	Yes
Centropomidae	Cu	*Centropomus undecimalis*	Common snook	Demersal	Enters brackish waters and freshwater	0–25	Yes +	
Cyclopsettidae	Sp	*Syacium papillosum*	Dusky flounder	Demersal	Sandy bottoms	10–140	Yes	Yes
Gerreidae	Eb	*Eugerres brasilianus*	Brazilian mojarra	Demersal	Soft bottoms	0–30	Yes	
Gobiidae	Go	*Gobionellus oceanicus*	Highfin goby	Demersal	Muddy and sandy bottoms (near estuaries)	0–40		
Merlucciidae	Mh	*Merluccius hubbsi*	Argentine hake	Dem-pel	Migrate inshore during spring and summer	20–800 (40–180)	HE	
Ophichthyidae	Og	*Ophicthus gomesii*	Shrimp eel	Demersal	Soft bottoms	0–450 (<90)		

Familiy	Code	Species	Common name	Habitat	Relevant information bottom and pelagic environment	Depth ranges (m)[a]	Fisheries target[b]	Bycatch
Ophidiidae	Rbr	*Raneya brasiliensis* (sin. *R. fluminensis*)	Cusk eel	Dem-pel	Uncommon in very coastal waters	30–150	Yes	
Percophidae	Pe	*Percophis brasiliensis*	Brazilian flat-head	Demersal	Sandy bottoms	0–80	Yes +	
Phycidae	Ubr	*Urophycis brasiliensis*	Squirrel cod-ling	Demersal	Sandy bottoms	0–250 (<180)	Yes +	
Pomatomidae	Psx	*Pomatomus saltatrix*	Bluefish	Pelagic	Migrate to warmer water during winter	0–200 (<30)	HE	
Sparidae	Ppg	*Pagrus pagrus*	Red porgy	Dem-pel	Rock, rubble, or sand bottoms	0–250 (80–180)	Yes +	
Syngnathidae	Lbl	*Leptonotus blainvil-leanus*	Deep-bodied pipefish	Demersal	Found in algal beds in coastal areas	0–250(<25)		
Trichiuridae	Tla	*Trichiurus lepturus*	Cutlassfish	Dem-pel	Muddy bottoms (enter estuaries)	0–350 (<250)	Yes	Yes
Triglidae	Psp	*Prionotus* sp.	Searobins	Demersal	Muddy and sandy bottoms			
Nomeidae	Csp	*Cubiceps* sp.	Driftfish	Pelagic	Not common in inshore waters			
Cynoglossidae	Ssp	*Symphurus* sp.	Tonguefish	Demersal	Muddy and sandy bottoms			
Cephalopod								
Loliginidae	Ds	*Doryteuthis sanpau-lensis*	Common long-finned squid	Dem-pel	Concentrates near the bottom during the day	0–250 (<60)	Yes +	Yes

(Continued)

Table 2.2 Cont'd

Familiy	Code	Species	Common name	Habitat	Relevant information bottom and pelagic environment	Depth ranges (m)[a]	Fisheries target[b]	Bycatch
	Dp	*Doryteuthis plei*	Slender inshore squid	Dem-pel	Congregate in large numbers near shore	0–370 (15–280)	Yes +	Yes
	Lbr	*Lolliguncula brevis*	Inshore squid	Dem-pel	Commonly in estuaries	0–20	Yes	Yes
Ommastrephidae	Ia	*Illex argentinus*	Argentine short-finned squid	Dem-pel	Sometimes congregate near seafloor	0–800 (80–400)	HE	
Sepiolidae	St	*Semirossia tenera*	Lesser bobtail squid	Demersal	Muddy and sandy bottoms	50–200		
Argonautidae	An	*Argonauta nodosa*	Knobby argonaut	Pelagic	Prefers to live near the surface level	20–500		
Octopodidade	Ot	*Octopus tehuelchus*	Tehuelche octopus	Benthic	Muddy and sandy bottoms	0–90 (<30)	Yes	
Eledonidae	Eg	*Eledone gaucha*	Octopus	Benthic	Muddy and sandy bottoms	40–600 (40–120)	Yes	
Crustacean								
Penaeidae	Alo	*Artemesia longinaris*	Marine shrimp	Benthic	Muddy and sandy bottoms	0–120 (<50)	Yes +	
	Pp	*Penaeus paulensis*	Pink shrimp	Benthic	Muddy and sandy bottoms	0–130 (35–50)	Yes +	
	Xk	*Xiphopenaeus kroyeri*	Atlantic seabob	Benthic	Muddy and sandy bottoms	0–70 (<30)	Yes +	

Familiy	Code	Species	Common name	Habitat	Relevant information bottom and pelagic environment	Depth ranges (m)[a]	Fisheries target[b]	Bycatch
Solenoceridae	Pm	*Pleoticus muelleri*	Argentine red shrimp	Benthic	Muddy and sandy bottoms	0–170 (5–25)	Yes +	
Sergestidae	Ppe	*Peisos petrunkevitchi*	Marine shrimp	Benthic	Soft bottoms	0–20 (10–20)		Yes
Mysidae	Na	*Neomysis americana*	Opossum shrimp	Benthic	Soft bottoms	0–240		
Diogenidae	Di	*Dardanus insignis*	Red brocade hermit	Benthic	Sandy bottoms	0–500 (25–190)		Yes
	Llo	*Loxopagurus loxocheles*	Marine crab	Benthic	Sandy bottoms	0–30 (15–25)		Yes

[a]The range in which the species are most abundant is given in parenthesis.
[b]HE: heavily exploited; OE: overexploited; Yes: reported in any type of artisanal or commercial fisheries; Yes +: important target for the commercial fleet.

Franciscanas feed mostly on demersal fish (mainly bottom-dwelling teleosts) and cephalopod species, followed by pelagic fish and crustaceans. Sciaenids account for most of the fish species; Loliginidae and Penaeidae for the squids and shrimps, respectively.

The franciscana lives in coastal areas, preferring up to the 30 m isobath (Danilewicz et al., 2009). In many regions, they occur in transitional zones like estuarine areas, which reflects the behavior of its demersal prey that seems to prefer muddy and sandy bottoms, many entering brackish water and estuaries. Their pelagic prey includes mostly schooling species (Table 2.2 and Fig. 2.3). The distribution of franciscana prey species according to depth shows that many species are abundant in waters less than 50 m deep (Fig. 2.3). Foraging in those habitats with higher prey abundance can increase franciscana chances of encountering prey. Therefore, franciscana seems to be an opportunistic consumer, feeding upon prey most frequently found in its habitat (e.g., Campos et al., 2020; Di Beneditto and Ramos, 2001; Pinedo, 1982). This behavior is reflected in the seasonal fluctuations of its diet, which coincide with the pattern of variation observed in the abundance of prey species through the year (e.g., Henning et al., 2017), mainly in the southern regions of its distribution (e.g., Bassoi et al., 2020; Paso-Viola et al., 2014).

Spatial dietary variation

The franciscana endemic distribution along the western South Atlantic ranges from warm waters in Itaúnas (18°25'S) (Espírito Santo, Brazil) to cold waters in Golfo San Matias (41°10'S) in Argentina (Crespo et al., 1998; Siciliano et al., 2002). Distinct features of the coastal areas along its distribution, such as continental shelf extent, current systems and boundaries, local productivity, upwellings, and the presence of estuaries and their associated distinct ecosystems influence the presence of the available prey. Therefore, differences in the prey species composition of the franciscana diet throughout its range are expected.

The northernmost regions (FMA I) are mainly influenced by warm oligotrophic waters carried by the Brazil Current (BC), and are also characterized by the upwelling of South Atlantic Central Waters, mainly in spring and summer months (Campos et al., 2000). Two main freshwater sources are present in these areas, the Doce and Paraíba do Sul rivers (Gonçalves and Carvalho, 2006; Pinto et al., 2015). There are three estuarine complexes within FMA II: Cananéia, Paranaguá estuarine complex and Babitonga Bay; but franciscanas only occupy the Paranaguá and Babitonga estuaries

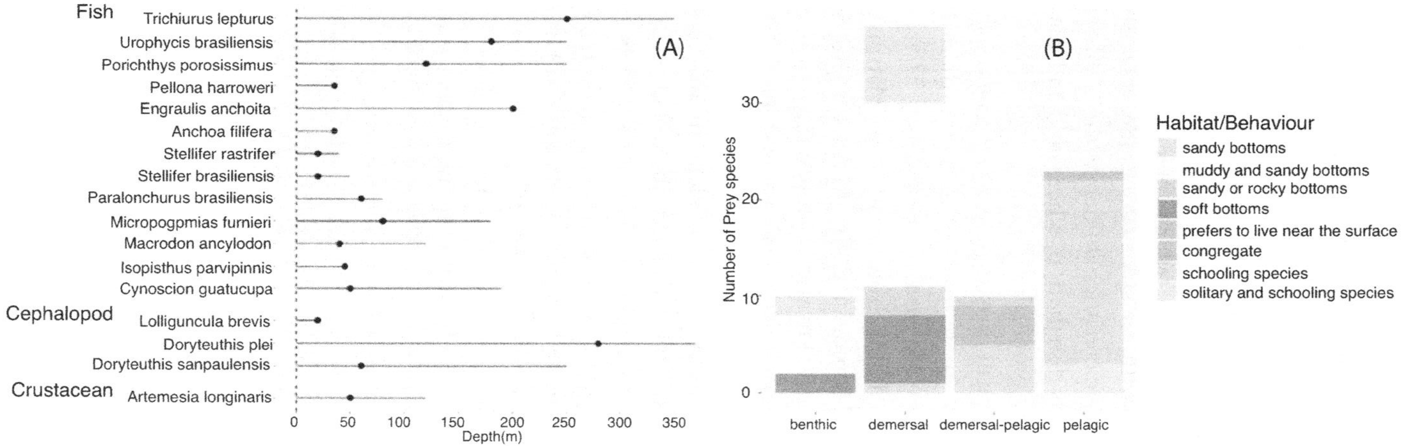

Fig. 2.3 ***Depth ranges of franciscana prey species (>5% FO, n = 17) throughout its distribution.*** The lines represent the maximum ranges and the dots, the depth in which the species are most abundant (A) and franciscana prey species habitat and behavior information throughout its distribution (B).

(Cremer and Simões-Lopes, 2005; Santos et al., 2009). The southernmost areas of franciscana distribution, FMAs III and IV, are mainly influenced by the Malvinas (Falklands) Current (MC), where cold waters of subantarctic origin (reaching 24°S throughout coastal waters in wintertime) meet warm waters of tropical origin carried by the BC (Campos et al., 1995). Coastal areas are also influenced by the freshwater discharge of the Patos Lagoon and the Río de la Plata estuaries, where franciscanas are found in both Uruguay and Argentina (within FMA IV) sides (Secchi, 2010) in the latter. These oceanographic characteristics influence the availability of distinct prey species along with its distribution, where cold and warm prey species are consumed in different proportions.

Species related to warm waters (e.g., *Doryteuthis plei, Stellifer* spp., *Cynoscion jamaicensis*) were more frequent in the northern FMAs, whereas cold water species (e.g., *Doryteuthis sanpaulensis, Micropogonias furnieri, Engraulis anchoita*) prevailed in the southern FMAs (Martins and Haimovici, 2017, 2020; Santos and Haimovici, 2001). Moreover, prey species that are more sensitive to cold and warm water temperatures (e.g., *Stellifer* spp. and *Pellona harroweri* vs *M. furnieri* and *Cynoscion guatucupa*), were not part of the franciscana diet in the southernmost and northernmost FMAs, respectively. These preferences were clearly evident in the modular network (Qw = 0.45). Indeed, three modules were identified in the network that could be associated with geographic regions of the FMAs: module 1 comprised FMAs Ib and II and 11 prey species, module 2 included FMAs III and IV and 12 prey and module 3 contained only FMA Ia and eight prey items (Fig. 2.4). Cold water prey species (*Doryteuthis sanpaulensis, Trichiurus lepturus*, and *Paralonchurus brasiliensis*) connect modules 1 and 2, while module 3 was connected to module 1 through warm water prey species (*Stellifer* sp., *Isopisthus parvipinnis*, and *Chirocentrodon bleeker*). In module 1, the squid *D. plei* and, to a lesser extent, the fish *P. harroweri* occupy central positions. Within module 2 *D. sanpaulensis* and *C. guatucupa* showed the strongest links with FMAs III and IV and *T. lepturus* and *P. brasiliensis* had strong connection to only FMA III. Almost all prey species within module 3 were connected to module 1, and only those prey identified as higher-level taxa (cephalopod, Dendrobrachiata and Engraulidae) were exclusively connected to FMA Ia. It is worth noting that stomach content data for this FMA is based on a single study that analyzed 18 stomachs (Table 2.1).

From northern to southern FMAs, there is a decrease in prey diversity, with fewer species representing more than 5% of occurrence in the franciscana diet (Fig. 2.5). This latitudinal variation is probably the reflection of

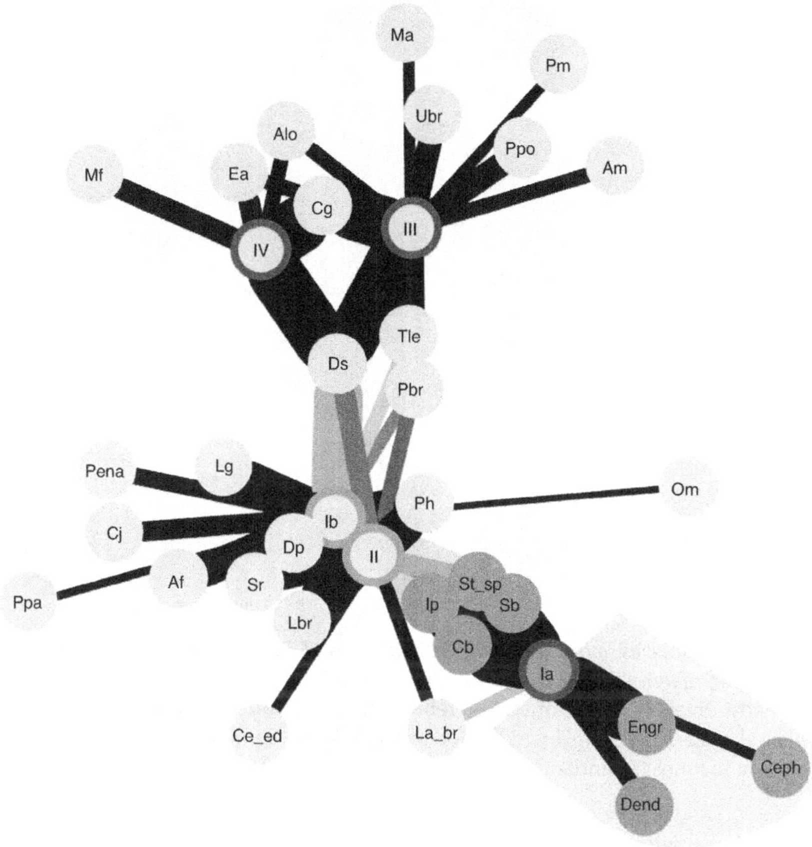

Fig. 2.4 ***Observed modular structure of the Fransiscana Management Areas (FMAs Ia, Ib, II, III and IV) and prey bipartite network.*** The nodes represent the FMAs and prey types and the links denote prey consumed by franciscanas of each FMA. Link thickness is proportional to the mean frequency of occurrence of the prey in the diet of franciscanas from each FMA. Modules 1 (blue), 2 (green), and 3 (orange) are shown. Black links connect prey to FMAs within the same module and colored links connect prey shared by two modules. Prey species abbreviations according to Table 2.2.

the latitudinal gradient of decreasing fish biodiversity between the northern warm areas and the southern colder areas (Pinheiro et al., 2018; Vilar et al., 2017). This biodiversity pattern is usually considered as derived from the overlap of tropical and temperate species around Rio de Janeiro region, which is the transition zone between the Tropical Atlantic and Temperate South America biogeographic realms (Floeter et al., 2008; Pinheiro et al., 2018; Spalding et al., 2007). On the other hand, the shallow coastal waters

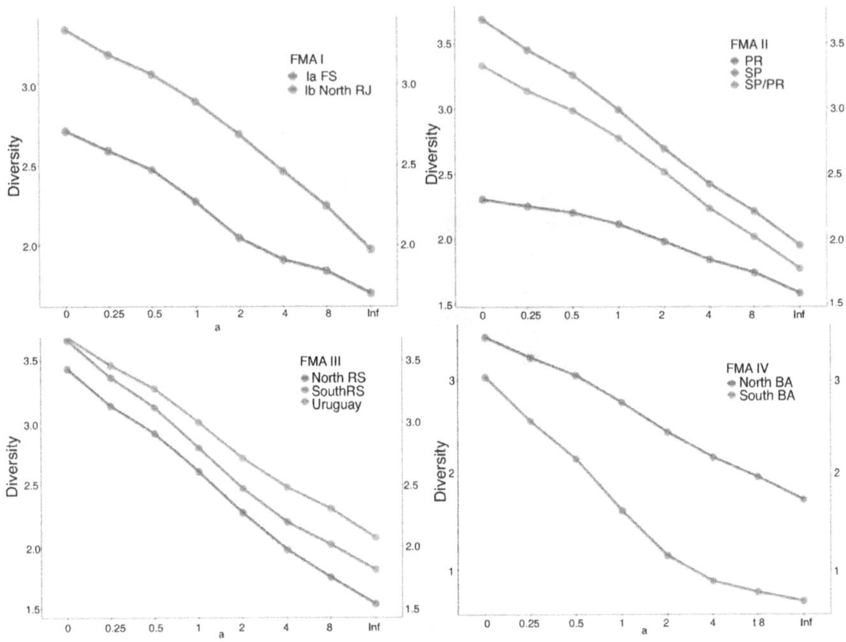

Fig. 2.5 ***Prey species richness and diversity profiles of Rényi comparing the franciscana geographic regions within the FMAs.*** Only the marine environment is contemplated. The Rényi diversity profile at scale 0 reflects species richness, at scale 1 the Shannon index, at scale 2 the Simpson index and at scale Inf the Berger-Parker index (the dominance of the most abundant prey species).

of FMAs III and IV represent an important spawning area for many demersal teleosts, with dense schools of recruits (<50 mm) and *C. guatucupa* is the most abundant teleost (Haimovici et al., 1996). These FMAs are important productive fishing areas of demersal resources, represented by fewer species with great abundance (spawning and recruitment grounds), which could explain the low diversity in the franciscana diet. Moreover, franciscanas seem to prey on more pelagic species in northern FMAs when compared to southern FMAs, and this could be explained by the great demersal fish assemblages in the southern areas, as previously mentioned.

In a small spatial scale, similar biodiversity differences could be found within each FMA (Fig. 2.5). For FMA I, prey diversity difference between Espirito Santo (ES - FMA Ia) and North RJ (FMA Ib) regions is more evident probably because of the gap in franciscana distribution between these areas (differences in the shelf extent and habitats, Amaral et al., 2018), but this should be looked at carefully since in the ES region the number of stomach contents analyzed is smaller. For FMA II, the species richness distinction between SP and Paraná (PR) coastal areas is clear despite the geographical distance between them. Prey species richness diverge among the geographic regions in the FMA III, and significant divergences in prey composition between South RS and North RS were also reported (Bassoi et al., 2020). Looking at FMA IV, the South BA region is noticeably less diverse in terms of reported prey items than North BA, but this should be taken with caution as South BA sample size is small. However, research studies for the area have shown important differences in the franciscana feeding regime between the nearest regions in this FMA (Denuncio et al., 2017; Rodriguez et al., 2002).

Our meta-analysis of franciscanas' isotopic values throughout the species distributional range has shown remarkable latitudinal patterns in both carbon and nitrogen stable isotopes, which might reflect the preference for isotopically different prey species, but also differences at the base of food webs. The lowest $\delta^{13}C$ values in teeth were observed in franciscanas from FMA Ia (-12.6‰ ± 0.5), whereas the highest values for the same tissue were observed in individuals from FMA III (-10.8‰ ± 0.6, Fig. 2.6). In

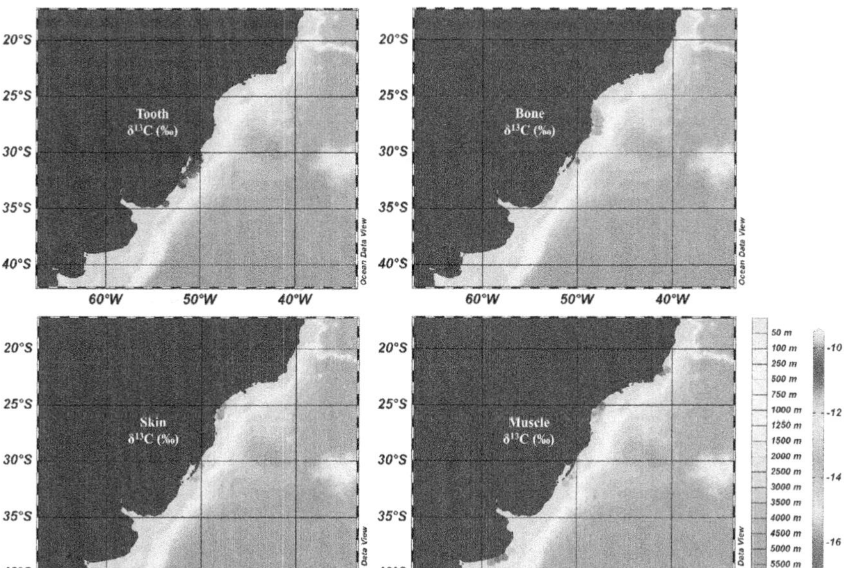

Fig. 2.6 ***Isotopic maps representing the patterns in carbon stable isotope values in the different tissues of franciscanas throughout the species' distributional range in the coastal waters of eastern South America.***

general, all tissues showed similar latitudinal patterns, with the lowest $\delta^{13}C$ values in the northernmost portion of franciscanas' distribution (FMA Ia), increasing southwards, reaching the highest values in individuals from FMA III, slightly decreasing again in franciscanas from FMA IV (Fig. 2.6). The influence of upwelling of cold South Atlantic Central Waters could be the cause of low carbon isotope values at the base of food webs in FMAs Ia and b (Troina et al., 2020a) as low $\delta^{13}C$ values are typically observed in deep-water dissolved inorganic carbon (McMahon et al., 2013). On the other hand, the influence of large estuaries (Patos Lagoon and Río de la Plata), characterized by extensive salt-marshes composed of ^{13}C-enriched C4 producers such as *Spartina* and *Ruppia* (Isacch et al., 2006), might be responsible for the high $\delta^{13}C$ values observed in FMAs III and IV. Nevertheless, the higher proportional consumption of ^{13}C-depleted pelagic prey species of franciscanas from FMAs Ia and Ib should also account for the lower carbon isotopic values observed in the northernmost areas.

Similar latitudinal increase was observed in $\delta^{15}N$ values, with franciscanas from FMAs III and IV showing between 2‰ and 6‰ increase in $\delta^{15}N$ in comparison to those from FMAs Ia, Ib, and II (Fig. 2.7). Such latitudinal

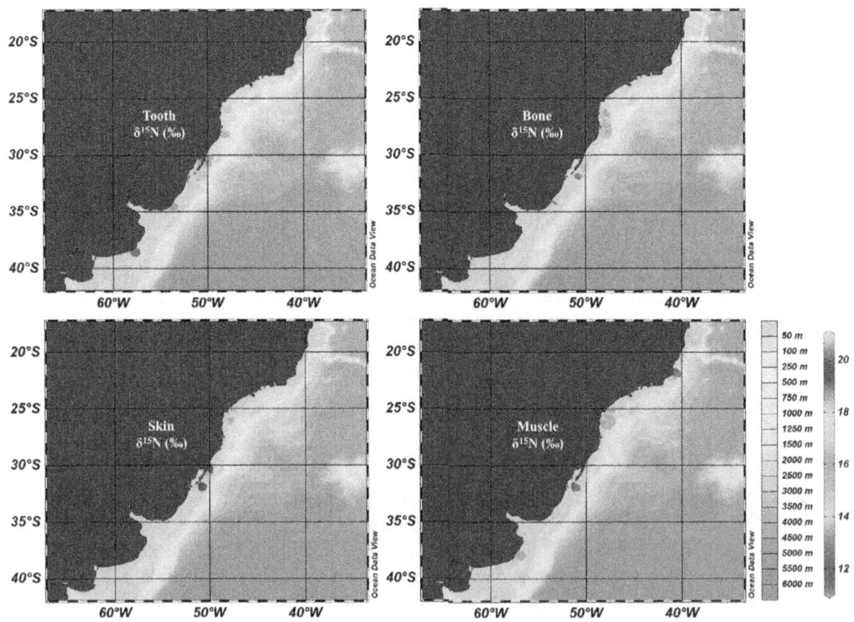

Fig. 2.7 ***Isotopic maps representing the patterns in nitrogen stable isotope values in the different tissues of franciscanas throughout the species' distributional range in the coastal waters of eastern South America.***

increase in $\delta^{15}N$ values does not seem to be due to differences in the trophic position among populations from different FMAs as, although the importance of prey species varies throughout franciscanas' latitudinal range, their diet mainly comprises secondary and tertiary consumer fish and squid species. Instead, the latitudinal increase in $\delta^{15}N$ values is most likely related to isotopic gradients at the base of the food webs.

The coastal water (CW) receiving low $\delta^{15}N$ nutrients from continental sources and the presence of the oligotrophic Tropical Water (TW), carried by BC (Knoppers et al. 2009), might be the cause of the relatively low $d^{15}N$ values found from franciscanas from the northernmost areas. N2-fixation by cyanobacteria and recycling N (low $\delta^{15}N$) are important in oligotrophic regions (Sigman et al., 2009). There is low isotopic fractionation during its uptake by phytoplankton (Minagawa and Wada, 1984; Montoya, 2008), thus producing low basal $d^{15}N$ values. On the other hand, the highest nitrogen isotope values were found in the southern areas (FMAs III and IV). The seasonal influence of the freshwater discharge from the Río de la Plata and, to a lesser extent, the Lagoa dos Patos estuary and the convergence of subantarctic shelf waters (SASW), and tropical and subtropical (Central) waters (STCW), result in high nutrients concentration and biological productivity (Ciotti et al., 1995). These special oceanographic conditions must be sources of high $\delta^{15}N$ nutrients fertilizing coastal food webs, producing high nitrogen isotope values that could be observed in predators foraging in these regions (Coletto et al., 2021; Troina et al., 2020b).

Temporal dietary variation

Most franciscana prey species are targeted in artisanal or commercial fisheries, and about 18% are heavily or overexploited; various prey species have also been reported as bycatch (Table 2.2). Marine mammals and fisheries interact indirectly through trophic pathways, competing for food resources (Beverton, 1985). Thus, fluctuation in fish abundances as a consequence of anthropogenic activities can affect the feeding habits of franciscana. For example, shifts in prey composition due to the depletion of important prey by fisheries, were reported for franciscanas in the southern coast of Brazil (Bassoi and Secchi, 1999). This type of ecological interaction with fisheries would be present throughout the franciscana distribution, although with different degrees of occurrence.

Three diet studies were available for the North RJ geographic region. The first study encompasses stomach content data for 89 individuals collected between 1986 and 1999 (reference 22 in Table 2.1). The second study reanalyzed those data adding information of 10 individuals collected during the 2000s (reference 25 in Table 2.1). There is almost no change in fish prey composition and occurrence when comparing both studies but interestingly the shift in the importance between the squids *D. sanpaulensis* and *D. plei* is evident. *D. sanpaulensis* was the main cephalopod in the diet before 2000; conversely, *D. plei* became the most ingested cephalopod in the second study, after adding only 10 more stomach samples. The exchange between these squid species ingested over time could be related to upwellings during intense and weaker El Niño events. Upwelling events are more pronounced in Cabo Frio system during strong ENSO, and an inverse situation occurs during a "weaker" El Niño with typical winter temperatures higher than 22°C (Coelho-Souza et al., 2012; Paes and Moraes, 2007). During the decades 1990 to 2010 there were three intense and three weaker El Niño events (Pinheiro et al., 2018), probably influencing the abundance of *D. sanpaulensis* which is one of the upwelling's most important fisheries target species (Valentin, 2001), and the warm water squid *D. plei* (Santos and Haimovici, 2001). This prey-switching behavior can be beneficial to franciscana, ensuring continuous food intake and foraging efficiency, through its ability to use alternative prey sources.

Similarly, franciscana feeding regime in the SP region (FMA II) showed an inverse trend in the ingestion of cephalopod species throughout time (Fig. 2.8); *D. sanpaulensis* decreasing and *D. plei* increasing in occurrence. The coast of SP is also highly influenced by El Niño events, which could explain these species shifting in occurrence. Depletion by fisheries is unlikely to be the reason affecting the squids' abundance in the area as only small-scale fisheries in coastal waters (<30 m) catch the squid *D. plei* (Rodrigues and Gasalla, 2008). Another possible explanation for the increase of *D. plei* in franciscanas' diet could be related to an increase in its abundance due to higher availability of food resources. Gasalla et al. (2010) studying the trophic role of *D. plei* in the coast of SP found that Engraulids (*Anchoa* spp.) were the second most important fish in the diet of this squid. Our results showed an increase in the occurrence of *Anchoa* spp. (Fig. 2.8) in franciscanas' diet through time, possibly reflecting the increase in abundance for this species which is an important prey resource for the squid. Other complex interactions could also be occurring as *D. plei* may potentially be a keystone species in the SP region (Gasalla et al., 2010).

Lolliguncula brevis showed a substantial increase in the diet of franciscanas over time. This squid species is found in shallow and low salinity waters in the estuarine region off the coast of SP (Araújo and Gasalla, 2018). The increase in the occurrence of *L. brevis* could be either due to franciscanas

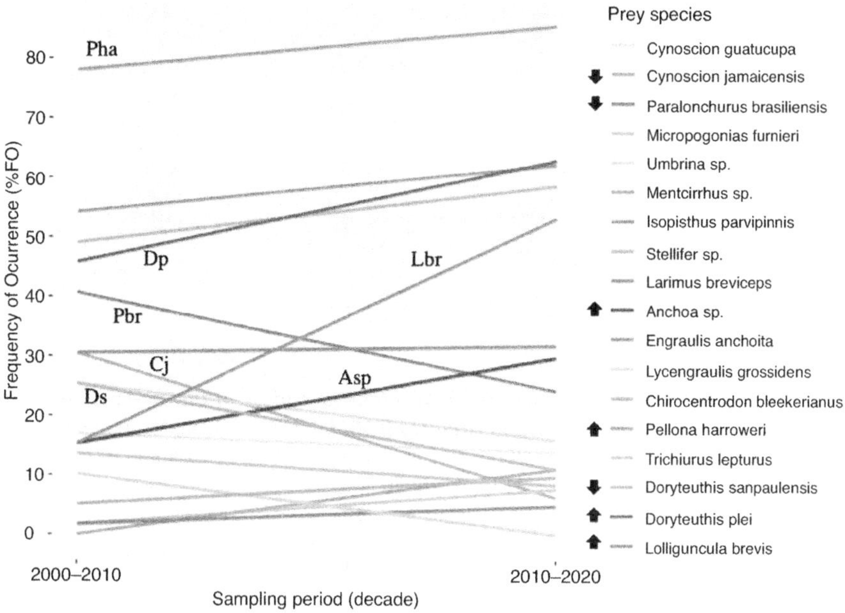

Fig. 2.8 ***Franciscana prey species (% FO) overtime for the geographic region SP (FMA II).*** $N = 58$ stomachs for 2000–10, and $N = 145$ for 2010–20.

eating and spending more time in shallower waters or to an increase of this prey in coastal zones because of freshwater discharge from the estuarine areas (e.g., La Niña events). Moreover, Gasalla and Rossi-Wongtschowski (2004) found that the heavy exploitation of weakfish (*Cynoscion jamaicensis*) by the bottom-trawlers along SP coast could result in the increment in squid biomass. Indeed, a decline in the frequency of occurrence of *C. jamaicensis* (Fig. 2.8) in the diet of the franciscanas was detected thus reinforcing a possible decline in the abundance of this important *L. brevis* predator. Nonetheless, whether these fluctuations in the importance of franciscana prey species in the SP geographic region were due to trophic interactions or resulting from oceanographic influences (or both), is still inconclusive.

Uruguay is the geographic region with the most representative studies and samples of franciscana stomach contents (see Table 2.1), for both marine and estuarine areas. One of the most evident prey species fluctuations for the marine area is the strong decrease in the occurrence of the striped weakfish (*Cynoscion guatucupa*) after the 1990s (Fig. 2.9), which coincided with the decline in the commercial catches (Gianelli and Defeo, 2017) of this heavily exploited species (see Table 2.2). Concomitantly, we observed higher occurrence over time of other prey species that are not

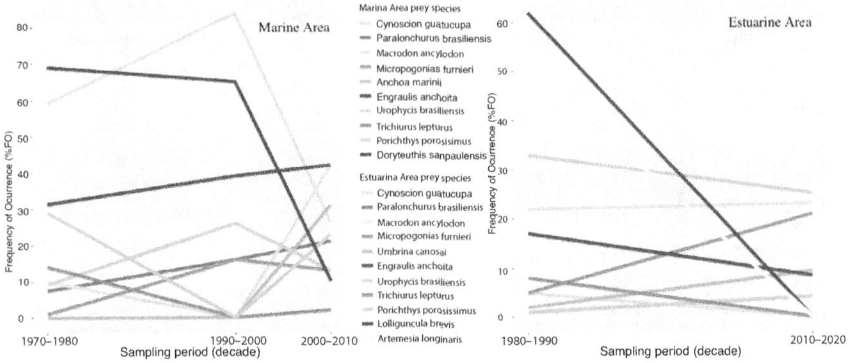

Fig. 2.9 ***Franciscana prey species (% FO) over time (decades) for the marine and estuarine areas of the geographic region Uruguay (FMA III).*** Marine Area: *N* = 633 stomachs for 1970–80, *N* = 110 for 1990–2000, and *N* ≈ 38 for 2000–10. Estuarine area: *N* = 157 stomachs for 1980–90, and *N* = 41 for 2010–20.

important targets for the Uruguayan fisheries, such as *Porichthys porossimus* and *Paralonchurus brasiliensis*. On the other hand, in the last decade, *M. furnieri* had occurred more frequently in the diet of franciscanas in the estuarine area, where it still has one of the main catch rates by fisheries (Vögler et al., 2020). This species is also important in the diet of franciscanas in the estuarine waters of Argentina (Rodriguez et al., 2002). Inversely, the shrimp prey (*Artemesia longinaris*) had an abrupt reduction in the dolphin diet, reflecting the continuous decline of this crustacean landings by the Uruguayan fleet after 2000 (Gianelli and Defeo, 2017). Clear shifts in the franciscana diet in parallel to changes in Uruguayan fisheries landings provide indirect evidence that this dolphin responded rapidly to variation in prey abundance. Notwithstanding, the decreases of squid species occurrence in the diet both for the estuarine area (near 60% *L. brevis*) after 3 decades, and for the marine area (55% *D. sanpaulensis*) after 2 decades, are still unclear as there is no information of fishing effort related to these species (target or bycatch).

Final considerations

The franciscana dolphin, as a generalist species, varies its diet according to the presence and abundance of its preferred prey items: demersal fish and coastal cephalopods. Our meta-analysis of published stomach content and stable isotopes information allowed us to recognize both broad and small spatial scale differences in franciscana diet throughout its distribution. Differences in environmental characteristics along the study area and the consequent biogeographic patterns of prey species were reflected in a diet mainly composed of warm water species in the northern FMAs (Ia, Ib, and II) and cold-water species in the two southernmost FMAs (III and IV). This latitudinal distribution is also reflected in the diversity patterns of the diet, which mirrors the peak in prey species richness in the regions encompassed by FMAs Ia and Ib, decreasing toward the south.

Isoscapes, derived from stable isotope values obtained from diverse tissue types from franciscanas, showed increasing trends in $\delta^{13}C$ and $\delta^{15}N$ values toward high latitudes, which reflected the distinct oceanographic conditions along with the distribution that generate distinct basal values which are transferred up the food web. Nevertheless, higher proportions of ^{13}C-depleted pelagic fish in the diet of franciscanas from northern FMAs could also play a role in the observed patterns.

Temporal analysis in prey frequency of occurrence revealed some interesting trends for important cephalopod and fish species. Fisheries overexploitation was previously pointed out as an important factor contributing to a different dietary proportion for franciscanas from FMAIII over time and this might be the case for the observed trend in main prey species of franciscanas from Uruguay too (Fig. 2.9). Nevertheless, shifts in the prey composition due to the higher frequency of natural events (e.g., Franco et al., 2020) might also alter the abundance of prey species in some areas. Natural changes in the abundance of some prey species of franciscana (e.g., keystone species such as *D. plei*) could also affect the abundance of other franciscana prey species, thus contributing to the observed changes in the frequency of occurrence of prey in FMAs Ib and II (Fig. 2.8).

The meta-analysis conducted also revealed gaps in the knowledge of the trophic ecology of this endangered cetacean. Further studies using a combination of methods are encouraged in order to better understand the feeding habits of this species, especially in less studied regions, such as FMA Ia. Long-term feeding databases are needed in order to identify trends in the composition and relative importance of prey that help to understand the effect of various natural and anthropogenic factors in the trophic ecology of the species throughout its distribution.

References

Amaral, K.B., Danilewicz, D., Zerbini, A., Di Beneditto, A.P., Andriolo, A., Alvares, D.J., et al., 2018. Reassessment of the franciscana Pontoporia blainvillei (Gervais & d'Orbigny, 1844) distribution and niche characteristics in Brazil. J. Exp. Mar. Biol. Ecol. 508, 1–12.

Araújo, C.C., Gasalla, M.A., 2018. Distribution patterns of loliginid squid paralarvae in relation to the oceanographic features off the South Brazil Bight (22°–25° S). Fisheries Oceanography 27, 63–75.

Artecona, F., De Maria, M., Bergamino, L., Szteren, D., 2019. A historical perspective of niche differentiation between two top predators in the Uruguayan coastal area. Wildlife Research 46, 136–144.

Barros, N.B., Clarke, M.R., 2002. Diet. In: Perrin, W.F., Würsig, B., Thewissen, J.G.M. (Eds.), Encyclopedia of Marine Mammals. Academic Press, San Diego, CA, pp. 323–327.

Bassoi, M., 1997. Feeding of franciscana, Pontoporia blainvillei (Gervais and D'Orbigny, 1844), accidentally caught in coastal gillneting in southern Brazil.

Bassoi, M., 2005. Feeding ecology of franciscana dolphin, Pontoporia blainvillei (Cetacea: Pontoporiidae), and oceanographic processes on the Southern Brazilian coast.

Bassoi, M., Secchi, E.R., 1999. Temporal variation in the diet of, franciscana, Pontoporia blainvillei (Cetacea, Pontoporiidae), as a consequence of fish stocks depletion off southern Brazil, Ecosystem Effects of Fishing ICES/SCOR Symposium, 51.

Bassoi, M., Shepherd, J.G., Secchi, E.R., Moreno, I.B., Danilewicz, D., 2020. Oceanographic processes driving the feeding ecology of franciscana dolphin off Southern Brazilian coast. Cont. Shelf Res. 201, 104–124.

Bassoi, M., Secchi, E., Danilewicz, D., Moreno, I., Santos, R., Shepherd, J., 2021. Intrapopulation variation in the diet of franciscana dolphin (Pontoporia blainvillei) off southern Brazil. Journal of the Marine Biological Association of the United Kingdom, 1–17. http://doi.org/10.1017/S0025315421000436.

Bastida, R., Rodríguez, D., Moreno, V., Pérez, A., Marcovecchio, J. and Gerpe, M., 1992. Varamientos de pequeños cetáceos durante el período 1984-1988 en el área de Mar del Plata (Provincia de Buenos Aires, Argentina). Anales, Tercera Reunión de Especialistas en Mamíferos Acuáticos de América del Sur, pp. 25–30.

Beckett, S.J., 2016. Improved community detection in weighted bipartite networks. R. Soc. Open Sci. 3, 140536.

Beverton, R.J.H., 1985. Analysis of Marine Mammal-Fisheries Interactions. Marine Mammals and Fisheries. George Allen & Unwin, London, pp. 3–33.

Bittar, V.T., Beneditto, Di, A.P.M., 2009. Diet and potential feeding overlap between Trichiurus lepturus (Osteichthyes: Perciformes) and Pontoporia blainvillei (Mammalia: Cetacea) in northern Rio de Janeiro, Brazil. Zoologia (Curitiba) 26, 374–378.

Brownell Jr, R.L., Ness, R., 1970. Preliminary notes on the biology of the Franciscana, Pontoporia blainvillei (Cetacea: Platanistidae), P: 23–28, Proc. Sixth Annu. Conf. on Biological Sonar and Diving Mammals. Palo Alto, CA. Stanford Res. Inst.

Brownell Jr, R.L. and Praderi, R., 1976. Present research and conservation problems with the franciscana, Pontoporia blainvillei in Uruguayan waters. In Advisory Committee on Marine Resources Research. Scientific Consultation on Marine Mammals, Bergen.

Brownell Jr, R.L., 1975. Progress report on the biology of the franciscana dolphin, Pontoporia blainvillei, in Uruguayan waters. Journal of the Fisheries Board of Canada 32, 1073–1078.

Brownell Jr., R.L., 1989. Franciscana, Pontoporia blainvillei (Gervais and D'Orbigny, 1844). 45–67. In Ridgway and Harrison (Eds). Handbook of Marine Mammals: River Dolphins and the Larger Toothed Whales (4). Academic Press, London and San Diego, CA.

Burmeister, H., 1867. Preliminary observations on the anatomy of Pontoporia blainvillei. In Proc. Zool. Soc. Lond. 1867, 484–489.

Burmeister, H., 1869. Fauna Argentina. Mammifera Pinnata Argentina, Pontoporia blainvillei. Anal. Mus. Publ. Bs. As. 1, 389–422.

Cabrera, A., Yepes, J., 1940. Mamíferos Sud-Americanos. Vida, costumbres y descripción. Hist. Natural Ediar. Buenos Aires., 370.

Campos, E.J.D., Gonçalves, J.E., Ikeda, Y., 1995. Water mass characteristics and geostrophic circulation in the South Brazil Bight – Summer of 1991. J. Geophys. Res. 100, 18537–18550.

Campos, E.J.D., Velhote, D., Silveira, I.C.A., 2000. Shelf break upwelling driven by Brazil Current cyclonic meanders. Geophys. Res. Lett. 27, 751–754.

Campos, L.B., Lopes, X.M., da Silva, E., Santos, M.C.O., 2020. Feeding habits of the franciscana dolphin (Pontoporia blainvillei) in south-eastern Brazil. Journal of the Marine Biological Association of the United Kingdom 100, 301–313.

Carvalho, C.D., 1961. Stenodelphis blainvillei na costa meridional do Brasil, com notas osteológicas (Cetacea, Platanistidae). Rev. Bras. Biol. 21, 443–454.

Ciotti, A.M., Odebrecht, C.O., Fillman, G., Moller Jr., O., 1995. Freshwater outflow and subtropical convergence influence on phytoplankton biomass on the southern Brazilian continental shelf. Cont. Shelf Res. 15, 1737–1756.

Clarke, M.R., 1986. Research on Dolphins, chapter Cephalopods in the diet of odontocetes. Claredon Press, Oxford.

Coelho-Souza, S.A., López, M.S., Guimarães, J.R.D., Coutinho, R., Candella, R.N., 2012. Biophysical interactions in the Cabo Frio upwelling system, Southeastern Brazil. Braz. J. Oceanogr. 60, 353–365.

Coletto, J.L., Botta, S., Fischer, L.G., Newsome, S.D., Madureira, L.S., 2021. Isotope-based inferences of skipjack tuna feeding ecology and movement in the southwestern Atlantic Ocean. Mar. Environ. Res. 165, 105246.

Cremer, M.J., Pinheiro, P.C., Simões-Lopes, P.C., 2012. Prey consumed by Guiana dolphin Sotalia guianensis (Cetacea, Delphinidae) and franciscanas dolphin Pontoporia blainvillei (Cetacea, Pontoporiidae) in an estuarine environment in southern Brazil. Iheringia. Série Zoologia 102, 131–137.

Cremer, M.J., Simões-Lopes, P.C., 2005. The occurrence of Pontoporia blainvillei (Gervais & d'Orbigny) (Cetacea, Pontoporiidae) in an estuarine area in southern Brazil. Revista Brasileira de Zoologia 22, 717–723.

Crespo, E.A., 2018. Franciscana dolphin: Pontoporia blainvillei. In: Würsig, B., Thewissen, J.G.M., Kovacs, K.M. (Eds.). Encyclopedia of marine mammals, 3rd ed. Academic Press, London, pp. 388–392.

Crespo, E.A., Harris, G., González, R., 1998. Group size and distributional range of the franciscana, Pontoporia blainvillei. Marine Mammal Science 14, 845–849.

Cunha, H.A., Medeiros, B.V., Barbosa, L.A., Cremer, M.J., Marigo, J., Lailson-Brito Jr., J., et al., 2014. Population structure of the endangered franciscana dolphin (Pontoporia blainvillei): reassessing management units. PLoS One 9, e85633.

Danilewicz, D., Secchi, E.R., Ott, P.H., Moreno, I.B., Bassoi, M., Borges-Martins, M., 2009. Habitat use patterns of franciscana dolphins (Pontoporia blainvillei) off southern Brazil in relation to water depth. Journal of the Marine Biological Association of the United Kingdom 89, 943–949.

Deagle, B.E., Tollit, D.J., 2007. Quantitative analysis of prey DNA in pinniped faeces: potential to estimate diet composition? Conservation Genetics 8, 743–747.

Denuncio, P., Paso-Viola, M.N., Machovsky-Capuska, G.E., Raubenheimer, D., Blasina, G., Machado, R., et al., 2017. Population variance in prey, diets and their macronutrient composition in an endangered marine predator, the franciscana dolphin. J. Sea Res. 129, 70–79.

Di Beneditto, A.P.M., Ramos, R.M.A., 2001. Biology and conservation of the franciscana (Pontoporia blainvillei) in the north of Rio de Janeiro, Brazil. Journal of Cetacean Research and Management 2, 185–192.

Di Beneditto, A.P.M., Siciliano, S., 2021. Hábitos alimentares da toninha (Pontoporia blainvillei) na costa norte do Estado do Rio de Janeiro, Brasil. Revista Ibero-Americana de Ciências Ambientais 12, 745–749.

Dormann, C.F., Gruber, B., Fründ, J., 2008. Introducing the bipartite package: analysing ecological networks. R News 8, 8–11.

Drago, M., Franco-Trecu, V., Segura, A.M., Valdivia, M., González, E.M., Aguilar, A., Cardona, L., 2018. Mouth gape determines the response of marine top predators to long-term fishery-induced changes in food web structure. Sci. Rep. 8, 1–12.

Fitch, J.E., Brownell Jr, R.L., 1971. Food habits of the franciscana Pontoporia blainvillei (Cetacea: Platanistidae) from South America. Bull. Mar. Sci. 21, 626–636.
Floeter, S.R., Rocha, L.A., Robertson, D.R., Joyeux, J.C., Smith-Vaniz, W.F., Wirtz, P., et al., 2008. Atlantic reef fish biogeography and evolution. J. Biogeogr. 35, 22–47.
Franco, B.C., Defeo, O., Piola, A.R., Barreiro, M., Yang, H., Ortega, L., et al., 2020. Climate change impacts on the atmospheric circulation, ocean, and fisheries in the southwest South Atlantic Ocean: a review. Clim. Change 162, 2359–2377.
Gasalla, M.A., Rodrigues, A.R., Postuma, F.A., 2010. The trophic role of the squid Loligo plei as a keystone species in the South Brazil Bight ecosystem. ICES J. Mar. Sci. 67, 1413–1424.
Gasalla, M.A., Rossi-Wongtschowski, C.L.D.B., 2004. Contribution of ecosystem analysis to investigating the effects of changes in fishing strategies in the South Brazil Bight coastal ecosystem. Ecol. Modell. 172, 283–306.
Gianelli, I., Defeo, O., 2017. Uruguayan fisheries under an increasingly globalized scenario: long-term landings and bioeconomic trends. Fish. Res. 190, 53–60.
Gonçalves, G.M., Carvalho, C.E.V., 2006. Particulate heavy metal dynamics in a tropical estuary under distinct river discharge and tidal regimes, Southeastern, Brazil. Journal of Coastal ResearchSpecial Issue 39, 1032–1035.
Haimovici, M., Martins, A.S., Vieira, P.C., 1996. Distribuição e abundância de peixes teleósteos demersais sobre a plataforma continental do sul do Brasil. Rev. Bras. Biol. 56, 27–50.
Henning, B., de Sa Carvalho, B., Pires, M.M., Bassoi, M., Marigo, J., Bertozzi, C., Araújo, M.S., 2017. Geographical and intrapopulation variation in the diet of a threatened marine predator, Pontoporia blainvillei (Cetacea). Biotropica 50, 157–168.
Isacch, J.P., Costa, C.S.B., Rodriguez-Gallego, L., Conde, D., Escapa, M., Gagliardini, D.A., Iribarne, O.O., 2006. Distribution of saltmarsh communities associated with environmental factors along a latitudinal gradient on the south-west Atlantic coast. J. Biogeogr. 33, 888–900.
Kindt, R., Coe, R., 2005. Tree diversity analysis: a manual and software for common statistical methods for ecological and biodiversity studies. World Agroforestry Centre, Nairobi.
Knoppers, B., Souza, W.F.L., Ekau, W., Figueiredo, A.G., Soares-Gomes, A., 2009. Interface Terra-Mar do Brasil. In: Pereira, R.C., Soares-Gomes, A. (Eds.), Biologia Marinha. Editora Interciência, Rio de Janeiro, pp. 529–553.
Lahille, F., 1899. Note sur les dimensiones du Stenodelphis blainvillei. Rev. Mus. La Plata 9, 389–392.
Lahille, F., 1905. Las ballenas de nuestros mares Revista del Jardin Zoologico de Buenos Aires11. Epoca, pp. 28–82.
Lajtha, K., Michener, R.H. (Eds.), 1994. Stable isotopes in ecology and environmental science. Blackwell, Oxford.
Lopes, X.M., 2012. Hábitos alimentares de toninha, Pontoporia blainvillei (Gervais e D'Orbigny, 1844) (Mammalia, Cetacea), no sul do estado de São Paulo e norte do Paraná, Brasil. Dissertação de Mestrado. Universidade Estadual Paulista (UNESP), Rio Claro. 94.
Machado, R., de Oliveira, L.R., Ott, P.H., Haimovici, M., Cardoso, L.G., Milmann, L., Romero, M.A., dos Santos, R.A., Borges-Martins, M., 2020. Trophic overlap between marine mammals and fisheries in subtropical waters in the western South Atlantic. Marine Ecology Progress Series 639, 215–232.
Martins, A.S., Haimovici, M., 2017. Seasonal mesoscale shifts of demersal nekton assemblages in the subtropical South-western Atlantic. Mar. Biol. Res. 13, 88–97.
Martins, A.S., Haimovici, M., 2020. Key prey indicates high resilience on marine soft bottom habitats. Mar. Environ. Res. 159, 104963.
McMahon, K.W., Hamady, L.L., Thorrold, S.R., 2013. A review of ecogeochemistry approaches to estimating movements of marine animals. Limnol. Oceanogr. 58, 697–714.
Minagawa, M., Wada, E., 1984. Stepwise enrichment of ^{15}N along food chains: further evidence and the relation between $\delta^{15}N$ and animal age. Geochim. Cosmochim. Acta 48, 1135–1140.

Montoya, J.P., 2008. Chapter 29 - Nitrogen stable isotopes in marine environments. In: Capone, D.G., Bronk, D.A., Mulholland, M.R., Carpenter, E.J. (Eds.), Nitrogen in the Marine Environment, second ed., Academic Press, pp. 1277–1302.

Oksanen, J., Blanchet, F.G., Kindt, R., Legendre, P., Minchin, P.R., O'hara, R.B., et al., 2020. Vegan: community ecology package. R package version 2.5–7.

Oliveira, M.R., Pinheiro, P.C. and Rosas, F.C.W., 1998. Ecologia alimentar de Sotalia fluviatilis e Pontoporia blainvillei acidentalmente capturados no litoral do Paraná. Abstracts, VIII Reunião de Trabalhos de Especialistas em Mamíferos Aquáticos da América do Sul, pp. 25–29.

Ott, P.H., 1994. Estudo da ecologia alimentar de Pontoporia blainvillei (Gervais e D'Orbigny, 1844) (Cetacea, Pontoporiidae) no litoral norte do Rio Grande do Sul, sul do Brasil. Dissertação de Monografia. Universidade Federal do Rio Grande do Sul. p. 69.

Paes, E.T., Moraes, L.E.S., 2007. A new hypothesis on the influence of the El Niño/La Niña upon the biological productivity, ecology and fisheries of the Southern Brazilian Bight. Pan-American Journal of Aquatic Sciences 2, 94–102.

Paitach, R.L., 2015. Ecologia alimentar e espacial da toninha (Pontoporia blainvillei) e do boto-cinza (Sotalia guianensis) na Baía de Babitonga, sul do Brasil. Dissertação de Mestrado. Universidade Federal de Santa Catarina, Florianópolis. p. 142.

Paso-Viola, M.N., Denuncio, P., Negri, M.F., Rodríguez, D., Bastida, R., Cappozzo, H.L., 2014. Diet composition of franciscana dolphin Pontoporia blainvillei from southern Buenos Aires, Argentina and its interaction with fisheries. Revista de Biología Marina y Oceanografía 49, 393–400.

Perez, J.E., Gingarelli, M., Beilis, A. and Corcuera, J., 1996. Alimentación del delfín Franciscana en el Sur de la Provincia de Buenos Aires, Argentina. Abstracts, VII Reunión de Trabajo de Especialistas en Mamíferos Acuáticos de América del Sur. Viña del Mar, p. 91.

Perez-Macri, G., 1987. Resultados preliminares del estudio de alimentación de Pontoporia blainvillei en aguas costeras argentinas. Abstracts, II Reunião de Trabalhos de Especialistas em Mamíferos Aquáticos da América do Sul. Rio de Janeiro, p. 107.

Pierce, G.J., Boyle, P.R., 1991. A review of methods for diet analysis in piscivorous marine mammals. Oceanography and Marine Biology: An Annual Review 29, 409–486.

Pinedo, M.C., 1982. Análise dos conteúdos estomacais de Pontoporia blainvillei (Gervais e D'Orbigny, 1844) e Tursiops gephyreus (Lahille, 1908) (Cetacea, Platanistidae e Delphinidae) na zona estuarial e costeira de Rio Grande, Brasil (Master's thesis). Universidade do Rio Grande, Rio Grande.

Pinheiro, H.T., Rocha, L.A., Macieira, R.M., Carvalho-Filho, A., Anderson, A.B., Bender, M.G., et al., 2018. South-western Atlantic reef fishes: Zoogeographical patterns and ecological drivers reveal a secondary biodiversity centre in the Atlantic Ocean. Diversity and Distributions 24, 951–965.

Pinto, W., Lima, G., Zanetti, J., 2015. Comparative analysis of models for times to series modeling and forecasting of scheme of average monthly streamflow of the Doce River, Colatina, Espirito Santo, Brazil. Ciência e Natura 26, 1–11.

Praderi, R., 1986. Comentarios sobre la distribución de Pontoporia blainvillei en aguas del Rio de La Plata. Actas I Reunión de Trabajo de Especialistas en Mamíferos Acuáticos de América del Sur, Buenos Aires, pp. 206–214.

R Core Team, 2020. R: A Language and Environment for Statistical Computing. R Foundation for Statistical Computing. Vienna, Austria. https://www.R-project.org/.

Rodrigues, A.R., Gasalla, M.A., 2008. Spatial and temporal patterns in size and maturation of Loligo plei and Loligo sanpaulensis (Cephalopoda: Loliginidae) in southeastern Brazilian waters, between 23oS and 27oS. Scientia Marina 72, 631–643.

Rodríguez, D., Rivero, L., Bastida, R., 2002. Feeding ecology of the franciscanas (Pontoporia blainvillei) in marine and estuarine waters of Argentina. Latin American Journal of Aquatic Mammals 1, 77–94.

Rupil, G.M., Barbosa, L., Marcondes, M.C.C., de Carvalho, B.M., Farro, A.P.C., 2019. Franciscana dolphin (Pontoporia blainvillei) diet from Northern Espírito Santo State coast, Brazil. Revista Biotemas 32, 2.

Santos, M.C., Oshima, J.E., Silva, E.D., 2009. Sightings of franciscana dolphins (Pontoporia blainvillei): the discovery of a population in the Paranaguá estuarine complex, southern Brazil. Braz. J. Oceanogr. 57, 57–63.

Santos, R.A., Haimovici, M., 2001. Cephalopods in the diet of marine mammals stranded or incidentally caught along southeastern and southern Brazil (21–34°S). Fish. Res. 52, 99–112.

Schlitzer, R., 2002. Interactive Analysis and Visualization of Geoscience Data with Ocean Data View. Computers and Geosciences 28, 1211–1218.

Schmiegelow, J.M.M. 1990. Estudos sobre cetáceos odontocetos encontrados em praias da região entre Iguape (SP) e Baía de Paranaguá (PR) (24°42'S - 25°08'S) com especial referência a Sotalia fluviatilis (Gervais, 1853) (Delphinidae). Tese de Mestrado. Universidade de São Paulo, São Paulo. p. 149.

Secchi, E.R., 2010. Life history and ecology of franciscana, Pontoporia blainvillei (Cetacea, Pontoporiidae). Biology, Evolution and Conservation of River Dolphins within South America and Asia. Nova Science Publishers Inc., Hauppauge, NY, pp. 301–321.

Secchi, E.R., Danilewicz, D., Ott, P.H., 2003. Applying the phylogeographic concept to identify franciscana dolphin stocks: implications to meet management objectives. Journal of Cetacean Research Management 5, 61–68.

Siciliano, S., Di Beneditto, A.P., Ramos, R., 2002. A toninha, Pontoporia blainvillei (Gervais & d'Orbigny, 1844) (Mammalia, Cetacea, Pontoporidae), nos estados do Rio de Janeiro e Espírito Santo, costa sudeste do Brasil: Caracterização dos hábitos e fatores de isolamento das populações. Boletim do Museu Nacional 476, 1–15.

Sigman, D.M., Karsh, K.L., Casciotti, K.L., 2009. Ocean process tracers: nitrogen isotopes in the ocean. In: Steele, J.H., Turekian, K.K., Thorpe, S.A. (Eds.), Encyclopedia of Ocean Sciences, second ed. Elsevier Ltd, San Diego, CA, pp. 40–54.

Spalding, M.D., Fox, H.E., Allen, G.R., Davidson, N., Ferdaña, Z.A., Finlayson, M.A.X., et al., 2007. Marine ecoregions of the world: a bioregionalization of coastal and shelf areas. Bioscience 57, 573–583.

Tellechea, J.S., Perez, W., Olsson, D., Lima, M., Norbis, W., 2017. Feeding habits of franciscana dolphins (Pontoporia blainvillei): echolocation or passive listening? Aquatic Mammals 43, 440–448.

Troina, G., Botta, S., Secchi, E.R., Dehairs, F., 2016. Ontogenetic and sexual characterization of the feeding habits of franciscanas, Pontoporia blainvillei, based on tooth dentin carbon and nitrogen stable isotopes. Marine Mammal Science 32, 1115–1137.

Troina, G.C., Dehairs, F., Botta, S., Di Tullio, J.C., Elskens, M., Secchi, E.R., 2020a. Skin $\delta^{13}C$ and $\delta^{15}N$ reveal spatial and temporal patterns of habitat and resource use by free ranging odontocetes from the southwestern Atlantic Ocean. Marine Biolology 167, 186.

Troina, G.C., Dehairs, F., Botta, S., Di Tullio, J.C., Elskens, M., Secchi, E.R., 2020b. Zooplankton-based $\delta^{13}C$ and $\delta^{15}N$ isoscapes from the outer continental shelf and slope in the subtropical western South Atlantic. Deep-Sea Research Part I 159, 103235.

Valentin, J.L., 2001. The Cabo Frio upwelling system, Brazil. In: Seeliger, U., Kjerfve, B. (Eds.), Coastal Marine Ecosystems of Latin America. Springer, Berlin and Heidelberg, pp. 97–105.

Van Erp, I., 1969. In quest of the La Plata dolphin. Pacific Discovery 22, 18–24.

Vilar, C.C., Joyeux, J.C., Spach, H.L., 2017. Geographic variation in species richness, rarity, and the selection of areas for conservation: An integrative approach with Brazilian estuarine fishes. Estuarine Coastal Shelf Sci. 196, 134–140.

Vögler, R., González, C., Segura, A.M., 2020. Spatio-temporal dynamics of the fish community associated with artisanal fisheries activities within a key marine protected area of the Southwest Atlantic (Uruguay). Ocean & Coastal Management 190, 105175.

CHAPTER 3

Macroparasites of franciscana: what can their coexistence say to conservationists?

Alessandra Rocha[a], Ana Luisa Valente[b], Juliana Marigo[c]
[a]Universidade Federal do Rio Grande, Rio Grande, Rio Grande do Sul, Brazil
[b]Departamento de Morfologia, Instituto de Biologia, Universidade Federal de Pelotas, Pelotas, Rio Grande do Sul, Brazil
[c]Laboratório de Patologia Comparada de Animais Selvagens, Departamento de Patologia, Faculdade de Medicina Veterinária e Zootecnia, Universidade de São Paulo, São Paulo, Brazil

Overview

Parasites represent a meaningful portion of Earth's diversity and, although usually considered as negative by humans, they have an important ecological role (Marcogliese, 2004). As in other vertebrates, parasites are common in marine mammals; thus, the host–parasite close association in the aquatic ecosystem may provide information about animals' behaviors, migratory routes, potential geographical movements, and stock identity (Aznar, 2001; Balbuena et al., 1995). On the other hand, marine mammals are sentinel organisms of the ocean's health as they are exposed to environmental stressors such as chemical pollutants, harmful algal biotoxins, and emerging or resurging pathogens that can negatively affect the host population, as well as those with zoonotic potential (Bossart, 2011; Gulland and Hall, 2007).

The wide term "parasite" can be divided into microparasites (e.g., virus, bacteria, protozoans) and macroparasites (e.g., arthropods and helminths) (Anderson, 1993). This chapter will focus on macroparasites as indicators of important ecological aspects related to the conservation of franciscana.

Historical studies

The early marine mammal studies in South America focused primarily on obtaining other biological data and samples, such as age and reproduction, feeding habits, anatomy, and osteology. Therefore, during necropsies or beach sampling, researchers were not necessarily looking for or paying

The Franciscana Dolphin
DOI: https://doi.org/10.1016/B978-0-323-90974-7.00021-5

attention to systematically collecting parasites and recording their lesions. This applied to the first records of franciscana's helminths, especially on the Uruguayan coast (Brownell, 1975). The gastrointestinal helminth fauna of franciscana was described later on in Argentina (Aznar et al., 1994) and southern Brazil (Andrade et al., 1997). Subsequent studies and the recorded species are listed in Tables 3.1 and 3.2.

Once the main cause of death of franciscana is bycatch (Crespo, 2018), in some areas, researchers were able to establish collaboration with fishermen and obtain fresh bycaught animals. In other areas, they had to mainly rely on stranded and more decomposed carcasses which made it difficult to analyze other organs aside from the gastrointestinal tract. Nevertheless, there is still a knowledge gap regarding various aspects of parasite infections, especially those related to their life cycle (Silveira et al., 2017). Parasitological studies have intensified along the Brazilian coast in the last 20 years, shedding light into franciscana population identities employing parasites as biological markers.

Parasite diversity

Host particularities

The helminth assemblage found in franciscana is apparently influenced by two factors: evolutionary (host–parasite association) and ecological (predator–prey interactions) (Aznar et al. 1994; Marigo et al., 2011, 2013).

Franciscans are part of the peculiar, endangered, and poorly known group of "river dolphins," which are morphologically and phylogenetically distinct from marine dolphins. Although considered a river dolphin, during its evolution, *Pontoporia* followed the marine waters and colonized the nearshore north and south coastal zone of the La Plata estuary (Hamilton et al., 2001). Aznar et al. (1994) mention that Pontoporiidae has adapted secondarily to freshwater habitats which could have represented significant habitat shifts for its helminth parasites and, therefore, evolutionary restrictions may be responsible for the relatively low richness and diversity of helminths found in the franciscana. For instance, franciscana and Guiana dolphin (*Sotalia guianensis*) are not closely phylogenetic related, which may partially explain why their parasite community differs, even though they occur in similar habitats.

Among cetaceans, franciscana is one with the largest number of small teeth, adapted to hold soft prey such as squid and small juvenile fish. The presence of only two stomachs and the absence of the anterior stomach

Table 3.1 Chronological records of parasitology studies on franciscana (*Pontoporia blanvillei*) and respective methodologies.

Year	Biogeographical region	Authors	Comments
1971	Uruguayan coast (northern region)	Pilleri	Methodology not explained by author
1972		Dailey and Brownell	Methodology not explained by authors
1975		Brownell	Methodology not explained by author
1976		Kagei et al.	Punctual records in one stomach analyzed
1984		Praderi	Only stomachs analyzed
1994	Argentinean coast (Buenos Aires Province)	Raga et al; Aznar et al.	Description of Hadwenius pontoporiae and first study with parasito-logical focus, with analysis of both small and large intestines
1996	Brazilian coast (Rio de Janeiro) northern region	Santos et al.	All organs were analyzed but no parasites were found
1997	Brazilian coast (Rio Grande do Sul) southern region	Andrade et al.	Study considering age and gender of the sampled dolphins. Methodological proposition for sampling the small intestine separately from the large
2001	Brazilian coast (Rio de Janeiro) northern region[a]	Di Beneditto and Ramos	All organs were analyzed but no parasites were found
2002	Brazilian coast (Paraná and São Paulo)	Marigo et al.	Used the Andrade et al. (1997) methodology
2010	Brazilian coast (Rio Grande do Sul) southern region[a]	Rocha	Analysis of stomach and whole intestines and statistical analyses by area and gender
2013	Uruguay	Rocha et al.	Only the intestines were analyzed as whole
2017	Brazilian coast (Rio Grande do Sul) northern region	Hoss et al.	Study focusing solely on stomach samples
2017	Brazilian coast (Santa Catarina)	Alves et al.	Used the Andrade et al. (1997) methodology. Only intestines were sampled
2018	Brazilian coast (São Paulo)[a]	Seabra	More samples of stomach and intestines. Used the Andrade et al. (1997) methodology
2021	Argentinian coast	Irigoitia et al.	First molecular record of *A. typica* in a franciscana

[a]Studies with overlapping sampled regions in different years.

Table 3.2 Comparison of identified gastrointestinal helminths in franciscana (*Pontoporia blanvillei*) specimens from Brazil, Uruguay, and Argentina.

		FMAII			FMAIII		FMAIV	
Helminths		**Brazil**				**Uruguay**	**Argentina**	**References**
	São Paulo	**Paraná**	**Santa Catarina**	**Santa Catarina**	**Rio Grande do Sul**			
Anisakis spp.	x	-	-	-	x	-	-	8,12
Anisakis typica	-	-	-	-	x	x	x	3,4,8,10,13
Anisakis simplex	-	-	-	-	x	-	x	5,8
Anisakis pegreffi	-	-	-	-	-	-	x	13
Contracaecum sp.[a]	-	-	-	-	-	x	x	1,2,8
Procamallanus sp.[a]	-	-	-	-	x	x	-	3,8
Raphidascaris sp.[a]	-	-	-	-	x	-	-	8,10
Hadwenius/Synthesium pontoporiae[b]	x	x	x	-	x	x	x	5,6,7,8,9,10,11,12
Pholeter gastrophilus[a]	-	-	-	-	-	-	x	5
Lecithochirium sp.[a]	-	-	-	-	x	-	-	8
Polymorphus (P.)/ Corynosoma cetaceum[b]	x	-	-	-	x	x	x	1,2,3,4,5,6,8,10,12
Polymorphidae sp. [a]	-	-	-	-	-	-	x	5
Corynosoma australe	-	-	-	-	x	-	-	6,8
Bolbosoma turbinella[a]	x	-	-	-	x	-	-	6,8,10,12
Andracantha sp.[a]	x	-	-	-	-	-	-	12

[a]Species with low infection levels and no sexual maturity evaluation in franciscana; thus, considered as accidental finding.
[b]Changes in nomenclature.
(x) = presence; (-) = absence
References: 1- Dailey and Brownell, 1972 (n = ?); 2- Brownell, 1975 (n = 191); 3- Kagei et al, 1976 (n = 1); 4- Praderi, 1984 (n = 157); 5- Aznar et al, 1994 (n = 46); 6- Andrade et al, 1997 (n = 53); 7- Marigo et al, 2002 (n = 41); 8- Rocha, 2010 (n = 32); 9- Rocha et al, 2013 (n = 11); 10- Hoss et al, 2017 (n = 34); 11- Alves et al, 2017 (n = 28); 12- Seabra, 2018 (n = 87); 13- Irigoitia et al, 2021 (n = 1).

may also be related to its peculiar diet (Pinedo, 1991). Differences in prey consumed within a small spatial range indicate that franciscana could spend long periods of time in the same area (Bassoi et al., 2020; Henning et al., 2017) and, therefore, be exposed to the same local parasite cycles. Consequently, due to their feeding habits, small home range, limited movement (Bassoi et al. 2020, Bordino et al., 2008), and evolutionary process, the overall helminth assemblage found in franciscana seems to be predictable and depauperate.

So far, at least 12 species have been recorded, most of them, seven species, the franciscana may be considered to be unsuitable hosts (accidental infection) for these parasites (see Table 3.2). The gastrointestinal tract harbors the majority of the franciscana's helminth fauna with three main groups: Digenea, Acanthocephala, and Nematoda. The two key species are the trematode *Synthesium pontoporiae* and the acanthocephalan *Corynosoma cetaceum*.

Ectoparasites and other arthropods

Only a few species of ectoparasites have been reported in franciscana: the barnacle *Xenobalanus globicipitis* (Cirripedia: Coronulidae) found attached to flukes and fins of three specimens (Brownell, 1975; Di Benedito and Ramos, 2001), and an unidentified necked-barnacle found on the lower jaw of a possibly old and sick female (Secchi, 2010). The barnacle *Conchoderma auritum* (Cirripedia: Lepadidae) was first recorded (fixed between teeth) in 1992 and has been found frequently in the same area, Santa Catarina and Rio Grande do Sul State, Brazil, after implementation of consistent beach monitoring (Henrique-Garcia et al., 2010; Silva and Castilho, 2020; Soto, 2001). Depending on the severity of infection, it may cause malnutrition and even death (Silva and Castilho, 2020). The isopods *Nerocila* sp. was reported on the skin of franciscana from Rio Grande do Sul state, Brazil, and northern Uruguay; whereas *Cirolana* sp. was observed in the blowhole and stomach of specimens from Uruguayan waters (Brownell, 1975; Secchi, 2010). *Nerocila* sp. is known to infest fishes and sharks, suggested by Brownell (1975) as transmitters of these parasites to franciscana while in the gillnet—once the analyzed specimens were bycaught. The relationship of *Cirolana* sp. and dolphins are not known; however, these isopods likely entered the blowhole of franciscana after death (Brownell, 1975) or through ingestion of infected prey (Danilewicz et al., 2002). Furthermore, Ferreira et al. (1998) observed *Riggia* sp. (Isopoda: Cymothoidae) in the vagina of one franciscana in Santos, São Paulo state, Brazil. Most likely, this infection was not a case of parasitism and, as occurs with *Cirolana* sp., the crustacean probably entered into the vagina after the animals' death (Danilewicz et al., 2002).

Endoparasites

Nematodes (roundworms)

Anisakis spp. (Nematoda, Anisakidae) (Dujardin, 1845) Nematodes of the genus *Anisakis* are parasites that, as adults, live in the digestive system of marine mammals and humans. Their life cycle also involves planktonic crustaceans as first intermediate hosts, while fish and squids are intermediate/paratenic hosts (Mattiucci et al., 2018).

Franciscana has been recorded to harbor *Anisakis typica, A. simplex,* and *Anisakis* spp. identified by morphology. These nematodes are mostly found in the main stomach, and in some records, in the esophagus, but this could be related to regurgitation and death of the host. Sexually immature forms are frequently found in the stomach and intestines, and are difficult to identify, once only mature males have the main morphological characteristics to differentiate between species (size proportion, and shape of male spicules) (Davey, 1971).

Molecular identification has recently identified nine species of *Anisakis* that differ in geographical distribution and definitive host range. Morphologically identified as *Anisakis simplex,* is now considered a complex of species: *Anisakis* (s1) *simplex* which includes *A. berlandi, A. pegreffii, and A. simplex* (s.s.), but only the previous two anisakids occur in austral regions (Irigoitia et al., 2021; Mattiucci et al., 2014, 2018).

Trematodes (flukes)

Syntesium pontoporiae, Hadwenius pontoporiae (Digenea, Brachycladiidae) (Raga et al. 1994; Marigo et al., 2008): *Synthesium pontoporiae* is franciscana host-specific, presenting high infection prevalence (over 50%). Although also found in the stomach, the main site of infection is the small intestine, where it does not seem to have a constant distribution, occurring only in the initial portion (first third) of the organ—a pattern described in many studies (Alves et al., 2017; Andrade, 1996; Aznar et al., 1995; Hoss et al., 2017; Marigo et al., 2002, Rocha, 2010; Seabra, 2018). These small hermaphrodite trematodes are easy to observe macroscopically but very delicate, potentially lost or damaged during intestinal analysis, especially if the professional is not aware of their anatomy or performs mucosal scraping.

The genus *Synthesium* has a worldwide distribution and occurs in a number of odontocetes, suggesting a long period of host–parasite interactions. Also, most digenean occurs in the upper intestines, where they browse on mucus, epithelial cells, and sometimes blood, confirming an older association in comparison to those species established in extra-intestinal sites, which could be relatively recent members that adopted new forms of nutrition (Fernández et al., 1998; Marigo et al., 2011).

Marigo et al. (2008), using specimens from six different locations in Brazil and a paratype taken off Argentina, redescribed this trematode, initially described as *Hadwenius pontoporiae*, incorporating more details to the original description and differentiating the parasite from related species within this genus. The main added features were related to details regarding the pyriform cirrus sac, the lobed testicles (described as an oval), and the spines on the tegument and cirrus, only visible with fresh samples (such loss probably due to the fact that the hosts were dead and frozen, an observation already reported by other authors—Adams and Rausch (1989) and (Gibson, 2005)). A valuable diagnostic feature of *S. pontoporiae* is the sucker ratio: oral x ventral sucker, which presented a 1:1.5–3.0 pattern (Marigo et al., 2008). All studied specimens were adults with eggs, and the measurements were similar to those specimens used in the first description, including the conserved characters of egg size and a variation in total body length (also observed in biometric variability studies of *Synthesium tursionis*) (Marigo et al., 2008; Fernandez et al., 1995). Recently, Seabra (2018) observed the parasite in 18% of the sampled mature specimens from São Paulo, Brazil, presenting both larger and smaller sizes than previously described, proposing three different morphotypes for the area.

In a phylogenetic study, the genus *Synthesium*, represented by *S. pontoporiae* and *S. tursionis*, was positioned among Brachycladidae; marine mammals' digenean parasites closely related to Acanthocolpidae, which are fish parasites (Marigo et al., 2011). It is suggested that the ancestor of the *Synthesium* species could have first adapted to other odontocete families and later on to *Pontoporia*, and that the evolutionary and biological peculiarities of franciscana may be responsible for the exclusivity of *S. pontoporiae* in this species (Marigo et al., 2013).

In a genetic intraspecific study testing nuclear and mitochondrial markers, the obtained sequences suggested a lack of population structure in *S. pontoporiae* and population expansion (Marigo et al., 2013).

Acanthocephala (thorny head worms)

Corynosoma cetaceum (Acanthocephala, Polymorphidae) (Johnston and Best, 1942; Schmidt and Dailey, 1971; Aznar et al., 1999): *Corynosoma cetaceum* was one of the first helminths recorded in franciscana. Initially, it was classified as *C. cetaceum* and subsequently, it was included in the genus *Polymorphus* due to the absence of genital spines (Schmidt and Dailey, 1971). Currently, it is once more classified as a *Corynosoma* species, and due to the adherence to a less restrictive definition of genital spines used by several authors, females of *P. cetaceum* could be considered as bearers of genital spines (Aznar et al., 1999). Articles have mentioned these three valid nomenclatures according to the time in which these parasites were recorded and published.

Unlike other species of Acanthocephala, the main infection site of this parasite is the gastric instead of the intestinal mucosa, where it is usually seen attached to the main and pyloric stomachs, and the duodenal ampulla (Aznar et al., 1999). Worms were easily visualized during macroscopic examinations because of their considerable size (approximately 0.7 mm long, with a broad anterior region), and natural yellowish to greenish coloration. The presence of a large copulatory bursa is visible in males. During the identification process, gravid females carrying a large number of eggs hamper the analysis of internal organs.

Corynosoma australe (Acanthocephala, Polymorphidae) (Johnston, 1937): Although present in a relatively high prevalence in the southern Brazilian coast, most of the specimens found in franciscana are immature; in contrast to what is seen in pinnipeds (mainly otariids), in which *C. australe* infections are characterized by high parasite burden and mature worms (Hernàndez-Orts et al., 2013; Pereira, 2012; Silveira et al., 2014). Thus, franciscana appear to be unsuitable hosts for *C. australe* (Andrade et al., 1997; Rocha, 2010). Fish ingested by franciscana, such as the striped weakfish (*Cynoscion guatucupa*), the banded croaker (*Paralonchurus brasiliensis*), and the white mouth croacker (*Micropogonias furnieri*), are known to be infected with *C. australe* acanthocisths (Pereira et al., 1993; Timi et al., 2005; Silveira et al., 2017). In southern Brazilian waters, these fish are also preyed on by otariids like the South American fur seal (*Arctocephalus australis*) and the South American sea lion (*Otaria flavescens*) (Falco, 2008; Machado et al., 2020).

Bolbosoma turbinella (Acanthocephala, Polymorphidae) (Diesing, 1851; Porta, 1908): This thorn-headed worm is common in whales and oceanic cetaceans (Pinto et al., 2004). Most specimens found in franciscana were immature and without complete somatic development. Nevertheless, some sexually developed females with still forming eggs were recorded in some hosts (Andrade, 1996; Hoss et al., 2017; Rocha, 2010).

Upon necroscopy, these helminths may be seen macroscopically in the stomach, although their preferential site is the large intestine. Immature individuals are not usually adhered to the mucosa, remaining free in the gastrointestinal tube and retained in the mesh during collection. Larger worms may be attached to the large intestine mucosa by the whole bulb inside and require careful removal through dissection under a stereoscopic microscope.

In the southern Brazilian coast, aside from the franciscana records, sexually mature worms were found in Brydes' whale (Valente et al., 2014). Larval forms were found in fish rough scad, *Trachurus lathami*, a species considered as a possible responsible for the infection in franciscanas, once it is part of their diet and a fish that lives both in oceanic and coastal waters (Silveira et al., 2017).

Cestoda

Unidentified cestode larvae were found attached to the outer intestine surface of a franciscana during the parasitological analysis of 193 individuals and were considered an accidental infection (Brownell, 1975).

Less common parasites

It is believed that most of the helminth species recorded in franciscana do not use this dolphin as a definitive host, because sexual maturity and the complete life cycle are not attained in this host. In some cases, helminths species typical of fish recorded in the gastrointestinal tract of dolphins may be considered a mix of information regarding parasites of recently ingested prey and the host. For instance, infection by the acanthocephalan *Andracantha* sp. and the trematode *Lecitochrium* sp. are considered accidental, caused by ingestion of infected prey, and, therefore, are not part of the host's parasitic fauna (Rocha, 2010; Seabra, 2018).

The trematode *Pholeter gastrophilus* is usually found in the stomach of odontocete cetaceans worldwide (Aznar et al., 2006), and the infection produced by this species apparently shows low specificity to their hosts (Fraija-Fernández et al., 2017). In franciscana, to this date, the only record was on the Argentinean coast, where three worms were found in a single dolphin (prevalence 5.9%) (Aznar et al., 1994). Records of this parasite in other cetaceans are known in areas of franciscana occurrence (Berón-Vera et al., 2008; Ebert et al., 2021), suggesting that the record in Argentina was an accidental infection.

Similarly, the nematodes *Contracaecum* sp., mostly found in pinnipeds and sea birds, and *Raphidascaris* sp. and *Procamallamus* found, respectively, in marine and fresh water fish, are considered accidental parasites (Table 3.2;

Hernandez-Ortiz et al., 2013; Machado-Pereira et al., 2017; Moravec et al., 2000; Pérez-i-García et al., 2015). To the authors' knowledge, no other parasites have been reported in the remaining organs.

Main factors influencing infection

Parasitological studies in franciscanas show that some helminths infect younger dolphins with increased intensity and prevalence. Juveniles were more infected with *S. pontoporiae* than mature individuals in the southern Rio Grande do Sul state, Brazil (Andrade et al., 1997). This could be due to the difference in food preferences between immature and mature dolphins, or age-related increased immune response (Andrade, 1996; Andrade et al., 1997).

In most studies, no significant difference was found between host gender and parasites (Andrade et al., 1997; Hoss et al., 2017; Marigo et al., 2002, Seabra, 2018). Once gastrointestinal parasites are transmitted to franciscana through consumed prey, such fact can be explained based on the similarity of male and female diets (Bassoi et al., 2021). Nevertheless, Rocha (2010) found that females harbor more *S. pontoporiae* than males and that females ingest more fish than males in the southern Rio Grande do Sul state, Brazil (Bassoi et al., 2021; Rocha, 2010). Although the life cycle of *S. pontoporiae* is still unknown, indirect evidence (i.e., lack of population structure) suggests that a mobile intermediate host such as a fish participates in its life cycle (Marigo et al., 2013) once franciscana has one of the smallest home ranges known among cetaceans (Bassoi et al., 2020; Bordino et al., 2008). This hypothesis should be further investigated, but if true, the numerical abundance of consumed prey could contribute to the higher *S. pontoporiae* infection levels found in females when compared to the number of parasites infecting males in that region (see Rocha, 2010 - appendix 1).

Temporal variation may be the greatest challenge in the interpretation of parasite infections. Such studies are not performed simultaneously and it is difficult to ensure that differences are due to the availability of infected prey in each study area. The patency period of each helminth infection is unknown, and recruitment (recent infection) could be misinterpreted between subsequent years. The influence of currents carrying intermediate hosts, migratory displacements of these prey, as well as changes in consumption rates on less available prey due to overexploitation of fishing stocks may partially contribute to temporal variability. Nevertheless, Rocha (2010) compared the helminth assemblage of franciscana in two periods,

15 years apart, in the coast of southern Rio Grande do Sul state, Brazil, and did not find significant differences, despite the overfishing evidence of important fish species preyed by franciscana in the region (Haimovici et al., 2006; Miranda and Haimovici, 2007; Vasconcellos and Haimovici, 2006).

Undoubtedly, in the case of franciscana, the geographical origin of the sampled animals is the main factor leading to variations in the levels of helminth infection. Latitudinal gradients have so far justified the occurrence or absence of certain species, as well as their infection levels. Such an issue has been the center of several studies focusing on a more ecological approach to the use of parasitological data as bioindicators and will be discussed in the next topic.

Parasites as biological tags and biogeographical distribution of parasite infections in franciscana

The close host–parasite association can provide important biological information on host populations, once parasites are frequently considered as biological tags for their hosts (Aznar et al., 2001). This concept was initially applied in fish, but can also be applied to marine mammals (Balbuena et al., 1995). The meaningful information about the use of parasites as biological tags in terms of marine mammal population management and conservation regards stock identity, geographical movements or their absence, and behavioral/social structure (Aznar et al., 2001; Balbuena et al., 1995).

The basic principle of this technique is that hosts will only be infected by a parasite species if they enter the parasite's endemic area; namely, a geographical area with suitable conditions to transmit the parasite (Mackenzie and Abaunza, 1998). With that in mind, it is possible to compare the summary of statistical infection levels (e.g., prevalence, mean intensity, abundance) of one or more parasite species between several host groups from different regions.

The ideal parasite to select as tag has to follow some general lines: (1) different infection levels in hosts from different regions; (2) the parasite species should be monoxene—with a complete life cycle involving just a single host; (3) be easily detected upon gross examination; (4) present a stable prevalence between seasons or years; and (5) the parasite should not cause disease or affect host behavior (Moser, 1991; MacKenzie, 1999; Catalano et al., 2014). Nevertheless, as a single parasite rarely attends all the desirable attributes, these guidelines cannot be considered as a rule, and adjustments are usually made (Catalano et al., 2014). The information provided by monoxene parasites is related to behavioral studies, once host-to-host contact is necessary. On the other hand, parasites indirectly transmitted or with complex life cycles that involve intermediate hosts, are linked to food webs and, therefore, give information about past and present host feeding grounds (Aznar et al., 2001).

The use of parasites as biological tags presents both advantages and limitations. The advantage of this method is the low cost, which can be well implemented in developing countries. Aside from that, the parasite selected as tag can often be used to identify host subpopulations that are distinguished by behavioral differences instead of genetic differences (MacKenzie and Abaunza, 2014). The weaknesses of this technique are sampling biases and the lack of information about the parasite's biology (Balbuena et al., 1995; MacKenzie and Abaunza, 2014).

It is important to note that parasites indicate ecological host stocks: distribution and abundance spatial patterns according to an ecological scale, indicating local food webs. It is different information than the one provided by genetic stocks. Ecological and genetic stocks may, or may not, have the same boundaries. For instance, several ecological host stocks can belong to a single genetic stock (Aznar et al., 2001).

Studies on the use of gastrointestinal helminths of franciscana as potential biological tags to investigate the existence of host stocks have been conducted since 1990 (Aznar et al., 1995). According to these studies, *Synthesium pontoporiae*, *Corynosoma cetaceum*, and *Anisakis* spp are the species of parasites recommended as biological markers of franciscana (Alves et al., 2017; Andrade et al., 1997; Aznar et al., 1995; Hoss et al., 2017; Marigo et al., 2002; Rocha, 2010; Seabra, 2018). All three species have indirect life cycles and are transmitted to franciscana through the food web. Also, they are easy to collect and identify, in addition to having no lethal effects on their host.

Synthesium pontoporiae is frequently indicated as the most useful biological tag of franciscana (Alves et al., 2017; Andrade et al., 1997; Aznar et al., 1995; Marigo et al., 2002; Rocha, 2010; Seabra, 2018). Although their life cycle is still unknown, *S. pontoporiae* as a digenetic trematode tends to be highly specific to the primary host, usually a mollusk, which can determine the presence of this parasite in a geographical area (MacKenzie and Abaunza, 2014; Kremnev et al., 2020). Furthermore, its exclusivity in franciscana, its high site of infection specificity, easy macroscopic observation (Aznar et al., 1994), and varying prevalence and infection levels in franciscana along with their occurring areas make these digenetic worms good markers for franciscana populations (Alves et al., 2017; Aznar et al., 1994; Andrade et al., 1997; Marigo et al., 2002; Rocha, 2010; Seabra, 2018). The different parasitological indexes corroborate with the stocks proposed (Cunha et al. 2020a; Secchi et al., 2003): absence of parasites in franciscana from FMA I (northern Rio de Janeiro state, Brazil) (Di Benedito and Ramos, 2001; Santos et al., 1996); the low intensity of infection in animals in FMA II (São Paulo e Paraná states, Brazil) (Marigo et al., 2002; Seabra, 2018); high intensity of infection in FMA III (south of Rio Grande do Sul state, Brazil and Uruguay) (Andrade et al., 1997; Rocha, 2010; Rocha et al., 2013); and the highest intensity of infections by *S. pontoporiae* in FMA IV (northern Argentina). Nevertheless, when the parasitological indexes of this trematode infecting franciscana from the estuarine waters of Babitonga bay (Santa Catarina state, Brazil) were compared with the infection levels of *S. pontoporiae* in dolphins from the adjacent coastal waters, the results changed. Dolphins from the estuary exhibited lower infection levels than individuals from oceanic waters (Alves et al., 2017), suggesting some degree of isolation by estuary dolphins and corroborating with molecular studies in these regions (Cunha et al., 2020b), as well with the stock subdivision proposed by Cunha et al., 2020a (see Fig. 3.1).

Nevertheless, although these preliminary data suggest different infection levels between areas, parasite morphometric (Marigo et al., 2008; Rocha, 2010) and molecular analysis (Marigo et al., 2013) revealed no geographical structuring and no gene restriction flows between FMAs, suggesting that *S. pontoporiae* belongs to a single population along with host occurrence (Marigo et al., 2013). The reason behind the unstructured population of *S.ontoporiae* was attributed to the intermediate hosts involved with the life cycle of the parasite, which is probably more mobile than the definitive host (Marigo et al., 2013).

Additionally, when more intestinal samples of franciscana from FMA II were analyzed, the results of different infection levels also changed and no longer supported the evidence of stock subdivision proposed by Secchi et al. (2003) (Marigo et al., 2013). Nevertheless, a recent study comprising 87 intestine samples from FMA II still reported low infection levels in comparison with animals from FMA III (Seabra, 2018). In spite of that, studies with larger sample sizes should be performed in all FMAs to better understand the role of *S. pontoporiae* as a biological marker of franciscana.

Corynosoma cetaceum has been pointed out as a potential biological tag of franciscana because they are easy to collect and less fragile than *S. pontoporiae,* and also because these acanthocephalan parasites have distinct

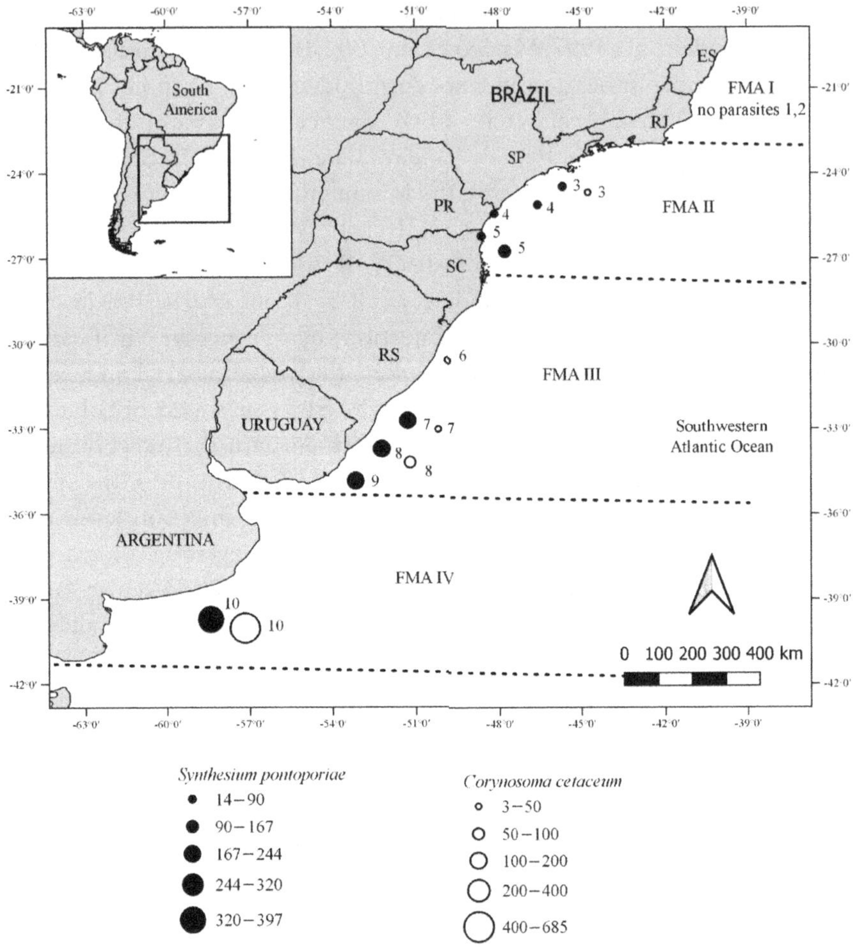

Fig. 3.1 ***Mean intensities of Synthesium pontoriae and Corynossoma cetaceum infections along Franciscana management areas (FMA) in Brazil, Uruguay, and Argentina.*** (Image created using QGIS software version 3.8.2 www.qgis.org). References: 1—Santos et al., 1996 (*n* = 42); 2—Di Beneditto and Ramos (2001) (*n* = 48); 3—Seabra (2018) (*n* = 87); 4—Marigo et al. (2002) (*n* = 41); 5—Alves et al. (2017) (*n* = 28); 6—Hoss et al. (2017) (*n* = 34); 7—Andrade et al. (1997) (*n* = 53); 8—Rocha (2010) (*n* = 32); 9—Rocha et al. (2013) (*n* = 11); 10—Aznar et al. (1994) (*n* = 46).

prevalence and intensity of infection according to host geographical distribution (Andrade et al., 1997; Aznar et al., 1994; Hoss et al., 2017). The infection pattern is similar to what is so far observed in *S. pontoporiae* with the infection levels increasing toward the south. Therefore, no parasites were found in franciscana from FMA I (northern Rio de Janeiro) (Di Benedito and Ramos, 2001; Santos et al., 1996), those from FMA II

(São Paulo) exhibit low infection levels (Seabra, 2018), hosts from FMA III (Rio Grande do Sul state, Brazil, and northern Uruguay) showed higher infection levels than infected franciscana from São Paulo, Brazil, but lower infection levels when compared with dolphins from FMA IV (Argentina) (Andrade et al., 1997; Aznar et al., 1994; Brownell, 1975; Hoss et al., 2017; Rocha, 2010) (Fig. 3.1).

Anisakid nematodes were suggested as the most suitable parasite to detect latitudinal movements of franciscana across regions (Aznar et al., 1995); however, recent satellite device studies proved the contrary, that the species has very small home ranges (Bordino et al., 2008). It has been suggested that water temperature shapes the distribution of *Anisakis typica* and of sibling species belonging to *A. simplex* (s. l.) complex (i.e., *A. pegreffii* and *A. berlandi*). Thus, *A. typica* is found in warm temperate and tropical waters, and *A. simplex* (s. l.) is related to cold temperate and polar waters in the Southwestern Atlantic Ocean (Davey, 1971; Mattiucci et al., 2018). The *Anisakis* spp. infection pattern in franciscana follows the same one of the previous *S. pontoporiae* and *C. cetaceum* worms, with the infection levels increasing along with the latitude. Thus, anisakids were not observed in franciscana from FMA I (Santos et al., 1996), while individuals from FMA II showed low infection levels (Seabra, 2018).

Franciscanas from FMA III were infected with *A. typica* (Andrade et al., 1997; Hoss et al., 2017; Rocha, 2010) and less often with *A. simplex* (only one specimen), whilst franciscana from the FMA IV were predominantly infected with *A. simplex* (Aznar et al., 1994; Rocha, 2010). Recently, *A. pegreffii* and *A. typica* were molecularly identified in one dolphin from Argentina (39°S, Irigoitia et al., 2021). Rocha (2010) suggested that the single *A. simplex* found in a franciscana from the southern Rio Grande do Sul state, Brazil, could be explained by a higher frequency of *Engraulis anchoita* occurrence in the diet of the hosts analyzed in that study. This fish is associated with cold waters masses (Lima and Castello, 1995) and is known to be part of the life cycle of *A. simplex* in Argentina (Berón-Vera et al., 2008). Thus, anisakid nematodes could provide clues about separated prey stocks along latitudinal gradients.

A difference in terms of the occurrence of less common parasites was also detected. The acanthocephalans *Andracantha* sp. was registered only in FMA II (Seabra, 2018), *Corynosoma australe* appeared in FMA III, solely in the southern Rio Grande do Sul state, Brazil (Andrade et al., 1997; Rocha, 2010), whereas *Bolbosoma turbinella* was found in FMA II and FMA III, but with higher parasitological indexes in the last one (Andrade et al., 1997;

Rocha, 2010; Seabra, 2018). The nematodes *Procamallanus* sp. and *Raphidascaris* sp. were found only in FMA III (Hoss et al. 2017; Kagei et al. 1976; Rocha 2010), while *Contracaecum* sp. appeared in FMA III and FMA IV (Aznar et al., 1995; Rocha, 2010). The digenetics *Pholether gastrophilus* and *Lecithochrium* sp. were registered just in FMA IV and FMA III, respectively (see Table 3.2). These differences regarding the occurrence of less common parasites could be related to the availability of intermediate hosts and abiotic factors in such areas.

Although those studies on franciscana helminth fauna were not carried out at the same period, due to the inherent difficulties in marine mammal sampling, the use of data from several parasites may compensate for such weakness and, in the case of franciscana, could provide interesting information about host populations. There are quantitative and qualitative differences between the helminth fauna across FMAs. The *S. pontoporiae, C. cetaceum, Anisakis* spp., and *B. turbinella* seem to have the same pattern of infection, with the parasitological indexes increasing toward south. Qualitative and quantitative differences were observed even in the same FMA, such as *C. australe,* which seems to infect francicana only in the southern Rio Grande do Sul state, Brazil, in FMA III, and *S. pontoporiae* infecting dolphins from the estuary of Babitonga Bay, Brazil, and adjacent oceanic coast in FMA II, respectively. All this information suggests the localized habitat use of franciscana, as seen in other studies (e.g., Bassoi et al., 2020; Bordino et al., 2008). Therefore, the use of parasites as indicators of host populations has potential, and when integrated with data across fields, can provide better information about the franciscana's stock's identity. This holistic approach can help stakeholders to develop management and conservation plans.

Methodological recommendations

Parasitological data may be useful to provide ecological information about their host when comparison among results of different studies is viable and the franciscana data compilation from more than four decades clearly showed such need. Studies carried out prior to Margolis et al. (1982) did not follow a standardized methodology to express the observed infection levels. In some cases, this lack of standardized results limited comparisons. Another limitation was the failure to inspect the intestines and/or stomachs in some areas. As some of the parasites have preferential infection sites, their absence or even reduction in numbers could be confused when the organ in question was not analyzed. The extraordinarily long small intestine of franciscana is also a reason for using different inspection methodologies. Although Aznar et al. (1994) have collected helminths from the whole organ, after verifying no expressive difference when collecting from proportional fragments, Andrade (1996) recommended a more quick and partial analysis using one-third of the total intestinal length. This methodology was used by subsequent studies, mostly in Brazil.

Considering that, at many times, the same researcher is not performing stomach and intestine content analysis, only items related to the animal's diet (otoliths and others) are of interest and gastric parasites can be underestimated or even lost if the examination is not performed by a skilled professional. Standardized and more detailed parasitological analyses are constantly recommended (although not always possible) and at all times depend on carcass decomposition state.

Host total body length, sex, and stage of decomposition (1–5 according to Geraci and Lounsbury, 2005) are basic information to be recorded; preferably the carcass should be in stages 2 or 3 of decomposition (fresh) according to this criteria. The first step is always to perform a careful inspection of the host searching for ectoparasites and all organs during necropsy. If there is a restriction of decomposition state, time, transportation, space to freeze or storage, at least the whole gastrointestinal tract, or stomach, and the first meters of the small intestines should be tied and collected for posterior analysis. In order to inspect all organs in one laboratory, appropriate sieves, stereomicroscope, and light are important and desirable.

Franciscana lacks the forestomach, which is found in other cetaceans (except beaked whales). Distinct longitudinal folds can be seen on the internal surface of the esophagus; microscopically clearly distinguishable in terms of color. Three compartments are demarcated by shallow grooves when viewed from the ventral side. The glandular main stomach is the largest compartment, nearly pear-shaped, and continuous from the esophagus, without any discernible sphincter. The smallest compartment, between the main and the pyloric stomach, is treated as the 'connecting channel' due to its tube-like passage. The pyloric stomach is a J-shaped elongated tubular structure, and at its end, there is an opening, with a sphincter that leads to the duodenal ampulla, which is demarcated from the pyloric stomach by a shallow groove (Yamazaki et al., 1974).

Each stomach chamber should be analyzed separately. Parasites recovered from the esophagus can be considered from the main stomach.

The differentiation between small and large intestines is made by observing the mesentery. Franciscana has no caecum and the ileum continues to the colon (which is near twice as thick), sustaining a different blood supply. In the ileum, vessels make an arcade and many vasa recta reach the ileum (perpendicularly), while in the colon, no (parallel) vasa recta can be seen. Internally, this transition area presents one evident bifurcation of the mucosa plica (Yamazaki et al., 1975).

After excluding the mesentery, the small intestine should be measured, divided into smaller segments and each compartment should be washed on a sieve with 150 micra mesh. All contents should be collected in petri dishes and inspected using a stereomicroscope (5 to 40x) to collect helminths.

In case the researcher decides to follow Andrade (1996), one should divide the small intestines into five equal parts and analyze each first third separately. The analysis of the whole small intestine could also be compared with the data provided by the partial analysis and the compartments considered equivalents to the partial analysis (e.g., Rocha, 2010). If available, the large intestines can be separated for the analysis (preferably).

The stomach and intestine walls should be examined, and the attached worms, collected. All parasites found should be collected and counted, at least from the most frequent species, allowing the determination of different parasite indexes, according to Bush et al. (1997). The parasitological indexes could be calculated in the software Quantitative Parasitology 3.0 (http://www.zoologia.hu/qp/qp.html). Parasites should be fixed in 70% alcohol for DNA extraction and morphological studies. Staining could follow Amato et al. (1991) for species identification based on taxonomic keys and specific descriptions.

Conclusion and general recommendations

The study of franciscana helminth fauna has constantly been included and recommended in workshops and action plans for this endangered species, especially in Brazil (Aznar et al., 1997; IBAMA, 1997, 2001; ICMBio, 2010; Reeves and Leatherwood, 1994). These studies have been showing population differences over the years and should continue to be performed, as it is a useful, simple, noninvasive, not expensive research that also provides health information.

It is always important to consider that parasite infection, prevalence, and mean intensity may vary through seasons, and differences in diet, age, gender and sample size of their hosts. Therefore, new standardized data can always be added to further understand these parameters.

There is a gap of knowledge in several biogeographic regions where there are records of partial data from the gastrointestinal tract. In the vast Brazilian coast, FMA I: the Espírito Santo and Rio de Janeiro state need new samplings; FMA II: the data from Paraná state, need to be updated with larger sample size; FMA III: in Santa Catarina state, information on stomach helminths is lacking, as well as southern region sampling; data is also incomplete for the northern coast of Rio Grande do Sul state, requiring greater sampling of intestines. In Uruguay, there is a need to repeat and standardize infection parameters for the stomach and analysis of a larger number of intestines both from franciscana of coastal waters and those residing in the estuary. In Argentina, FMA IV, franciscana occur along approximately 600 km of the Argentine coast and the only data available is from a more central area, being necessary to sample franciscanas further south of this point and individuals in estuaries.

In addition to presenting existing information, this chapter aimed at encouraging new researchers to advance in this area of great conservation relevance, where information is still lacking. New studies are fundamental when considering the ecology of a host with restricted distribution and whose survival is closely related to the level of habitat degradation. The coastal marine ecosystem is highly vulnerable to anthropogenic action and data renewal could provide a different panorama in every subsequent decade.

Acknowledgment

We thank all research groups and fishermen that contributed to the parasitological studies of franciscana. We also thank Ana Carolina Ewbank and Andrea Mondino for the revision of the draft of this chapter.

References

Adams, A.M., Rausch, R.L., 1989. A revision of the genus *Orthosplanchnus* Odhner, 1905 with consideration of the genera *Odhneriella* Skriabin, 1915 and *Hadwenius* Price, 1932 (Digenea: Campulidae). Can. J. Zool. 67, 1268–1278.

Alves, A.K.M., Souza, F.A.L., Marigo, J., Cremer, M.J., 2017. Intestinal helminths of franciscana (*Pontoporia blainvillei*) and Guiana dolphins (*Sotalia guianensis*) from the north coast of Santa Catarina state. Brazil. Oecol. Aust. 21 (1), 83–87.

Amato, J.F.R., Boeger, W.A., Amato, S.B., 1991. Amato, J.F.R. (Ed.), Protocolo para Coleta e Processamento de Parasitos de Pescado. Seropédica, Imprensa Universitária Universidade Federal do Rio, de Janeiro, p. 78.

Anderson, R.M., 1993. Epidemiology in Modern Parasitology (Ed. by FEG Cox), pp. 75-116. Blackwell Oxford.

Andrade, A.L.V., 1996. Comunidade componente de helmintos gastrointestinais da Franciscana, *Pontoporia blainvillei* (Cetacea, Pontoporiidae) no Rio Grande do Sul, Brasil, e sua utilização como marcador biológico na identificação de estoques MSc Thesis. Universidade do Rio Grande, Rio Grande, p. 98.

Andrade, A., Pinedo, M.C., Pereira, Jr, J., 1997. The gastrointestinal helminths of franciscana, *Pontoporia blainvillei*, in southern Brazil. Ann. Rep. Int. Whaling Comm. 47, 669–673.

Aznar, F.J., Balbuena, J.A., Raga, J.A., 1994. Helminths communities of Pontoporia blainvillei (Cetacea: Pontoporiidae) in Argentinian waters. Can. J. Zool. 72, 1–5.

Aznar, F.J., Raga, J.A., Corcuera, J., Monzón, F., 1997. Los parasitos, un posible método para diferenciar stocks y detectar migraciones em las franciscanas.EF/2/DT1. Anais do 2[0] Encontro sobre coordenação de pesquisa e manejo de franciscana/editado por M.C. Pinedo e A.S. Barreto, Rio Grande, Ed FURG, pp. 13–22.

Aznar, F.J., Bush, A.O., Fernández, M., Raga, J.A., 1999. Constructional morphology and mode ofattachment of the trunk of *Corynosoma cetaceum* (Acanthocephala: Polymorphidae). J. Morphol. 241 (3), 237–249.

Aznar, F.J., Fognani, P., Balbuena, J.A., Pietrobelli, M., Raga, J.A., 2006. Distribution of *Pholeter gastrophilus* (Digenea) within the stomach of four odontocete species: the role of the diet and digestive physiology of hosts. Parasitology 133 (3), 369–380.

Aznar, F.J., Raga, J.A., Corcuera, J., Monzón, F., 1995. Helminths as biological tags for franciscana (*Pontoporia blainvillei*) (Cetacea:Pontoporidae) in Argentinian and Uruguayan waters. Mammalia 59 (3), 427–435.

Aznar, F., Balbuena, J.A., Fernández, M., Raga, J., 2001. Living Together: The Parasites of Marine Mammals. Marine Mammals). Springer, Boston, MA, pp. 385–423.

Balbuena, J.A., Aznar., F.J., Fernandez, M., Raga, J.A., 1995. Parasites as indicators of social structure and stock identity of marine mammals. In: Blix, A.S., Walloe, L., Ultang, O. (Eds.), Proceedings of the international Symposium on the Biology of Marine Mammals in the east Atlantic. Whales, seals, fish and man: developments in marine biology. Elesevier Science publishere Ltd., Amsterdam, p. 720.

Bassoi, M., Sheperd, J.G., Secchi, E.R., Moreno, I.B., Danilewicz, D., 2020. Oceanographic processes driving the feeding ecology of franciscana dolphin off Southern Brazilian coast. Cont. Shelf Res. 201, 104124.

Bassoi, M., Secchi, E., Danilewicz, D., Moreno, I., Santos, R., Shepherd, J., 2021. Intrapopulation variation in the diet of franciscana dolphin (*Pontoporia blainvillei*) off southern Brazil. J. Marine Biological Assoc. United Kingdom 101 (3), 621–637. doi:10.1017/S0025315421000436.

Berón-Vera, B., Crespo, E.A., Raga, J.A., 2008. Parasites in stranded cetaceans of Patagonia. J. Parasitol. 94 (4), 946–948.

Bossart, G.D., 2011. Marine mammals as sentinel species for oceans and human health. Vet. Pathol. 48 (3), 676–690.

Bordino, P., Wells, R.S., Stamper, M.A., 2008. Satellite tracking of Franciscana Dolphins *Pontoporia blainvillei* in Argentina: preliminary information on ranging, diving and social patterns, International Whaling Commission Scientific Committee Meeting SC60/SM14 2008 Jun. IWC.

Brownell Jr, R.L., 1975. Progress report on the biology of the franciscana dolphin *Pontoporia blainvillei*, in Uruguayan waters. J. Fish. Res 32 (7), 1073–1078.

Bush, A.O., Lafferty, K.D., Lotz, J.M., Shostak, A.W., 1997. Parasitology meets ecology on terms: Margolis et al. Revisited. J. Parasitol. 83 (4), 575–583.

Catalano, S.R., Whittington, I.D., Donnellan, S.C., Gillanders, B.M., 2014. Parasites as biological tags to assess host population structure: guidelines, recent genetic advances and comments on a holistic approach. IJP:PAW 3 (2), 220–226.

Crespo, E.A., 2018. Franciscana dolphin: *Pontoporia blainvillei*. Encyclopedia of marine mammals. Academic Press, pp. 388–392.

Cunha, H.A., Gariboldi, M.C., Mendez, M., Secchi, E.R., Oliveira, L.R., Ott, P., Trres-Flores, J.P., Farro, A.P.C., 2020a. Review on the franciscana stock structure and Franciscana Management Areas (FMA). IWC: SC/68B/SDDNA/07.

Cunha, H.A., Dias, C.P., Alvarenga, L.C., Wells, R.S., Cremer, M.J. Microscale population structure and kinship analyses suggest philopatry of both sexes in franciscana (Pontoporia blainvillei), 2020b. IWC: SC/68B/SDDNA/04.

Dailey, M.D., Brownell Jr, R.I., 1972. A check list of marine mammal parasites. In: Ridgway, S.H. (Ed.), Mammals of the Sea, Biology and Medicine. Charles C. Thomas. Springfield, Illinois, pp. 528–589.

Danilewicz, D., Rosas, F., Bastida, R., Marigo, J., Muelbert, M., Rodriguez, D., et al., 2002. Special issue on the biology and conservation of franciscana. LAJAM 1, 25–42.

Davey, J.T., 1971. A revision of the Genus *Anisakis* Dujardin, 1845 (Nematoda: Ascaridata). J. Helminthol. 45, 51–72.

Di Beneditto, A.P.M., Ramos, R.M.A, 2001. Biology and conservation of the franciscana (*Pontoporia blainvillei*) in the north of Rio de Janeiro State. Brazil. J. Cetacean Res. Manag. 3, 185–192.

Ebert, M.B., Valente, A.L.S., Cremer, M.J., da Silva, R.J., 2021. *Pholeter gastrophilus* (Trematoda: Heterophyidae), a parasite of dolphins from Brazilian waters. Biologia (Bratisl) 14 1-0.

Falco, A.L., 2008. Composição alimentar do leão-marinho do sul, *Otaria flavescens*, no litoral sul do RS: avaliação histórica e ontogenética. Dissertação de mestrado. Universidade Federal do Rio Grande 100.

Fernández, M., Balbuena, J.A., Pertusa, J.F., Raga, J.A., 1995. Biometric variability of *Hadwenius tursionis* (Marchi, 1873) (Digenea, Campulidae) from the intestine of the bottlenose dolphin *Tursiops truncatus* (Montagu, 1821). Syst. Parasitol. 30 (1), 67–76.

Fernández, M., Aznar, F.J., Latorre, A., Raga, J.A., 1998. Molecular phylogeny of the families Campulidae and Nasitrematidae (Trematoda) based on mtDNA sequence comparison. Int. J. Parasitol. 28, 767–775.

Ferreira, H.S., Cunha, C.M., da, Gonzalez, M.M.B, 1998. Endoparasitismo de franciscana *Pontoporia blainvillei* (Cetacea, Platanistidae) por *Riggia* sp. (Isopoda, Cymothoidae). VIII Reunião de Trabalhos de Especialistas em Mamíferos Aquáticos da América do Sul. held in Olinda, Brazil (p. 80). Rio de Janiero. Brazil: Status of Aquatic Mammals in Latin America (SOLAMAC).

Fraija-Fernández, N., Fernández, M., Lehnert, K., Raga, J.A., Siebert, U., Aznar, F.J., 2017. Long-distance travellers: phylogeography of a generalist parasite, *Pholeter gastrophilus*, from cetaceans. PLoS One 12 (1), e0170184.

Geraci, J.R., Lounsbury, V.J., 2005. Marine mammals ashore: a field guide for strandings. 2nd Edition. National Aquarium Baltimore, MD.

Gibson, D.I., 2005. Family Brachycladiidae Odhner, 1905. In: Jones, A., Bray, R.A., Gibson, D.I. (Eds.). Keys to the Trematoda, Vol. 2. CAB International and the Natural History Museum, London, pp. 641–652.

Gulland, F.M.D., Hall, A.J., 2007. Is marine mammal health deteriorating? Trends in the global reporting of marine mammal disease. EcoHealth 4, 135–150.

Haimovici, M., Vasconcellos, M., Kalikoski, D.C., Abdalah, P., Castello, J.P., Hellebrant, D., 2006. Diagnóstico da pesca no Rio Grande do Sul Eds. In: Isaac, V., Martins, S.A., Haimovici, M., Andriguetto, J.M. (Eds.), A Pesca Marinha e Estuarina do Brasil no Início do Século XXI: recursos, tecnologias, aspectos socioeconômicos e institucionais. Publicação do Especial do Projeto: Uso e Apropriação de Recursos Costeiros—Modelo Gerencial da Pesca. Instituto do Milenio/CNPq Editora Universitaria UFPA, Belem, pp. 157–180.

Hamilton, H., Caballero, S., Collins, A.G., Brownell, R.L.Jr., 2001. Evolution of river dolphins. Proc. Royal Soc. London Series B 268, 549–556.

Henning, B., Carvalho, B.S., Pires, M.M., Bassoi, M., Marigo, J., Bertozzi, C., et al., 2017. Geographical and intrapopulation variation in the diet of a threatened marine predator, *Pontoporia blainvillei* (Cetacea). Biotropica 50, 157–168.

Henrique-Garcia, J., Barreto, A.S., Secchi, E.R., 2010. Presença de *Conchoderma auritum* (Linnaeus, 1769) (Crustacea, Cirripedia) em *Pontoporia blainvillei* (Mammalia, Cetacea,

Pontoporiidae) no litoral sul do Brasil. Anais 14ª RT Reunião de Trabalhos de Especialistas em Mamíferos Aquáticos da América do Sul. Florianópolis. 2010.

Hernández-Orts, J.S., Montero, F.E., Juan-García, A., García, N.A., Crespo, E.A., Raga, J.A., Aznar, F.J., 2013. Intestinal helminth fauna of the South American sea lion *Otaria flavescens* and fur seal *Arctocephalus australis* from northern Patagonia. Argentina. J. Helminthol. 87, 336–347.

Hoss, D., Valente, A.L., Ott, P.H., 2017. Gastrointestinal helminths of the franciscana *Pontoporia blainvillei* (Mammalia: Cetartiodactyla) in the Northern coast of Rio Grande do Sul State. Southern Brazil. Oecol. Aust. 21 (1), 72–82.

IBAMA, 1997. Mamíferos Aquáticos do Brasil: Plano de Ação – Versão I. Brasília, IBAMA (Instituto Brasileiro do Meio Ambiente e dos Recursos Naturais Renováveis, Ministério do Meio Ambiente), p. 80.

IBAMA, 2001. Mamíferos Aquáticos do Brasil: Plano de Ação – Versão II. Brasília, IBAMA (Instituto Brasileiro do Meio Ambiente e dos Recursos Naturais Renováveis, Ministério do Meio Ambiente), p. 706.

ICMBio, 2010. Plano de ação nacional para conservação do pequeno cetáceo Toninha: *Pontoporia blainvillei*, Ana Paula di Beneditto... (et al), organizadores C.C.R. Campos, D. S. Danilewicz, S. Siciliano. Brasilia: Instituto Chico Mendes de Conservação da Biodiversidade, ICMBio, 2010. p. 76.

Irigoitia, M.M., Palomba, M., Braicovich, P.E, Lanfranchi, A.L., Denuncio, P.E., Gana, J.C, M., M,S., 2021. Genetic identification of *Anisakis* spp. (Nematoda: Anisakidae) from cetaceans of the Southwestern Atlantic Ocean: ecological and zoogeographical implications. Parasitol. Res. 120 (5), 1699–1711.

Kagei, N., Tobayama, T., Nagasaki, Y., 1976. On the helminthum of franciscana *Pontoporia blainvillei*. Sci. Rep. Whales Res. Inst. 28, 161–166.

Kremnev, G., Gonchar, A., Krapivin, V., Knyazeva, O., Krupenko, D., 2020. First elucidation of the life cycle in the family Brachycladiidae (Digenea), parasites of marine mammals. Int. J. Parasitol 50, 997–1009.

Lima, I.D., Castello, J.P., 1995. Distribution and abundance of South-west Atlantic anchovy spawners (*Engraulis anchoita*) in relation to oceanographic processes in the southern Brazilian shelf. Fish. Oceanogr. 4 (1), 1–16.

Machado, R., Oliveira, L.R., Ott, P.H., Haimovici, M., Cardoso, L.G., Millman, L., Romero, M.A., Santos, R.A., Borges-Martins, M., 2020. Trophic overlap between marine mammals and fisheries in subtropical waters in the western South Atlantic. MEPS 639, 215–232.

Machado-Pereira, E., Pereira, J., Secchi, E., Müller, G., Valente, A.L., 2017. *Contracaecum ogmorhini* Johnston & Mawson, 1941 (Nematoda: Anisakidae) in southern sea lion, *Otaria flavescens* (Shaw, 1800) from southern Rio Grande do Sul State, Brazil. Pan-Am. J. Aquat. Sci. 12, 227–233.

Mackenzie, K., 1999. Parasites as biological tags in population studies of marine organisms. Qatar Univ. Sci J. 19, 117–127.

MacKenzie, K., Abaunza, P., 1998. Parasites as biological tags for stock discrimination of marine fish: a guide to procedures and methods. Fish. Res. 38, 45–56.

MacKenzie, K., Abaunza, P., 2014. Parasites as biological tags. Stock Identification Methods. Applications in Fisheries Science, Academic Press, 185–203.

Marcogliese, DJ., 2004. Parasites: Small Players with Crucial Roles in the Ecological Theater. EcoHealth 1, 151–164.

Margolis, L., Esch, G.W., Holmes, J.C., Kuris, A.M., Schad, G.A., 1982. The use of ecological terms in parasitology (Report of an Ad Hoc Committee of the American Society of Parasitologists). The J. Parasitol. 68, 131–133.

Marigo, J., Rosas, F.C.W., Andrade, A.L.V., Oliveira, M.R., Dias, R.A., Catão-Dias, J.L., 2002. Parasites of *Pontoporia blainvillei* from São Paulo and Paraná States. Brazil. LAJAM 1 (1), 115–122.

Marigo, J., Vicente, A.C.P., Valente, A.L.S., Measures, L., Santos, C.P., 2008. Redescription of *Synthesium pontoporiae* n. comb. with notes on *S. tursionis* and *S. seymouri* n. comb. (Digenea: Brachycladiidae Odhner, 1905). J. Parasitol. 94, 505–514.
Marigo, J., Thompson, C.C., Santos, C.P., Iñiguez, A.M., 2011. The *Synthesium* Brachycladiidae Odhner, 1905 (Digenea) association with hosts based on nuclear and mitochondrial genes. Parasitol. Int. 60 (4), 530–533.
Marigo, J., Cunha, H.A., Bertozzi, C.P., Souza, S.P., Rosas, F.C.W., Cremer, M.J., Barreto, A.S., Oliveira, L.R., Cappozo, H.L., Valente, A.L.S., Santos, C.P., Vicente, A.C.P., 2013. Genetic diversity and population structure of *Synthesium pontoporiae* (Digenea, Brachycladiidae) linked to its definitive host stocks, the endangered Franciscana dolphin, *Pontoporia blainvillei* (Pontoporiidae) off the coast of Brazil and Argentina. J. Helmint. 89 (1), 19–27.
Mattiucci, S., Cipriani, P., Webb, S.C., Paoletti, M., Marcer, F., Bellisario, B., Gibson, D.I., Nascetti, G., 2014. Genetic and morphological approaches distinguish the three sibling species of the *Anisakis simplex* species complex, with a species designation as *Anisakis berlandi* n. sp. for *A. simplex* sp. C (Nematoda: Anisakidae). J. Parasitol. 100, 199–214.
Mattiucci, S., Cipriani, P., Levsen, A., Paoletti, M., Nascetti, G., 2018. Molecular epidemiology of *Anisakis* and anisakiasis: an ecological and evolutionary road map. Adv. Parasitol. 99, 93–263.
Miranda, L.V., Haimovici, M., 2007. Changes in the population structure, growth and mortality of striped weakfish *Cynoscion guatucupa* (Scianidae, Teleostei) of southern Brazil between 1976 and 2002. Hydrobiologia 589, 69–78.
Moravec, F., Salgado-Maldonado, G., Caspeta-Mandujano, J., 2000. Three new *Procamallanus* (Spirocamallanus) species from freshwater fishes in Mexico. J. Parasitol. 86, 119–127.
Moser, M., 1991. Parasites as biological tags. Parasitol. Today 7 (7), 182–185.
Pereira, E.M., 2012. Identificação da comunidade componente de helmintos, gastrointestinais hepáticos, pulmonares, cardíacos e renais de *Otaria flavescens* (Leão-marinho-do-sul), no litoral sul do Brasil Pereira. MSc Thesis. Instituto de Biologia. Universidade Federal de Pelotas, p. 70.
Pereira Jr, J., Neves, L.F.M., 1993. *Corynosoma australe* Johnston, 1937 (Acanthocephala: Polymor-phidae) em *Micropogonias furnieri* (Desmarest, 1823) (Perciformes: Sciaenidae) do litoral do Rio de Grande do Sul. Com. Mus. Ciênc. PUCRS 6, 51–61.
Pérez-i-García, D., Constenla, M., Carrassón, M., Montero, F.E., Soler-Membrives, A., González-Solís, D., 2015. *Raphidascaris* (*Raphidascaris*) macrouri n. sp. (Nematoda: Anisakidae) from two deep-sea macrourid fishes in the Western Mediterranean: Morphological and molecular characterizations. Parasit. Int. 64 (5), 345–352.
Pilleri, G., 1971. On the La Plata dolphin *Pontoporia blainvillei* on the Uruguayan coast. Invest. Cet. 3, 69–73.
Pinedo, M.C., 1991. Development and variation of the franciscana, *Pontoporia blainvillei*. PhD. Thesis. University of California, Santa Cruz, p. 406.
Pinto, R.M., Muniz-Pereira, L.C., Alves, V.C., Siciliano, S., 2004. First report of a helminth infection for bryde's whale, *Balaenoptera edeni* Anderson, 1878 (Cetacea, Balaenopteridae). LAJAM 3 (2), 167–170.
Praderi, R., 1984. Mortalidad de Franciscana, *Pontoporia blainvillei*, en pesquerias artesanales de tiburón de la costa Atlântica Uruguaya. Pages 259-272 in Actas, III Reunión Iberoamericana de Conservacion y Zoologia de Vertebrados, 15-19 November, 1992, Buenos Aires.
Raga, J.A., Balbuena, J.A., Dailey, M.D., 1994. *Hadwenius pontoporiae* sp. n. (Digenea: Campulidae) from the intestine of franciscana (Cetacea: Pontoporiidae) in Argentinian waters. J. Helminthol. Soc. Washington 61 (1), 45–49.
Reeves, R., Leatherwood, S., 1994. Dolphins, porpoises and whales: 1994–1998 Action Plan for the Conservation of Cetaceans. Gland, IUCN (International Union for Conservation of Nature).

Rocha, A.A., 2010. Comunidade componente de helmintos gastrointestinais de *Pontoporia blainvillei* (Gervais & d'Orbigny, 1844) do litoral sul do Rio Grande do Sul relação com a dieta e determinação de estoques ecológicos MSc Thesis. Instituto de Oceanografia da Universidade Federal do Rio Grande, p. 133.

Rocha, A.A., Franco-Trecu, V., Valente, A.L.S., Secchi, E.R., Pereira Jr, J., 2013. *Synthesium pontoporiae* (Digenea: Brachycladiidae) em franciscanas (*Pontoporiae blainvillei*) provenientes do Uruguay. ST: número 13, pôster-oral. XIII Congreso de Ciencias del Mar-COLACMAR. 27-31th october. Punta del Este, Uuruguay.

Santos, C.P., Rohde, K., Ramos, R., Di Beneditto, A.P., 1996. Helminths of cetaceans on the Southeastern coast of Brazil. J. Helminthol. Soc. Washington 63 (1), 149–152.

Seabra, C.B., 2018. Helmintos gastrointestinais de toninha, *Pontoporia blainvillei* (Cetartiodactyla, Pontoporiidae), do litoral de São Paulo, Brasil MSc Thesis. Universidade Estadual Paulista, Campus do Litoral Paulista, Instituto de Biociências, p. 60.

Secchi, E.R., Danilewicz, D., Ott, P.H., 2003. Applying the phylogeographic concept to identify franciscana dolphin stocks: implications to meet management objectives. J. Cetacean Res. Manag. 5 (1), 61–68.

Secchi, E.R., 2010. Review on the threats and conservation status of franciscana, *Pontoporia blainvillei* (Cetacea, Pontoporiidae). In: Shostell, J.M., Ruiz-Garcia, M. (Eds.). Biology, Evolution and Conservation of River Dolphins within South America and Asia, Vol. 1. Hauppauge, NY: Nova Science Publishers Inc, pp. 323–339.

Schmidt, G.D., Dailey, M.D., 1971. Zoogeography and the generic status of *Polymorphus (Polymorphus) cetaceum* (Johnston et Best, 1942) comb. n. (Acanthocephala). Proceedings of the Helminthological Society of Washington 38, 137.

Silveira, T.L.R., 2014. Avaliação de componentes da dieta de *Pontoporia blainvillei* Gervais & d'Orbigny, 1844 (Cetacea, Pontoporiidae) como possíveis hospedeiros intermediários ao parasitismo por helmintos MSc Thesis. Universidade Federal de Pelotas. Instituto de Biologia, p. 87.

Silva, N., Castilho, P.V., 2020. Avaliação de ocorrência de cracas pedunculares (*Conchoderma auritum*) em toninhas (*Pontoporia blainvillei*) no litoral Sul de Santa Catarina. Trabalho de Conclusão, Engenharia de Pesca, Centro de Educação Superior da Região Sul, da Universidade do Estado de Santa Catarina, Laguna, Santa Catarina. p. 25.

Silveira, T., Remião, M.H., Robaldo, R.B., Freitas, K., Valente, A.L, 2017. Prey species of franciscana *Pontoporia blainvillei* as paratenic hosts of helminths. Dis. Aquat. Org. 123 (3), 251–256.

Soto, J.M.R., 2001. First record of a rabbit-eared barnacle, *Conchoderm auritum* (Linnaeus, 1767) (Crustacea, Cirripedia), on the teeth of the La Plata dolphin, *Pontoporia blainvillei* (Gervais & D'Orbigny, 1844) Cetacea, Platanistoidea). Mare Magnum 1, 172–173.

Timi, J.T., Luque, J.L., Sardella, N.H., 2005. Parasites of *Cynoscion guatucupa* along South American Atlantic coasts: evidence for stock discrimination. J. Fish Biol. 67, 1603–1618.

Valente, A.L., Echenique, J.Z., Pereira, E., Schild, A.L.P., 2014. First record of *Bolbosoma turbinella* (Acantocephala: Polymorphydae) in bryde's whale in southern coast of Brazil. X Congreso de la Sociedad Latinoamericana de Especialistas en Mamíferos Acuáticos (SOLAMAC).

Vasconcellos, M., Haimovici, M., 2006. Status of white croaker *Micropogonias furnieri* exploited in southern Brazil according to alternative hypotheses of stock discreetness. Fish. Res. 80, 196–202.

Yamasaki, F., Takahashi, K., Kamiya, T., 1974. Digestive tract of la Plata dolphin, *Pontoporia blainvillei*. I. Oesophagus and Stomach. Okajimas Folia Anat. Jpn. 51, 29–52.

Yamasaki, F., Takahashi, K., Kamiya, T., 1975. Digestive tract of la Plata dolphin, *Pontoporia blainvillei*. II. Small and Large Intestines. Okajimas Folia Anat. Jpn. 52, 1–26.

CHAPTER 4

Pathologies and diseases

Ana Paula Frederico Rodrigues Loureiro Bracarense, Isabela Guarnier Domiciano
Laboratório de Patologia Animal, Universidade Estadual de Londrina, Londrina, Paraná, Brazil

Respiratory diseases

Pulmonary changes are the most frequent alterations observed in stranded franciscana analyzed in PMP-BS (Simba, 2021). From 352 animals, 296 (84%) had lung lesions and death associated with respiratory dysfunctions. Similarly, a previous study reported 91.7% of death causes as asphyxia (Ruoppolo, 2003). *Postmortem* examination revealed pulmonary congestion and mild to severe edema as the main findings (Fig. 4.1). In addition, a frothy and clear fluid in the tracheal lumen and lung hemorrhage were frequently described. Microscopically, moderate to severe alveolar edema and congestion were observed (Simba, 2021). These changes associated with superficial and deep bruises on the skin (mouth, fins, and tail), including nodes of fishing gears, are commonly interpreted as bycatch (Domiciano et. al., 2016) (Fig. 4.2).

In 23.6% of the cases that asphyxia was considered the cause of death, the occurrence of pulmonary lesions was observed, mainly pneumonia. A diversity of pneumonic patterns has been diagnosed, as interstitial and granulomatous pneumonia. *Pseudomonas aeruginosa* and Micrococcus sp. have been isolated from thoracic cavity fluid in two animals with chronic pneumonia (Ruoppolo, 2003).

Interstitial pneumonia is characterized by damage to alveolar septa. Grossly, the lungs are enlarged, with a rubbery consistency. Furthermore, rib imprints are commonly observed in lungs (Fig. 4.3); the lesions are diffuse, but predominantly dispersed on the caudal area. Microscopically, the interalveolar septa are thickened by the proliferation of smooth muscle and fibroblasts along with infiltration with lymphocytes, neutrophils, and macrophages.

A report case described the husbandry and medical management of an injured orphan calf entangled in a fishing gear (Baldassin et al., 2007). The animal died and the gross and histological lesions revealed interstitial pneumonia, but the cause was not determined. Another case included a neonate stranded

The Franciscana Dolphin
DOI: https://doi.org/10.1016/B978-0-323-90974-7.00008-2

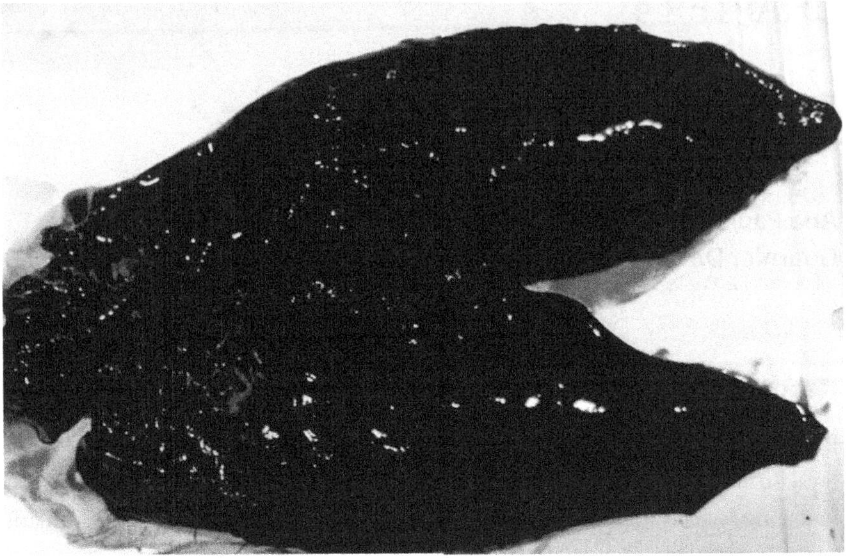

Fig. 4.1 ***Diffuse lung congestion.*** *Source: Simba, 2021.*

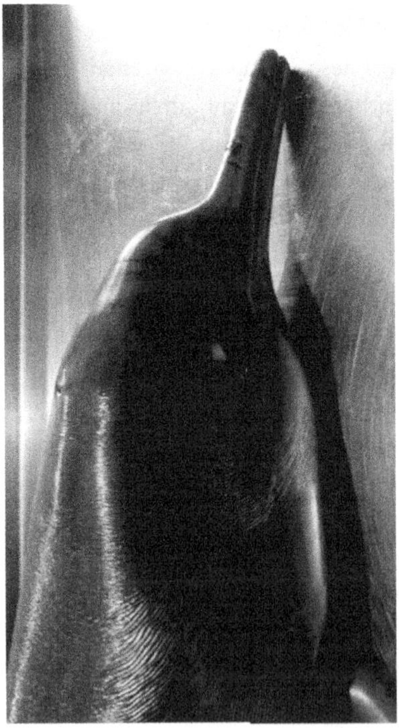

Fig. 4.2 ***Impression from gillnet entanglement at the head.*** *Source: Simba, 2021*

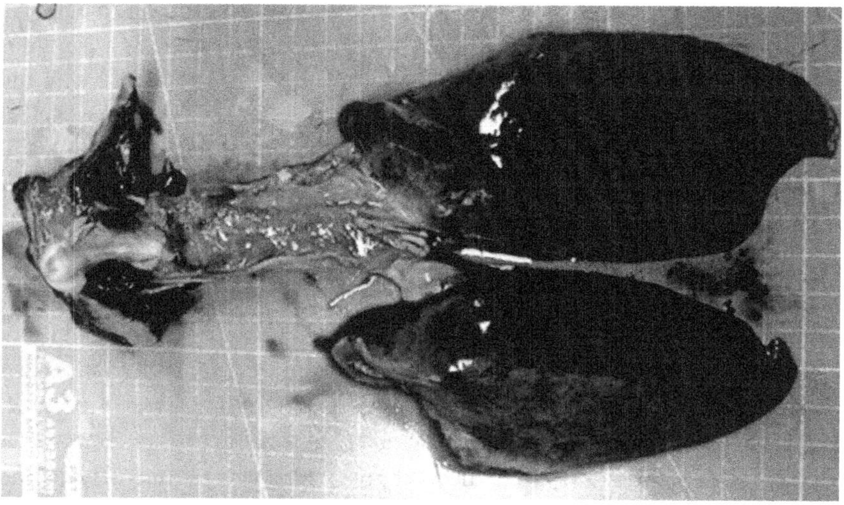

Fig. 4.3 ***The lungs are diffusely firmer than normal, showing a rubbery texture and ribs impression.*** *Source: Simba, 2021*

alive, but the animal died after intensive care, and exams revealed severe pneumonia associated with *Enterobacter kobei* (Fuentes-Castillo et al., 2021). The genetic analysis indicated *E. kobei,* a multidrug-resistance pathogen, with possible negative implications for wild endangered populations, humans, and associated ecosystems (Fuentes-Castillo et al., 2021). Interstitial pneumonia was reported in an adult female specimen associated with *Toxoplasma gondii* infection (Simba, 2021). Cysts containing bradyzoites were observed in lungs, pulmonary and mediastinal lymph nodes, and liver associated with lymphohistiocytic infiltrate and areas of necrosis (Fig. 4.4).

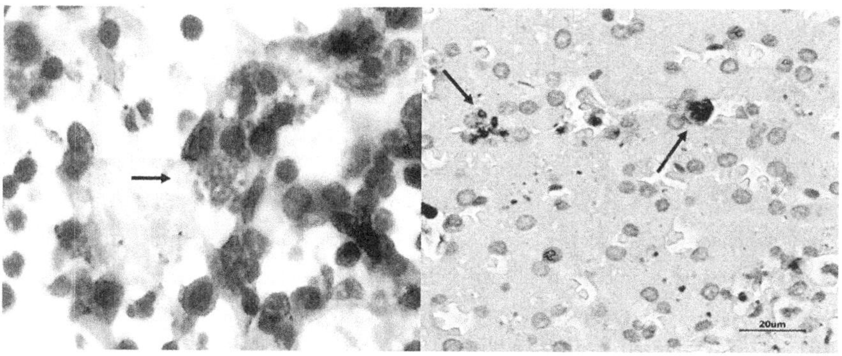

Fig. 4.4 ***A-Lungs showing bradyzoites in the alveolar parenchyma (arrow).*** Hematoxylin-eosin staining. Objective 40x. Laboratory PatAnimal, São Paulo. B- Liver showing tachyzoites and a cyst immunostained for *T. gondii.* Immunoperoxidase. Bar 20 μm. *Source*: Laboratory of Animal Pathology, Universidade Estadual de Londrina, Paraná.

Chronic pneumonia associated with cholesterol deposits was diagnosed in 16.7% (10/60) of franciscanas stranded or bycaught. Grossly, findings included multifocal coalescent firm areas (1–10 mm), distributed along the peripheral edges of lungs. Less than 10% of the lungs were affected. The granulomatous areas were pale yellow and homogeneous. Microscopically, multifocal areas containing cholesterol clefts circumscribed by multinucleated giant cells and abundant foamy macrophages were observed. The etiology of this type of pneumonia remains unsolved (Ruoppolo et al., 2010).

Necrotic pneumonia associated with *Aspergillus* spp. infection was reported in a *P. blainvillei* stranded in Argentina. Extensive areas of necrosis were observed macroscopically, while histologically no granulomas were noticed. Special histological stains revealed septate hyphae presenting dichotomous branching at an acute angle within the necrotic areas (Romano et al., 2020). Furthermore, suppurative necrotic pneumonia was reported in a fetus associated with pulmonary edema and intralesional cocci and bacilli (Simba, 2021).

Aspiration pneumonia associated with the inspiration of exogenous material was observed in two cases. The animals were found alive and carried to rehabilitation centers. Probably the administration of diets through a gastric catheter contributed to the lesion (Ruoppolo, 2003). Another case of aspiration pneumonia was reported in a calf associated with the presence of epithelial cells and foamy macrophages suggesting aspiration of meconium and amniotic liquid (Simba, 2021).

Gastrointestinal diseases

Infection by *Helicobacter* species was related to gastric erosions and ulcers in captive and wild cetaceans (Harper et al., 2000). *Helicobacter* spp. DNA was identified in gastric juice from autopsy samples in *P. blainvillei* from La Plata River, showing 99.4% identity with *H. cetorum* (Goldman et al., 2011).

Parasites are routinely observed in the gastrointestinal tract of *P. blainvillei*, and some species indicate biological and ecological aspects of host stocks, specific niche, and low pathogenicity owing to host-parasite coevolution, but diagnosis of infection is scarce. Gastric ulceration has been associated with stress and parasite infection within the gastric mucosa in marine mammals (Sweeney and Ridgway, 1975). In bycatch, *P. blainvillei* off southern Brazil, mild gastritis associated with adults *Polymorphus* (*P.*) *cetaceum* (Acanthocephala, Polymorphidae) fixed to the mucosa, caused macroscopically low, long linear to discoid depression. Dysplasia, necrosis in the main gastric chamber, and hyperplasia, mild inflammation in pyloric, and connecting channel mucosa was observed microscopically (Silva and Cousin, 2006a). Furthermore, grossly round, edematous, pale lesions and petechiae in the mucosa, and thickened submucosa and muscularis layers. Microscopically, eosinophilic granulomas and fibrosis in the main gastric chamber were diagnosed, caused by adult fixation or larvae migration of *Anisakis typica* Diesing, 1860 (Nematoda - Anisakidae) (Silva and Cousin, 2004).

Additionally, a parasitic infection of acanthella and adult *Bolbosoma turbinella* Diesing, 1851 (Acanthocephala — Polymorphidae) was observed in large intestines in bycatch animals off southern Brazil. Grossly, round, yellow, edematous, and suppurative lesions were associated with parasitic infection, diagnosed as severe miofibrino purulent abscesses microscopically. The acanthella migration may cause transmural enteritis and scarce healing process, while adults may cause mucosa and submucosa infection owing to parasitic fixation (Silva and Cousin, 2006b). Conversely, the intestinal trematode *Hadwenius pontoporiae* was observed in 15 of the 21 (71.42%) *P. blainvillei*, and no gross lesions were associated with the presence of the parasites (Danilewicz et al., 2002). It is important to note that parasites may detach from adhered tissue after host death and migrate to lumen owing to *postmortem* anoxia and/or carcasses refrigeration.

Plastic debris ingestion was observed in 28.1% (30/106) and 15.7% (14/89) of bycatch animals off northern Argentina and southeastern Brazil, respectively. No perforation, ulceration, or obstructions were macroscopically observed in the digestive tracts, owing to the small size and number of fragments found (Denuncio et al., 2011; Di Beneditto and Ramos, 2014). Nevertheless, the presence of bigger fragments and the effect of plastic digestion cannot be ruled out as a potential cause of death (Denuncio et al., 2011).

Liver diseases

Hepatic changes have been reported in franciscanas stranded off southern and southeastern Brazil. Cytoplasmic vacuolation of hepatocytes, fibrosis, and hepatitis were the main findings observed (4/60, 2/60, 1/60, respectively) (Ruoppolo, 2003). In another study, in the same region, hyaline inclusions (62%), lymphoplasmacytic portal hepatitis (34%), and steatosis (13%) were the most frequent lesions. Hyaline inclusions were detected more frequently in bycatch animals and were associated with acute stress, cate-

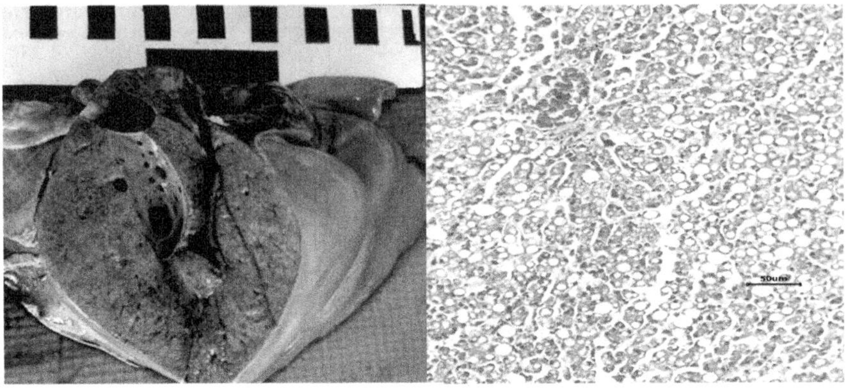

Fig. 4.5 ***Left- the liver is enlarged, diffusely pale.*** *Source: Simba, 2021.* Right- liver, hepatocytes showing diffuse cytoplasmic vacuolation. HE. Scale bar 50 μm. *Source: Laboratory of Animal Pathology, UEL.*

cholamines release, and vascular constriction. The cause of hepatitis was not established in this study; however, an infectious or toxic agent is suspected, considering the portal location. Steatosis affected mainly calves and juvenile specimens, while hepatitis was observed in all ages (Gonzalez-Viera, 2012). Steatosis can be related to nutritional disturbance and/or associated with another disease (Baldassin et al., 2007; Gonzalez-Viera, 2012). Degenerative hepatic process and acute interstitial pneumonia were observed in a calf that died after intensive care off southeastern Brazil. Necropsy revealed friable liver of light pink color and mild to severe diffuse micro-and macrovacuolization of hepatocytes which can be related to nutritional disturbance (Baldassin et al., 2007). From 2015 to 2020, 428 *P. blainvillei* were analyzed and the most frequent change in the liver was cytoplasmic vacuolation of hepatocytes (Simba, 2021). Severe steatosis was observed in a calf entangled in fishing nets in its rostrum and in a pregnant animal (Fig. 4.5).

Renal diseases

The kidneys of cetaceans have specialized structures to the marine environment such as reniculate morphology and increased thickness of the medullary region. In Brazil, a study conducted between 1996 and 2011 revealed that franciscana is commonly affected by cystic diseases (Gonzales-Viera et al., 2015). Simple, well-demarcated cysts were observed in a random distribution only in the cortex. The cysts ranged from 0.2 to 0.7 cm and were filled by a gelatinous substance. Microscopically, the animals showed

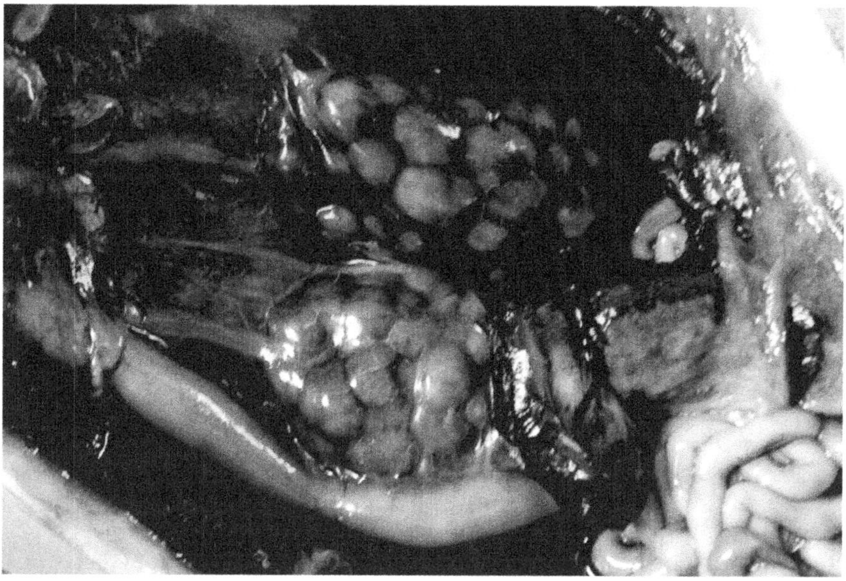

Fig. 4.6 ***Kidneys showing diffuse pale color.*** The microscopically tubular acute injury was diagnosed.

a diversity of changes including membranous and membranoproliferative glomerulonephritis, lymphoplasmacytic nephritis, glomerulosclerosis, lipidosis, simple cysts, and a tubular adenoma.

Acute tubular injury (ATI) is frequently diagnosed in stranded specimens in PMP-BS. From 54 animals with a fresh carcass condition analyzed between 2015 and 2019, 25 (46%) showed this change microscopically (Simba, 2021). The ATI is a reversible condition characterized by tubular cell degeneration and an increased number of pyknotic nuclei resulting in the functional loss (Cianciolo and Mohr, 2016). In severe cases, the kidneys show a pale color in the cortical region (Fig. 4.6). Microscopically, tubular epithelial degeneration and tubular necrosis were the main findings. Hyaline casts within the tubular lumen are observed because of epithelial death and detachment, and loss of proteins (Fig. 4.7). Tubular calcification was rarely seen and it was associated with tubular necrosis.

Lymphoid system diseases

Reports concerning lesions in the lymphoid system are scarce in the literature. A newborn calf captured in fishing gears showed granulomatous thymitis, lymphoid reactive hyperplasia with sinus histiocytosis in lymph

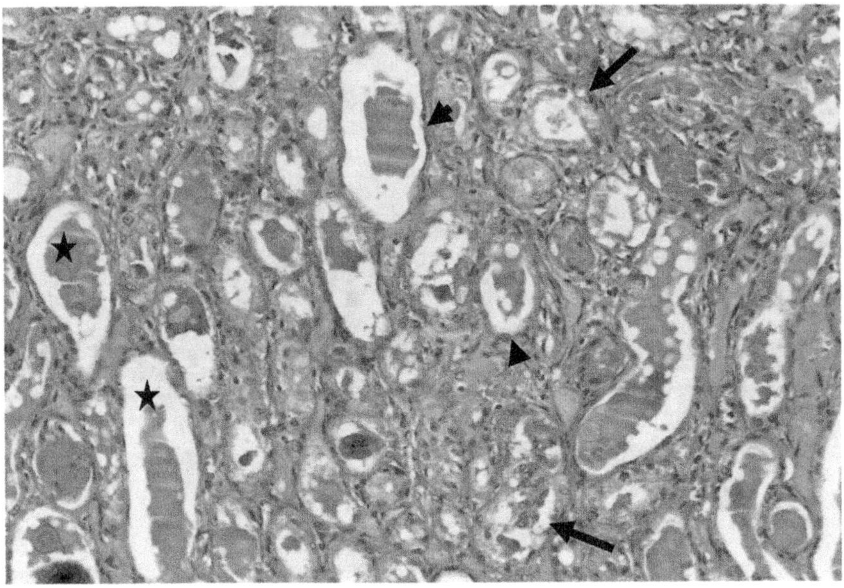

Fig. 4.7 ***Kidney showing acute tubular injury.*** Multifocally tubules show epithelial cells with cytoplasmatic vacuolation (arrows) and pyknotic nuclei (arrowhead). Multiple hyaline casts are observed within a tubular lumen. HE. 20x objective. *Source*: Laboratory of Animal Pathology, Universidade Estadual de Londrina, Paraná.

nodes, and hemodynamic alterations associated with the asphyxiation process. Although asymptomatic—no marked lesions—*Brucella* spp. was identified by molecular diagnosis in lymph nodes (Sarmiento et al., 2019), reinforcing a zoonotic potential infection.

Lymphoid depletion affecting multiple lymph nodes was frequently observed in *P. blainvillei* stranded in southern and southeastern Brazil (Simba, 2021). Focal to multifocal necrosis in pulmonary, prescapular, and mesenteric lymph nodes have been reported in animals presenting concomitant pneumonia (Marutani, 2020). Lymphoid depletion was primarily related to infectious diseases as morbillivirus infection (Diaz-Delgado et al., 2019); however, in these animals, no viral infection was detected so far.

Musculoskeletal diseases

Neonatal and juvenile marine mammals are frequently affected by emaciation. This condition can be related to the mothers (decreased prey resources, agalactia, mastitis, trauma) or the calves (dystocia and neonatal weakness, hypothermia, infection) (Colegrove, 2018). Emaciation was reported as the

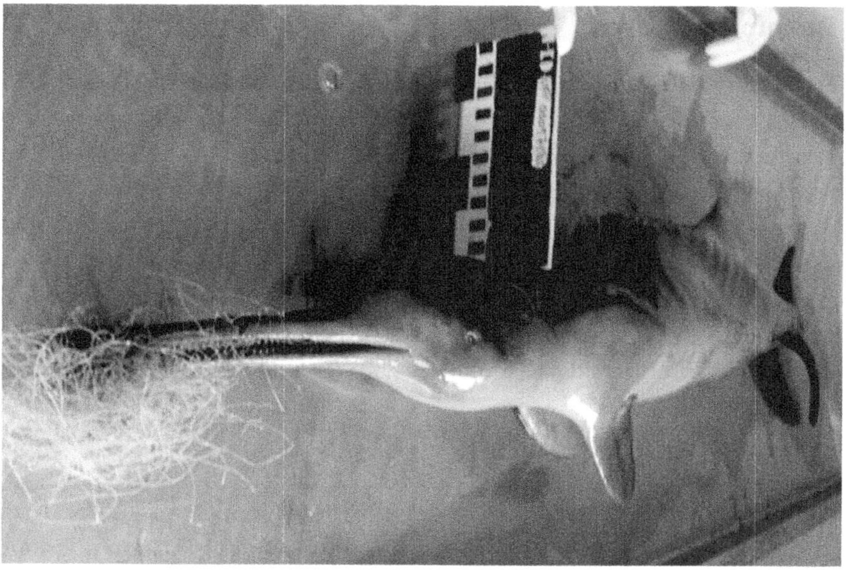

Fig. 4.8 ***Cachectic animal with fishing lines encircling the rostrum.*** Severe loss of skeletal and adipose tissue. Scapular bones and ribs are well-defined under the skin.

cause of death in one *P. blainvillei* dependent calf. It was characterized by generalized muscular and adipose tissue atrophy, resulting in postcranial depression and well-defined rib impressions under the skin, and empty stomachs (Domiciano et al., 2016). Severe emaciation and hepatic lipidosis were observed in a stranded *P. blainvillei* dolphin analyzed in PMP-BS (Simba, 2021). The animal showed fishing lines encircling the rostrum, preventing its opening (Fig. 4.8). The hepatic lipidosis is probably related to lipid metabolic disturbance owing to a lack of feed intake.

The *P. blainvillei* is characterized by a long rostrum, and occasional reports of lesions and fractures are recorded. In Argentina, mild to moderate beak deviation was identified in 12% (29/239) of stranded or bycaught animals and only two cases showed a complex curvature that may have precluded feeding (Denuncio et al., 2016). The cause of those abnormalities (e.g., natural or anthropogenic) was unknown, but some cases are probably associated with lesions caused by fishing gear entrapment (Denuncio et al., 2016). In Brazil, similar findings were observed in <1% of animals (Ott et al., 1996). Studies of pathological conditions on teeth included caries-like lesions (König, 2013) and geminated teeth (König, 2013; Pinedo et al.,1991).

Systemic diseases

Septicemia may have caused the death of a stranded neonate female after intensive care, owing to strains of *Pseudomonas aeruginosa* cultured from its peripheral blood (Danilewicz et al., 2002). Furthermore, *Micrococcus* spp. was isolated in a young male from the thoracic fluid and *Staphylococcus* spp. was isolated from the lung, diagnosed with fibrinous pleuropneumonia, lymphoadenomegaly of the nodes associated with the respiratory tract. This animal had a five cm scar on its abdomen, and the wound was possibly caused by a puncturing object that had ruptured the liver, crossed the diaphragm, and reached the thoracic cavity (Danilewicz et al., 2002).

Further considerations

The knowledge about *P. blainvillei* diseases is incipient. The cause of death of animals described in published data was primarily asphyxiation in fishing gear. Therefore, information about pathological processes, etiological diagnosis, and immunology of these animals, added to stranded alive or dead animals, may improve the health assessment of this endangered species.

References

Baldassin, P., Werneck, M.R., Barbosa, C.B., Gallo, B.M.G., Gallo, H., Walsh, M., 2007. Veterinary treatment of an injured wild Franciscana dolphin calf (Pontoporia blainvillei, Gervais & d'Orbigny, 1844). LAJAM 6 (2), 185–187.

Cianciolo, R.E., Mohr, F.C., 2016. Urinary system. In: Maxie, M.G., Jubb, K.V.F., Kennedy, P.C., Palmer, N.C. (Eds.), Pathology of Domestic Animals, vol. 1 6th ed. Elsevier, St Louis, p. 376–464.

Colegrove, K., 2018. Noninfectious diseases. In: Gulland, F.M.D., Dierauf, L.A., Whitman, K.L. (Eds.), In: CRC Handbook of Marine Mammal Medicine3rd ed. Published by CRC Press (Taylor & Francis), Boca Raton, FL, pp. 267–296.

Danilewicz, D., Rosas, F., Bastida, R., Marigo, J., Muelbert, M., Rodríguez, D., Brito, Jr. J.L., Ruoppolo, V., Ramos, R., Bassoi, M., Ott, P.H., Caon, G., Rocha, A.M., Catão-Dias, J.L., Secchi, E.R., 2002. Report of the working group on biology and ecology. LAJAM 1 (1), 25–42.

Denuncio, P., Bastida, R., Dassis, M., Giardino, G., Gerpe, M., Rodríguez, D., 2011. Plastic ingestion in Franciscana dolphins, Pontoporia blainvillei (Gervais and d'Orbigny, 1844), from Argentina. Mar. Pollut. Bull. 62, 1836–1841.

Denuncio, P., Panebianco, M.V., Del Castillo, D., Rodríguez, D., Cappozzo, H.L., Bastida, R., 2016. Beak deviations in the skull of Franciscana dolphins Pontoporia blainvillei from Argentina. Dis. Aquat. Org. 120 (1), 1–7.

Di Beneditto, A.P.M., Ramos, R.M.A., 2014. Marine debris ingestion by coastal dolphins: what drives differences between sympatric species? Mar. Pollut. Bull. 83 (1), 298–301.

Diaz-Delgado, J., Groch, K.R., Sierra, E., Sacchini, S., Zucca, D., Quesada-Canales, O., Arbelo, M., Fernandez, A., Santos, E., Ikeda, J., Carvalho, R., Azevedo, A.F., Lailson-Brito, J., Flach, L., Ressio, R., Kanamura, C.T., Sansone, M., Favero, C., Porter, B.F.,

Centelleghe, C., Mazzariol, S., Di Renzo, L., Di Francesco, G., Di Guardo, G., Catao-Dias, J.L., 2019. Comparative histopathologic and viral immunohistochemical studies on CeMV infection among Western Mediterranean, Northeast-Central, and Southwestern Atlantic cetaceans. PLoS One 14 (3), e0213363.

Domiciano, I.G., Domit, C., Broadhurst, M.K., Koch, M.S., Bracarense, A.P., 2016. Assessing disease and mortality among small cetaceans stranded at a world heritage site in Southern Brazil. PLoS One 11 (2), e0149295.

Fuentes-Castillo, D., Sellera, F.P., Goldberg, D.W., Fontana, H., Esposito, F., Cardoso, B., Ikeda, J., Kyllar, A., Catão-Dias, J.L., Lincopan, N., 2021. Transbound. Emerg. Dis. http://doi.org/10.1111/tbed.13980.

Goldman, C.G., Matteo, M.J., Loureiro, J.D., Almuzara, M., Barberis, C., Vay, C., Catalano, M., Heredia, S.R., Mantero, P., Boccio, J.R., Zubillaga, M.B., Cremaschi, G.A., Solnick, J.V., Perez-Perez, G.I., Blaser, M.J., 2011. Novel gastric helicobacters and oral campylobacters are present in captive and wild cetaceans. Vet. Microbiol. 152 (1-2), 138–145. http://doi.org/10.1016/j.vetmic.2011.04.023.

Gonzales-Viera, O.A., 2012. Patologia comparada das hepatopatias e nefropatias em cetáceos do Brasil. (In Portuguese). São Paulo, Brazil. Faculdade de Medicina Veterinária e Zootecnia, Universidade de São Paulo, Master in Animal Science. 136 f.

Gonzales-Viera, O., Ruoppolo, V., Marigo, J., Carvalho, V.L., Groch, K.R., Bertozzi, C.P., Takakura, C., Namiyama, G., Vanstreels, R.E., Catão-Dias, J.L., 2015. Renal lesions in cetaceans from Brazil. J. Comp. Pathol. 152 (4), 345–354. http://doi.org/10.1016/j.jcpa.2015.02.004.

Harper, C.M., Dangler, C.A., Xu, S., Feng, Y., Shen, Z., Sheppard, B., Stamper, A., Dewhirst, F.E., Paster, B.J., Fox, J.G., 2000. Isolation and characterization of a Helicobacter sp. from the gastric mucosa of dolphins, Lagenorhynchus acutus and Delphinus delphis. Appl. Environ. Microbiol. 66 (11), 4751–4757. http://doi.org/10.1128/aem.66.11.4751-4757.2000.

König, F.F., 2013. Morfologia dentária de Pontoporia blainvillei (Cetacea, Pontoporiidae) do Rio Grande do Sul: variações, anomalias e patologias. (in Portuguese). Pelotas, Brazil. Universidade Federal de Pelotas, Monografia (Trabalho de conclusão de curso). 52f.

Marutani, V.H.B., 2020. Caracterização histopatológica, imuno-histoquímica e molecular de doenças emergentes em cetáceos no litoral do Paraná (in Portuguese). Londrina, Brazil: Universidade Estadual de Londrina, Master in Animal Science, p. 71.

Pinedo, M.C., 1991. Development and variation of the Franciscana, Pontoporia blainvillei PhD Thesis. University of California, Santa Cruz, CA, p. 406.

Romano, L.A., Klosterhoff, M.C., Medeiros, A.F.F., Pedrosa, V.F., 2020. Pulmonary Aspergillis in the dolphin (Pontoporia blainvillei). Braz J. Vet. Pathol. 13 (2), 524–526. B978-0-323-88668-0.00014-010.24070/bjvp.1983-0246.v13i2p524-526.

Ruopollo, V., 2003. Patologia comparada de cetáceos e pinípedes. (in Portuguese). São Paulo, Brazil. Faculdade de Medicina Veterinária e Zootecnia, Universidade de São Paulo. Master in Experimental and Compared Pathology, p. 131.

Ruoppolo, V., Vanstreels, R.E.T., Marigo, J., Catão-Dias, J.L., 2010. Unusual incidence of chronic pneumonia associated with cholesterol deposits in stranded and bycaught franciscanas Pontoporia blainvillei. Dis. Aquat. Org. 92, 75–81. https://doi.org/10.3354/dao02263.

Sarmiento, A.M.S., Carvalho, V.L., Díaz-Delgado, J., Ressio, R.A., Fernandes, N.C.C.A., Guerra, J.M., Sacristán, C., Groch, K.R., Silvestre-Perez, N., Ferreira-Machado, E., Costa-Silva, S., Navas-Suaréz, P., Meirelles, A.C.O., Favero, C., Marigo, J., Bertozzi, C.P., Colosio, A.C., Marcondes, M.C.C., Cremer, M.J., Silva, N.S., Ferreira Neto, J.S., Keid, L.B., Soares, R., Sierra, E., Fernández, A., Catão-Dias, J.L., 2019. Molecular, serological, pathological, immunohistochemical and microbiological investigation of Brucella spp. in marine mammals of Brazil reveals new cetacean hosts. Transbound. Emerg. Dis. 66, 1674–1692.

Silva, R.Z., Cousin, J.C.B., 2004. Anormalidade gástrica parasitária em Pontoporia blainvillei (Cetacea, Platanistoidea, Pontoporiidae) da região litorânea do Rio Grande do Sul. Brasil. Biociências 12 (2), 99–113.

Silva, R.Z., Cousin, J.C.B., 2006a. Ancoragem de Polymorphus (P.) cetaceum (Acanthocephala, Polimorphidae) nos compartimentos estomacais de Pontoporia blainvillei (Cetacea, Platanistoidea, Pontoporiidae) da região litorânea do Rio Grande do Sul. Brasil. Biociências 14 (2), 156–167.

Silva, R.Z., Cousin, J.C.B., 2006b. Anormalidade intestinal parasitária em Pontoporia blainvillei (Cetacea, Platanistoidea, Pontoporiidae) da região litorânea de Rio Grande. RS, Brasil. Biociências 14 (1), 37–46.

SIMBA, 2021 Sistema de Monitoramento da Biota Aquática (SIMBA). https://simba.petrobras.com.br/simba/web/. (Accessed 10 July 2021).

Sweeney, J.C., Ridgway, S.H., 1975. Common diseases of small cetaceans. J. Am. Vet. Med. Assoc. 167 (7), 533–540.

CHAPTER 5

The life history of franciscana dolphins

Daniel Danilewicz[a], Pablo Denuncio[b], Eduardo R. Secchi[c], Gabriela Tanios[d]

[a]Grupo de Estudos de Mamíferos Aquáticos do Rio Grande do Sul, Torres, Rio Grande do Sul, Brazil
[b]Instituto de Investigaciones Marinas y Costeras, Mar del Plata, Buenos Aires, Argentina
[c]Laboratório de Ecologia e Conservação da Megafauna Marinha, Instituto de Oceanografia, Universidade Federal do Rio Grande, Rio Grande, Rio Grande do Sul, Brazil
[d]Programa de Pós-graduação em Zoologia, Universidade Estadual de Santa Cruz, Ilhéus, Brazil

Introduction

Life history concerns how organisms grow, reproduce and survive (Stearns, 1992). From the classical optimization approach, a species′ life history is shaped by the interaction of extrinsic and intrinsic factors (Roff, 1992; Stearns, 2000). Extrinsic factors can be exemplified by the ecological impacts on growth, reproduction, and survival (for example, how environment shapes reproductive seasonality), while intrinsic factors are the tradeoffs among life-history traits and lineage-specific constraints (why offspring size is fixed at 1 in cetaceans). When describing the life history traits of a species, its constraints, flexibilities, and variation, the outcome of several evolutionary and ecological processes can be uncovered.

Understanding life traits that shape the population dynamics and variation in the demography of threatened species is critical for their conservation. In a broader spectrum, it is the framework to assess how a species can respond to stressors, to predict the consequences of human-related impacts, and to guide decision-making processes (e.g., Frederiksen et al., 2014; Wade, 1998). The long-term monitoring of certain reproductive parameters that are subject to density-dependent effects (e.g., age at sexual maturity (ASM)) may also provide indirect clues about population trajectory relative to resources availability (Perrin and Reilly, 1984). Moreover, it has been demonstrated that life-history traits also have the potential for predicting species vulnerability to climate change in mammals (Jackson et al., 2021).

DOI: https://doi.org/10.1016/B978-0-323-90974-7.00012-4

The franciscana, *Pontoporia blainvillei* (Gervais and d'Orbigny, 1844), is the only extant member of the Family Pontoporiidae and the most threatened small cetacean in the Western South Atlantic (Secchi et al., 2021). It is distributed in coastal waters of Brazil, Uruguay and Argentina (Amaral et al., 2018) and its range is divided into four main Franciscana Management Areas (FMAs) with subdivisions (Cunha et al., 2014; Secchi et al., 2003) (Fig. 5.1). The high and long-standing mortality due to by-catch in gillnet fisheries along its distribution has led the franciscana to be classified as "Vulnerable" in the IUCN Red List (Zerbini et al., 2017). In

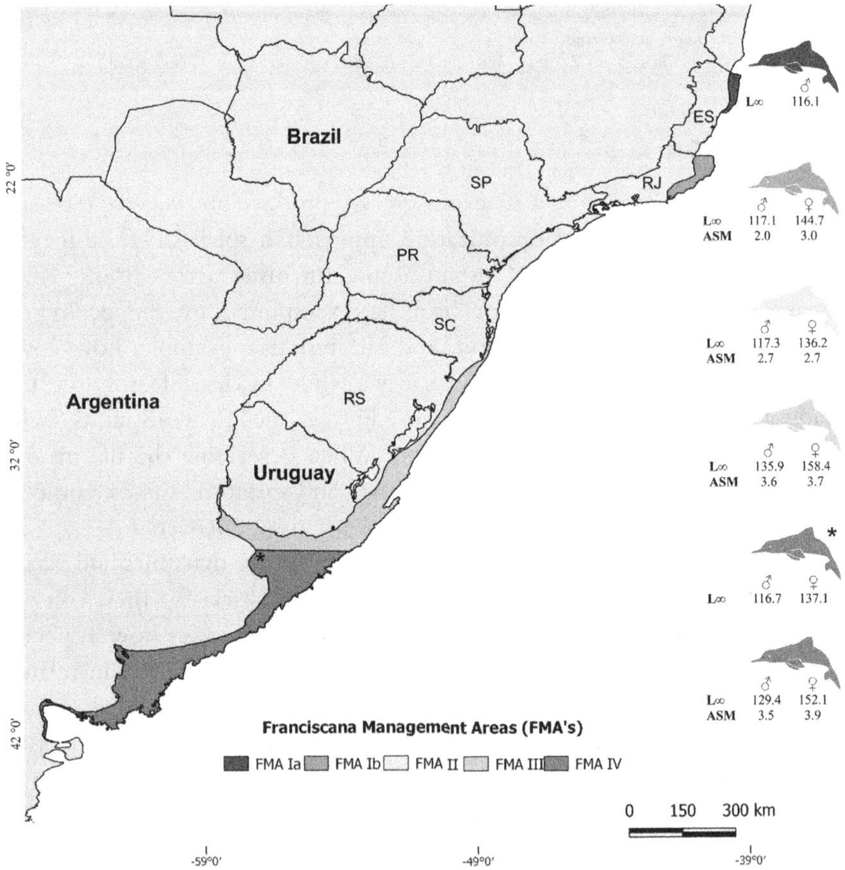

Fig. 5.1 ***Map of the distribution range of franciscana,*** **Pontoporia blainvillei*****, evidencing the four Franciscana Management Areas (FMAs - FMA I subdivided in FMA Ia and Ib).*** The most updated information on asymptotic length (L∞) and average age at sexual maturity (ASM) of both sexes are presented. Parameters for franciscanas from estuarine area in Argentina is representa by *.

Brazil, the species is officially considered "Critically Endangered" (MMA, 2014), while in Uruguay it has Conservation Priority (Gonzalez et al., 2013) and in Argentina it is listed as Vulnerable (Denuncio et al., 2019).

The first franciscana life history studies began in Uruguay in the late 1960s as a result of the collaborative work of Uruguayan, Japanese, and American researchers. The large number of animals by-caught in Uruguay at that time, together with cooperation with fishers, allowed researchers to have access to quality biological material. After some preliminary reports (Brownell, 1975; Harrison and Brownell, 1971), two fundamental articles were published by Kasuya and Brownell (1979) and Harrison et al. (1981). These articles brought unprecedented and robust information based on a large sample size and were responsible for all the baseline knowledge of *Pontoporia*'s life history until the early 2000s, when new studies were published (Danilewicz, 2003; Danilewicz, et al., 2004; Di Beneditto and Ramos, 2001; Rosas and Monteiro-Filho, 2001). More recently, contributions in other geographic areas have emerged (e.g., Panebianco et al., 2012, 2016; Silva et al., 2020).

In this chapter, we provide a comprehensive review of the life history of the franciscana, integrating information from 45 years of studies on reproduction, growth, and survival. Moreover, unpublished data on aerial observations and calf strandings from beach monitoring programs in Brazil were incorporated to contribute to reproductive behavior and birth seasonality, respectively. We also discuss the current evidence about the species´ mating system and make recommendations regarding future research to address areas where understanding is limited or incomplete.

Age, growth, and sexual dimorphism

Age and growth

Age is an important parameter to better understand wildlife populations and several biological parameters at individual and populational levels. Age in franciscanas is determined through the counting of growth layer groups (GLGs) present in the dentine and cementum of the teeth (Kasuya and Brownell, 1979; Pinedo and Hohn, 2000). A model that allows accurate and precise age estimates was developed by Pinedo and Hohn (2000) and proved to be useful for franciscanas in many study areas (e.g., Botta et al., 2010; Conversani et al., 2020). A GLG was defined to comprise one unstainable and one stainable layer. Age can be read both from the cementum and dentine, although dentine GLGs are deposited in much less quantity and are only useful in determining age in the first three years.

The maximum age of female and male franciscana was 21 and 20 years old, respectively (Pinedo and Hohn, 2000; Silva et al., 2020). The age calculated from incidentally caught and washed ashore animals suggests that only a small fraction of the population lives more than 12–14 years, irrespective of their geographic location (Denuncio et al., 2018; Di Beneditto and Ramos, 2001; Kasuya and Brownell, 1979; Pinedo, 1994; Pinedo and Hohn, 2000; Secchi et al., 2003b).

Growth curves are available for franciscanas in all FMAs, fitting the most frequently used growth models in cetaceans: von Bertalanffy and Gompertz (Hohn et al., 1996; Mattson et al., 2006; Zullinger et al., 1984). Table 5.1 summarizes information on length intervals and asymptotic length for both sexes for all FMAs. Franciscanas are larger in their southern range (Rio Grande do Sul State, Brazil; Uruguay and Argentina—FMAs III and IV) followed by animals from the northern range (Rio de Janeiro and the Espírito Santo States, Brazil—FMA I) and are smaller in the middle of its range (São Paulo, Paraná, and the Santa Catarina States, Brazil—FMA II). In Argentina (FMA IV), the northern forms seem to be smaller than the southern (Denuncio et al., 2018). The asymptotic length for the estuarine dolphins from Bahía Samborombón (north FMA IV) was estimated at 116.7 cm for males and 137.1 cm for females, whereas the value was estimated at the nearby marine area for females was 135.8 cm. For specimens from southern FMA IV, asymptotic lengths for males and females were 129.4 cm and 152.1 cm, respectively (Negri, 2010).

Bergmann's Rule (Mayr, 1970) states that animals of a species inhabiting colder regions are larger. In franciscanas there is not a clear clinal (latitudinal) pattern in size (total length) of individuals from the north (warm) to south (cold) areas of distribution, according to the asymptotic length estimations (Ramos et al., 2002; Secchi et al., 2021). It was soon evident that the largest franciscanas were those from the FMA III (Rio Grande do Sul State, Brazil, and Uruguay) and the smallest those from FMA II (São Paulo and Paraná, Brazil), regardless of the sex (Pinedo, 1991; Barbato et al., 2012). For males, however, a clearer clinal pattern is observed with franciscanas particularly large in the FMAs III and south IV, being between 14% and 19% larger than in the northern FMAs (see Table 5.1).

As in other mammals, physical maturity in franciscanas occurs years after the attainment of sexual maturity (Botta, 2005). Estimates of mean length at physical maturity in franciscana populations are very close to the asymptotic length, representing 96%–100% of this last parameter.

Table 5.1 Summary of information on size at birth, asymptotic length, and length range of franciscana, *Pontoporia blainvillei*, separated by franciscana management areas (FMA) and sample area.

FMA	Sample area	size (m/f)	Size at birth (cm) m/f	Asymptotic TL (cm) m	f	Length range (cm) m	f	SSD	References
FMAIa	Espirito Santo	13/1	–	116.1	–	–	–		Ramos et al. (2002)
FMAIb	Rio de Janeiro	39/43	71.1	117.1	144.7	78–120	74–147.5	1.24	Ramos et al. (2000)
FMAIb	Rio de Janeiro	60/57	–	121.9	145.6	66–130	74–147.5	1.19	Di Beneditto and Ramos (2001)
FMAIb	Rio de Janeiro	27/19	–	122	140.9	–	–	1.15	Ramos et al. (2002)
FMAII	São Paulo	13/10	–	120.5	134.0	–	–	1.11	Ramos et al. (2002)
FMAII	São Paulo/Paraná	23/18	–	113.3	129.1	–	–	1.14	Barreto and Rosas (2006)
FMAII	São Paulo/Paraná	73/87	71.2/71.6	117.0 ± 1.45	135.8 ± 2.36	59–125	66.5–146	1.16	Conversani et al. (2020)
FMAIII	Rio Grande do sul	23/18	73.4	–	–	76–147	75–171		Botta et al. (2010)
FMAIII	Rio Grande do sul	59/48	–	129.8–130.6	146.3–152.6	–	–	1.13	Barreto and Rosas (2006)
FMAIII	Rio Grande do sul	–/97	– /73.4	–	158.4	–	75–171		Danilewicz (2003)
FMAIII	Rio Grande do sul	47/–	–	135.9	–	76–147	–		Danilewicz et al. (2004)
FMAIII	Uruguay	97/121	70–85[a]	133.3	153	84–147	85–171	1.15	Kasuya and Brownell (1979)
FMAIV	North Argentina, estuarine	42	67.9/68.8	116.7	137.1	–	–	1.17	Denuncio et al. (2018)
FMAIV	South Argentina,marine	66	– /67	–	135.8	–	–		Denuncio et al. (2018)
FMAIV	South Argentina,marine	65	81.6/79.6	129.4	152.1	–	–	1.18	Negri (2010)
FMAIV	South Argentina,marine	15/20	–	125.9	158.4	85–141	73–158	1.26	Botta et al. (2010)

m, male; f, female, SSD, sexual size dimorphism. [a]Suggested, not estimated. (–) Not estimated.

Size at birth

Size at birth is applied to estimate the time of birth, the gestation period and may be a starting point for developing a growth curve (Perrin and Reilly, 1984). The size at birth of franciscanas was first presented for Uruguay, where Kasuya and Brownell (1979) concluded that the body length at birth is between 70 and 85 cm, probably in the low 70 cm range. The later estimates in Brazil and Argentina fluctuates around 70 cm and agreed to, with some exceptions, the range sizes that these authors had proposed (see Table 5.1 for details).

In Argentina (FMA IV) it is possible to present size at birth in a finer scale within the FMA that reflects differences between the northern and smaller forms against the southern and larger ones. Estuarine franciscanas showed a size at birth of 68.8 and 67.9 cm for males and females respectively (Denuncio et al., 2018). Similar results were obtained in the nearby marine area (67.0 cm for females). Size at birth for the nearby marine FMA IV was also calculated by the Borjesson and Read (2003) overlap method, resulting in 60.7 cm, even smaller than the length at time zero obtained from growth curves. A high frequency of live ashore calves with similar size and with umbilical cord remains reinforces a smaller size at birth in this area.

Sexual size dimorphism

Reversed sexual size dimorphism (RSSD) (*i.e.*, females larger than males) occurs throughout the species' geographic distribution (Pinedo, 1991; Ramos et al., 2002). Considering the asymptotic length of several populations, adult females are on average 17% larger than adult males (estimates range from 13% to 26%) (Table 5.1). RSSD is also present in skull and body morphometry (Higa et al., 2002; Barbato et al., 2012, Wells et al., 2013, Botta et al., 2006), where in all cases females are the larger sex. Larger females might represent an evolutionary advantage for small species by allowing gestation and birth of larger, even slightly, calves (Danilewicz et al., 2004; Ralls 1976, 1977). Larger calves potentially have a higher likelihood of survival due to several factors including a relatively lower metabolic rate and a lower surface/volume ratio, which reduces heat loss (Secchi et al., 2021). Supporting this hypothesis is the evidence that most odontocete with RSSD (*Pontoporia blainvillei*, *Cephalorhynchus hectori*, *C. commerssoni*, *Phocoena phocoena*, *P. sinus*) are the smallest species within the group.

Franciscana reproductive cycle

Ovulation time, characteristics, and conception

The franciscana is an annual breeding and monoestrous species (Harrison et al., 1981; Danilewicz, 2003, Panebianco et al., 2016). Although multiple ovulations have been recorded, particularly in young adult females, most animals only ovulate once in a season (Harrison et al., 1981; Panebianco et al., 2016). The morphometrics, cytology, and histology of ovaries and its corpora lutea (CL) and albicantia (CA) are described in detail by Harrison et al. (1981). CL does not increase in size as gestation progresses and there is a considerable individual variation in its size in different stages of pregnancy (Danilewicz, 2003). There is strong evidence that CAs do not persist throughout life in franciscanas, being fully reabsorbed after 4 years (Harrison et al., 1981; Panebianco et al., 2016). Consequently, CAs have a limited potential for reconstructing individual reproductive lifetime histories in this species. The ovulation rate was estimated at 0.39 for a sample in Argentina.

The tendency of a much higher activity of the left ovary is well documented in franciscanas from Uruguay and Rio Grande do Sul (FMA III) (Harrison et al., 1981; Danilewicz, 2003). This asymmetrical ovarian activity may be a populational character since in Argentina (FMA IV) and Paraná/São Paulo (FMA II) both ovaries ovulate at similar rates (Panebianco et al., 2016, Rosas and Monteiro-Filho, 2001; Silva et al., 2020).

The ovulation period is determined when recent ovulation is detected in the ovary through the presence of a new ruptured follicle or a CL in a female carrying no embryo or fetus. Franciscana ovulations are clearly concentrated in the austral summer, about two-thirds occurring from December to February (Danilewicz, 2003; Harrison et al., 1981; Silva et al., 2020). Consequently, most conceptions also occur in this period. Nevertheless, it is important to note that studies in different sites have evidenced that "out of peak" ovulations and conceptions take place in the winter and autumn (Danilewicz, 2003; Harrison et al., 1981). A proportion of those animals that ovulate outside the main reproductive period may fail to conceive and ovulate a second time near the peak of breeding (Harrison et al., 1981).

Seasonal variation in testicular mass in mature males has not been detected (Danilewicz et al., 2004; Kasuya and Brownell, 1979; Panebianco et al., 2012). That indicates that males do not follow the reproductive seasonality presented by females and probably are able to mate year-round, explaining some few fetuses displaced in the populational fetal growth curves (Danilewicz, 2003; Harrison et al., 1981). Danilewicz et al. (2004) suggest that the small energy investment in producing sperm all over the year, due to the small testicular mass, may be an evolutionary advantage for male franciscanas in case of the appearance of off-season reproductive females.

Gestation length

The gestation period is one of the life history parameters with less variation among mammal populations (Kiltie, 1982). No direct data on franciscana gestation length through the monitoring of pregnant females kept in captivity is available. Therefore, the duration of gestation has been estimated through indirect mathematical methods that utilize the empirical concept that fetal growth is directly related to the gestation period (*e.g.*, Hugget andWiddas, 1951; Calder, 1982).

Estimates of franciscana gestation length range from 10.2 to 11.2 months. Both estimates available for FMA III are very similar, lasting 10.8–11 months in Uruguay and 11.2 months in Rio Grande do Sul (Danilewicz, 2003; Kasuya and Brownell, 1979). Estimates for other areas in Brazil are slightly shorter, lasting 10.6 months in Paraná/southern São Paulo (FMA II) (Rosas and Monteiro-Filho, 2001) and 10.2–10.4 months in northern Rio de Janeiro (FMA Ib) (Ramos, 1997). Whether this difference is the result of a gestation length a few weeks shorter in the northern range, possibly linked to a smaller franciscana body size, or whether it is due to differences in sample sizes/methods remains to be further investigated. In this context, it is known that the linear regression given by Perrin et al. (1977) and applied by Ramos tends to produce shorter gestation length estimates than the Huggett and Widdas (1951) method (used in FMA III), even with the same dataset. Regardless of these considerations, franciscana gestation length is at the lower part of a continuum of values for cetaceans that ranges from ~10.5 months in the smaller odontocetes such as harbor porpoises, *Phocoena phocoena* (Read, 1990), to 18 months in the sperm whale (Harrison, 1969).

Birth period

Franciscana presents a typical birth-pulse pattern in most of its range, with births starting in early spring (October, sometimes late September) and decreasing gradually until the end of austral summer (March) (Danilewicz, 2003; Denuncio et al., 2013). Although some variation exists between areas, the period from October to December invariably comprises the vast majority of births. The analysis of data on stranding frequency of animals smaller than 75 cm carried out for this chapter confirms the same birth pattern across all FMA II. It has been postulated that franciscanas from northern Rio de Janeiro (FMA Ib) are birth-flow breeders, with no spring/summer peak (Di Beneditto and Ramos, 2001; Ramos et al., 2000). The sex ratio at birth is 1:1 (Brownell, 1984).

Lactation period

The duration of lactation may be considered the most variable component of the reproductive cycle in odontocete cetaceans (Perrin and Reilly, 1984). Lactation lasts a minimum of nine months in Uruguay (Harrisson et al., 1981). In Argentina, based on the range of dates for calves with milk in their stomachs, lactation duration was estimated at a minimum of 7 months (or 28 weeks) (Denuncio et al., 2013). Ramos (1997) estimated a lactation period of 8.4 months for northern Rio de Janeiro. The aerial photographic record of several mother-calf pairs over the years in Ubatuba, Brazil, accompanied by a small juvenile, probably a yearling, raises the hypothesis that perhaps in some cases the calf of the previous year may be able to accompany its mother for some time after lactation or the birth of the new calf. The genetic profile of franciscanas by-caught in the same net also indicates that some groups can be composed of half-siblings and their mother (Valsecchi and Zanellato, 2003).

Considering the estimates presented above about the time length for each stage of franciscana reproduction, the following generalization about its reproductive cycle can be made. The complete franciscana reproductive cycle, from ovulation and conception to calf weaning, should last an average of 20 months. In the case of females that give birth every 2 years, it would be expected a resting period of about 4-6 months until the next ovulation, and in the case of females giving birth within two years in a row, the conception of the second offspring would take place between the first and fifth month of lactation of the previous offspring, depending on the first calf's birth date.

Population reproductive parameters

Sexual maturity

Age and size at sexual maturity are reproductive parameters critical to the understanding of the ecology and conservation of a population. Sexual maturity definition in female cetaceans is the presence of at least one ovulation corpora (CA or CL) in the ovary indicating that the animal has ovulated at least once (Perrin and Reilly, 1984). For males, criteria are more subjective and depend on verifying the presence of spermatozoa and spermatids in the gonads. Nevertheless, studies on male franciscanas have pointed out that the testes morphometrics can provide very confident indirect indexes to determine sexual maturity without the need for histological examination (Danilewicz et al., 2004; Panebianco et al., 2012).

Franciscanas have one of the lowest ages at the attainment of sexual maturity among cetaceans. Franciscanas mature when they are between 2 and 5 years of age, and the average ASM ranges from 2.7 to 3.9 years, depending on the population and method applied (Table 5.2). There are some cases of 1-year-old sexually mature franciscanas (*e.g.* Silva et al., 2020). As the species present reproductive seasonality of births, it is likely that these cases (1 year old mature) are animals that matured very close to completing their second year of life but with the second growth layer not yet fully formed or undetected in the teeth.

As a natural consequence of the interpopulation variation in body size and reversed sexual size dimorphism, there are considerable differences in length and weight at sexual maturity along with the franciscana range and between sexes (Table 5.2). Sexual bimaturism, that is, sex differences in the time of onset of sexual maturity, is not present in franciscanas in spite of the sexual dimorphism in body size. On average, within the same population, both sexes mature at the same age. The differential energetic investment in growth between sexes seems not to have a high influence on the onset of sexual maturity. Considering the age of first reproduction (AFR), life span, and calving intervals, a female franciscana might produce from four to eight offspring in her lifetime (Danilewicz, 2003).

Pregnancy rate and calving interval

Annual pregnancy rate (APR) is computed as the proportion of mature females in the sample that is pregnant, including those simultaneously pregnant and lactating, divided by gestation length in years. In practical terms, it informs us how many females are reproducing each year and is a useful biological parameter to understand the potential growth of a population. APR is also considered the reciprocal value for calving interval (CI = 1/APR). In this sense, APR is functional to estimate the period between births: the higher the proportion of pregnant mature females in the population, the shorter the calving interval. As discussed by Perrin and Reilly (1984), APR estimates may be very sensitive to sampling biases. For example, differential vulnerability to by-catch by pregnant females, differential habitat use by

Table 5.2 Information about age at sexual maturity (ASM), length at sexual maturity (LSM), oldest immature and youngest mature, smaller mature and largest immature of *Pontoporia blainvillei* females and males, separated by franciscana management areas (FMA) and sample area.

FMA	AREA	ASM (years)		Sample size	LSM (cm)		Sample size	Oldest immature		Youngest mature		Smallest mature		Largest immature		References
		M	F	M/F	M	F	M/F	M	F	M	F	M	F	M	F	–
FMA Ib	Rio de Janeiro	2	3	39/43	115	130	39/43	–	–	2	3	113	130	–	–	Ramos et al. (2000)
FMA II	São Paulo/ Paraná	4–5	4–5	22/18	112–116(e)	122/ 126(e)	22/18	–	–	–	–	110	125	112	114	Rosas and Monteiro-filho (2002)
FMA II	São Paulo	2.7 ± 0.4(a) 2(b)	2.7 ± 0.1(a) 2.4(b)	90/72 90/72	114(b)	128.5(b)	90/70	3	2	1	2	–	125	–	100	Silva et al. (2020)
FMA III	Rio Grande do sul	–	3.7 ± 0.36(d) 3.5(b)	–/72 –/72	–	138.9 ± 3.14(f)	–/94	–	6	–	3	–	133	–	149.9	Danilewicz (2003)
FMA III	Rio Grande do sul	3.6 ± 0.47(d) 3(b)	–	47/– 47/–	128.2 ± 1.49(f)	–	110/–	5	–	2	–	120.5	–	137.5	–	Danilewicz et al. (2004)
FMA III	Uruguay	2–4	2.7(c)	25/86	131.4(c)	140.3(c)	48/40	2.1	4	2	2	121	137	137	146	Kasuya and Brownell (1979)
FMA IV	Argentina	–	3.92 ± 0.09(d) 3.42(b)	–/31 –/31	–	133.5 ± 11.11(f)	–/31	–	4	–	3	–	–	–	–	Panebianco et al. (2016)
FMA IV	Argentina	3.54 ± 0.17(d) 2.92 ± 0.30(b)	–	28/– 28/–	126.19 ± 0.18(f) 126.27 ± 3.97(b)	–	38/– 40/–	3	–	5	–	119.3	–	135.2	–	Panebianco et al. (2012).

The method applied: (a) Dabin et al. (2008). (b) Logistic regression. (c) Linear regression (50% mature). (d) DeMaster (1978). (e) 50% mature. M, male. F, female. (–) not estimated.

reproductive condition, small sample sizes, or even undetected pregnancies have the potential to introduce severe biases in estimates (Perrin and Reilly, 1984).

Franciscana APR was estimated as 0.66 (95% CI = 0.47 – 0.85) for Rio Grande do Sul, Brazil (FMA III), which translates to a calving interval of 1.5 years (Danilewicz, 2003). Since franciscanas are seasonal breeders, this CI indicates that about half of the mature females are reproducing annually, while the other half are reproducing every 2 years in southern Brazil. Considerably lower APRs were estimated for Argentina (APR: 0.36; 95% CI: 0.1 - 0.65) (Panebianco et al., 2016) and southeastern Brazil (São Paulo) (APR = 0.36 ± 0.11) (Silva et al., 2020). The calving interval resulting from an APR of 0.36 is 2.8 years, which means that more than half of mature females would be giving birth every three years or more. This calving interval does not seem to match the life history of a species with such an early maturation and short lifespan as the franciscana. In fact, Panebianco et al. (2016) credited the lower APR estimated for Argentina as a result of possible sampling biases. Discrepancies in sample size between studies should also be considered. In any case, efforts to improve pregnancy rate estimates all along the franciscana range should be made.

New methodological approaches may be useful to evaluate alternative pregnancy indexes for franciscanas. Images taken from unmanned aerial vehicles can be applied in small populations, or in areas where the species may be more easily observed and monitored, such as Babitonga Bay and Ubatuba. Photographic data from Ubatuba sampled just before the start of the birth period allowed the identification of females in advanced stages of pregnancy accompanied by a likely calf from the previous year (Fig. 5.2), while some pregnant females had only the company of one or two adults. In the first case, it may be assumed that the female experienced a one-year calving interval. In the second case, the female may either have experienced a calving interval longer than 1 year or have lost the offspring in the first months and subsequently became pregnant again. However, a careful sampling design should be implemented in order to avoid biases of the same nature as the traditional APR.

Longevity and senescence

Franciscana has one of the shortest life spans among all cetaceans. It lives up to around 20 years of age, based on the maximum observed age, with no apparent difference between the sexes. The oldest female and male

Fig. 5.2 ***A franciscana group composed of a pregnant female (left) accompanied by a yearling (center, <75% the female´s length, shorter rostrum) and a likely adult of unknown sex (right).*** Group recorded on December 7, 2016 in Ubatuba, Brazil (FMA 2). Credit: Daniel Danilewicz.

franciscana were 21 and 20 years old, respectively (Pinedo and Hohn, 2000; Silva et al., 2020). Based on the age structure of several studies, only a few individuals live longer than 12 years. Nevertheless, the age structure observed in these studies is very likely influenced by the high non-natural mortality, due to bycatch in fisheries, which reduces the chances to observe individuals with higher expectancy.

Reproductive senescence and postreproductive lifespan are two similar concepts and are often used interchangeably. Reproductive senescence, however, corresponds to a decline in demographic rates (e.g., pregnancy rates) with increasing age and is quite widespread in mammals (Lemaître et al., 2020). On the other hand, postreproductive lifespan occurs if females live a substantial proportion of their lives after their last reproductive event. Postreproductive stages are rare in mammals and confined to a limited number of species, such as humans, killer whales, and short-finned pilot whales (Ellis et al., 2018).

Franciscanas do not show any evidence of senescent ovaries (Danilewicz, 2003; Kasuya and Brownell, 1979; Panebianco et al., 2016; Rosas and Monteiro-Filho, 2002; Silva et al., 2020) and most likely females reproduce until late life, not presenting a long postreproductive lifespan. Unfortunately, data on reproductively active older females are scarce. A

15-year-old franciscana was reported to be lactating (Silva et al., 2020), and another 19-year-old with recent ovulation (Danilewicz, pers. obs.).

There is still no information about a decline in franciscana reproductive rates with age, although if they follow the general mammalian pattern it would be expected to increase to a peak, then stay relatively constant during adulthood before declining in late life (Emlen, 1970). In bottlenose dolphins, for example, there is evidence that older females reproduced at lower rates, suggesting an age-related decrease in reproductive fitness (Fruet et al., 2015).

Survival

The survival rate is the probability of survival of an individual from one age to the next. Estimating survival rates that are free of bias or assumptions is difficult for long-lived species (*e.g.*, Eberhardt, 1985). For cetaceans, survival has been estimated from several different types of data and approaches, including (1) age-distribution of catch data during commercial whaling (*e.g.*, Ohsumi and Masaki, 1975); (2) age-distribution of animals killed in drive fishery (Kasuya and Marsh, 1984); (3) individual-based mark-recapture models of photo-identified animals (see several articles in Hammond et al., 1990; Fruet et al., 2015b); (4) using a range of life tables of several species with similar life histories (*e.g.*, Barlow and Boveng, 1991; Caswellet al., 1998; Secchi, 1999); and (5) fitting Siler (1979) competing-risk models to age-at-death data (Stolen and Barlow, 2003). All these approaches rely on assumptions that, in many cases, are difficult to satisfy.

For several reasons, the use of mark-recapture techniques is unsuitable for franciscana. Age distribution data of bycatch has not been used so far for any cetacean species because of the unequal vulnerability of different age classes. Beachcast data are typically biased, as naturally-killed dolphins and bycatch are mixed in unknown proportions (Secchi and Fletcher, 2004). These authors developed a method that combines age distribution data of bycatch and beachcast animals to identify and adjust the bias. Once the bias was adjusted, the age distribution data were fitted by the Siler model and survival rates of franciscana were estimated. Results were compared to those obtained from an alternative approach of using life tables from similar species to estimate survival rates of franciscana (Secchi and Fletcher, 2004; Secchi 2006). This method consists of using survival schedules from other species (herein called model species) to represent franciscana's survival schedules.

All selected model species had the same litter size (=1) and a calving interval varying from 1 to 3 years. The exception to this was the bottlenose dolphin, which was included in order to have one cetacean amongst the model species (Secchi and Fletcher, 2004). To increase comparability between the model species and franciscana, the time scales of the model species' survivorship curves were rescaled according to franciscana's AFR. After rescaling, 1 year in the life of a model species with mean AFR at 4 years, for example, (e.g., zebra, buffalo, fur seal) becomes 1.125 years. Thus survivorship to age 1 for franciscana is equivalent to survivorship to age 0.889 for the model species.

Rescaled survivorship schedules and curves of the five model species are presented in Table 5.3 and Fig. 5.3. A Monte Carlo sample of 50 survivorship curves generated from randomly sampling the survival probabilities corresponding to the rescaled model survivorship curves fills the space where we expect the franciscana survivorship curve to be (Fig. 5.4). The mean stage-specific survival rates estimated from the 5000 bootstrap resamples of franciscana survivorship schedules were 0.673 (SE = 0.052), 0.883 (SE = 0.023), and 0.87 (SE = 0.016) for calves, juveniles and adult franciscanas using four model species (zebra, buffalo, northern fur seal, and bottlenose dolphin) and 0.708 (SE = 0.052), 0.90 (SE = 0.022), 0.86 (SE = 0.016) when Dall's sheep is included. The mean stage-specific survival rates estimated by fitting the Siler model to franciscana age-at-death data of beachcast and from Monte Carlo methods using model species life tables are similar (Table 5.3).

The two methods resulted in identical mean survival rates for juvenile franciscana (i.e., 0.91 and 0.90). The values for adults were also

Table 5.3 Stage-specific survival rates mean (and SE) for franciscana, *Pontoporia blainvillei*, estimated from model life-tables of mammal species with similar life history and from franciscana species-specific age-at-death data.

Stage-class	Calves	Juveniles	Adults
Life-Table (w/o Dall's sheep)	0.67 (0.052)	0.88 (0.023)	0.87 (0.016)
Life-Table (with Dall's sheep)	0.71 (0.052)	0.90 (0.022)	0.86 (0.016)
Age-at-death	0.73[a] (a) 0.74[a] (b) (0.027)	0.91 (0.023)	0.854 (0.015)

[a]After Approximations: a—square of adult S; b—ratio calf/non-calf S.
The later values are after adjustment for calves and adult mean survival rates.

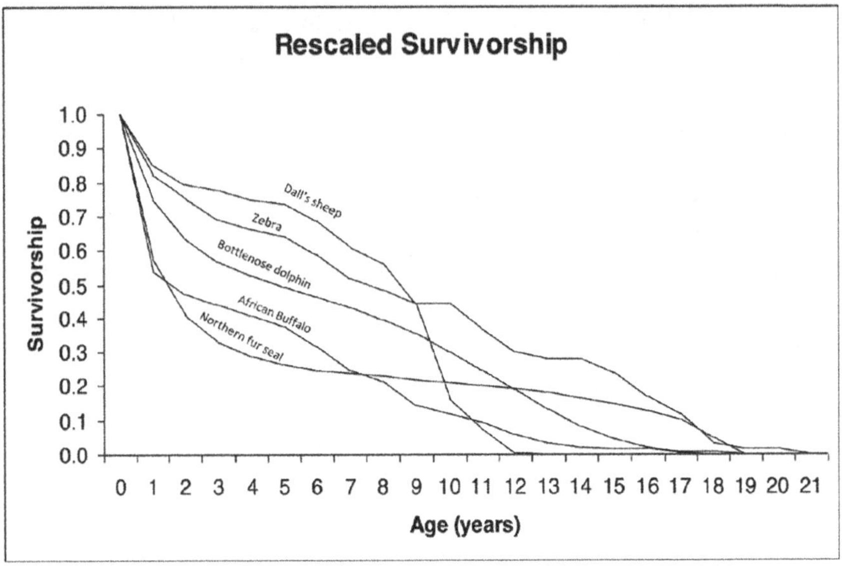

Fig. 5.3 ***Rescaled survivorship curves of the five species used as model life tables to estimate plausible range of survivorship schedules for franciscana.***

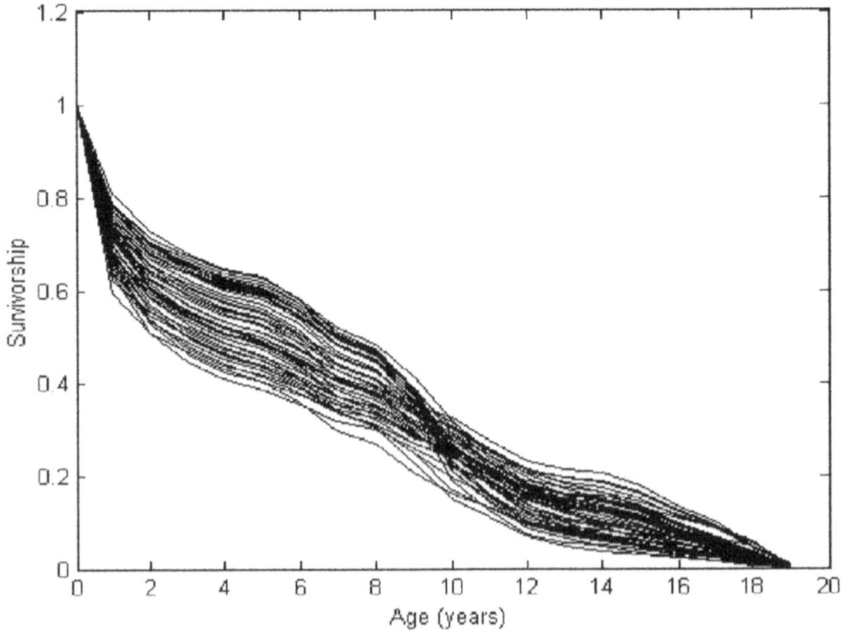

Fig. 5.4 ***A Monte Carlo sample of 50 survivorship curves generated from a random sample of rescaled model survivorship curves.***

identical when the age-at-death data was truncated at age 14 (i.e., 0.854 and 0.86). This suggests that beachcast age-at-death data, after careful treatment to reduce bias effects, may be useful to estimate approximate survival rates.

Estimated survival rates of franciscana are lower than other small cetaceans (e.g., Hector's dolphins – Cameron et al., 2016) and much lower than medium size and large cetaceans (e.g., killer whales – Olesiuk et al., 1990). This seems biologically reasonable as there is strong evidence that body mass is positively correlated with survival in mammals (e.g., Millar and Zammuto, 1983). Hector's dolphins however have the similar body mass to franciscanas. The explanation for the higher survival rate in Hector's dolphin is related, to some extent, to its much lower reproductive potential. A female Hector's dolphin first reproduces at approximately 8 years of age and produces one offspring every two to three years. A female franciscana, on the other hand, has its first calf at an age of approximately 3–4 years and reproduces every 1 or 2 years.

Franciscana mating system: assembling a complex puzzle

Understanding franciscana´s mating strategy and social structure is a complex puzzle that has been put together by many hands over the years. It is currently possible to present a hypothetical scenario with some confidence, although many fundamental questions remain open. The scientific tools employed to understand this important aspect of the species biology and evolution include gonad and body morphometrics, genetics, satellite telemetry, and behavioral observations by helicopter and unmanned aerial vehicles (Costa-Urrutia et al., 2012; Danilewicz et al., 2004; Rosas and Monteiro-Filho, 2002; Valsecchi and Zanelatto, 2003; Wells et al., 2013).

The assessment process was initiated as elimination of alternatives in the broad spectrum of mating systems of mammals (Danilewicz et al., 2004). The first studies to address franciscana mating system were published in the early 2000s when observations of franciscanas in the wild were still very rare and restricted to two study sites (San Blas and Babitonga bays). At that time, the idea of the existence of some sort of monogamy in the franciscana's mating system was hypothesized based exclusively on the theoretical field evaluating anatomic evidence, with no direct evidence of free-ranging dolphins to corroborate this assertion (Danilewicz, 2000; Danilewicz et al., 2004; Rosas and Monteiro-Filho, 2001). Male gonad anatomy was the starting point. The most remarkable characteristic of franciscana testis is that they are much smaller than they should be for a mammal its size. By plotting franciscana testis weight in the general regression that describes the body/testis weight relationship in mammals (Kenagy and Trombulak, 1986), it was evidenced that they have testis on average six times lighter than predicted (Danilewicz et al., 2004). That feature alone precluded the existence of sperm competition, a mating behavior that occurs in many delphinids (*e.g.*, Murphy et al., 2005; Rosas and Monteiro-Filho, 2002). Testes are relatively smaller in species presenting monogamy or extreme polygyny, where a single male copulates with many/all females of a group. Nevertheless, extreme polygyny requires a body size dimorphism towards males, a degree of intrasexual competition and aggressive behavior, and many times the presence of secondary sexual characteristics. None of these characteristics are present in the franciscana, pointing to monogamy as the most likely hypothesis, despite its rarity in mammals.

In parallel, molecular ecology was pointing in a similar direction and raising interesting questions. The contribution of genetics to this issue comes largely from the analysis of the relationship between individuals that were incidentally captured together in the same gillnet, assuming that such a sample represents a group in nature. Valsecchi and Zanelatto (2003) reported on a group of four individuals sampled in the same bycatch event and revealed that it was composed of a mother, two offspring, and a putative father of the youngest offspring. Costa-Urrutia et al. (2012) also report the presence of half-siblings in the same groups, evidencing a lack of dispersal from the natal group, neither by female nor by male offspring. It was then suggested that the franciscana live in groups of related individuals on the basis of a matrilineal structure and, at least in some cases, the father of one offspring may still be accompanying the mother-calf group. However, the lack of relatedness between one adult male and a calf of the same group reported by Mendez et al. (2010) suggests that a permanent family unit may not be the rule and added complexity to this theme. The fact that half-siblings have been recorded in the same group also suggests that, in spite of the mate guarding, the franciscana reproductive pair formation may not last longer and male/father exchange occurs.

Mate guarding is a common behavioral tactic to overcome sperm competition (Møller and Birkhead, 1989) and is predicted to be one of the driving forces for the evolution of monogamy (Brotherton and Komers, 2003). Franciscana mate-guarding behavior may occur both with males accompanying females with and without a calf. The application of satellite telemetry to study franciscana association brought more evidence for the mate-guarding/monogamy hypothesis. Through the deployment of satellite-linked tags on franciscanas in Bahia San Blas, Argentina, Wells et al. (2013) demonstrated that genetically unrelated female–male pairing occurs for extended periods, suggesting the formation of reproductive pairs. The authors reported close proximities locations between one adult male and female during March–May, a period when most adult females in the population have already ovulated and conceived, suggesting the possibility of post-copulatory guarding.

The great majority of mother-calf groups monitored by helicopter or UAVs present escorts, usually one or two adult franciscanas of unknown sex (Fig. 5.5). On some occasions, it has been observed one calf moving from a close association to an adult towards another, suggesting the existence extramaternal care in the species (see Cremer et al., this volume). Understanding the sex and kinship relationship of these escorts with the mother and the calf would be enlightening. Whether it is a mate-guarding male, an older female that remained in the group, or another possibility, this information is key to further understand franciscana social structure.

In summary, multimethodological approach suggests that adult franciscanas form monogamous reproductive pairs of unknown duration, where a

Fig. 5.5 ***A group of franciscanas,*** **Pontoporia blainvillei,** ***photographed in Ubatuba, Brazil, during the birth season.*** A mother in the front with its < 3 months calf at left, a yearling/juvenile, and an adult escort of unknown sex in the rear.

male performs mate guarding. Mate guarding can extend after conception and, in some cases, the father can remain with the mother-calf pair after the offspring birth, although it is not clear if it is the species´ general pattern. A matrilineal society where offspring remain for some time in the same group as their mother is proposed.

Perspectives

The understanding of the franciscana´s general biology and life history is one of the best among small cetaceans. However, some aspects clearly still deserve attention and research effort. Franciscanas from the small and isolated population in Espírito Santo (FMA Ia) are definitely a conservation priority and their biological parameters are the least known along with all the species range. Efforts to estimate basic age-length reproductive and growth parameters from the existing carcasses are urgently needed.

There are still considerable uncertainties about pregnancy rates in franciscanas across their range. The three existing estimates are contradictory and it would be essential to refine them, as well as calculate this important index for areas where it is not available.

Reproductive and growth parameters of franciscanas from Uruguay are well known from specimens collected in the 1970s. However, this time span of about half a century without information by itself demands an update, mainly related to parameters that may indirectly reflect the population status. In this context, ASM estimates should be encouraged to be updated throughout FMA III, where there is a historical series of biological data that allow temporal comparisons and an assessment of the effects of by-catch mortality on populational parameters.

Although there is substantial information on the biological parameters in Argentina, in a small-scale lens, there is a temporal and geographical discrepancy in the information. For instance, in the area where about 70% of by-catch occurs and where it is proposed to be a distinct population (FMA IV b, sensu Cunha et al., 2014), available information on the reproductive biology of franciscana is absent. In the same area, a deeper analysis of the growth parameters is also needed, especially for males.

References

Amaral, K.B., Danilewicz, D., Zerbini, A., Di Beneditto, A.P., Andriolo, A., Alvares, D.J., Secchi, E., Ferreira, E., Sucunza, F., Borges-Martins, M., de Oliveira Santos, M.C., Cremer, M., Denuncio, P., Ott, P.H., Moreno, I.B., 2018. Reassessment of the franciscana

Pontoporia blainvillei (Gervais & d'Orbigny, 1844) distribution and niche characteristics in Brazil. J. Exp. Mar. Bio. Ecol. 508, 1–12. https://doi.org/10.1016/j.jembe.2018.07.010.

Barbato, B.H.A., Secchi, E.R., Beneditto, Di, A.P.M., R., R.M.A., B., C., M., J., B., P., K, P.G., 2012. Geographical variation in franciscana (Pontoporia blainvillei) external morphology. J. Mar. Biol. Assoc. United Kingdom 92, 1645–1656. https://doi.org/10.1017/S0025315411000725.

Barlow, J., Boveng, P., 1991. Modeling age-specific mortality for marine mammal populations. Mar. Mammal Sci. 7, 50–65. https://doi.org/10.1111/j.1748-7692.1991.tb00550.x.

Barreto, A.S., Rosas, F.C.W., 2006. Comparative growth analysis of two populations of Pontoporia blainvillei on the Brazilian coast. Mar. Mammal Sci. 22, 644–653. https://doi.org/10.1111/j.1748-7692.2006.00040.x.

Börjesson, P., Read, A.J., 2003. Variation in timing of conception between populations of the harbor porpoise. J. Mammal. 84, 948–955. https://doi.org/10.1644/BEM-016.

Botta, S., 2005. Reprodução e crescimento dos machos de toninha (Pontoporia blainvillei) do Rio Grande do Sul, Brasil. Universidade Federal do Rio Grande, Rio Grande.

Botta, S., Muelbert, M.M.C., Secchi, E.R., 2006. Morphometric relationships of franciscana dolphin, Pontoporia blainvillei (Cetacea), off Rio Grande do Sul coast, southern Brazil. Lat. Am. J. Aquat. Mamm. 5, 117–123. https://doi.org/10.5597/lajam00102.

Botta, S., Secchi, E.R., Muelbert, M.M.C., Danilewicz, D., Negri, M.F., Cappozzo, H.L., Hohn, A.A., 2010. Age and growth of franciscana dolphins, Pontoporia blainvillei (Cetacea: Pontoporiidae) incidentally caught off southern Brazil and northern Argentina. J. Mar. Biol. Assoc. United Kingdom 90, 1493–1500. https://doi.org/10.1017/S0025315410001141.

Brotherton, P.N.M., Komers, P.E., 2003. Mate guarding and the evolution of social monogamy in mammals. In: Reichard, U.H., Boesch, C. (Eds.), Monogamy: Mating Strategies and Partnerships in Birds, Humans and Other Mammals. Cambridge University Press, Cambridge, pp. 42–58.

Brownell Jr., R.L., 1975. Progress report on the biology of the franciscana dolphin, Pontoporia blainvillei, in Uruguayan Waters. J. Fish. Res. Board Can. 32, 1073–1078.

Brownell Jr., R.L., 1984. Review of reproduction in platanistid dolphins. Rep. Int. Whal. Commn. 6, 149–158.

Calder, W.A., 1982. The pace of growth: an allometric approach to comparative embryonic and post-embryonic growth. J. Zool. 198, 215–225.

Cameron, C., Barker, R., Fletcher, D., Slooten, E., Dawson, S., 2016. Modelling Survival of Hector's Dolphins around Banks Peninsula, New Zealand. J. Agric. Biol. Environ. Stat. 4, 126–135.

Caswell, H., Brault, S., Read, A.J., Smith, T.D., 1998. Harbor porpoise and fisheries: an uncertainty analysis of incidental mortality. Ecol. Appl. 8, 1226–1238. https://doi.org/10.1890/1051-0761(1998)008[1226:HPAFAU]2.0.CO;2.

Conversani, V.R.M., Barbosa, R.A., Santos, M.C.D.O., Silva, D.F., Hohn, A.A., 2020. Age and growth of franciscana, Pontoporia blainvillei, and Guiana, Sotalia guianensis, dolphins from southeastern Brazil. Mar. Mamm. Sci., 1–15. https://doi.org/10.1111/mms.12763.

Costa-Urrutia, P., Abud, C., Secchi, E.R., Lessa, E.P., 2012. Population genetic structure and social kin associations of franciscana dolphin, Pontoporia blainvillei. J. Hered. 103, 92–102. https://doi.org/10.1093/jhered/esr103.

Cunha, H.A., Medeiros, B.V., Barbosa, L.A., Cremer, M.J., Marigo, J., Lailson-Brito, J., Azevedo, A.F., Solé-Cava, A.M., 2014. Population structure of the endangered franciscana dolphin (Pontoporia blainvillei): reassessing management units. PLoS One 9, 1–8. https://doi.org/10.1371/journal.pone.0085633.

Danilewicz, D., 2000. Biologia reprodutiva e padrões de uso de habitat da toninha, Pontoporia blainvillei (Mammalia, Cetacea), no litoral do Rio Grande do Sul. Pontificia Universidade Católica do Rio Grande do Sul, p. 90.

Danilewicz, D., Secchi, E.R., Ott, P.H., Moreno, I.M., 2000. Analysis of the age at sexual maturity and reproductive rates of franciscana (Pontoporia blainvillei) from Rio Grande do Sul, southern Brazil. Com. Mus. Ciên. Tecnol. PUCRS. 13, 89–98.

Danilewicz, D., 2003. Reproduction of female franciscana (Pontoporia blainvillei) in Rio Grande do Sul, southern Brazil. Lat. Am. J. Aquat. Mamm. 2, 67–78. https://doi.org/10.5597/lajam00034.

Danilewicz, D., Claver, J.A., Péres Carrera, A.L., Secchi, E.R., Fontoura, N.F., 2004. Reproductive biology of male franciscanas (Pontoporia blainvillei) (Mammalia: Cetacea) from Rio Grande do Sul, southern Brazil. Fish. Bull. 102, 581–592.

Denuncio, P., Negri, M.F., Bastida, R., Rodríguez, D., 2018. Age and growth of franciscana dolphins from northern Argentina. J. Mar. Biol. Assoc. United Kingdom 98, 1197–1203. https://doi.org/10.1017/S0025315417000765.

Denuncio, P., Viola., P., M.N., C.-.S., I., C., H.L., R., D., M,M.A, 2019. Pontoporia blainvillei. En: SAyDS–SAREM (eds.) Categorización 2019 de los mamíferos de Argentina según su riesgo de extinción. Lista Roja de los mamíferos de Argentina. https://doi.org/10.31687/SaremLR.19.181.

Denuncio, P.E., Bastida, R.O., Danilewicz, D., Morón, S., Rodríguez-Heredia, S., Rodríguez, D.H., 2013. Calf chronology of the franciscana dolphin (Pontoporia blainvillei): Birth, onset of feeding, and duration of lactation in coastal waters of Argentina. Aquat. Mamm. 39, 73–80. https://doi.org/10.1578/AM.39.1.2013.73.

Di Beneditto, A.P.M., Arruda Ramos, R.M., 2001. Biology and conservation of the franciscana (Pontoporia blainvillei). J. Cetacean Res. Manage. 3, 185–192.

Eberhardt, L.L., 1985. Assessing the dynamics of wild populations. J. Wildl. Manage. 49, 997–1012. https://doi.org/10.2307/3801386.

Ellis, S., Franks, D.W., Nattrass, S., Cant, M.A., Bradley, D.L., Giles, D., Balcomb, K.C., Croft, D.P., 2018. Postreproductive lifespans are rare in mammals. Ecol. Evol. 8, 2482–2494. https://doi.org/10.1002/ece3.3856.

Emlen, J.M., 1970. Age specificity and ecological theory. Ecology 51, 588–601.

Frederiksen, M., Lebreton, J.D., Pradel, R., Choquet, R., Gimenez, O., 2014. Identifying links between vital rates and environment: a toolbox for the applied ecologist. J. Appl. Ecol. 51, 71–81. https://doi.org/10.1111/1365-2664.12172.

Fruet, P.F., Daura-Jorge, F.G., Möller, L.M., Genoves, R.C., Secchi, E.R., 2015. Abundance and demography of bottlenose dolphins inhabiting a subtropical estuary in the Southwestern Atlantic Ocean. J. Mammal. 96, 332–343. https://doi.org/10.1093/jmammal/gyv035.

Gervais, P. and d'Orbigny A., 1844. Mammalogie. Bulletin de la Société Philomathique, Paris, pp. 38–40.

González, E.M., Martínez-Lanfranco, J.A., Juri, E., Rodales, A.L., Botto, G., Soutullo, A., 2013. Pontoporia blainvillei. Base de datos de espécies. https://www.dinama.gub.uy/especies.

Hammond, P.S., Mizroch, S.A., Donovan, G.P., 1990. Individual recognition of cetaceans: use of photo-identification and other techniques to estimate population parameters. International Whaling Commission, Cambridge, (Special issue 12). p. 440.

Harrison, R.J., 1969. Reproduction and reproductive organs. In: H.T. Andersen (ed.). The Biology of Marine Mammals, 51. Academic Press, N.Y. and London. pp. 253–342.

Harrison, R.J., Brownell, R.L.Jr, 1971. The gonads of the South American dolphins, Inia geoffrensis, Pontoporia blainvillei, and Sotalia fluviatilis. J. Mammal. 52, 413–419. https://doi.org/10.2307/1378683.

Harrison, R.J., Bryden, M.M., McBrearty, D.A., Brownell, R.L.,Jr., 1981. The ovaries and reproduction in Pontoporia blainvillei (Cetacea:Platanistidae). J. Zoology 193, 563–580.

Higa, A., Hingst-Zaher, E., Vivo, M., 2002. Size and shape variability in the skull of Pontoporia blainvillei (Cetacea: Pontoporiidae) from the Brazilian coast. Lat. Am. J. Aquat. Mamm. 1. https://doi.org/10.5597/lajam00018.

Hohn, A.A., Read, A.J., Fernandez, S., Vidal, O., Findley, L.T., 1996. Life history of the vaquita, Phocoena sinus (Phocoenidae, Cetacea). J. Zool. 239, 235–251.
Hugget, A., St., G., Widdas, W.F., 1951. The relationship between mammalian fetal weight and conception age. J. Physiol. 114, 3063 17.
Jackson, J., Le Coeur, C., Jones, O.R., 2021. Life-history predicts global population responses to the weather in the terrestrial mammals. BioRxiv 1–22. https://doi.org/10.1101/2021.04.22.440896.
Kasuya, T., Brownell, R.L.J., 1979. Age determination, reproduction, and growth of the Franciscana Dolphin Pontoporia blainvillei. Sci. Reports Whales Res. Inst.Tokyo 31, 45–67.
Kasuya, T., Marsh, H., 1984. Life history and reproductive biology of the short-finned pilot whale, Globicephala macrorhynchus, off the Pacific Coast of Japan. Reprod. whales, dolphins porpoises. Proc. Conf. La Jolla, CA 1981, 259–310.
Kenagy, G.J., Trombulak, S.C., 1986. American Society of Mammalogists size and function of mammalian testes in relation to body size. J. Mammal. 67, 1–22.
Kiltie, R.A., 1982. Intraspecific variation in mammalian gestation period. J. Mammal. 63, 646–652.
Lemaître, J.F., Ronget, V., Gaillard, J.M., 2020. Female reproductive senescence across mammals: a high diversity of patterns modulated by life history and mating traits. Mech. Ageing Dev. 192. https://doi.org/10.1016/j.mad.2020.111377.
Mattson, M.C., Mullin, K.D., Ingram, G.W., Hoggard, W., 2006. Age structure and growth of the bottlenose dolphin (Tursiops truncatus) from strandings in the Mississippi sound region of the north-central Gulf of Mexico from 1986 to 2003. Mar. Mammal Sci. 22, 654–666. https://doi.org/10.1111/j.1748-7692.2006.00057.x.
Mayr, E., 1970. Populations, species and evolution. The Belknap Press, Cambridge.
Mendez, M., Rosenbaum, H.C., Wells, R.S., Stamper, M.A, Bordino, P., 2010. Genetic evidence highlights serious demographic impacts of by-catch in cetaceans. PLoS One 5 (12), e15550. http://doi.org/10.1371/journal.pone.0015550.
Millar, J.S., Zammjto, R.M., 1983. Life Histories of Mammals: An Analysis of Life Tables. Ecology, 64, 631–635. https://doi.org/10.2307/1937181.
MMA, 2014. Lista Nacional Oficial de Espécies da Fauna Ameaçadas de Extinção – Mamíferos, Aves, Répteis, Anfíbios e Invertebrados Terrestres. Portaria MMA no. 444, de 17 de dezembro de 2014. Brazil.
Møller, A.P., Birkhead, T.R., 1989. Copulation behavior in mammals: evidence that sperm competition is widespread. Biol. J. Linn. Soc. 38, 119–131.
Murphy, S., Collet, A., Rogan, E., 2005. Mating strategy in the male common dolphin (Delphinus delphis): What gonadal analysis tells us. J. Mammal. 86, 1247–1258. https://doi.org/10.1644/1545-1542.
Negri M.F., 2010. Estudio de la biologıa y ecologıa del delfiin franciscana, Pontoporia blainvillei, y su interacción con la pesquería costera en la provincia de Buenos Aires. PhD thesis. University of Buenos Aires, Buenos Aires, Argentina.
Ohsumi, S., Masaki, Y., 1975. Biological parameters of the antarctic minke whale at the virginal population level. J. Fish. Res. Board Canada 32, 995–1004. https://doi.org/10.1139/f75-119.
Olesiuk, P.F., Bigg, M.A., Ellis, G.M., 1990. Life history and population dynamics of resident killer whales (Orcinus orca) in the coastal waters of British Columbia and Washington State. Rep. - Int. Whal. Comm. Spec. Issue 12, 209–243.
Panebianco, M.V., Del Castillo, D.L., Denuncio, P.E., Negri, M.F., Bastida, R., Failla, M., Vitullo, A.D., Cappozzo, H.L., 2016. Reproductive biology of female franciscana dolphins (Pontoporia blainvillei) from Argentina. J. Mar. Biol. Assoc. United Kingdom 96, 831–840. https://doi.org/10.1017/S002531541500137X.
Panebianco, M.V., Negri, M.F., Cappozzo, H.L., 2012. Reproductive aspects of male franciscana dolphins (Pontoporia blainvillei) off Argentina. Anim. Reprod. Sci. 131, 41–48. https://doi.org/10.1016/j.anireprosci.2012.02.005.

Perrin, W.F., Holts, D.B., Miller, R.B., 1977. Growth and reproduction of the eastern spinner dolphin, a geographical form of Stenella longirostris in the eastern tropical Pacific. Fish. Bull. 75, 725–750.

Perrin, W.F., Reilly, S.B., 1984. Reproductive parameters of dolphins and small whales of the family Delphinidae. Reprod. whales, dolphins porpoises. Proc. Conf. La Jolla, CA 1981, 97–133.

Pinedo, M.C., 1991. Development and variation of the franciscana, Pontoporia blainvillei PhD Thesis. University of California, Estados Unidos, Santa Cruz, p. 406.

Pinedo, M.C., 1994. Impact of incidental fishery mortality on the age structure of Pontoporia blainvillei in southern Brazil and Uruguay. Report of the International Whaling Commission, pp. 261–264.

Pinedo, M.C., 1991. Development and variation of the franciscana, Pontoporia blainvillei Doctoral Thesis. University of California, Santa Cruz, p. 406.

Pinedo, M.C., Hohn, A.A., 2000. Growth layer patterns in teeth from the franciscana, Pontoporia blainvillei: developing a model for precision in age estimation. Mar. Mammal Sci. 16, 1–27. https://doi.org/10.1111/j.1748-7692.2000.tb00901.x.

Ralls, K., 1976. Mammals in which females are larger than males. Q. Rev. Biol. 51, 245–276. https://doi.org/10.1086/409310.

Ralls, K., 1977. Sexual dimorphism in mammals: avian models and unanswered questions. Am. Nat. 111, 917–938. https://doi.org/10.1086/283223.

Ramos, R., 1997. Determinação de idade e biologia reprodutiva de Pontoporia blainvillei e de Sotalia fluviatilis (Cetacea: Pontoporiidae e Delphinidae) no Norte do Rio de Janeiro MSc Thesis. Universidade Estadual do Norte Fluminense, Campos, p. 95.

Ramos, R.M.A., Di Beneditto, A.P.M., Wille Lima, N.R., 2000. Growth parameters of Pontoporia blainvillei and Sotalia fluviatilis (Cetacea) in northern Rio de Janeiro, Brazil. Aquat. Mamm. 26, 65–75.

Ramos, R.M.A., Di Beneditto, A.P.M., Siciliano, S., Santos, M., Zerbini, A., Bertozzi, C., Vicente, A., Zampirolli, E., Alvarenga, F.S., Lima, N.R.W., 2002. Morphology of the franciscana (Pontoporia blainvillei) off southeastern Brazil: sexual dimorphism, growth and geographic variation. Lat. Am. J. Aquat. Mamm. 1, 129–144. https://doi.org/10.5597/lajam00017.

Read, A.J., 1990. Age at sexual maturity and pregnancy rates of harbour porpoises Phocoena phocoena from the Bay of Fundy. Can. J. Fish. Aquat. Sci., 47, 561–565.

Roff, D.A., 1992. The Evolution of Life Histories. Chapman and Hall, New York, NY.

Rosas, F.C.W., Monteiro-Filho, E.L.A., 2001. Reproductive parameters of Pontoporia blainvillei (Cetacea, Pontoporiidae), on the coast of São Paulo and Paraná States. Brazil. Mammalia 66, 231–245. https://doi.org/10.1515/mamm.2002.66.2.231.

Rosas, F.C.W., Monteiro-Filho, E.L.A., 2002. Reproduction of the estuarine dolphin (Sotalia guianensis) on the coast of Paraná, southern Brazil. J. Mammal. 83, 507–515. https://doi.org/10.1644/1545-1542(2002)083<0507:ROTEDS>2.0.CO;2.

Secchi, E.R., 2006. Modeling the population dynamics and viability analysis of franciscana (Pontoporia blainvillei) and Hector's dolphins (Cephalorhynchus hectori) under the effects of bycatch in fisheries, parameter uncertainty and stochasticity Ph.D. Thesis. University of Otago, Dunedin.

Secchi, E.R., Cremer, M.J., Danilewicz, D., Lailson-Brito, J., 2021. A Synthesis of the Ecology, Human-Related Threats and Conservation Perspectives for the Endangered Franciscana Dolphin. Front. Mar. Sci., 8, 1.

Secchi, E.R., Danilewicz, D., Ott, P.H., 2003. Applying the phylogeographic concept to identify franciscana dolphin stocks: implications to meet management objectives. J. Cetacean Res. Manag. 5, 61–68.

Secchi, E.R., Fletcher, D., 2004. Estimating survival rates of franciscana by fitting the Siler model to data on age-at-death of beachcast and bycatch and by a modeling approach using life tables of similar species: a comparison. International Whaling Commission

Scientific Committee Paper, Technical Paper SC/56/SM20 presented to the Scientific Committee of the International Whaling Commission. 29 June – 08 July, Sorrento, Italy.

Secchi, E.R., 1999. Taxa de crescimento potencial intrínseco de um estoque de franciscanas, Pontoporia blainvillei (Gervais & D'Orbigny, 1844) (Cetacea, Pontoporiidae) sob o impacto da pesca costeira de emalhe Master Thesis. Fundação Universidade Federal do Rio Grande, Rio Grande.

Siler, W., 1979. A competing-risk model for animal mortality. Ecology 60, 750–757. https://doi.org/10.2307/1936612.

Silva, D.F., Barbosa, R.A., Conversani, V.R.M., Botta, S., Hohn, A.A., Marcos, M.C., 2020. Reproductive parameters of franciscana dolphins (Pontoporia blainvillei) of Southeastern Brazil. Mar. Mammal Sci. 36, 1291–1308. https://doi.org/10.1111/mms.12720.

Stearns, S.C., 1992. The Evolution of Life Histories. Oxford University Press, Oxford.

Stearns, S.C., 2000. Life history evolution: Successes, limitations, and prospects. Naturwissenschaften 87, 476–486. https://doi.org/10.1007/s001140050763.

Stolen, M.K., Barlow, J., 2003. A model life table for bottlenose dolphins (Tursiops truncatus) from the Indian River Lagoon system. Florida, U.S.A. Mar. Mammal Sci 19, 630–649. https://doi.org/10.1111/j.1748-7692.2003.tb01121.x.

Valsecchi, E., Zanelatto, R.C., 2003. Molecular analysis of the social and population structure of the franciscana (Pontoporia blainvillei): conservation implications. J. Cetaceans Res. Manag. 5, 69–75.

Wade, P.R., 1998. Calculating limits to the allowable human-caused mortality of cetaceans and pinnipeds. Mar. Mamm. Sci. 14, 1–37.

Wells, R.S., Bordino, P., Douglas, D.C., 2013. Patterns of social association in the franciscana, Pontoporia blainvillei. Mar. Mammal Sci. 29, 1–9. https://doi.org/10.1111/mms.12010.

Zerbini, A.N., Secchi, E., Crespo, E., Danilewicz, D., and Reeves, R. 2017. Pontoporia blainvillei (errata version published in 2018). The IUCN Red List of Threatened Species 2017: e.T17978A123792204: https://dx. doi.org/10.2305/IUCN.UK.20173.RLTS. T17978A50371075.en. (Accessed date: 15.apr.2022).

Zullinger, E.M., Ricklefs, R.E., Redford, K.H., Mace, G.M., 1984. Fitting sigmoidal equations to mammalian growth curves. J. Mammal. 65, 607–636. https://doi.org/10.2307/1380844.

CHAPTER 6

Genetic diversity, population structure, and phylogeography

Haydée Andrade Cunha
Laboratório de Mamíferos Aquáticos e Bioindicadores, Faculdade de Oceanografia, Universidade do Estado do Rio de Janeiro, Rio de Janeiro, Brazil

Introduction

In the Middle Miocene, some 20–15 Million years ago (Mya), a vast portion of the South American continent was covered by seawater (Hoorn, 1993; Räsänen et al., 1995). Two epicontinental water bodies, formed by marine transgressions, were named the Paranense and the Pebasian Seas. The Paranense Sea opened to the Atlantic Ocean through the mouth of the La Plata River, and the Pebasian Sea through the mouth of Lake Maracaibo, and also possibly through the mouth of the current Amazon River (Fig. 6.1). Many authors believe that the Pebasian and Paranense seas were connected at least during part of the Miocene. These epicontinental seas became the habitat for an ancestral odontocete, which adapted and thrived in this environment for millions of years. Still during the Miocene marine waters retreated and were gradually replaced by the Paraná, Orinoco, and Amazon river basins. This event caused the definitive isolation between the ancestors of botos (genus *Inia*), which remained in the Orinoco and Amazon basins, and the ancestors of the franciscana, which were confined to the Paraná basin. Each of those lineages followed its own evolutionary trajectory in the ensuing 17 million years (Cunha et al., 2011). Eventually, the franciscana lineage reached the Atlantic Ocean through the La Plata River Estuary, and began colonizing the coast northwards and southwards.

These geological and evolutionary events provide a plausible explanation for the phylogenetic relationship between *Inia geoffrensis* and *Pontoporia blainvillei*, revealed by DNA analyses (Cassens et al., 2000; Hamilton et al., 2001). The evolutionary scenario just described is also supported by fossil data and molecular clock divergence estimates (Cunha et al., 2011; Hamilton et al., 2001) (Fig. 6.1). It is thus the starting point of this chapter,

The Franciscana Dolphin
DOI: https://doi.org/10.1016/B978-0-323-90974-7.00018-5

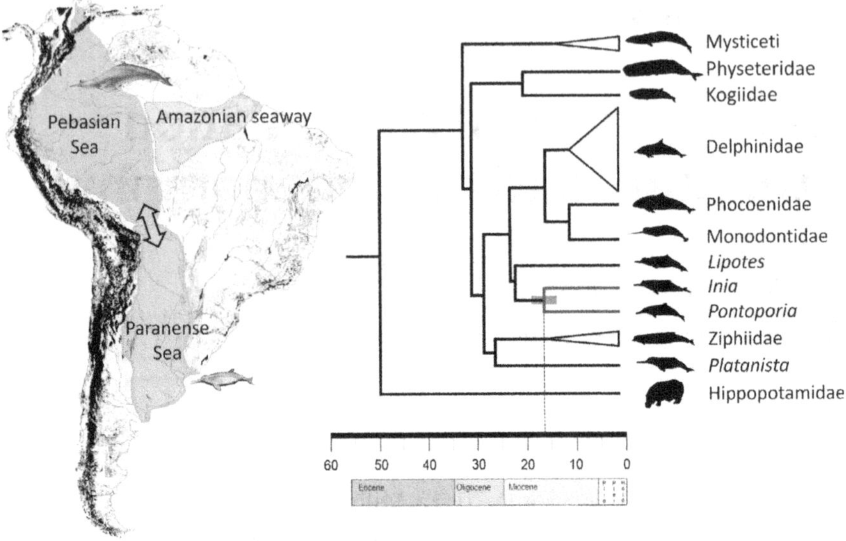

Fig. 6.1 ***Phylogenetic tree with molecular dating of the divergence between the franciscana*** **(Pontoporia blainvillei)** ***and its closest living relative, the Amazon boto*** **(Inia geoffrensis)*****, obtained with complete mitogenomes and calibrated with fossil data (Cunha et al. 2011); and the paleogeography of South America in Middle Miocene (20–15 Mya).***

which aims to describe how DNA data have been used to infer the evolutionary past of franciscanas from the moment they reversed to life in seawater, to the formation of their present-day populations, and to discuss how such information is useful to understand population structure and the vulnerability of each population and of the species as a whole.

Genetic diversity and molecular markers

Genetic diversity is the raw matter of evolution. Natural selection, which is one of the evolutionary forces, can only act when there are genetic variants with differential effect on survival and/or reproduction. Those variants may determine if a population will be able to cope with a new pathogen, new predator, new competitor, new environment. It is, ultimately, the evolutionary potential of a species, and as such, should be maintained to insure the species' persistence both in space and time. Genetic diversity is originated by the slow process of mutation, coupled with recombination, and therefore a species' present-day genetic diversity was accumulated over uncountable generations. Its loss, however, can be fast. The decline in populations' size and their isolation cause a reduction of genetic diversity and have been accelerated for some decades in all endangered species, further increasing their extinction risk (Allendorf and Luikart, 2006; Frankham et al., 2002).

Besides its importance for the evolution and mid and long-term persistence of any species, genetic diversity is a powerful tool to assess many intraspecific parameters that are relevant for conservation. Genetic data can be used to detect populations' geographic limits, to estimate gene flow (migration) among them, and to infer (relative) population size, amongst its more important applications. Genetic studies employ one or more molecular markers, which are portions of the mitochondrial or nuclear DNA that have a mutation rate compatible with the evolutionary timeframe of interest. In intraspecific studies dealing with population delimitation in cetaceans, molecular markers are usually sequences of the mitochondrial DNA (mtDNA) control region and microsatellite (nuclear DNA) genotypes. With the advent of high-throughput sequencing technologies, sequences of the entire mitochondrial genome (mitogenomes) and genotypes assessed over thousands of nuclear loci (e.g., SNP obtained by RAD-seq) will probably dominate population genetics studies in the near future. Their advantages are the much larger amount of data and genome coverage compared to traditional molecular markers, and the only drawback is cost.

Population genetics and phylogeography

Population genetics is a discipline that studies, among other things, the genetic differentiation of populations. This is done, in a simplified way, by measuring genetic diversity (nucleotide diversity in the case of mtDNA, and heterozygosity in the case of microsatellites) and testing if it is statistically different from that expected if all individuals were randomly mating (i.e., panmixia). Deviations are interpreted to mean restrictions to gene flow (nonrandom mating, lack of panmixia) because the traditionally used markers are neutral. Neutral markers are not influenced by natural selection, they only reflect the action of the other evolutionary forces: genetic drift and gene flow (the last force, mutation, is negligible in the timeframe of interest).

Analyses of population differentiation, also known as population structure (or genetic population structure), aim to provide a picture of the existing populations, as if we could see the invisible barriers between them. Those barriers restrict gene flow to less than would exist between the two areas, if mating was random, and are not necessarily physical. By defining population boundaries, genetic data guide the collection of data for abundance estimates, and other population parameters, such as natality and mortality rates, which are all basic population information for any endangered species. Gene flow can also be used as a proxy for immigration and emigration, the interchange of individuals between two populations. It can, then, inform how each population depends on its own natality and mortality rates to persist.

The importance of defining populations is a long-standing issue in the fields of species conservation and management, even before genetic data were available. The term "stock" is a jargon traditionally used in literature about management of explored species and is frequently used interchangeably with "population" in the sense used by population geneticists (i.e., to indicate genetically differentiated populations).

While population structure provides a picture of present-day populations, phylogeography investigates how and when they were formed, and it is thus a helpful complimentary approach. Phylogeography is the study of the principles and processes, historical and contemporary, that shaped the geographical distribution of the genealogical lineages of a species, or between closely related species (Avise et al., 1987). Just as evolutionary forces (mutation, genetic drift, gene flow, and natural selection) do, demographic events (such as founding and extinction of local populations, population expansion or decline, dispersal, and vicariance) leave signatures in a species' genealogical history. Phylogeography aims at deciphering those signatures to understand how those factors influenced the distribution of genetic diversity in a species in time and space (Cunha and Solé-Cava, 2012).

The phylogeography of franciscanas

Now we return to franciscanas, exactly to the point in time when they reached the Atlantic Ocean through the La Plata River. This scenario was proposed by Hamilton et al. (2001) as the most likely explanation for marine franciscanas being the sister taxon to *Inia geoffrensis*, the Amazon boto, as demonstrated by phylogenetic analyses (Cassens et al., 2000; Hamilton et al., 2001). But the plausibility of this scenario is much increased when population genetic data are taken into consideration, because the La Plata estuary and adjacent areas have the highest genetic diversity found in franciscanas (Cunha et al., 2014). This indicates that franciscanas have lived there for longer than anywhere because genetic diversity accumulates slowly, as discussed before. From there, franciscanas dispersed southwards and northwards. Franciscanas from the southern portion of the distribution exhibit relatively high genetic diversity (Gariboldi et al., 2015, 2016; Mendez et al., 2010a; Mendez et al., 2008), which may be a result of a recent split from the original population, or prolonged restricted contact following an old splitting event. Contrastingly, franciscanas to the north show a gradient of decreasing genetic diversity (Cunha et al., 2014), with low genetic variability in the northernmost limit (Espírito Santo: Cunha et al., 2014; de Oliveira et al., 2020).

In the most geographically comprehensive genetic study to date, Cunha et al. (2014) detected a strong divergence between the mtDNA lineages of franciscanas from the north (states of Espírito Santo and Northern Rio de Janeiro), compared to the south (Southern Rio de Janeiro southwards). The authors proposed that the two lineages should be regarded as evolutionarily significant units (ESU) *sensu* Moritz (1994), to acknowledge their distinctiveness and emphasize the need to preserve each ESU. In a recent study, this finding was reexamined through the inclusion of a larger number of sequences and application of phylogeographic analyses (Nara et al.). The authors doubled the number of mtDNA sequences, including more sequences from the northern portion, from where few had been analyzed previously. With this increment, they were able to confirm a doubtful haplotype sequenced before and found new haplotypes that indicate a stronger connection between franciscanas from Espírito Santo and northern Rio de Janeiro, and those from southern Rio de Janeiro to the central coast of Santa Catarina (including São Paulo and Paraná). This resulted in a new suggestion of the oldest evolutionary break in the species, that is now believed to have separated a northern group (ancestors of franciscanas that now inhabit the area from northern Santa Catarina to Espírito Santo) and a southern group (ancestors of franciscanas from southern Santa Catarina to Argentina), around 2.7 Mya (Nara et al.). The next split would have been in the northern group, dividing the franciscanas from Espírito Santo and northern Rio de Janeiro from the others. A little afterwards, the southern group also split, giving rise to the present-day populations. The divergence between Espírito Santo and northern Rio de Janeiro would be the most recent (1.0–0.1 Mya). These results put the microevolution of franciscanas into a temporal perspective, enabling the formulation of hypotheses about geological events that may have driven the distribution of genealogical lineages in the species. In this context, the authors discuss Pleistocene events that may have played a role in the micro-evolution of franciscanas (Nara et al.).

Population structure

As mentioned above, the delimitation of populations (or stocks) is of paramount importance for endangered species. The goal of population structure analyses is to detect the number of genetically differentiated populations and their limits. Basically, genetic data provide a means of assessing the interchange of individuals between populations, using gene flow as a proxy. The amount of emigration and immigration could also be measured using ecological approaches, but they are unfeasible, time-consuming, or too expensive for most species.

Genetic differentiation is expected to be detectable when populations exchange less than one migrant per generation for many generations. As a consequence, each genetically differentiated population should be managed independently, because when migration is negligible, the fate of each population will be determined solely by its natality and mortality rates. Moreover providing this basic demographic information, very useful to guide ensuing research and conservation actions, genetic differentiation *per se* is relevant, as it is the evolutionary potential of a species. Each population's exclusive genetic diversity includes adaptive variation that can be decisive for the persistence of the species when faced with changes in its selection regime. Examples are genetic resistance to a new pathogen, capacity to adapt when environmental conditions change, to prey on different species, to avoid a new predator or human-related threat, or to breed earlier.

Genetic data have been successfully used to aid in the delimitation of franciscana populations for more than two decades. Secchi et al. (1998) was the first published genetic study, which compared franciscanas from Rio Grande do Sul and northern Rio de Janeiro and found strong differentiation, indicating at least two populations. These data, coupled with results from Valsecchi and Zanelatto (2003), who analyzed sequences from Paraná, and with then still unpublished results from Lazaro et al. (2004), was integrated to other lines of evidence (morphology, life-history traits, parasites, and contaminant load) to formulate the first proposal of Management Units for the species, termed Franciscana Management Areas (FMA, Secchi et al., 2003). Originally, four FMA were proposed (Fig. 6.2).

This scheme was corroborated by subsequent studies, including different datasets and nuclear microsatellites along with mtDNA sequences (Ott, 2002), and only mtDNA (Lazaro et al., 2004). Ott (2002) found evidence of differentiation between FMA I, II and III, and that the limit between FMA II and III would lie further to the north (from ~29°20´S to ~27°35´S). These findings were posteriorly corroborated by Santos (2011) with a larger sample size. Lázaro et al. (2004) detected differentiation between FMA I, III and IV, and was the only study to report genetic differentiation between FMA III and IV.

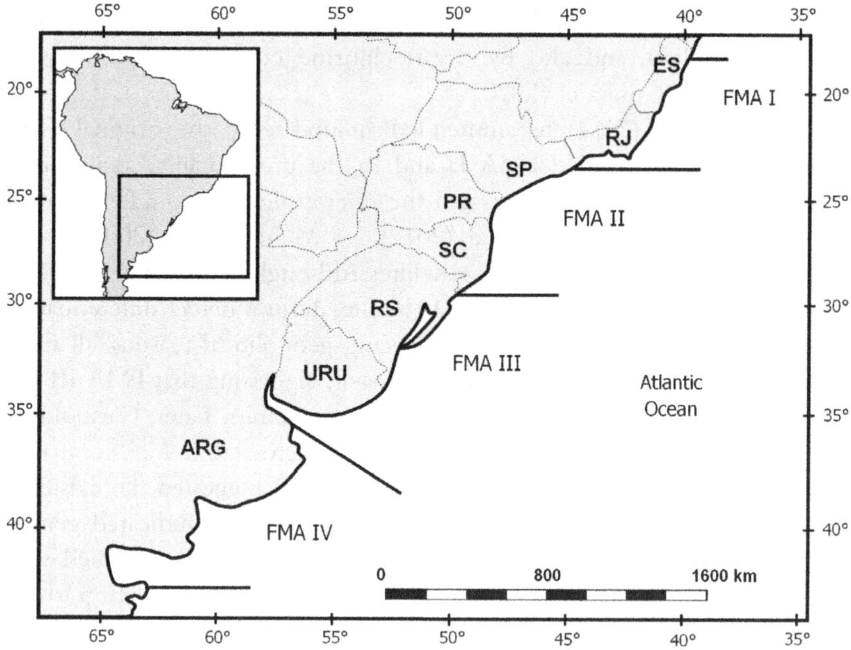

Fig. 6.2 ***The original four FMA scheme proposed by Secchi et al. (2003) based in preliminary genetic data and other lines of evidence (morphology, life history traits, parasites and contaminant load).*** *FMA*, franciscana management areas.

In the following years, Costa-Urrutia et al. (2012) and Mendez et al. (2008, 2010a) provided evidence of microscale genetic differentiation within FMA III and FMA IV, respectively, based on mitochondrial and microsatellite data. Those studies did not compare FMA III and IV, or with the other FMA. However, they were important by revealing the existence of local, genetically differentiated populations within two FMA, suggesting that the FMA scheme could be insufficient to protect them.

The original FMA proposal of Secchi et al. (2003) was refined later with the inclusion of samples from throughout the species' distribution (Cunha et al., 2014). Using mtDNA sequences, the authors detected genetic differentiation between franciscanas from Espírito Santo and northern Rio de Janeiro, which were split as FMA Ia and Ib, respectively. The authors also suggested that FMA II could comprise two populations (FMA IIa: southern Rio de Janeiro and northern São Paulo; and FMA IIb: central São Paulo to northern Santa Catarina), although data was not conclusive due to the small sample size in some of these areas. Recently, the hypothesis of differentiation between FMA IIa and IIb was strengthened by increased genetic sampling, and also by organochlorine compounds profiles and stable isotopes (Cunha et al., 2020a).

Cunha et al. (2014) recommended an update of the original FMA proposal, which included FMA Ia and Ib, the provisional recognition of FMA IIa and IIb, and incorporated the microscale genetic differentiation presented by Costa-Urrutia et al. (2012) and Mendez et al. (2008, 2010a) based on both mtDNA and microsatellites. Although Cunha et al. (2014), as well as Ott (2002) and Santos (2011) before, did not detect differentiation between FMA III and IV, and considering geopolitical reasons, all those authors followed the precautionary approach, suggesting that FMA III and IV should be considered distinct Management Units. Later, Gariboldi et al. (2015, 2016), analyzing both mtDNA and microsatellite loci, confirmed that FMA IV was not a single panmitic unit, and suggested the existence of further population subdivision within it. Their data indicated genetic differentiation in the southern portion of FMA IV, and combined with previous results by Mendez et al. (2008, 2010a) could mean that up to five populations exist within FMA IV.

The abovementioned studies were thoroughly reviewed by Cunha et al. (2020b). The authors concluded that the available data for subdivision within FMA III and IV were still limited. Regarding FMA III, the results found by Costa-Urrutia et al. (2012) were considered insufficient to warrant subdivision within the southern portion of that FMA, because a comprehensive microsatellite analysis of spatially representative samples from the La Plata River estuary, and adjacent coasts of Argentina, Uruguay and southern Brazil was still lacking.

Cunha et al. (2020b) also recognized that none of the genetic studies in FMA IV included a geographically comprehensive analysis with minimum sampling size in all sampling localities. Thus, like in the case of FMA III just described, the reported population structure could be different if all samples were analyzed together, or with increased sampling and geographic coverage. On the other hand, the available data indicate genetic distinctiveness over much of FMA IV, and ignoring them could have negative consequences to the conservation of franciscanas in the area. The authors then decided that the populations detected by Mendez et al. (2008, 2010a) and Gariboldi et al. (2016) should be regarded provisionally as Management Units, and recommended a comprehensive analysis using mtDNA and microsatellites to provide a better assessment of the genetic differentiation within FMA IV.

More recently, Cunha et al. (2020c) reported genetic differentiation of the franciscanas from Babitonga Bay (Santa Catarina) and adjacent coastal waters only 20Km away, using mtDNA and microsatellites. Surprisingly, this difference is detectable in the macroscale mtDNA analyses done by Nara and colleagues (Nara et al.). They were able to recover nine of the currently recognized or putative populations, including Babitonga, FMA IIa and IIb as distinct populations. That study could also find evidence of genetic differentiation between FMA III and IV.

Considering all the available information, Cunha et al. (2020b) recommended an updated subdivision of FMA, as summarized in Fig. 6.3. Notwithstanding this scheme, it is important to highlight that even if local, genetically differentiated populations do not correspond to stocks or FMAs, their existence must not be neglected. This is the case of the populations from Babitonga Bay, from the inner La Plata Estuary, and possibly some of the local populations in FMA IV. At least in some of these cases there is convincing evidence of a fine-scale differentiation that reflects isolation of local, demographically independent populations. Some are probably vulnerable due to unsustainable non-natural mortality rates. Recognition of their genetic distinctiveness may be important to prevent the extirpation of a local population that may retain adaptive variation.

Philopatry and its relevance to population structure

Genetic studies have suggested that the social structure of franciscanas is matriarchal (Costa-Urrutia et al., 2012; Mendez et al., 2010b; Valsecchi and Zanelatto, 2003). Matriarchal structure may reflect female philopatry, which was in fact proposed for franciscanas (Mendez et al., 2010b). Philopatry of females is a general pattern in mammals and has been used to explain population structure in many cetaceans. In most mammals, including cetaceans, female philopatry is frequently coupled with male biased dispersal (Connor, 2000; Greenwood, 1980). Female philopatry is usually inferred by a stronger degree of structure detected with mitochondrial markers compared to nuclear loci (i.e., microsatellites), due to the matrilinear inheritance of mtDNA (Möller, 2012; Prugnolle and de Meeus, 2002).

According to evolutionary theory, dispersal, and ultimately gene flow, is necessary to counterweight the effects of genetic drift and inbreeding and maintain genetic variation within populations. At the same time, gene flow is required to homogenize populations and prevent their differentiation (Wright, 1978). Philopatry means that a species does not disperse as frequently as if mating was random. This results in a lack of panmixia

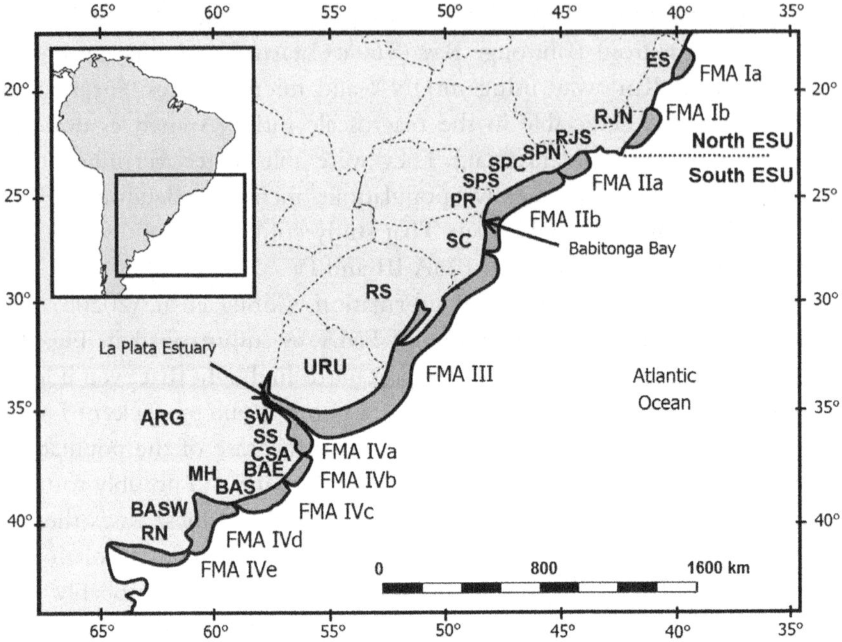

Fig. 6.3 ***Updated FMA proposal, considering all the available genetic information (Cunha et al. 2020b). See text for details.*** *FMA,* franciscana management areas.

and, consequently, in population structure. Thus philopatry, like all other phenomena that restrict dispersal and random mating (such as habitat selection, presence of strong physical barriers, distance, etc.), may lead to population differentiation (Avise, 2004; Wright, 1978). Female philopatry, along with habitat specialization, could explain the genetic differentiation of franciscanas among geographically close localities in FMA III and FMA IV (Costa-Urrutia et al., 2012; Gariboldi et al., 2015, 2016; Mendez et al., 2008, 2010a) and would have direct consequences for conservation.

Notwithstanding the possibility of male-mediated gene flow, the lack or very infrequent dispersal of philopatric females means that the female core of each population must be protected. Under this circumstance, female breeders will only be replaced through recruitment, which depends on the survival of young females up to sexual maturity, and beyond. In extreme cases, the loss of this female core will unavoidably lead to the extirpation of the population, without the possibility of recolonization—because even if males disperse, they cannot found a new population on their own. This reasoning explains why females are considered the most relevant demographic component, and why extra care must be taken in the case of species with female philopatry. It also justifies why the definition of management units should be based on the population structure detected with mtDNA when females are the philopatric sex (Dizon et al., 1997; Taylor et al., 2000).

In a recent study, Cunha et al. (2020c) analyzed kinship in franciscanas from Babitonga Bay and nearby coastal areas, besides the already mentioned microscale analyses in the same areas. The authors concluded that females are philopatric, but went a step further, by proposing that both female and male franciscanas tend to remain in their natal population after reaching sexual maturity. The philopatry of both sexes would favor the genetic differentiation of populations even at a small geographic scale, and increase the risk of inbreeding, especially when human interference intensifies the isolation and decline of populations.

Kinship analyses revealed that in Babitonga over 60% of franciscana dyads (every two analyzed franciscanas) were related, while individuals collected outside the bay, used as a control group, were not. And interestingly, kinship in Babitonga was high regardless of the sex and included dyads involving females and males (Cunha et al., 2020c). Relatedness among females was expected, but the frequent presence of related adult males within their natal population was an unusual finding and suggested lack of male-biased dispersal in the species.

Cunha et al. (2020c) reviewed all previous genetic studies that analyzed kinship in franciscanas, and verified that data concerning male dispersal is fragmentary and ambiguous. There are two reported cases of juvenile males traveling with probable mother and aunt, respectively (Valsecchi and Zanelatto, 2003, Costa-Urrutia et al., 2012), but there are also some cases of adult males accompanying females that were unrelated to them (Mendez et al., 2010b; Wells et al., 2013). Valsecchi and Zanelatto (2003) analyzed four franciscanas that entangled in the same net and suggested that they travel in family groups, and that males could display at least short-term paternal care, but doubted the existence of a longer bond between males and their offspring and respective mothers. Mendez et al. (2010b) analyzed 12 franciscana social groups and concluded that they are matrilineally oriented but include unrelated reproductive males that could form temporary or longer-lasting bonds to sexually mature females of the group. Costa-Urrutia et al. (2012) was the largest of those studies, including 21 social groups. Similarly to Mendez et al. (2010b), the authors proposed that the species' basic social unit is the family group, structured in matrilines.

They also suggested that males may remain in their natal group, but their evidence was weak.

Although those previous studies provide some insights about male dispersal and are not incompatible with their hypothesis, Cunha et al. (2020c) notice that they evaluated kinship at group level, which is not ideal for detecting population-level phenomena like philopatry or dispersal of one sex. They also highlight that dispersal from the natal population (i.e., philopatry), not from the natal group, is important to conservation because it will translate into immigration and gene flow. In Babitonga, kinship analyses suggest that both females and males are philopatric because even if males do disperse from one family group to another, both sexes still remain in their natal site, and consequently in their natal population.

The hypothesis of philopatry of both sexes deserves further testing, but seems to be in accordance with genetic data from other populations. Philopatry of both sexes would be able to generate microscale population differentiation, and would produce concordant structure patterns between mitochondrial and nuclear markers. Both situations are met in the case of franciscanas: microscale structure has been verified in several localities (Mendez et al., 2008, 2010a; Costa-Urrutia et al., 2012; Gariboldi et al., 2015, 2016; Cunha et al., 2014, 2020a, 2020c), and concordant structure between mtDNA and microsatellites have been found in Babitonga Bay (Cunha et al., 2020c) and in the studies by Ott (2002) and Santos (2011). As discussed above, the confirmation of this hypothesis would mean that franciscana populations are more isolated than previously anticipated, and therefore need even more protection.

How population structure can assist in the conservation of franciscanas

As has been discussed in this chapter, data about genetic population structure can help in two ways. The delimitation of populations' geographic boundaries provided by genetic data is fundamental to assist in the sampling design of ecological and natural history studies that will assess basic demographic parameters such as abundance, natality, and mortality rates. These parameters are still unknown for many franciscana populations. Those data will indicate which populations are more vulnerable and require more immediate and effective actions.

At the same time, population structure also guides the definition of management units. All genetically differentiated populations, as distinct

Management Units, need measures to minimize unnatural mortality rates and maintain or increase birth and survival rates. Immigration should be assumed as an extremely infrequent event in this case, insufficient to compensate mortality and to promote the colonization of extirpated local populations. In addition, genetic erosion, inbreeding, and ultimately loss of local, genetically differentiated populations could result in reduction of the adaptive potential of this critically endangered species, an irreversible and unwanted outcome that could compromise its persistence.

References

Allendorf, F.W., Luikart, G., 2006. Conservation and the Genetics of Populations, 1st ed. Blackwell Publishing, New Jersey.

Avise, J., 2004. Molecular Markers, Natural History and Evolution, 2nd ed. Sinauer Associates, Massachussets.

Avise, J., Arnold, J., Ball, R., Lambt, T., Neigel, J., Carol, A., et al., 1987. Intraspecific phylogeography: the mitochondrial DNA bridge between population genetics and systematics. Annu. Rev. Ecol. Syst. 18, 489–522.

Cassens, I., Vicario, S., Waddell, V.G., Balchowsky, H., Van Belle, D., Ding, W., et al., 2000. Independent adaptation to riverine habitats allowed survival of ancient cetacean lineages. Proc. Natl. Acad. Sci. USA 97, 11343–11347.

Connor, R.C., 2000. Group living in whales and dolphins. In: Mann, J., Tyack, P.L., Whitehead, H. (Eds.), Cetacean Societies: Field Studies of Dolphins and Whales. Chicago University Press, Chicago, pp. 199–218.

Costa-Urrutia, P., Abud, C., Secchi, E.R., Lessa, E.P., 2012. Population genetic structure and social kin associations of franciscana dolphin, *Pontoporia blainvillei*. J. Hered. 103, 92–102.

Cunha, H.A., Bisi, T.L., Bertozzi, C.P., Dias, C.P., Santos-Neto, E.B., Manhães, B.M., et al., 2020a. Population differentiation of endangered franciscanas in Southeastern Brazil: new genetic, contaminant and stable isotope data support subdivision of FMAII (SC/68B/SDDNA/05). 68th Annual Meeting of the International Whaling Comission Scientific Commitee. Available at: https://archive.iwc.int/pages/search.php?search=!collection29952.

Cunha, H.A., Dias, C.P., Alvarenga, L.C., Wells, R.S., Cremer, M.J., 2020b. Microscale population structure and kinship analyses suggest philopatry of both sexes in franciscanas (Pontoporia blainvillei) (SC/68B/SDDNA/04). 68th Annual Meeting of the International Whaling Comission Scientific Commitee. Available at: https://archive.iwc.int/pages/search.php?search=!collection29952.

Cunha, H.A., Gariboldi, M.C., Mendez, M., Secchi, E.R., Oliveira, L.R., Ott, P., et al., 2020c. Review on franciscana stock structure and Franciscana Management Areas (FMA) (SC/68B/SDDNA/07). 68th Annual Meeting of the International Whaling Comission Scientific Commitee. Available at: https://archive.iwc.int/pages/search.php?search=!collection29952.

Cunha, H.A., Medeiros, B.V., Barbosa, L.A., Cremer, M.J., Marigo, J., Lailson-Brito, J., et al., 2014. Population structure of the endangered franciscana dolphin (*Pontoporia blainvillei*): reassessing management units. PLoS One 9 (1), e85633.

Cunha, H.A., Moraes, L.C., Medeiros, B.V., Lailson-Brito Jr., J., da Silva, V.M.F., Solé-Cava, A.M., et al., 2011. Phylogenetic status and timescale for the diversification of *Steno* and *Sotalia* dolphins. PLoS One 6.

Cunha, H.A., Solé-Cava, A.M., 2012. Análise filogeográfica. In: Matioli, S.R. (Ed.), Biologia Molecular e Evolução. Editora Holos, São Paulo.

de Oliveira, V.K.M., Faria, D.M., Cunha, H.A., dos Santos, T.E.C., Colósio, A.C., Barbosa, L.A., et al., 2020. Low genetic diversity of the endangered franciscana (*Pontoporia blainvillei*) in its northernmost, isolated population (FMAIa, Espírito Santo, Brazil). Front. Mar. Sci. 7, 608276.

Dizon, A., Perrin, W., Amos, W., Baker, C., Chivers, S., Costa, A., et al., 1997. Report of the workshop. In: Dizon, A.E., Chivers, S.J., Perrin, W.F. (Eds.), Molecular Genetics of Marine Mammals (Special Publication Number 3). Society for Marine Mammalogy, University of California, CA, USA.

Frankham, R., Ballou, J.D., Briscoe, D.A., 2002. Introduction to Conservation Genetics, 1st ed. Cambridge University Press, Cambridge.

Gariboldi, M.C., Túnez, J.I., Dejean, C.B., Failla, M., Vitullo, A.D., Negri, M.F., et al., 2015. Population genetics of franciscana dolphins (*Pontoporia blainvillei*): introducing a new population from the southern edge of their distribution. PLoS One 10, 1–15.

Gariboldi, M.C., Túnez, J.I., Failla, M., Hevia, M., Panebianco, M.V., Paso Viola, M.N., et al., 2016. Patterns of population structure at microsatellite and mitochondrial DNA markers in the franciscana dolphin (*Pontoporia blainvillei*). Ecol. Evol. 6, 8764–8776.

Greenwood, P.J., 1980. Mating systems, philopatry and dispersal in birds and mammals. Anim. Behav. 28, 1140–1162.

Hamilton, H., Caballero, S., Collins, A.G., Brownell, R.L., 2001. Evolution of river dolphins. Proc. R. Soc. London Ser. B-Biological Sci. 268, 549–556.

Hoorn, C., 1993. Marine incursions and the influence of Andean tectonics on the Miocene depositional history of northwestern Amazonia: results of a palynostratigraphic study. Palaeogeogr. Palaeoclimatol. Palaeoecol. 105, 267–309.

Lazaro, M., Lessa, E.P., Hamilton, H., 2004. Geographic genetic structure in the franciscana dolphin (*Pontoporia blainvillei*). Mar. Mammal Sci. 20, 201–214.

Mendez, M., Rosenbaum, H.C., Bordino, P., 2008. Conservation genetics of the franciscana dolphin in Northern Argentina: population structure, by-catch impacts, and management implications. Conserv. Genet. 9, 419–435.

Mendez, M., Rosenbaum, H.C., Subramaniam, A., Yackulic, C., Bordino, P., 2010a. Isolation by environmental distance in mobile marine species: molecular ecology of franciscana dolphins at their southern range. Mol. Ecol. 19, 2212–2228.

Mendez, M., Rosenbaum, H.C., Wells, R.S., Stamper, A., Bordino, P., 2010b. Genetic evidence highlights potential impacts of by-catch to cetaceans. PLoS One 5, e15550.

Möller, L.M., 2012. Sociogenetic structure, kin associations and bonding in delphinids. Mol. Ecol. 21, 745–764.

Moritz, C., 1994. Defining evolutionarily-significant-units for conservation. Trends Ecol. Evol. 9, 373–375.

Nara, L., Cremer, M.J., Farro, A.P.C., Colósio, A.C., Barbosa, L.A., Bertozzi, C.P., Secchi, E.R., Pagliani, B., Costa-Urrutia, P., Gariboldi, M.C., Lazoski, C., Cunha, H.A., (in press). Phylogeography of the endangered franciscana dolphin: timing and geological setting of the evolution of populations. J. Mamm. Evol.

Ott, P.H., 2002. Diversidade genética e estrutura populacional de duas espécies de cetáceos do Atlântico Sul Ocidental: *Pontoporia blainvillei e Eubalaena australis*. Universidade Federal do Rio Grande do Sul, Porto Alegre.

Prugnolle, F., de Meeus, T., 2002. Inferring sex-biased dispersal from population genetic tools: a review. Heredity 88, 161–165.

Räsänen, M.E., Linna, A.M., Santos, J.C.R., Negri, F.R., 1995. Late Miocene tidal deposits in the Amazonian foreland basin. Science 269, 386–390.

Santos, E.V., 2011. Estrutura populacional e história filogeográfica da toninha (*Pontoporia blainvillei*). Pontifície Universidade Católica do Rio Grande do Sul, Porto Alegre, RS, Brazil.

Secchi, E.R., Danilewicz, D., Ott, P.H., 2003. Applying the phylogeographic concept to identify franciscana dolphin stocks implications to meet management objectives. J. Cetacean Res. Manag. 5, 61–68.

Secchi, E.R., Wang, J.Y., Murray, B.W., Rocha-Campos, C.C., White, B.N., 1998. Population differentiation in the franciscana (*Pontoporia blainvillei*) from two geographic locations in Brazil as determined from mitochondrial DNA control region sequences. Can. J. Zool. Can. Zool. 76, 1622–1627.

Taylor, B.L., Chivers, S.J., Sexton, S., Dizon, A.E., 2000. Evaluating dispersal estimates using mtDNA data : comparing analytical and simulation approaches. Conserv. Biol. 14, 1287–1297.

Valsecchi, E., Zanelatto, R.C., 2003. Molecular analysis of the social and population structure of the franciscana (*Pontoporia blainvillei*): conservation implications. J. Cetacean Res. Manag. 5, 69–75.

Wells, R.S., Bordino, P., Douglas, D.C., 2013. Patterns of social association in the franciscana *Pontoporia blainvillei*. Mar. Mammal Sci. 29, 520–528.

Wright, S., 1978. Evolution and the Genetics of Populations. The University of Chicago Press, Chicago.

CHAPTER 7

Abundance estimates of franciscana dolphins (*Pontoporia blainvillei*): a review and future recommendations

Alexandre Novais Zerbini[a,b], Artur Andriolo[c], Marta Jussara Cremer[d], Enrique A. Crespo[e], Daniel Danilewicz[f], Camila Domit[g], Federico Sucunza[f]

[a]Cooperative Institute for Climate, Ocean and Ecosystem Research (CICOES), University of Washington and Marine Mammal Laboratory, Alaska Fisheries Science Center, Seattle, WA, United States
[b]Marine Ecology and Telemetry Research, Seabeck, WA, United States
[c]Laboratório de Ecologia Comportamental e Bioacústica, Universidade Federal de Juiz de Fora, Juiz de Fora, MG, Brazil
[d]Laboratório de Ecologia e Conservação de Tetrápodes Marinhos e Costeiros, Universidade da Região de Joinville, São Francisco do Sul, Santa Catarina, Brazil
[e]Centro Nacional Patagonico, Consejo Nacional de Investigaciones Científicas y Técnicas, Puerto Madryn, Argentina
[f]Grupo de Estudos de Mamíferos Aquáticos do Rio Grande do Sul, Torres, Rio Grande do Sul, Brazil
[g]Laboratório de Ecologia e Conservação, Universidade Federal do Paraná, Pontal do Sul, Paraná, Brazil

Introduction

The franciscana (*Pontoporia blainvillei*) is one of the smallest cetaceans and the sole living member of the family Pontoporiidae. Franciscanas are endemic of the eastern coast of South America and are found in shallow waters (typically less than 50 m of depth) in Brazil (from about 18°S), Uruguay, and northern Argentina (up to about 41°S) (Crespo et al., 1998; Siciliano et al., 2002; do Amaral et al. 2018). The distribution is discontinuous in the northern range of the species (states of Espírito Santo [ES] and Rio de Janeiro [RJ] in Brazil), where two isolated populations are found (Siciliano et al., 2002; do Amaral et al. 2018). Starting in southern Rio de Janeiro, the distribution of the franciscana in the remainder of the range appears to be continuous and include the states of São Paulo (SP), Paraná (PR), Santa Catarina (SC), and Rio Grande do Sul (RS) in Brazil, the whole coast of Uruguay and the provinces of Buenos Aires, Rio Negro and Chubut in Argentina (Crespo et al., 1998; Secchi, 2010a). The species is also found in at least four enclosed

DOI: https://doi.org/10.1016/B978-0-323-90974-7.00009-4

bays or estuarine areas, three in Brazil (Cremer and Simões-Lopes, 2005; Santos et al., 2009; Weyn, 2016) and one in Argentina (Failla et al., 2012).

The franciscana is considered the most threatened small cetacean in South America (Secchi, 2010b) and is listed as *Vulnerable* in the IUCN Red List of threatened species (Zerbini et al., 2017). High levels of mortality of fransciana dolphins associated with gillnet fisheries have been documented throughout the species range for at least 50 years (Ott et al., 2002; Secchi et al., 2021), making bycatch the greatest threat to this species (Secchi, 2010b). Currently, it is estimated that a few thousand animals are killed incidentally by fisheries and there is evidence that bycatch is largely unsustainable (Anonymous, 2016; Ott et al., 2002; Secchi, 2010b). Habitat degradation, including chemical and noise pollution, is another important threat, but their effects on franciscana populations are poorly understood and are likely to vary regionally.

Estimation of density and abundance is key to the management of wildlife as they provide baseline information on the size of a population at the time the estimate is computed. Conservation actions often rely on some knowledge of the status of a population and are more effectively implemented if quantitative determination on whether the population is declining, increasing, or stable is available. Therefore, times series of abundance estimates can be useful in monitoring population trends. Knowledge of abundance also has other applications, including understanding the ecological roles of a species in the ecosystem. For example, population estimates can be converted into estimates of biomass and used as input parameters in ecosystem modeling research.

In the early 2000s, the range of the franciscana was divided into four "franciscana management areas" (FMAs) (Secchi et al., 2003), two in Brazil (FMA I and II), one shared between Brazil and Uruguay (FMA III), and a fourth in Argentina (FMA IV). These areas were determined based on the phylogeographic approach developed by Dizon et al. (1992) and took into consideration information on population dynamics, morphology, molecular genetics, contaminants, and parasite loads (Secchi et al., 2003). Additional research on population structure provided evidence of population substructure within these FMAs (Cunha et al., 2014) and led to changes in some of the originally proposed FMA boundaries (Anonymous, 2016). Currently, five FMAs are recognized, FMA Ia (ES), FMA Ib (northern coast of RJ), FMA II (southern RJ, SP, PR, and norther SC), FMA III (southern SC and RS in Brazil and the whole coast of Uruguay), and, finally, FMA III (Argentina). Additional details on the current FMA boundaries are provided in Cunha et al. (2022).

Even though human-induced threats to the franciscana dolphin have been recognized for many decades, estimates of abundance to assess the potential effects of these threats have only been consistently developed in the past 25 years. However, there are still only a few reliable estimates of abundance for the franciscana and many do not cover the whole range of the currently recognized FMAs. Also, no empirical estimates of trends in abundance for franciscana exist and further efforts are needed to assess the rate of population change for all management areas. In this chapter, we provide an overview of what is known about estimates of abundance of franciscanas, including a historical perspective, improvements in survey methodology, current estimates for the various FMAs, and a discussion about future work that can improve our ability to estimate abundance and potentially monitor trends in the abundance of this threatened species.

Early years

While many methods to estimate the abundance of cetaceans exist, the most used ones are line transect sampling and capture-recapture (Hammond et al., 2021). Capture-recapture methods can only be applied to species where individual identification (e.g., through photo-identification or genotyping) is possible. With a few exceptions (e.g., franciscanas in Babitonga Bay, Brazil, Sartori et al., 2017), franciscana individuals are not typically recognizable via natural marks and are not easy to be biopsy sampled. Therefore, abundance estimates must rely largely on methods that do not require individual identification. Given their relatively broad distribution patterns and social characteristics (e.g., franciscanas typically live in small groups and are relatively widely distributed throughout their habitats), line transect surveys have long been considered the most suitable method to estimate the abundance of franciscanas (Crespo et al., 2002). Abundance surveys conducted to date have used fixed-wing manned aircraft to sample large geographical areas (e.g., Crespo et al., 2010; Danilewicz et al., 2010; Secchi et al., 2001; Sucunza et al., 2020a), but small boat-based line-transect surveys have also been in more limited regions (Bordino et al., 2004; Cremer and Simões-Lopes, 2008; Weyn, 2016).

Surveys to estimate the abundance of franciscanas started in the mid-1990s and early 2000s (Bordino et al., 2004; Cremer and Simões-Lopes, 2008; Crespo et al., 2010; Secchi et al., 2001). Additional details on the estimates computed during these studies are presented in section "Franciscana abundance estimates" below. Here we provide some background information on the relevance of these studies as the first step toward computing estimates of density and abundance of the franciscana.

The first survey to estimate franciscana abundance was conducted in southern RS, Brazil in 1996 (Secchi et al., 2001). In addition to providing an estimate for this species, this research exposed franciscana scientists to aerial line transect methods and set the stage for the development of a number of studies in the subsequent years. Perhaps most important, this study fostered international collaboration across franciscana scientists to discuss the best methods to develop abundance surveys within the range of the species, including survey design, data collection, and data analysis. This study also brought an important lesson, quickly learned by all franciscana researchers. At the end of the study, the single-engine aircraft used in the survey was forced to ditch nearly 5 miles offshore due to engine and fuel problems, almost claiming the lives of three young (and now prominent) franciscana scientists (D. Danilewicz, P.H. Ott, and E.R. Secchi) and their pilot. Fortunately, the airplane crew survived the crash and was nearly miraculously rescued by the Brazilian armed forces. This incident made it clear that offshore surveys should be conducted under rigorous safety protocols and should avoid the use single-engine aircraft (Crespo et al., 2002; Secchi et al., 2001), a standard followed until today for aerial surveys of franciscanas and other marine species.

Estimates of density and abundance were also obtained through small boat and aerial surveys within the range of the franciscana FMA II and in FMA IV (Argentina). Boat surveys were conducted in coastal waters off the coast of the Buenos Aires Province to estimate seasonal density and abundance (Bordino et al., 2004) and in Babitonga Bay, State of Santa Catarina, Brazil (Cremer and Simões-Lopes, 2008). Aerial surveys sampled a broad geographical range within the range of FMA IV, including offshore habitats off the Argentinian Provinces of Buenos Aires and Rio Negro (Crespo et al., 2010), including Golfo San Matias, which is considered the southern limit of the franciscana distribution (Crespo et al., 1998).

These pioneer studies in FMA II, FMA III and FMA IV demonstrated the feasibility of using line-transect surveys to estimate the abundance of franciscanas and emphasized the urgent need for additional surveys to monitor trends, especially considering the high level of anthropogenic threats the species had been and continue to be subjected to. These studies also revealed that improvements in survey design and implementation would be required to produce reliable estimates of abundance (IWC, 2005).

Methodological improvements and computation of correction factors

In 2004, the International Whaling Commission (IWC) Scientific Committee reviewed the status of the franciscana and provided recommendations to improve the accuracy of abundance estimates computed in future surveys, including expanding survey coverage within franciscana habitats and developing correction factors to minimize bias in density and abundance estimates (IWC, 2005). Surveys conducted since the mid-2000s attempted to address these recommendations. Improvements in survey design included more even sampling effort and greater sampling coverage of the area for which estimates of abundance were desired (Crespo et al., 2020; Cunha et al., 2020; Danilewicz et al., 2010, 2020; Sucunza et al., 2020a, b; Zerbini et al., 2010). These surveys were specific to different FMAs and are discussed further in the next section. In this section, we will describe research on the development of correction factors to minimize bias in franciscana abundance estimates, with a focus on those computed for large-scale aerial surveys. Boat surveys have been applied only to small geographic areas within the range of the species (Bordino et al., 2004; Cremer and Simões-Lopes, 2008; Weyn, 2016), but correction factors to address bias for this type of survey platform have not yet been developed.

Line transect surveys assume that the probability of detecting objects on the survey trackline is certain (an assumption known as $g(0) = 1$) (Buckland et al., 2001), but this assumption is often violated in aerial surveys for marine mammals (Laake and Borchers, 2004). This "visibility bias" ($g[0] < 1$, Laake et al., 1997) leads to underestimation of abundance. Marsh and Sinclair (1989) coined the terms "perception" and "availability" bias to differentiate two forms of visibility bias. Perception bias occurs when dolphins are available to be seen but are missed by the observers while availability bias corresponds to animals that are missed because they are submerged (and thus not available to be seen). Another important assumption of line transect methods is that the number of individuals in a group is accurately estimated during surveys. Because of the relatively high speed of the aircraft and because franciscanas are small and difficult to see from fast-moving platforms, group sizes are often negatively biased.

Early attempts to estimate availability bias relied on dive parameters of franciscanas observed from shore-based stations (2 m above the sea level) and vessels in Bahía Anegada, Argentina (Bordino et al., 1999). Surfacing and dive times estimated at 1.2 and 21.7 s, respectively, were used to correct for availability bias in aerial surveys of franciscanas in FMA III and FMA

IV (Secchi et al. 2001, Crespo et al. 2010, Danilewicz et al. 2010) using a method developed by Barlow et al. (1988) for harbor porpoises. Estimates of availability in these studies varied between 0.28 and 0.36 because different aircraft were used in each survey, and estimates of this parameter are typically platform-specific. Because surface and dive intervals estimated by Bordino et al. (1999) were computed from a surface platform, it is unlikely they apply for aerial surveys because of the different fields of view and perspective from observers on an aircraft. Observers in a low surface platform typically see the animals when they are breaking the surface of the water, whereas in an aircraft, dolphins can often be seen when they are close to the surface but still underwater. Therefore, the proportion of the time a group of franciscanas is visible is greater for an observer in an aircraft than one on a boat or on land. This leads to a negative bias in the estimation of availability bias and, consequently, a positive bias in the estimation of abundance.

The need to develop estimates of availability bias applicable to franciscana aerial surveys was addressed by Sucunza et al. (2018; 2020c). These authors computed the availability of franciscanas in two regions with contrasting environmental conditions, Babitonga Bay (shallow and more turbid waters) and Ubatuba (deeper and more clear waters). Dive parameters were obtained from a helicopter flying at an altitude consistent with that flown during aerial surveys (500 ft). Although surfacing and dive intervals (Table 7.1) were significantly longer in Ubatuba than in Babitonga Bay, the proportion of time at the surface did not vary significantly between the study areas ($P = 0.32$), and thus a combined availability bias estimate of 0.39 (CV = 0.02) was computed for both areas combined using the method proposed by Laake et al. (1997).

Table 7.1 Dive parameters and availability estimated from aerial platforms to correct aerial surveys of franciscana.

Location	Surface time	Dive time	Proportion at surface	Availability[a]
Babitonga Bay	16.1 (SE = 9.7)	39.8 (SE = 29.1)	0.36 (SE = 0.23)	0.39 (CV = 0.03)
Ubatuba	38.8 (SE = 13.1)	77.3 (SE = 20.0)	0.34 (SE = 0.09)	–
Total	23.70 (SE = 15.33)	52.33 (SE = 31.75)	0.35 (SE = 0.19)	0.39 (CV = 0.02)

[a]Estimated assuming window time (w[0]) is 6 s, consistent with the aircraft, and survey altitude (500 ft) and speed (100 kts) used in aerial surveys for franciscana abundance estimation in FMA Ia, FMA Ib, FMA II and FMA III (Danilewicz et al., 2010, 2020; Sucunza et al., 2020a, b, d; Zerbini et al., 2010).

Sucunza et al. (2020a), using mark-recapture distance sampling methods (Laake and Borchers, 2004), estimated that between 13% and 23% of the available franciscana groups were missed by the observers (perception bias) during aerial surveys in FMA II. This is the only empirical estimate of perception bias for franciscanas.

Independent boat and aerial surveys were concurrently carried out in southern Brazil to assess potential bias in aerial surveys for franciscanas (Sucunza et al., 2020c; Zerbini et al., 2011). Estimates of density and group sizes computed for the boats were assumed to be accurate (i.e., g[0] was assumed unit because surveys were conducted in good weather conditions [sea state = 0 – 2 in the Beaufort scale] and no responsive movement was documented). A correction factor (CF = 4.42, CV = 0.04) for visibility and group size bias for aerial surveys was developed as the ratio of the density estimated by boats (D = 2.99 ind/km^2, CV = 0.23) and by the aircraft (D = 0.68 ind./km^2, CV = 0.28) (Sucunza et al., 2020c). In addition, in this study, generalized linear models were used to assess differences in group sizes estimated from the boats and the airplane and indicated that boat group size estimates are 33% greater than those computed from an airplane (Sucunza et al., 2020c). This suggests that abundance estimates that are not corrected from group size bias, are negatively biased.

Franciscana abundance estimates

Abundance estimates exist for all FMAs, mostly through aerial surveys. In some cases, the estimates comprised a portion of the FMA range and in others, they covered the entirety of the known distribution of the species within each management area. Details of FMA-specific estimates are provided below.

FMA Ia

Franciscanas in FMA Ia are distributed along with the northern coats of Espírito Santo State (ES), Brazil, and are likely isolated from all other populations (Cunha et al., 2014; Danilewicz et al., 2012; do Amaral et al., 2018; Siciliano et al., 2002). Aerial surveys to estimate abundance were conducted in 2011 and again in 2018 (Danilewicz et al., 2012; Sucunza et al., 2020b). During the 2011 surveys, franciscana groups were recorded as far south as Santa Cruz (19°58′) (Danilewicz et al., 2012), but in recent years franciscana sightings were only seen about 20 km north of this location despite an increase of up to 30 times in survey effort in the region (Sucunza et al., 2020b). Results from aerial surveys suggest that, at least during the summer,

Table 7.2 Estimates of franciscana abundance in FMA Ia in 2018.

Area	*D*	*N* (CV)	95% CI	References
FMA Ia North	0.14	200 (0.61)	69–503	Sucunza et al. (2020b)
FMA Ia South	0.35	395 (0.58)	81–938	Sucunza et al. (2020b)
Whole FMA I	0.23	595 (0.44)	203–1200	Sucunza et al. (2020b)

franciscanas in FMA Ia are distributed in a relatively small area (173 km of coastline) off ES, between Conceição da Barra (18°35′S) and Santa Cruz (19°56′S). An estimate of abundance could not be computed for the 2011 survey because no franciscana groups were detected on-effort that year (Danilewicz et al., 2012). An estimate was computed for the 2018 survey, which divided the survey area into two regions: north and south. The effort consisted of a set of parallel track lines allocated perpendicular to the coastline as far as 16.5 km from shore. A total of 2986 km was surveyed and 17 franciscana groups (41 individuals, mean group size = 2.41, CV = 0.14) were seen within 8 km from shore. Abundance corrected for visibility and group size biases (Correction Factor [CF] = 4.42, CV = 0.04; Sucunza et al., 2020c) was estimated at 200 (CV = 0.61) and 395 (CV = 0.58) dolphins for the northern and southern areas, respectively. Population size in the whole FMA Ia was estimated at 595 individuals (CV = 0.44) (Table 7.2).

FMA Ib

FMA Ib corresponds to the northern coast of Rio de Janeiro State (RJ) in southeastern Brazil. Aerial surveys to assess distribution and abundance have been carried out in December 2011, March 2017, and August 2019, and abundance was computed for 2011 and 2017 using line-transect methods (Danilewicz et al., 2012; Danilewicz et al., 2020). Similar to FMA Ia, franciscanas in FMA Ib also show a restricted distribution, within only 125 km of linear coastline between São João da Barra (21°56'S) and Carapebus (22°16'S). Sightings were detected as far as 15.1 n. miles from shore, but most of the sightings (70%) were seen within 4 n. miles from shore (Danilewicz et al., 2020). In 2011, a total of 1,010 km was surveyed and 15 franciscana groups were detected on effort. Average group size was estimated at 2.54 individuals (CV = 0.18). In 2017, the survey effort totaled 2,150 km and 29 sightings were recorded on effort. Average group size was estimated at 2.5 individuals (CV = 0.46). Density and abundance corrected for visibility and group size bias using the Sucunza et al. (2020c) correction factor were estimated, respectively, at 0.3 ind./km^2 and 1,692

Table 7.3 Estimates of franciscana abundance in FMA Ib.

Year	*D*	*N* (CV)	95% CI	References
2011	0.3	1,692 (0.47)	706–4,055	Danilewicz et al. (2020)
2017	0.316	1,280 (0.44)	566–2,894	Danilewicz et al. (2020)

individuals (CV = 0.47) in 2011 and at 0.316 ind./km^2 and 1,280 individuals (CV = 0.44) in 2017 (Table 7.3). Density estimates were similar between the two years, but the two abundances are not directly comparable because the area surveyed was slightly different across years.

FMA II

FMA II ranges from the southern coast of RJ to central SC an area of approximately 1,000 km of linear coastline. Abundance estimates for this management area (Table 7.4) have been computed for the whole stock range and also for two small estuarine regions, one in Babitonga Bay, SC, believed to represent an isolated population (Cunha et al., 2020; Dias et al., 2013; Sartori et al., 2017) and another for Paranaguá Estuarine Complex (Weyn, 2016), though the population identity of franciscanas found in this region is unclear.

The only estimate of franciscana abundance for the whole of FMA II was computed from an aerial survey conducted in the summer of 2008/9, which covered the range of that management area within the 30 m isobath (Zerbini et al., 2010). However, changes in the boundary between FMA II and FMA III in 2015, the existence of two putative "sub-populations" within FMA II (named FMA IIa and FMA IIb, Cunha et al., 2014), and the estimation of new correction factors led to the reanalysis of the original data (Sucunza et al., 2020a). Revised estimates were computed using mark-recapture distance sampling methods (Laake and Borchers, 2004) to estimate perception bias and were corrected for availability bias using correction factors developed by Sucunza et al. (2018). Estimates were provided for the whole of FMA II (6,827, CV = 0.33, Table 7.4) and for the putative subpopulations (FMA IIa = 1,915, CV = 0.32; FMA IIb = 4,353, CV = 0.24) as defined by Cunha et al. (2014). This estimate is not corrected for group size bias and is therefore negatively biased.

Within FMA II, there are independent abundance estimates for Babitonga Bay, SC, where a small population occurs year-round. Evidence from telemetry and genetics studies suggests that franciscanas in Babitonga Bay are isolated from those in adjacent coastal areas (Dias et al., 2013;

Table 7.4 Estimates of abundance of franciscanas computed for FMA II.

Area	Year	*D*	*N* (CV)	95% CI	References	Comments
Whole FMA II	2008/9	0.362	8,525 (0.34)	4,434–16,390	Zerbini et al. (2010)	Outdated due to changes in FMA boundaries and computation of new correction factors for visibility bias
Whole FMA II	2008/9	0.33	6,827 (0.26)	4,127–11,294	Sucunza et al. (2020a)	Estimate for updated FMA II boundaries
FMA IIa	2008/9	0.59	1,915 (0.32)	1,034–3,546	Sucunza et al. (2020a)	Estimate for a putative subpopulation within FMA II
FMA IIb	2008/9	0.25	4,353 (0.24)	2,728–6,946	Sucunza et al. (2020a)	Estimate for a putative subpopulation within FMA II
Babitonga Bay	2000-2003 (mid-point: 2002)	0.318	50 (0.30)	28–89	Cremer and Simões-Lopes (2008)	Population believed to be isolated from franciscanas found in adjacent coastal habitats (see text above)
Babitonga Bay	2011	2.99	49 (0.22)	–	Sucunza et al. (2020c)	Same general region as the estimate by Cremer and Simões-Lopes (2008) but different survey area. Estimates not comparable to that study due to different sampling area
Babitonga Bay	2011–2013		52-82		Sartori (2014)	Abundance computed using capture-recapture methods form individually identified dolphins
Paranaguá Estuarine Complex	2012–2013	0.16	198 (0.31)	109–361	Weyn (2016)	Population identity relative to franciscanas outside of the bay is unknown

Cunha et al., 2022). Boat-based surveys were conducted with line transect methods between 2000 and 2003 and estimated an abundance of 50 individuals (CV = 0.30), which is assumed here to correspond to 2002, the mid-year of the surveys (Cremer and Simões-Lopes, 2008). Another boat-based estimate of 49 individuals (CV = 0.23) for this area was computed in 2011 (Sucunza et al., 2020c), but the area surveyed was different (smaller) relative to the 2000-2003 surveys. While the area sampled in the most recent survey corresponded to the main area of occurrence of franciscanas, this estimate may not be comparable to the one computed by Cremer and Simões-Lopes (2008). Seasonal abundance of franciscanas in Babitonga Bay was estimated using mark-recapture methods in 2011 and 2013 (ranged from 52 to 82 individuals) and, as expected, indicated that abundances did not vary significantly across seasons (Sartori, 2014).

A boat-based line transect estimate of 198 franciscanas (CV = 0.31, Table 7.4) was computed for the Paranaguá Estuarine Complex, in Paraná state, during 2012–2013 (Weyn, 2016). It is unclear whether franciscanas in this estuarine complex correspond to a separate population (such as the one in Babitonga Bay) or whether they represent animals that seasonally occur in the estuarine area but are part of the larger FMA II population.

FMA III

FMA III includes the southernmost Brazilian states (SC and RS) and Uruguay. Estimates of abundance for this management area were computed in 1996, 2004, and 2014 (Danilewicz et al., 2010; Secchi et al., 2001; Sucunza et al., 2020c). For all these estimates, sampling occurred in the Brazilian portion of the range of the stock. Surveys to assess franciscana abundance have never been conducted in Uruguay.

The first abundance estimate for FMA III was computed from aerial surveys conducted in 1996 (Secchi et al., 2001) (Table 7.5). In this study, 1,428 km were surveyed in a small coastal area (435 km^2) between the coastline and 9.3 km from shore in a region located approximately in the center of the range of the stock (close to the mouth of Patos Lagoon). A total of 29 sightings (35 individuals, average group size = 1.16, CV = 0.07) were made on effort and density within the survey area corrected for availability (but not perception or group size) bias was estimated at 0.657 ind./km^2 (CV = 0.34, Secchi et al., 2001). This density estimate was extrapolated to the (nonsurveyed) range of the stock (both southern Brazil and Uruguay, total area of approximately 64,000 km^2) to compute an estimate of abundance of nearly 42,000 individuals (CV = 0.34).

Table 7.5 Density and abundance estimates of franciscanas computed within FMA III.

Area	Year	*D*	*N* (CV)	95% CI	References	Comments
FMA III	1996	0.65	42,078 (0.34)	33,047–53,572	Secchi et al. (2001)	No correction for perception and group size bias. Density was estimated in an area representing a small fraction of the range of the stock
Rio Grande do Sul	2004	0.51	6,839 (0.32)	3,709–12,594	Danilewicz et al. (2010)	No corrections for perception bias, and group size bias. The survey did not cover the whole range of FMA III
Rio Grande do Sul and south of Santa Catarina	2014	0.31	9,651 (0.24)	6,013–15,490	Sucunza et al. (2020d)	Corrected for visibility and group size bias. Does not represent the whole range of FMA III

A second survey conducted in RS in 2004 expanded the sampling area of Secchi et al. (2001) and used a different survey design (Danilewicz et al., 2010) (Table 7.5). A total of 1,256 km was surveyed along the coast of RS with track lines going as far offshore as 13 n. miles and covering an area of 13,341 km^2. Thirty-one franciscanas were seen in 25 groups (average group size = 1.36 [CV not available] for groups seen on effort and 2.0 [CV not available] for groups seen off effort). Density within the survey area was corrected for availability bias and estimated at 0.51 franciscanas/km^2 (CV = 0.32). The total estimated abundance within the survey area was 6,839 franciscanas (CV = 0.32). The sensitivity of abundance estimates to different values of expected group size, and correction factors for availability bias and for perception bias (using an estimate for harbor porpoise in Washington State, United States, Laake et al., 1997) were examined. This sensitivity analysis led to estimates of density ranging from 0.205 to 3.17 ind./km^2 (Danilewicz et al., 2010), which implied point estimates of franciscana abundance could range between 2,700 and 43,000 individuals in the study area.

In the mid-2010s, the boundary between FMA II and FMA III was shifted northwards and the latter now includes the southern portion of SC state (Anonymous, 2016). A new aerial survey conducted in 2014 covered the range of FMA III within Brazilian waters (Sucunza et al., 2020c). In this study, parallel tracklines were allocated perpendicular to the shore from the coastline up to the 50 m isobaths, resulting in a comprehensive sampling of the franciscana habitat in southern Brazil. A total of 3853 km was surveyed on effort and 84 franciscana groups were recorded (167 individuals, mean group size = 1.99, SE = 0.91, median = 2, range = 1–5). Density corrected for visibility bias and group size bias (Sucunza et al., 2020c) was estimated at 0.31 individuals/km^2 (CV = 0.24, 95% CI = 0.19–0.50). Density extrapolated to the survey region yielded an estimate of 9,651 individuals (CV = 0.24, 95% CI = 6,013–15,490) (Table 7.5).

Neither of the estimates of abundance presented above is comparable (e.g., to compute trends in density/abundance) because of differences in survey areas, observers, and survey platforms. In addition, they represent only a fraction of the total range of FMA III, which is shared between Brazil and Uruguay. A major difficulty in estimating the abundance of FMA III lies in the fact that waters off Uruguay have never been surveyed.

FMA IV

FMA IV includes the coast of Argentina, from the La Plata River down to the northern coast of Golfo San Matías. Estimates of abundance for the Province of Buenos Aires were computed by Bordino et al. (2004) using small boats. A total of 28 surveys were conducted in good visibility conditions (Beaufort 0–2); 200 transects totaling 1,363 km of survey effort were completed in four nearshore areas along the coast. Group size was estimated at 2.3 individuals (CV = 0.39). The average density across these areas was 0.38 ind./km^2 (CV = 0.30) and abundance within the survey area was estimated at 206 individuals (CV = 0.30). Density in the survey area (562 km^2) was extrapolated to the range of the species along the Province of Buenos Aires to estimate a population of 31,110 individuals (Table 7.6, Bordino et al., 2004). This study also investigated seasonal variance in density and while there was evidence that franciscanas were found closer to shore in the summer/spring, density estimates between summer/spring and winter seasons were not statistically different.

Crespo et al. (2010) conducted aerial surveys between the Province of Buenos Aires and the northern coast of Golfo San Matías, Rio Negro Province in 2003 and 2004. The survey area was divided into four strata, two in the north (a coastal stratum—0–30 m isobath—and an offshore stratum—30–50 m isobath) and two in the south (also coastal and offshore).

Table 7.6 Estimates of density and abundance for franciscana in FMA IV.

Area	Year	*D*	*N* (CV)	95% CI	References	Comments
Argentina—Buenos Aires Province	1996–1998/ 2000–2003	0.38	31,350 (-)	15,262–47,850	Bordino et al. (2004)	Density in the survey area was extrapolated to the unsurveyed portion of the pronvince of Buenos Aires
Argentina—North, coastal stratum (0–30 m)	2003/4	0.377	8,279 (0.21)	4,904–13,960	Crespo et al. (2010)	Density in the surveyed area was extrapolated to the unsurvey portion of the stratum
Argentina—North offshore stratum (30–50 m)	–	0.178	470 (0.37)	–	Crespo et al. (2010)	Not extrapolated to unsurvey areas and not included in the total estimate
Argentina—South (coastal stratum (0–30 m)	2003/4	0.197	5,896 (0.61)	1,928–17,999	Crespo et al. (2010)	Density in the surveyed area was extrapolated to the unsurvey portion of the stratum
Argentina—Total Coastal Stratum	2003/4	0.273	14,175 (0.30)	7,900–25,100	Crespo et al. (2010)	Overall abundance was calculated only for the coastal stratum (that is it's the sum of 8,279 and 5,896)
Argentina—North, coastal stratum (0–30 m)	2019	0.608	13,356 (0.28)	7,737–22,821	Crespo et al. (2020)	Density in the surveyed area was extrapolated to the unsurvey portion of the stratum

Four sets of zigzag tracklines were allocated across these regions and the area where these tracklines were allocated represented just a fraction of the total area of each stratum. Surveys were conducted in three periods, two in the summer (February 2003 and 2004) and one in the fall (April/May 2003). Overall, 6,634 km were flown and a total of 101 franciscanas were observed in 71 sightings (average group size = 1.43). Density and abundance were computed using standard line transect sampling methods and corrected for availability bias. Because there were no seasonal differences in estimates of density, data from autumn surveys were used because of better survey conditions in that season. In the northern regions (off the northern Buenos Aires Province) corrected density was higher in the coastal stratum (0.377 ind/km^2, CV = 0.21), but declined further offshore (0.178 ind./km^2, CV = 0.37). Due to the small sample size, density was computed for both coastal and offshore stratum in the south (0.197 ind./km^2, CV = 0.61). Abundance was computed for the northern coastal stratum by extrapolating density computed in the surveyed area (about 1/3 of the total area) to the whole stratum (Table 7.6). For the northern offshore stratum, abundance was calculated only for the area covered by the survey (that is, no extrapolation to unsurveyed areas) because the sampling region on this stratum represented a much smaller proportion of the total stratum area. In the south, abundance was estimated only for the coastal stratum by extrapolating density computed in the covered area to the whole area of the stratum (Table 7.6).

A new series of aerial surveys to estimate the abundance of franciscana off Argentina started in October 2019. They were designed to sample similar areas to those described in Crespo et al., (2010). However, the surveys were interrupted because of poor weather conditions and could not be continued during the Coronavirus pandemic. Therefore, to date, only a portion of the survey area was covered (Crespo et al., 2020). The October 2019 survey was able to cover the northern coastal stratum of Crespo et al. (2010), but at a different season (spring in the new survey, autumn in Crespo et al., 2010). Estimates of density and abundance are presented for comparison with the 2003/4 surveys (Crespo et al. 2020). A total of 1,532 km of trackline were covered and 41 franciscana groups were seen on effort (mean group size = 1.34, CV = 0.11). Density/abundance corrected for availability bias was estimated at 0.608 ind./km^2/13,356 individuals (CV = 0.28) in the northern coastal stratum (Table 7.6). While the 2003/4 and 2019 surveys are not entirely comparable, the new study suggests that the density of franciscanas in the study area is higher than previously thought (Crespo et al., 2020).

Conclusion and recommendations

Surveys to estimate the abundance of franciscanas started in the late 1990s. Early surveys suffered from logistical limitations, including the lack of appropriate survey platforms, which precluded wide spatial coverage of the range of the franciscana in each stock. Additional problems included inexperience of observers and lack of correction factors for known sources of bias in cetacean aerial surveys (e.g., perception bias). These limitations were recognized by the IWC SC in the last review of the status of the franciscana in 2004, when recommendations for improvements were made (IWC, 2005). More recent estimates (from the late 2000s to the present) have used improved survey and analytical methods. In addition, many studies have been carried out to better understand the magnitude of the different sources of bias in aerial surveys for franciscanas, including visibility and group size bias (Zerbini et al., 2011; Sucunza et al., 2018; Sucunza et al., 2020c). While correction factors to account for availability and perception bias have been applied to many franciscana aerial surveys, correction for group size bias has not been widely used. Because negative bias in this parameter can be relatively large (~30%), correction of franciscana abundance estimates to account for group size bias should always be considered.

Estimates of abundance in FMA Ia (Sucunza et al., 2020b) and Ib (Danilewicz et al., 2012; 2020) are likely the most reliable among all FMAs because the surveys were designed to evenly sample the whole habitat of the franciscana in these areas. In addition, the estimates have been corrected for visibility and group size bias. The populations in FMA Ia and FMA Ib are among the smallest franciscana populations. Two estimates are available for FMA Ib (2011, Danilewicz et al., 2012; Danilewicz et al., 2020) and density estimates are not statistically different across the two surveys. The survey regions in the two years overlapped to a great extent, but because abundance was computed for different areas, they are not directly comparable. Attempts to compute abundance for comparable areas should be made if future surveys allow for trends in abundance to be assessed.

There is only one range-wide estimate of abundance for FMA II (Sucunza et al., 2020a) and this estimate is likely biased low because no correction for group size bias has been applied. The second survey in FMA II was ongoing by the time this document was being prepared. Results from this survey could potentially be useful for comparison with estimates from 2008/9. Multiple boat/aerial surveys have been conducted within Babitonga Bay, but because the area covered differed over time, trends in abundance for this region have not yet been obtained. A review of these data could be conducted to evaluate whether estimates of abundance for comparable areas would be applicable for the estimation of trends as more information is obtained. Franciscanas in Babitonga Bay are possibly genetically distinct from franciscanas in FMA II (Cunha et al., 2020). In such a case, this would be the smaller franciscana population, with only around 50 individuals (Cremer and Simões-Lopes, 2008).

While multiple estimates of abundance have been computed within FMA III, surveys have never covered the whole range of that management area. The estimate of density computed for a 1996 survey in RS, Brazil (Secchi et al., 2001) was extrapolated for the range of FMA III (Brasil and Uruguay) and an estimate of abundance was presented. Because the region where sampling occurred was small (less than 1% of the range of the species within this management area), extrapolation to the whole range could lead to severe bias and the estimate was deemed not appropriate as an estimation of the abundance for franciscanas in FMA III (IWC, 2005). Other surveys conducted in RS (Danilewicz et al., 2010; Sucunza et al., 2020d) covered a wider geographic area, but still only a portion of the FMA III range. Therefore, these correspond to partial estimates for this management area. Because surveys in RS in 1996, 2004, and 2014 used different aircraft and/or observers with different experience levels, estimates are not comparable for estimation of trends in abundance, even at more regional scales. No surveys to estimate the abundance of franciscanas in Uruguay have been conducted to date; thus there is no information about the density of this species in a relatively large portion of FMA III.

Estimates of abundance were computed for a portion of FMA IV in Argentina in 1996–2003 (Province of Buenos Aires, Bordino et al., 2004) and in 2004 (Provinces of Buenos Aires and Rio Negro, Crespo et al., 2010). These estimates were examined by the IWC SC during the last review of the status of the franciscana and logistical and analytical limitations were identified (IWC, 2005). These were mostly related to lack of correction for perception bias and extrapolation of density from surveyed areas to un-surveyed areas. Partial results of a survey conducted in 2019 suggest that the abundance of franciscanas off the Province of Buenos Aires are higher than previously thought or may have increased since the 2004 survey (Crespo et al., 2020). Because the 2019 estimates are preliminary and for only a portion of the study area, the continuation of the surveys may provide additional information to better evaluate the abundance of franciscanas in FMA IV.

Estimates of abundance are key to guiding the management and conservation of threatened species. While improvements have been made in surveys to compute a range-wide abundance of franciscanas in many areas (e.g., FMA Ia, Ib, and II), new surveys are needed in areas not yet covered or incompletely covered (FMA III and IV). In particular, surveys in waters off Uruguay should be given the highest priority given that this region corresponds to one of the primary habitats for this species and an area with historically large bycatch rates (e.g., Secchi 2010). Because Uruguay represents just a portion of FMA III, simultaneous surveys should also sample southern Brazil in order to obtain an estimate for the whole management area. Surveys in Argentina (FMA IV) should also receive higher priority because this region has not been fully sampled. In Argentina, attempts should be made to cover not only the coastal (marine) region but also the estuarine waters of the Rio de La Plata which are also believed to be an important habitat for the franciscana. Because comparable surveys have not been repeated in most of the FMAs, empirical estimates of trends have not been computed. Future studies should attempt to standardize sampling methods so that estimates of density and abundance are comparable, and trends are assessed.

Consideration should be given to alternative methods to estimate abundance and trends for franciscana. The use of spatial modeling approaches (Hedley et al., 2004; Miller et al., 2013) to analyze existing survey data is encouraged. Spatial models can be used to assess the relationship between density and environmental covariates and may produce more precise estimates of density and abundance (e.g., Bortolotto et al., 2017). Also, in certain circumstances, spatial models can be used to predict abundance in regions (e.g., FMA IV) where logistical limitations preclude consistent sampling of more remote habitats (Mannocci et al., 2016). Because of the relatively large availability of biological samples (e.g., carcasses from bycatch), genetic methods (e.g., close-kin capture-recapture, Bravington et al. (2016) could also be a cost-effective alternative to estimate the abundance of franciscanas. Moreover, the feasibility of acoustic methods (e.g., from towed arrays or bottom set instruments) (Gibb et al., 2018) for future estimation of seasonal and interannual relative abundance and trends of franciscanas should be considered.

References

Anonymous, 2016. Report of the VIII workshop for the research and conservation of the franciscana (Pontoporia blainvillei), Paper SC/66b/SM05 presented to the IWC Scientific Committee, May 2016. Bled,Slovenia, 26 (unpublished).

Barlow, J., Oliver, C.W., Jackson, T.D., Taylor, B.L., 1988. Harbor porpoise, Phocoena phocoena, abundance estimation for California, Oregon, and Washington: II. Aerial surveys. Fish. Bull. 86, 433–444.

Bordino, P., Albareda, D., Fidalgo, G., 2004. Abundance estimation of Franciscana dolphin Pontoporia blainvillei from boat surveys in Buenos Aires, Argentina, Paper SC/56/SM13 presented to the IWC Scientific Committee, July 2004. Sorrento,Italy, p. 12 (unpublished).

Bordino, P., Thompson, G., Iñíguez, M., 1999. Ecology and behaviour of the franciscana (Pontoporia blainvillei) in Bahía Anegada. Argentina. J. Cetac. Res. Manag. 1 (2), 213–222.

Bortolotto, G.A., Danilewicz, D., Hammond, P.S., Thomas, L., Zerbini, A.N., 2017. Whale distribution in a breeding area: spatial models of habitat use and abundance of western South Atlantic humpback whales. Mar. Ecol. Progr. Series 585, 213–227.

Bravington, M.V., Skaug, H.J., Anderson, E.C., 2016. Close-Kin Mark-Recapture. Statistical Sci. 31 (2), 259–274.

Buckland, S.T., Anderson, D.R., Burnham, K.P., Laake, J.L., Borchers, D.L., Thomas, L., 2001. Introduction to Distance Sampling: Estimating Abundance of Biological Populations. Oxford University Press, New York, NY vi+xv+p. 432.

Cremer, M.J., Simoes-Lopes, P., 2005. The occurrence of Pontoporia blainvillei (Gervais and d'Orbigny) (Cetacea, Pontoporiidae) in an estuarine area in southern Brazil. Revta. Bras. Zool. 22 (3), 717–723.

Cremer, M.J., Simoes-Lopes, P., 2008. Distribution, abundance and density estimates of franciscanas, Pontoporia blainvillei (Cetacea, Pontoporiidae) in Babitonga Bay, southern Brazil. Revta. Bras. Zool. 25, 397–402.

Crespo, E.A., Coscarella, M.A., Arias, M., Sureyo, N., 2020. Abundance estimation of franciscana abundance by aerial surveys in Buenos Aires Pronvince. Argentina Paper SC/68B/ASI/03_rev1 presented to the IWC Scientific Committee, 9.

Crespo, E.A., Harris, G., González, R., 1998. Group size and distributional range of the franciscana, Pontoporia blainvillei. Mar. Mamm. Sci. 14 (4), 845–849.

Crespo, E.A., Pedraza, S.N., Grandi, M.F., Dans, S.L., Garaffo, G.V., 2010. Abundance and distribution of endangered franciscana dolphins in Argentine waters and conservation implication. Mar. Mamm. Sci. 26, 17–35.

Crespo, E.A., Secchi, E.R., Dalla Rosa, L., Kinas, P.G., Danilewicz, D., Bordino, P., 2002. Report of the working group on abundance estimates. LAJAM 1 (Special Issue on the Biology and Conservation of the Franciscana), 65–66.

Cunha, H.A., Dias, C.P., Alvarenga, L.C., Wells, R.S., Cremer, M.J. 2020. Microscale population structure and kinship analysis suggest philopatry of both sexes in franciscanas (Pontoporia blainvillei). Paper SC/68B/SDDNA/04 presented to the IWC Scientific Committee. p. 22.

Cunha, H.A., Medeiros, B.V., Barbosa, L., Cremer, M., Marigo, J., Lailson Brito Jr, J., Azevedo, A., Sole-Cava, A.M., 2014. Population structure of the endangered franciscana dolphin (Pontoporia blainvillei): reassessing management units. PLoS One 9 (1), e85633.

Danilewicz, D., Moreno, I.B., Ott, P.H., Tavares, M., Azevedo, A.M., Secchi, E., Andriolo, A., 2010. Abundance estimation for a threatened population of franciscana dolphins in southern Brazil: uncertainties and conservation considerations. J. Mar. Biol. Assoc. UK 90 (Special issue 8), 1649–1657.

Danilewicz, D., Sucunza, F., Ott, P.H., Ferreira, E., Perez, M.S., Berchieri, N., Alvarez, D., Andriolo, A., Secchi, E.R., Flores, P.A.d.C., Farro, A.P., Martins, A., Zerbini, A.N., 2020. Abundance and distribution of franciscanas (Pontoporia blainvillei) in norhtern Rio de Janeiro (FMA Ib), Brazil. Paper SC/68B/ASI/07_rev1 presented to the IWC Scientific Committee. p. 14.

Danilewicz, D., Zerbini, A.N., Andriolo, A., Secchi, E.R., Sucunza, F., Ferreira, E., Denuncio, P., Flores, P.A.C., 2012. Abundance and distribution of an isolated population of franciscana dolphin (Pontoporia blainvillei) in southeastern Brazil: red alert for FMA I?, Paper SC/64/SM17 presented to the IWC Scientific Committee, June 2012. PanamaCity, 11 (unpublished).

Dias, C.P., Cremer, M.J., Azevedo, A.F., Cunha, H.A, 2013. Baixa diversidade mitocondrial sugere filopatria das toninhas Pontoporia blainvillei (Gervais & D'Orbigny 1844) da Baía da Babitonga, SC, Paper read at II Encontro de Mastozoologia do Estado do Rio de Janeiro, at Rio de Janeiro. Brazil.

Dizon, A.E., Lockyer, C., Perrin, W.F., Demaster, D.P., Seisson, J., 1992. Rethinking the stock concept: a phylogeographic approach. Conserv. Biol. 6, 24–36.

do Amaral, K.B., Danilewicz, D., Zerbini, A., Di Beneditto, A.P., Andriolo, A., Alvares, D.J., Secchi, E., Ferreira, E., Sucunza, F., Borges-Martins, M., de Oliveira Santos, M.C., Cremer, M., Denuncio, P., Ott, P.H., Moreno, I.B, 2018. Reassessment of the franciscana Pontoporia blainvillei (Gervais & d'Orbigny, 1844) distribution and niche characteristics in Brazil. J. Exp. Mar. Biol. Ecol. 508, 1–12. https://doi.org/10.1016/j.jembe.2018.07.010.

Failla, M., Seijas, V.A., Espósito, R., Iñíguez, M.A., 2012. Franciscana dolphins, Pontoporia blainvillei, of the Río Negro Estuary, Patagonia, Argentina. Mar. Biod. Records 5, e102.

Gibb, R., Browning, E., Glover-Kapfer, P., Jones, K.E., 2018. Emerging opportunities and challenges for passive acoustics in ecological assessment and monitoring. Meth. Ecol. Evol. 10 (2), 169–185.

Hammond, P.S., Francis, T.B., Heinemann, D., Long, K.J., Moore, J.E., Punt, A.E., Reeves, R.R., Sepúlveda, M., Sigurðsson, G.M., Siple, M.C., Víkingsson, G., Wade, P.R., Williams, R., Zerbini, A.N., 2021. Estimating the abundance of marine mammal populations. Front. Mar. Sci. 8, 1316.

Hedley, S., Buckland, S.T., Borchers, D., 2004. Spatial distance sampling models. In: Buckland, S.T., Anderson, K.P., Burnham, K.P., Laake, J., Borchers, D., Thomas, L. (Eds.), Advanced Distance Sampling. Oxford University Press, Oxford, p. 595, pp.48-70.

IWC, 2005. Report of the Scientific Committee. Annex L. Report of the Sub-Committee on Small Cetaceans. pp. 307-17.

Laake, J., Borchers, D., 2004. Methods for incomplete detection at distance zero. In: Buckland, S.T., Anderson, K.P., Burnham, K.P., Laake, J., Borchers, D., Thomas, L. (Eds.), Advanced Distance Sampling. Oxford University Press, Oxford, p. 595, pp.108-89.

Laake, J.L., Calambokidis, J., Osmek, S.D., Rugh, D.J., 1997. Probability of detecting harbour porpoise from aerial surveys: Estimating g(0). J. Wild. Manage. 61 (1), 63–75.

Mannocci, L., Roberts, J.J., Miller, D.L., Halpin, P.N., 2016. Extrapolating cetacean densities to quantitatively assess human impacts on populations in the high seas. Conserv. Biol. 31 (3), 601–614.

Marsh, H., Sinclair, D.F., 1989. Correcting for visibility bias in strip transect aerial surveys for aquatic fauna. J. Wild. Manage. 53 (4), 1017–1024.

Miller, D.L., Burt, M.L., Rexstad, E.A., Thomas, L., 2013. Spatial models for distance sampling data: recent developments and future directions. Meth. Ecol. Evol. 4, 001–010.

Ott, P.H., Secchi, E.R., Moreno, I.B., Danilewicz, D., Crespo, E.A., Bordino, P., Ramos, R., Di Beneditto, A.P., Bertozzi, C., Bastida, R., Zanelatto, R., Perez, J.E., Kinas, P.G., 2002. Report of the working group on Fishery Interactions. LAJAM 1(Special Issue on the Biology and Conservation of the Franciscana), 55–64.

Santos, M.C.O., Oshima, J.E.F., Da Silva, E., 2009. Sightings of Franciscana dolphins (Pontoporia blainvillei): the discovery of a population in the Parangua estuarine complex, southern Brazil. Braz. J. Ocean. 57 (1), 57–63.

Sartori, C.M., 2014. Fotoidentificação, abundância e área de vida de toninhas, Pontoporia blainvillei, na baía da Babitonga. Universidade da Região de Joinville, Santa Catarina, Brasil, p. 88.

Sartori, C.M., Paitach, R.L., Cremer, M.J., 2017. Photo-identification of franciscanas (Pontoporia blainvillei) in Babitonga Bay, Santa Catarina State. Brazil. J. Cetac. Res. Manage. 16, 49–55.

Secchi, E.R., 2010. Life history and ecology of franciscana, Pontoporia blainvillei (Cetacea, Pontoporiidae). In: Ruiz-García, M., Shostell, J. (Eds.), Biology, Evolution and Conservation of River Dolphins within South America and Asia. Nova Science, New York, NY, pp. 301–322.

Secchi, E.R., 2010. Review on the threats and conservation status of franciscana, Pontoporia blainvillei (Cetacea, Pontoporiidae). In: Ruiz-García, M., Shostell, J. (Eds.), Biology, Evolution and Conservation of River Dolphins within South America and Asia. Nova Science, New York, NY, pp. 323–339.

Secchi, E.R., Cremer, M.J., Danilewicz, D., Lailson-Brito, J., 2021. A Synthesis of the Ecology, Human-Related Threats and Conservation Perspectives for the Endangered Franciscana Dolphin. Front. Mar. Sci. 8, 896.

Secchi, E.R., Danilewicz, D., Ott, P.H., 2003. Applying the phylogeographic concept to identify franciscana dolphin stocks: implications to meet management objectives. J. Cet. Res. Manage. 5 (1), 61–68.

Secchi, E.R., Ott, P.H., Crespo, E.A., Kinas, P.G., Pedraza, S.N., Bordino, P., 2001. A first estimate of franciscana (Pontoporia blainvillei) abundance off southern Brazil. J. Cet. Res. Manage. 3 (1), 95–100.

Siciliano, S., Di Beneditto, A.P., Ramos, R.M.A., 2002. A toninha, Pontoporia blainvillei (Gervais and d'Orbigny, 1844) (Mammalia, Cetacea, Pontoporiidae), nos estados do Rio de Janeiro e Espírito Santo, costa sudeste do Brasil: caracterizaçoes dos habitats e fatores de isolamento das populaçoes. Bol. Mus. Nac. Zool. 476, 1–15.

Sucunza, F., Danilewicz, D., Andriolo, A., Azevedo, A.F., Secchi, E.R., Zerbini, A.N., 2020a. Distribution, habitat use, and abundance of the endangered franciscana in southeastern and southern Brazil. Mar. Mamm. Sci. 36 (2), 421–435.

Sucunza, F., Danilewicz, D., Cremer, M., Andriolo, A., Zerbini, A.N., 2018. Refining estimates of availability bias to improve assessments of the conservation status of an endangered dolphin. PLoS One 13 (3), e0194213.

Sucunza, F., Danilewicz, D., Ott, P.H., Neves, M., Berchieri, N., Farro, A.P., Martins, A., Zerbini, A.N., 2020b. Population size and IUCN Red Listing of the isolated northern population of the franciscana (Pontoporia blainvillei). Paper SC/68B/ASI/05 presented to the IWC Scientific Committee. p. 24.

Sucunza, F., Danilewicz, D., Secchi, E.R., Andriolo, A., Cremer, M., Flores, P.A.d.C., Ferreira, E., Alves, L.C.P.d.S., Castro, F.R., Pretto, D., Sartori, C.M., Schulze, B., Denuncio, P., Perez, M.S., Ott, P.H., Zerbini, A.N. 2020c. Assessing bias in aerial surveys for threatened cetaceans: results from experiments conducted with the franciscana (Pontoporia blainvillei). Paper SC/68B/ASI/04_rev1 presented to the IWC Scientific Committee. p. 32.

Sucunza, F., Danilewicz, D., Cremer, M., Ferreira, E., Denuncio, P., 2020d. Abundance of the endangered franciscana in southern Brazil. Paper SC/68B/ASI/06 presented to the IWC Scientific Committee, April-May 2020. p.17.

Weyn, M., 2016. Population density of Pontoporia blainvillei in an estuarine area at a World Heritage Site, Southern Brazil. Universidade Federal do Paraná, Brazil, p. 36.

Zerbini, A.N., Danilewicz, D., Secchi, E.R., Andriolo, A., Cremer, M., Flores, P.A.C., Ferreira, E., Alves, L.C.P.d.S., Sucunza, F., de Castro, F.R., Pretto, D., Sartori, C.M., Schulze, B., Denuncio, P., Laake, J.L., 2011. Assessing bias in abundance estimates from aerial surveys to improve conservation of threatened franciscana dolphins: preliminary results from a survey conducted off southern Brazil, Paper SC/63/SM9 presented to the IWC Scientific Committee. Tromsø, Norway, 13 (unpublished).

Zerbini, A.N., Secchi, E., Crespo, E., Danilewicz, D., Reeves, R., 2017. Pontoporia blainvillei, Franciscana. The IUCN Red List of Threatened Species 2017: e.T17978A50371075. p. 16.

Zerbini, A.N., Secchi, E.R., Danilewicz, D., Andriolo, A., Laake, J.L., Azevedo, A., 2010. Abundance and distribution of the franciscana (Pontoporia blainvillei) in the Franciscana Management Area II (southeastern and southern Brazil), Paper SC/62/SM7 presented to the IWC Scientific Committee, June 2010. Agadir, Morocco, 14 (unpublished).

CHAPTER 8

Behavior and movement patterns of franciscana dolphins

Marta Jussara Cremer[a], Daniel Danilewicz[b], Federico Sucunza[b], Leonardo G. Berninsone[c], Renan Lopes Paitach[a], Randall S. Wells[d]
[a]Laboratório de Ecologia e Conservação de Tetrápodes Marinhos e Costeiros, Universidade da Região de Joinville, São Francisco do Sul, Santa Catarina, Brazil
[b]Grupo de Estudos de Mamíferos Aquáticos do Rio Grande do Sul, Torres, Rio Grande do Sul, Brazil
[c]Fundación AquaMarina, Centro de Estudios en Ciencias Marinas, Pinamar, Buenos Aires, Argentina
[d]Chicago Zoological Society's Sarasota Dolphin Research Program, % Mote Marine Laboratory, Sarasota, Florida, United States

Introduction

Behavior patterns expressed by animals are strongly influenced by ecological features, and marine mammals are no exception to this rule (Wells et al., 1999). The way animals respond to variations in the environment, in turn, depends on the morphological and physiological aspects of each species, determined by their genetics. However, unlike the species of mammals that maintain ties with the terrestrial environment, studies on the behavior of cetaceans are especially challenging. Cetaceans spend their entire lives at sea, and most of their time underwater, which limits our ability to observe and recognize individuals, making studies of their behavior exceptionally difficult.

In general, coastal species or populations of cetaceans tend to be better known than pelagic species, as long-term studies that include aspects related to behavior are more viable in environments close to land (Mann and Karniski, 2017). Despite being a very coastal species, knowledge about franciscana dolphin (*Pontoporia blainvillei*) behavior is very scarce. Franciscanas live in the coastal waters of the western South Atlantic, between northern Espírito Santo, Brazil (18°25′S–30°42′W) (Siciliano, 1994), and northern Patagonia, Argentina (42°35′S–64°48′W) (Crespo et al., 1998), mainly in waters beyond the surf zone and out to the 30 m

The Franciscana Dolphin
DOI: https://doi.org/10.1016/B978-0-323-90974-7.00014-8

isobath (Amaral et al., 2018; Crespo et al., 2010; Danilewicz et al., 2010; Secchi et al., 1997). The small size of franciscanas and their grayish brown color that matches the surrounding waters are important factors that explain why the existing knowledge about the behavior of the species is scarce, but it is probable that the discrete behavior on the surface is the main factor (Bordino et al., 1999; Cremer and Simões-Lopes, 2005). The simple detection of the species on the sea is a challenge in itself. The fact that the species' habitat is also characterized by having turbid waters in most of its range also contributes to making franciscanas' detection and/or observation very difficult.

Most aspects related to franciscanas' behavior have been addressed based on inferences made from biological parameters, from stomach content analysis, for example (Bassoi et al., 2020; Rosas and Monteiro-Filho, 2002), or from opportunistic observations. More recently, the use of novel methodologies and equipment, such as satellite-linked transmitters or acoustic recording devices, has contributed to obtaining more detailed data on movements, distribution, and habitat use patterns (Paitach, 2021; Wells et al., 2022). Moreover, recent observation efforts utilizing helicopters and drones as sampling platforms have added some new glimpses of the behavioral patterns of this elusive species (Sucunza et al., 2018).

This chapter aims to gather information available in the literature as well as unpublished data involving different aspects related to the behavior of franciscanas, providing a synthesis of existing knowledge.

Group composition and behavior

Franciscana is a social species, with individuals observed in close proximity of each other, moving in the same direction and in an apparent association. They are generally observed in groups of 2–5 animals, reaching up to 10 individuals (Bordino et al., 1999; Cremer and Simões-Lopes, 2005; Crespo et al., 1998; Danilewicz et al. 2009; Sucunza et al. 2018) (Fig. 8.1). Nevertheless, these groups frequently associate with others forming aggregations of up to 20–50 animals (e.g., Cremer and Simões-Lopes, 2005). Larger groups have been proposed to be related to aggregations of smaller social units and should be associated with foraging activities at least in winter (Bordino et al., 1999), and in Brazil are also observed in spring and summer.

It is interesting to note that these large aggregations are not tight or in rigid formations such as echelons, as observed in many delphinids, but rather sparser. Fig. 8.2 shows two groups containing mother–calf pairs using

Fig. 8.1 ***Group of franciscanas seen from a boat.*** *Credit: Projeto Toninhas/UNIVILLE.*

Fig. 8.2 ***Two franciscana groups containing mother, calf, and one adult escort, not directly interacting with each other but using the same area in Ubatuba, Brazil.*** These two groups belonged to an aggregation of about 20 individuals who were using an area of 1 km^2. *Credit: Diego Alvares/Gemars.*

the same area. Solitary animals seem to be rare, although their detectability should be naturally lower. Although the risk of predation is the most common reason for animals to live in groups (Connor, 2000), in the case of franciscanas the ability to capture prey may be the main factor. In Baía Babitonga, Brazil, a predator-free area, many groups in close proximity may form large aggregations of about 40 dolphins in a relatively small area where feeding is frequently observed (Cremer and Simões-Lopes, 2005; Cremer et al., 2018).

Genetic analyses of individuals incidentally caught in the same net or stranded within less than 1 km on the same day in Uruguay, suggest natal group philopatry (Costa-Urrutia et al., 2012), which can be considered rare in mammals (Connor, 2000). The few exceptions in cetaceans include killer whales (*Orcinus orca*) (Baird, 2000) and pilot whales (*Globicephala melas*) (Amos et al., 1993), and data suggest that franciscanas can be included in this restricted group.

It is possible that the basic social unit for franciscanas could be the family group. Data indicate that some adult males stay in their social unit beyond the reproductive period and the existence of half-sibling relationships within groups (Costa-Urrutia et al., 2012; Valsecchi and Zanelatto, 2003). Valsecchi and Zanelatto (2003) suggested that the permanence of the adult male in the group could be related to a mate-guarding behavior, associated with a low probability to find reproductive partners in the population. However, considering that probably the franciscana was an abundant species throughout its distribution, it is possible that the hypothesis proposed by Costa-Urrutia et al. (2012) would be more plausible. These authors suggest that the adult male in the group could provide extra parental care, increasing the reproductive success of the species. In fact, on three occasions during aerial surveys in Ubatuba, Brazil, small calves were recorded in very close association with an adult (side by side) swimming away from this animal and starting a new close association with another adult of the same group; however it should be noted that the sexes of the adults were not known.

Information on mother–calf pairs from Ubatuba indicates that in most cases (93%) there is a third or fourth animal escorting them, at least during the first three months after birth. These escorts usually swim behind the mother and its calf, as the mother assumes the front position in 75% of the records. The relationship of the escorts within the mother–calf pair still is not understood, although preliminary analysis on total body size measurement and rostrum size/shape from aerial photographs suggest that in some groups containing four animals, the two "escorts" are a juvenile and an adult

Fig. 8.3 ***Franciscana group containing four animals: a mother and calf and two escorts.*** These two escorts may be a juvenile and an adult franciscana. *Credit: Federico Sucunza/ Gemars.*

franciscana (Fig. 8.3). A comprehensive assessment of the sex and maturity status of these "escorts" in mother–calf groups, as well as their genetic relationship, is key for clarifying the social structure of the franciscana. It is fair to speculate that the juvenile might be the last calf of the same female, a scenario that matches the evidence that half-siblings can be by-caught in the same gillnet (Valsecchi and Zanellato, 2003). It is interesting to note that in those situations, where a mother–calf pair is accompanied by a juvenile and an adult, the latter invariably takes the last position in the group and performs broader and more investigative movements in semicircles while the mother–calf pair swims in a more linear pattern. Groups containing two mother–calf pairs were never observed in Ubatuba, at least in the first three months after the birth period initiates (October–December) (Silva et al., 2020).

Similarly to what is observed for other species of mammals, franciscana calves tend to remain close to their mother in the first months of life, which is related to the need for feeding, protection, and learning processes (Fig. 8.4). The analysis of stomach contents of stranded calves indicate that lactation can last up to eight months (Denuncio et al., 2013; Ramos et al., 2000), but the presence of food items in the stomach of younger calves indicates that weaning is gradual (Di Beneditto and Ramos, 2001). However, considering

Fig. 8.4 ***Franciscana calf swimming close to her likely mother.*** *Credit: Projeto Toninhas/ UNIVILLE.*

the hypothesis that males contribute to parental care, it is possible that the offspring will not remain only at the mother's side throughout this period.

Two records of epimeletic behavior were reported in the literature for franciscanas. The first case was described in Uruguayan waters by Pilleri (1984) and the second one was described in South Brazil by Cremer et al. (2006). In both cases, the behavior was directed to newborns, and the records were made from the analysis of newly dead animals found stranded on the beach. The hypothesis presented by both authors was that there was an attempt to remove the calf from the net by an adult, producing the large number of marks registered on the calves' bodies, which apparently were produced by the teeth of an adult. Di Beneditto et al. (2001) suggest that tooth marks on the body of specimens incidentally caught in fishing nets in Southeastern Brazil are related to rescue attempts made by one or more members of the group.

Nicks in the dorsal fin of small cetaceans are typically acquired over time and are commonly produced by bites from conspecifics (Würsig and Jefferson, 1990). Nicks are the main marks used for individual recognition and were present in around 40% of the individuals in Baía Babitonga population (Sartori et al., 2017) (Fig. 8.5). They are very small, proportional

Fig. 8.5 ***Group of franciscanas in Baía Babitonga; on the left is an individual with nicks on the dorsal fin and on the right one individual without marks.*** *Credit: Projeto Toninhas/ UNIVILLE.*

to the small teeth characteristic of the species, and suggest that social interaction includes bites from conspecifics, although we do not know in what behavioral context they occur. Scratches, another type of mark very common in small cetaceans and related to social interactions, are rare in franciscanas.

Mating system

There is no direct evidence on the mating system for most cetacean species in the field due to intrinsic difficulties related to the identification of the sex of individuals, the observation of copulatory behavior, or even the availability of samples from a large number of individuals within a population, that allowed assessment of the degree of kinship, and these limitations apply to franciscanas. However, many works suggest that franciscanas have a single-male breeding system (serial monogamy) based on indirect evidence (Costa-Urrutia et al., 2012; Danilewicz et al., 2004; Mendez et al., 2010; Panebianco et al., 2012; Wells et al., 2013), which can be considered as uncommon in cetaceans. They are the reversed length sexual dimorphism, the absence of secondary sexual characteristics, and the small size of the testes, which makes sperm competition unlikely (Danilewicz et al., 2004; Rosas and Monteiro-Filho, 2002; Secchi et al., 2021). In Baía Babitonga, Brazil, one adult male and female were captured and tagged in the same group and photographed together in 73% of their subsequent records, for at least eight months (Wells et al., 2022). In Argentina, tagged, unrelated adult males and females were together consistently or repeatedly over several of months (Wells et al. 2013).

Surface behavior

Franciscana is a discreet species, with rare aerial events and reduced dorsal exposure during emergence (Cremer and Simões-Lopes, 2005). In general, the long rostrum first projects out of the water, followed quickly by the back which moves with only a slight curvature (Bordino et al., 1999; Cremer and Simões-Lopes, 2005). The dorsal fin barely cuts the waterline, emerging and submerging almost vertically (Fig. 8.6). Mean surface time was estimated at 1.42 s (95% CI: 0.91–1.93) in Baía Babitonga (Actis et al., 2018) and 1.2 s in Baia Anegada (Bordino et al., 1999). Synchronous breathing may be a useful proxy for studying cetacean behavior. After evaluating the influence of anthropogenic and biological variables on this behavioral display in the dolphin populations, Actis et al. (2018) concluded that only the distance between animals within a pair was positively related to breathing synchrony for franciscana dolphins.

Only three aerial behavior events were observed: forward leap, tail slap, and spyhop (Cremer and Simões-Lopes, 2005). Forward leap and tail slap were observed during fishing activity. Spyhop was characterized by the accentuated exposure of the head out of the water, with the animal looking toward the boat, indicating curiosity (Fig. 8.7).

Dive behavior

Available information on franciscana diving behavior has come from a variety of sources, including shore- and helicopter-based observations, recovery of carcasses in fishing nets, and satellite-linked, time-depth-recording tags. Franciscana dives tend to be brief compared to those of other small cetaceans. Bordino et al. (1999) observed franciscanas from shore in Bahía San Blas, Argentina, and reported a mean dive time of 21.7 s. Observers in a helicopter measured franciscana dive intervals, considering the period when no members of a group were visible, and found significantly different mean values for franciscanas in Baía Babitonga (39.77 s) and Ubatuba (77.26 s) (Sucunza et al., 2018, 2021). Nevertheless, it is interesting to note

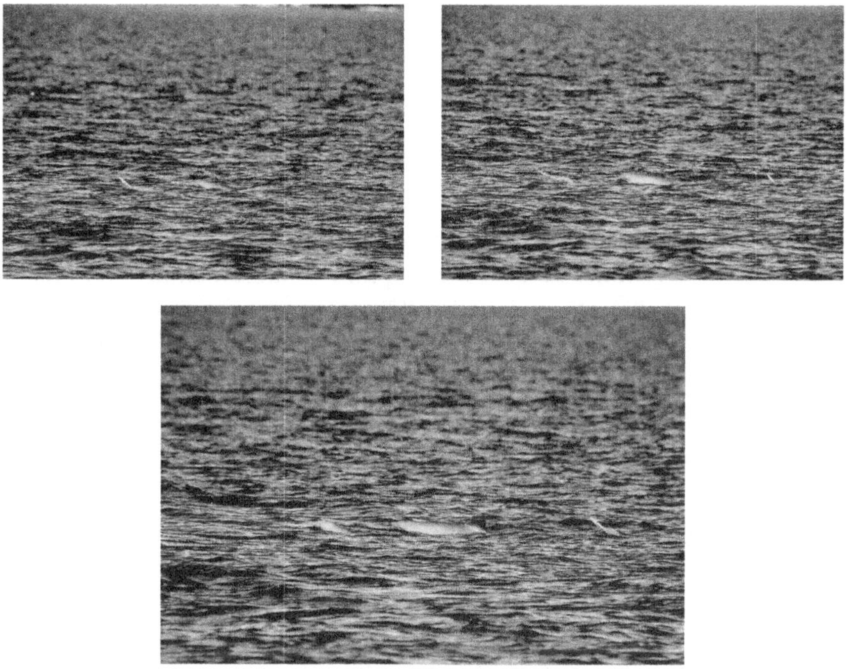

Fig. 8.6 ***A sequence of franciscana surface behavior.*** *Credit: Projeto Toninhas/UNIVILLE.*

Fig. 8.7 ***Franciscana looking toward the boat during research activities.*** *Credit: Projeto Toninhas/UNIVILLE.*

that despite the difference in the mean dive time between the areas, the proportion of time that a group remained close to the surface and available to be seen from the helicopter did not vary significantly (Sucunza et al., 2021). This proportion of time was observed to positively increase with the size of the group (Sucunza et al., 2018). These observations were designed to estimate availability bias for use as a correction factor for franciscana aerial surveys, rather than for defining true surface-dive cycles and may overestimate dive times because the dolphins could sometimes continue to be seen when below the surface. In addition, these observations reflect the diving behavior of the group as the unit rather than the individuals.

Dive patterns have been described for different populations of franciscana, with some similarities among them. It is interesting to note that Bordino et al. (1999) reported that franciscanas exhibited sequences of three to four short dives, followed by a long dive, in Argentina. During the helicopter surveys in Brazil (Sucunza et al., 2018, 2021), franciscanas in a group were visible near the surface (but not above it) between sequential short dives followed by a long dive, when all individuals of the group were not visible, corroborating with the pattern described by Bordino et al. (1999). In addition, the proportion of time that a group of franciscanas remains available to be seen from a helicopter was similar between different regions of Brazil (i.e., Baía Babitonga and Ubatuba), and indicate a similar surface-dive pattern of franciscana groups.

Franciscana habitat consists of a relatively narrow strip of shallow coastal and inshore waters, and at all of the locations where franciscana dive depths have been reported, the animals have been shown to use the entire water column, from the surface to the seafloor. In southern Brazil, fixed bottom-set gillnets caught males and females of all age classes in depths of up to 35 m, assuming that the dolphins were not caught in the nets closer to the surface during deployment or retrieval (Danilewicz et al., 2009). Tracking and time-at-depth data from satellite-linked tags used at two sites in Argentina and one in Brazil showed that while franciscanas spent nearly all of their time at depths less than 10 m, and mostly less than five m, maximum dives were sufficient to reach the seafloor in the deepest portions of their home ranges (Wells et al., 2022). Franciscanas occasionally dove to 10–15 m in or near Bahía Samborombón, Argentina, to 30–35 m in or near Bahía San Blas, Argentina, and to 15–20 m in Baía Babitonga, Brazil. The use of the entire water column is also evidenced considering the behavior of prey species found in franciscana stomach contents (Bassoi et al., 2020; Danilewicz et al., 2002). Research in Baía Babitonga, Brazil, has shown that the dolphins feed on bottom-dwelling fish, with the epi-benthic rake star-drum (*Stellifer rastrifer*) reaching nearly 70% of relative importance in their diet (Cremer et al., 2012).

Feeding behavior

Knowledge about the feeding behavior of the species comes mainly from studies of stomach contents, through the analysis of items resistant to digestion, and reveals a strong opportunistic tendency. Franciscanas prey upon neritic species, both pelagic and demersal (Bassoi et al., 2020; Bittar and Di Beneditto, 2009; Cremer et al., 2012; Rodriguez et al., 2002; Rupil et al., 2019), and mainly on juveniles or small-sized fish and cephalopod species (~10.0–15.0 cm length) (Danilewicz et al., 2002; Di Beneditto and Ramos, 2001) (Fig. 8.8).

Feeding behavior can be influenced by characteristics such as sex and age of individuals. In São Paulo State, southeastern Brazil, cephalopods are consumed more by females than males, and authors suggest an ontogenetic diet shift (Henning et al., 2017). The groups exhibit behaviors that indicate some level of cooperation in capturing prey (Fig. 8.9). Off the southern Brazilian coast, dietary studies also suggest that the individuals occupy small ranges for long periods of time (Bassoi et al., 2020).

Fig. 8.8 ***Franciscana capturing a small fish.*** *Credit: Projeto Toninhas/UNIVILLE.*

Fig. 8.9 ***A group of three franciscanas foraging in Ubatuba: fast circles around their own axis and partial leaps are often observed when franciscanas are catching fish on the surface.*** *Credit: Daniel Danilewicz/Gemars.*

Calves began to ingest solid food very early, within around 40–60 days, although lactation can be prolonged, lasting for at least 7–9 months (Denuncio et al., 2013; Kasuya and Brownell, 1979; Ramos et al., 2000), which suggests gradual weaning. Di Beneditto and Ramos (2001) recorded food items in a 79 cm 2-month male. Some studies indicate an ontogenetic diet shift (Henning et al., 2017), and the main prey captured by the calves are very small fish and shrimps, which are probably easier to capture (e.g., Danilewicz et al., 2002; Denuncio et al., 2013; Rodríguez et al., 2002).

It is widely recognized that during foraging, dolphins considerably increase the emission rate of echolocation click sounds as they approach prey until they reach extremely fast rates (so-called "feeding buzzes") just before capture (Wisniewska et al., 2012, 2014). The recognition of the presence of feeding buzzes from passive acoustic monitoring (PAM) devices is therefore a cue that contributes to the study of franciscana feeding behavior, which due to its elusive characteristics becomes very difficult to be observed and quantified visually. In a study carried out using PAM devices, a higher proportion of feeding buzzes was observed in relation to the other acoustic signals detected, both in estuarine (68%) and open sea (58%) environments (Paitach, 2021), than reported in studies with *Phocoena phocoena* (27%) and *Tursiops truncatus* (4%) (Nuuttila et al., 2013). This indicates that the feeding rate of franciscanas is extremely high, suggesting that they have high metabolic demand, or that it takes more effort to meet metabolic needs with such small prey items, which makes them more vulnerable to anthropogenic disturbances that reduce the availability of their prey, such as overfishing and habitat degradation (Paitach, 2021).

Interspecific interactions

There are no records of cooperative interactions between franciscanas and other species of marine mammals. However, similarly to what is recorded for other cetacean species, interactions between seabirds and franciscanas are common and are related to fishing activities in Baía Babitonga, Brazil. In this area, small groups of the neotropical cormorant (*Phalacrocorax brasilianus*) and the brown booby (*Sula leucogaster*) have been recorded interacting with franciscana groups (Cremer and Simões-Lopes, 2005). Terns of the genus *Thalasseus* show the highest number of interactions recorded in the area, involving groups of 1–15 birds, while the groups of franciscanas included 2–17 individuals (Cremer and Simões-Lopes, 2005) (Fig. 8.10). The terns flew over the franciscana groups, diving frequently toward the water and performing an intense vocalization, and the initiative to interact was always

Fig. 8.10 ***Interaction between a tern (Thalasseus sp.) and a franciscana group during fishing.*** *Credit: Projeto Toninhas/UNIVILLE.*

displayed by the terns. In Baía Babitonga, the fishing behavior of the terns is so characteristic that they can be used as an indicator for the location of franciscana groups. Over 10 years of franciscana observations in this area, the presence of interacting terns was recorded in 4.3%–46.7% of the groups each year. Beyond Baía Babitonga, only in Argentina there are records of interactions with seabirds, in this case, three species of the genus *Larus* (Bordino et al., 1999). In all the cases, the interactions seem to be commensalistic.

Despite living in direct sympatry with the Guiana dolphin (*Sotalia guianensis*) in much of its distribution, mixed groups have never been documented over 20 years of research in Baía Babitonga (Cremer et al., 2018), where resident populations of both species live. Although no direct interactions have been recorded, studies carried out using Passive Acoustic Monitoring - PAM shows that the intensity of occurrence of Guiana dolphins is one of the main factors affecting the habitat use of franciscanas in Baía Babitonga; they avoid areas where Guiana dolphins are present (Paitach, 2021).

Predators

The main predators of franciscanas are sharks and killer whales (*Orcinus orca*), but there is no information on the frequency of these events, not even on what this represents at the population level. Information about predation was mainly obtained by analyzing the stomach contents of sharks (*Galeocerdo cuvieri*, *Notorynchus cepedianus*, and *Sphyrna* spp.) (Di Beneditto, 2004; Monzón et al, 1994; Praderi, 1985) and killer whales (Ott and Danilewicz, 1998; Padula et al., 2022). Only one predation event by a killer whale has been documented on the southeast coast of Brazil (Santos and Netto, 2005). The presence of bite marks on the body of franciscanas found dead on the beaches is rare. This may be associated with the small size of the species, which possibly does not survive the attack of a predator, with a large part of the body being ingested.

There is little evidence that sharks influence the habitat use of franciscanas. Lucifora et al (2005) suggested that the pattern of habitat use of franciscanas in Anegada Bay might be affected by *N. cepedianus,* especially during summer/spring season when sharks prey on franciscana calves. In South Brazil, franciscanas of different ages, sizes, sex, and reproductive class are homogeneously distributed according to water depth (Danilewicz et al., 2009). But Failla et al. (2012) suggest that the presence of killer whales in the Rio Negro Estuary, in Argentina, could be responsible for the removal of franciscanas from certain areas. It has been suggested that franciscanas may have modified the frequency of their sound to reduce the overlap with the hearing capability of killer whales (Morisaka and Connor, 2007), producing narrow-band high-frequency (NBHF) stereotyped clicks (Madsen et al., 2005). This is known as the "acoustic crypsis" hypothesis, which suggests that predation pressure is the evolutionary driver of this characteristic, considering that the NBHF clicks are above the auditory range of potential predators such as the killer whales *Orcinus orca* (Szymanski et al., 1999). However, this hypothesis is still very controversial.

Reaction to boats and aircraft

Franciscanas have shown minimal behavioral reactions to planes and helicopters. During experiments in Babitonga Bay, a Robinson R44 helicopter hovered at 500 ft over franciscana groups during surface dive cycles, and in only 2% of all recorded cycles, a response (e.g., an abrupt dive) was observed (Sucunza et al., 2018). A clear response of individuals to the helicopter's shadow was observed during surveys at 500 ft with both a Robinson R44 and a Bell Jet Ranger in Ubatuba, with all individuals of a group diving abruptly when the shadow just passed over it at different distances between the helicopter and the group.

Although no controlled experiments on franciscana reactions to drones were carried out, no apparent reactions have been detected in Ubatuba,

Brazil, where sampling occurs from 5 to 50 m height flights. In Baía Babitonga, no specific analyzes were carried out in this regard either, but during drone overflights to estimate the health status of animals by photogrammetry and obtaining outreach videos, occasional reactions were noted in situations with the drone below 10m and no reaction above that height. The observed reactions were swimming acceleration and deep diving, sometimes synchronously by the group members.

Unlike what is observed for many species of dolphins, franciscanas do not swim at the bow of boats and do not usually get close. However, they can remain in areas with high traffic by small boats, such as fishing boats and speedboats, which does not represent the absence of impacts for the species. In fact, this is a very characteristic condition in the coastal zone, where there is greater traffic of small boats. On the other hand, franciscanas totally avoid areas with the movement of ships and large vessels, which is possibly related to the intense and constant noise generated by them, as is the case of harbor areas (Holz, 2014).

Bordino et al. (1999) suggest the use of sailboats for behavior studies on the species, as they cause minimal changes in behavior. However, this type of vessel has maneuvering restrictions, which can make it difficult to keep track of the animals. In Baía Babitonga, an enclosed area with large tidal flats, the use of sailboats is hardly viable, and studies have been carried out for several years using small boats with outboard motors (Cremer et al., 2018).

Movement patterns and site fidelity

Monitoring the movement of tagged individuals

Franciscanas at all three sites in Argentina and Brazil where satellite-linked tags were deployed during 2006–2013 demonstrated very localized movements, with home range size varying from site to site (Wells et al., 2022). Small tags (SPOT or SPLASH, Wildlife Computers, Redmond, WA, USA) attached to the dorsal fin during brief catch-and-release efforts transmitted location and other data for 7-258 days. Seven dolphins were tracked in Bahía Samborombón, Argentina, in 2006 and 2010, eight were tracked in Bahía San Blas, Argentina, in 2007 and 2008, and six were tracked in Baía Babitonga, Brazil, in 2011 and 2013. None of the dolphins ranged far from their tagging site.

The dolphins in Bahía Samborombón exhibited two ranging patterns, overlapping at the sandbanks at the mouth of the bay. Some of the dolphins used the shallow waters of the interior of southwestern Bahía Samborombón more than the open coastal waters to the east and south of the bay, with the sandbanks and deeper waters (>5 m deep) immediately offshore of the bay forming the eastern extent of their ranges, while others used the coastal, somewhat deeper (5–10 m deep) waters to the east and south of the bay to a greater extent, offshore of Cabo San Antonio and nearby sandbanks. The mean home range size for franciscanas with more than 25 locations was 132.9 km^2 (± 102.8 km^2 *SD*, range = 51–312 km^2, n = 5), with a mean core area of 23.4 km^2 (± 14.0 km^2 *SD*, range = 12–46 km^2, n = 5) (Wells et al., 2022).

The franciscanas tagged in Bahía San Blas occupied ranges that were nearly three times greater in size and included different habitats as compared to those in Bahía Samborombón, and they also exhibited two overlapping ranging patterns. Some of the dolphins remained inside the bay in shallows and channels, while others ranged through shallower interior waters as well as moving through deep channels and coastal waters to the north and south of Bahía San Blas. Mean home range size for dolphins with more than 25 locations was 463.5 km^2 (± 344.1 km^2 SD, range = 49–1013 km^2, n = 7), with a mean core area of 79.3 km^2 (± 47.8 km^2 SD, range = 16–151 km^2, n = 7) (Wells et al., 2022).

At the other end of the spectrum, the franciscanas of Baía Babitonga had unusually small ranges relative to those of other franciscanas and other small cetaceans. They exhibited a single-ranging pattern that involved using only a small portion of the bay in an area that included shallow flats and adjacent channels. Mean home range size for franciscanas with more than 25 locations was 6.4 km^2 (± 0.7 km^2 SD, range = 5.5–7.0 km^2, n = 4), with a mean core area of 1.6 km^2 (± 0.4 km^2 SD, range = 1.1–2.0 km^2, n = 4) (Wells et al., 2022). The small home range of the franciscanas of this population is strongly related to a high site fidelity, according to information obtained through photo-identification (Cremer et al., 2018; Sartori et al., 2017). Despite the challenge of obtaining high-quality photographs of the individuals' dorsal fin, 23 franciscanas have already been cataloged, representing about 40% of this population.

At all three sites, the dolphins moved widely through their home ranges on a daily basis, without any indication of migratory patterns. However, as described elsewhere in this chapter, movements within their home ranges appeared to be related to tidal fluxes. At each tagging site, individuals displayed a high degree of range overlap; however, there were considerable differences in ranging behavior among the three sites. Possible influences on ranging pattern variation include (1) prey availability, (2) availability of shallow habitat preferred by the species, (3) coastal development, (4) predator presence, and (5) possible competition with other dolphins.

In Bahía Samborombón and Baía Babitonga, artisanal fishing efforts overlap extensively with the franciscana ranges, leading to dolphin losses in gillnets. Extensive areas of shallow waters including sand and mudflats exist at all three sites, but in Baía Babitonga, the development of harbor facilities for large ships appears to have limited habitat available to franciscanas to a small portion of the bay. Predators may be a more important factor in the more open waters of Bahía San Blas and Bahía Samborombón, and extensive use of shallow flats may provide some protection from exposure to large sharks and killer whales (Bastida and Rodriguez, 2005; Lucifora, 2003; Lucifora et al., 2002, 2005), while dolphin predators are no longer seen in Baía Babitonga (Cremer, 2015; Gerhardinger et al., 2020).

Movements relative to tides

Tidal cycles play a central role in the dynamics of coastal environments, particularly in bays and estuaries. Franciscana movements relative to tides were analyzed in three bays in which the species occurs, with similar patterns observed, but with different intensities. The first observations were made from fixed points in Anegada/San Blas Bay (Argentina), and it was noted that the movement of franciscanas is not by chance, with an association with the direction and speed of tidal currents, predominating movements with the currents (Bordino, 2002). Later, in the study conducted by Wells et al. (2022), which tracked 15 franciscanas with satellite-linked transmitters in the bays of Samborombón and San Blas, it was verified that in fact, the animals moved further into the bays and north during flood tides and further east and south, closer to the mouth of the bays during ebb tides. In Bahia San Blas, where the daily excursion of the tides is greater, the movements were more evident (Wells et al., 2022).

In Baía Babitonga, visual records made from boats, covering predefined routes (surveys) over 10 years (2004–2014), allowed the identification of franciscana distribution patterns that show movements to be a function of the tidal cycle (Paitach et al., 2017). During ebb tide they are subtly distributed more toward the only inlet channel of the bay (although still far from the connection with the open sea), and in the opposite direction during flood tide, thus following the tidal flow. During ebb tide, franciscanas also tend to remain closer to the central islands, characterized by rocky shores, probably following the movement of their prey (Paitach et al., 2017). However, when analyzing the effect of the tides on the movement of six individuals tracked by satellite in Baía Babitonga, compared to Argentinean bays, there were no statistically significant differences (Wells et al., 2022). According to the authors, this result is possibly related to the spatial scale of distribution of the Brazilian population, which is much more restricted than the others.

In general, franciscanas apparently tend to move predominantly in the same direction of the tidal current. It is possible that this is associated with an energy-saving strategy, allowing a greater speed of movement with less effort, or even as a passive drag effect on animals by the current flow. Also, such movements are probably related in some way to hunting strategies, where the tidal cycle influences prey availability and capture potential, affecting franciscanas habitat use (Bordino, 2002; Paitach et al., 2017; Wells et al., 2022). Movements with the current are more commonly reported for other dolphin species, although countercurrent movements also occur (e.g., Fury and Harrison, 2011; Lin et al., 2013; Mendes et al., 2002; Shane, 1990). There are still cases in which the movements are apparently independent of the tidal state, as is the case of the Guiana dolphin sympatric of the franciscana population in Baía Babitonga (Paitach et al., 2017). The effect of tides on dolphin movement can vary among species and populations and understanding such effects for franciscanas may be useful to understand their susceptibility to incidental captures, contributing to planning management strategies.

Diel patterns

Diel patterns of behavior and habitat use of franciscanas were studied in only two locations, in the State of Santa Catarina, Brazil, which represent distinct habitats of occurrence of the species, one estuarine and the other in the open sea (Paitach, 2021). Through PAM, a diel pattern in the foraging/feeding behavior was observed. In both environments, a greater intensity of franciscana foraging cues was verified at night, demonstrating that this is the period when the individuals are most intensely involved with foraging/feeding, although they also do so at a high frequency during the daytime (Paitach, 2021). In the coast of São Paulo State, Brazil, Barcellos and Santos (2021) also observed a higher proportion of high repetition franciscanas echolocation signals, which is an indicator of foraging/feeding, but they did not carry out specific analysis in this regard. There are no more published data on diel echolocation patterns in the franciscana. Higher foraging/feeding frequency during the night has already been observed for other dolphin species (e.g., Carlström 2005; Todd et al., 2009).

However, this is not a rule and different populations may have patterns entirely opposite, probably associated with prey activity (Elwen et al., 2006; Leeney et al. 2011).

Diel movements in search of areas with high prey availability can be stereotyped with daily repetition. In Babitonga Bay, a spatio-temporal systematic study of franciscanas was carried out (Paitach, 2021). There was a clear cyclic pattern in the occurrence of franciscanas across the study area. In the areas where their occurrence was more intense (core area), they were more likely to occur during the morning. The rest of the day, the population was more dispersed to other areas, with seasonal variations from the areas used in other periods. It is interesting that the core area of the population has a predominant substrate of sand, a relationship that had already been reported in the past studies from visual observation (Paitach et al., 2017). However, when we look at the foraging/feeding behavior at dusk and dawn, there was an increase in the use of adjacent muddy bottom areas, demonstrating that these areas are also important and this has implications for population management. A very similar result was observed for the harbor porpoise in the Moray Firth, Scotland, where only sandy banks were identified as important feeding areas, but when the diel cycles were included in that investigation, then adjacent muddy areas were also found to be important feeding habitats for them at night (Williamson et al., 2017). Temporal multiscale approaches have been shown to be very useful in studies of behavior and habitat use of highly mobile species that explore dynamic habitats, such as the franciscanas (González-García et al., 2018).

Implications for the conservation of the species

The franciscana is a social species and, despite the challenges related to the study of behavior in the field, the data obtained to date strongly suggest that the species forms family groups, and has a serial monogamous system, which is very rare among mammals. These characteristics further exacerbate some of the threats to the species, such as entanglement in fishing nets and the degradation or loss of their coastal habitats. On the other hand, some aspects of the species' behavior seem to be more flexible, which may confer greater resilience in terms of responses to some types of changes in franciscana habitats, such as feeding behavior. All data point to the need for concern about this being a sensitive species, and that it should receive special attention in permitting of human activities in its habitat, and in marine spatial planning processes.

References

Actis, P.S., Danilewicz, D., Cremer, M.J., Bortolotto, G.A., 2018. Breathing synchrony in franciscana (*Pontoporia blainvillei*) and Guiana dolphins (*Sotalia guianensis*) in Southern Brazil. Mar. Mamm. Sci. http://doi.org/10.1111/mms.12480.

Amaral, K.B., Danilewicz, D., Zerbini, A., Di Beneditto, A.P., Andriolo, A., Alvares, D.J., et al., 2018. Reassessment of the franciscana *Pontoporia blainvillei* (Gervais & d'Orbigny, 1844) distribution and niche characteristics in Brazil. J. Exp. Mar. Biol. Ecol. 508, 1–12. http://doi.org/10.1016/j.jembe.2018.07.010.

Amos, B., Schlotterer, C., Tautz, D., 1993. Social structure of pilot whales revealed by analytical DNA profiling. Science 260, 670–672.

Baird, R.W., 2000. The killer whale: foraging specialization and groups hunting. In: Mann, J., Connor, R.C., Tyack, P., Whitehead, H. (Eds.), Cetacean Society: Field Study of Dolphins and Whales. The University of Chicago Press, Chicago.

Bassoi, M., Sheperd, J.G., Secchi, E.R., Moreno, I.B., Danilewicz, D., 2020. Oceanographic processes driving the feeding ecology of franciscana dolphin off Southern Brazilian coast. Cont. Shelf Res. 201, 104124. http://doi.org/10.1016/j.csr.2020.104124.

Bastida, R.O., Rodríguez, D., 2005. Marine Mammals of Patagonia and Antarctica. Vázquez Mazzini Editores, Buenos Aires.

Barcellos, D.D., Santos, M.C., de, O., 2021. Echolocation characteristics of franciscana dolphins (*Pontoporia blainvillei*). Mar. Mamm. Sci. 2021, 1–11. doi:10.1111/mms.12796.

Bittar, V.T., Di Beneditto, A.P.M, 2009. Diet and potential feeding overlap between *Trichiurus lepturus* (Osteichthyes: perciformes) and *Pontoporia blainvillei* (Mammalia: cetacea) in northern Rio de Janeiro. Brazil. Zoologia 26, 374–378.

Bordino, P., Thompson, G., Iñiguez, M., 1999. Ecology and behaviour of the Franciscana *Pontoporia blainvillei* in Bahia Anegada. Argentina. J. Cet. Res. Manag. 1, 213–222.

Bordino, P., 2002. Movement pattern of franciscana dolphins (*Pontoporia blainvillei*) in Bahia Anegada. Argentina. Lat. Am. J. Aq. Mamm. 1, 71–76.

Carlström, J., 2005. Diel variation in echolocation behavior of wild harbor porpoises. Mar. Mamm. Sci. 21, 1–12.

Connor, R.C., 2000. Group living in whales and dolphins. In: Mann, J., Connor, R., Tyack, P.L., Whitehead, H. (Eds.), Cetacean Societies: Field Studies of Dolphins and Whales. The University of Chicago Press, Chicago.

Costa-Urrutia, P., Abud, C., Secchi, E.R., Lessa, E.P., 2012. Population genetic structure and social kin associations of franciscana dolphin, *Pontoporia blainvillei*. J. Hered. 103, 92–102. http://doi.org/10.1093/jhered/esr103.

Cremer, M.J., Simões-Lopes, P.C., 2005. The occurrence of *Pontoporia blainvillei* (Gervais & d'Orbigny) (Cetacea, Pontoporiidae) in an estuarine area in southern Brazil. Revta. Bras. Zool. 22, 717–723.

Cremer, M.J., Pinheiro, P.C., Simões-Lopes, P.C., 2012. Prey consumed by Guiana dolphin *Sotalia guianensis* (Cetacea, Delphinidae) and franciscana dolphin *Pontoporia blainvillei* (Cetacea, Pontoporiidae) in an estuarine environment in southern Brazil. Iheringia. Série Zool. 102, 131–137. http://doi.org/10.1590/s0073-47212012000200003.

Cremer, M.J., Hardt, F.A.S., Tonello Júnior, A.J., 2006. Evidence of epimeletic behavior involving a *Pontoporia blainvillei* calf (Cetacea, Pontoporiidae). Biotemas 19, 83–86.

Cremer, M.J., Holz, A.C., Sartori, C.M., Schulze, B., Paitach, R.L., Simões-Lopes, P.C., 2018. Behavior and ecology of endangered species living together: long-term monitoring of resident sympatric dolphin populations. In: Rossi-Santos, M.R., Finkl, C.W. (Eds.), Advances in Marine Vertebrate Research in Latin America. Springer, Berlin. http://doi.org/10.1007/978-3-319-56985-7_17.

Cremer, M.J., 2015. Os mamíferos marinhos da Baía da Babitonga. Editora da Univille, Joinville.

Crespo, E.A., Harris, G., González, R., 1998. Group size and distributional range of the franciscana, *Pontoporia blainvillei*. Mar. Mamm. Sci. 14, 845–849. http://doi.org/10.1111/j.1748-7692.1998.tb00768.x.

Crespo, E.A., Pedraza, S.N., Grandi, M.F., Dans, S.L., Graffo, G.V., 2010. Abundance and distribution of endangered Franciscana dolphins (*Pontoporia blainvillei*) in Argentine waters and conservation implications. Mar. Mammal Sci. 26, 17–35. http://doi.org/10.1111/j.1748-7692.2009.00313.x.

Danilewicz, D., Rosas, F., Bastida, R., Marigo, J., Muelbert, M., Rodriguez, D., Lailson Brito, Jr., J., Ruoppolo, V., Ramos, R., Bassoi, M., Ott, P., Caon, G., Monteiro da Rocha, A., Catão-Dias, J.L., Secchi, E.R, 2002. Report of the working group on biology and ecology. LAJAM 1 (Special Issue 1), 25–42.

Danilewicz, D., Claver, J.A., Carrera, A.L.P., Secchi, E.R., Fontoura, N.F., 2004. Reproductive biology of male franciscanas (*Pontoporia blainvillei*) (Mammalia: Cetacea) from Rio Grande do Sul, southern Brazil. Fish. Bull. 102, 581–592.

Danilewicz, D., Secchi, E.R., Ott, P.H., Moreno, I.B., Bassoi, M., Martins, M.B., 2009. Habitat use patterns of franciscana dolphin (*Pontoporia blainvillei*) off southern Brazil in relation to water depth. J. Mar. Biol. Assoc. UK 89, 943–949.

Danilewicz, D., Moreno, I.B., Ott, P.H., Tavares, M., Azevedo, A.F., Secchi, E.R., et al., 2010. Abundance estimates for a threatened population of franciscana dolphins in southern coastal Brazil: uncertainties and management implications. J. Mar. Biol. Assoc. UK 90, 1649–1657. http://doi.org/10.1017/s0025315409991482.

Denuncio, P.E., Bastida, R.O., Danilewicz, D., Morón, S., Rodríguez-Heredia, S., Rodríguez, D.H., 2013. Calf chronology of the franciscana dolphin (*Pontoporia blainvillei*): birth, onset of feeding, and duration of lactation in coastal waters of Argentina. Aq. Mamm. 39, 73–80. http://doi.org/10.1578/am.39.1.2013.73.

Di Beneditto, A.P.M., Ramos, R.M.A., Lima, N.R.W., 2001. Sightings of *Pontoporia blainvillei* (Gervais & D'Orbigny, 1844) and *Sotalia fluviatilis* (Gervais, 1853) (Cetacea) in Southeastern Brazil. Braz. Arch. Biol. Tech. 44, 291–296.

Di Beneditto, A.P.M., 2004. Presence of franciscana dolphin (*Pontoporia blainvillei*) remains in the stomach of a tiger shark (*Galeocerdo cuvieri*) captured in Southeastern Brazil. Aq. Mamm. 30, 311–314.

Elwen, S.H., Meÿer, M.A., Best, P.B., Kotze, P.G.H., Thornton, M., Swanson, S., 2006. Range and movements of female Heaviside's dolphins *Cephalorhynchus heavisidii* as determined by satellite telemetry. J. Mammal. 87, 866–877.

Failla, M., Seijas, V.A., Esposito, R., Iñíguez, M.A., 2012. Franciscana dolphins, *Pontoporia blainvillei*, of the Río Negro Estuary, Patagonia, Argentina. Mar. Biodivers. Rec. 5, e102.

Fury, C.A., Harrison, P.L., 2011. Seasonal variation and tidal influences on the use of two Australian estuaries by bottlenose dolphins (*Tursiops aduncus*). Est. Coastal Shelf Science 93, 389–395.

Gerhardinger, L.C., Herbst, D.F., Cunha, S.M.B., Costa, M.D.P., 2020. Diagnóstico da ictiofauna do Ecossistema Babitonga. Revta Cepsul - Biod. Cons. Mar. 9, eb2020001.

González-García, L., Pierce, G.J., Autret, E., Torres-Palenzuela, J.M., 2018. Multi-scale habitat preference analyses for Azorean blue whales. PLoS One 13, e0201786.

Henning, B., Carvalho, B.S., Pires, M.M., Bassoi, M., Marigo, J., Bertozzi, C., et al., 2017. Geographical and intrapopulation variation in the diet of a threatened marine predator, *Pontoporia blainvillei* (Cetacea). Biotropica 50, 157–168. http://doi.org/10.1111/btp.12503.

Holz, A.C., 2014. Caracterização do ambiente acústico subaquático e uso de habitat pela toninha, *Pontoporia blainvillei*, na Baía da Babitonga, litoral norte de Santa Catarina, Brasil [Unpublished Masther thesis]. University of the Region of Joinville, Joinville.

Kasuya, T., Brownell, R.L., 1979. Age determination, reproduction, and growth of the franciscana dolphin, *Pontoporia blainvillei*. Sci. Rep. Whal. Res. Inst. 31, 45–67.

Leeney, R.H., Carslake, D., Elwen, S.H., 2011. Using static acoustic monitoring to describe echolocation behaviour of heaviside's dolphins (*Cephalorhynchus heavisidii*) in Namibia. Aquat. Mamm. 37, 151–160.

Lin, T., Chou, L., Akamatsu, T., 2013. Tidal influences on the habitat use of Indo-Pacific humpback dolphins in an estuary. Mar. Biol. 160, 1353–1363.

Lucifora, L.O., 2003. Ecología y conservación de los grandes tiburones costeros de Bahía Anegada, Provincia de Buenos Aires, Argentina [unpublished PhD thesis]. Universidad Nacional de Mar del Plata, Mar del Plata.

Lucifora, L.O., Menni, R.C., Escalante, A.H., 2002. Reproductive ecology and abundance of the sand tiger shark, *Carcharias taurus*, from the southwestern Atlantic. ICES J. Mar. Sci. 59, 553–561.

Lucifora, L.O., Menni, R.C., Escalante, A.H., 2005. Reproduction, abundance and feeding habits of the broadnose sevengill shark *Notorynchus cepedianus* in north Patagonia, Argentina. Mar. Ecol. Prog. Series 289, 237–244.

Madsen, P., Carder, D., Beedholm, K., Ridgway, S., 2005. Porpoise clicks from a sperm whale nose—convergent evolution of 130 kHz pulses in toothed whale sonars? Bioacoustic 15, 195–206.

Mann, J., Karniski, C., 2017. Diving beneath the surface: long-term studies of dolphins and whales. J. Mamm. 98, 621–630.

Mendes, S., Turrell, W., Lütkebohle, T., Thompson, P., 2002. Influence of the tidal cycle and a tidal intrusion front on the spatio-temporal distribution of coastal bottlenose dolphins. Mar. Ecol. Prog. Series 239, 221–229.

Mendez, M., Rosembaum, H.C., Subramaniam, A., Yackulic, C., Bordino, P., 2010. Isolation by environmental distance in mobile marine species: molecular ecology of franciscana dolphins at their southern range. Mol. Ecol. 19, 2212–2228. http://doi.org/10.1111/j.1365-294x.2010.04647.x.

Monzón, F., Perez, J., Varela, G., et al., 1994. Indices de predacion de tiburones sobre delfines franciscana en la costa argentina, Abstracts, VI Reunião de Trabalhos de Especialistas em Mamíferos Aquáticos da América do Sul. 24-28 October. Florianópolis, 101.

Morisaka, T., Connor, R.C., 2007. Predation by killer whales (*Orcinus orca*) and the evolution of whistle loss and narrow-band high frequency clicks in odontocetes. J. Evol. Biol. 20, 1439–1458.

Nuuttila, H.K., Meier, R., Evans, P.G.H., Turner, J.R., Bennell, J.D., Hiddink, J.G., 2013. Identifying foraging behaviour of wild bottlenose dolphins (*Tursiops truncatus*) and harbour porpoises (*Phocoena phocoena*) with static acoustic dataloggers. Aquat. Mamm. 39, 147–161.

Ott, P.H., Danilewicz, D., 1998. Presence of franciscana dolphins (*Pontoporia blainvillei*) in the stomach of a killer whale (*Orcinus orca*) stranded in southern Brazil. Mammalia 62, 605–609.

Padula, A.D., Gana, J.C.N., Giardino, G.V., De Leon, M.C., Elissamburu, A., Rodriguez, D.H., Denuncio, P., 2022. First record of orca predation on franciscana dolphins (*Pontoporia blainvillei*) in Argentina. LAJAM 17 (1), 1–5. doi:10.5597/lajam00283.

Paitach, R.L., 2021. Padrões de uso de habitat e comportamento de toninhas: abordagens acústicas para monitoramento e conservação de uma espécie ameaçada [Habitat use and behavior patterns of franciscana dolphins: acoustic approaches for monitoring and conservation of a threatened species] [Unpublished PhD thesis]. Federal University of Santa Catarina, Florianópolis.

Paitach, R.L., Simões-Lopes, P.C.A., Cremer, M.J., 2017. Tidal and seasonal influences in dolphin habitat use in a southern Brazilian estuary. Sci. Mar. 81, 49–56.

Panebianco, M.V., Negri, M.F., Cappozzo, H.L., 2012. Reproductive aspects of male franciscana dolphins (*Pontoporia blainvillei*) off Argentina. Anim. Rep. Sci. 131, 41–48. http://doi.org/10.1016/j.anireprosci.2012.02.005.

Pilleri, G., 1984. Epimeletic behaviour in cetacea: intelligent or instinctive? Invest. Cetacea 16, 30–48.

Praderi, R., 1985. Relaciones entre *Pontoporia blainvillei* (Mammalia: Cetacea) y tiburones (Selachii) de águas Uruguayas. Comm. Zool. Museo de Historia Natural 11, 1–19.

Ramos, R.M.A., Di Beneditto, A.P.M., Lima, N.R.W., 2000. Growth parameters of *Pontoporia blainvillei* in northern Rio de Janeiro. Brazil. Aq. Mamm. 26, 65–75.

Rodríguez, D.H., Rivero, L., Bastida, R.O., 2002. Feeding ecology of the franciscana (*Pontoporia blainvillei*) in marine and estuarine waters of Argentina. Lat. Am. J. Aq. Mamm. 1, 77–94.

Rosas, F.C.W., Monteiro-Filho, E.L.A., 2002. Reproductive parameters of *Pontoporia blainvillei* (Cetacea, Pontoporiidae), on the coast of São Paulo and Paraná States. Brazil. Mammalia 66, 231–245.

Rupil, G.M., Barbosa, L., Marcondes, M.C.C., Carvalho, B.M., Farro, A.P.C., 2019. Franciscana dolphin (*Pontoporia blainvillei*) diet from Northern Espírito Santo State coast. Brazil. Rev. Biotemas 32, 87–96.

Santos, M.C.O., Netto, D.F., 2005. Killer whale (*Orcinus orca*) predation on a franciscana dolphin (*Pontoporia blainvillei*) in Brazilian waters. Lat. Am. J. Aq. Mamm. 4, 69–72.

Sartori, C.M., Paitach, R.L., Cremer, M.J., 2017. Photo-identification of franciscanas (*Pontoporia blainvillei*) in Babitonga Bay, Santa Catarina State. Brazil. J. Cet. Res. Manage. 16, 49–55.

Secchi, E.R., Zerbini, A.N., Bassoi, M., Dalla Rosa, L., Möller, L.M., Rocha-Campos, C.C., 1997. Mortality of franciscanas, *Pontoporia blainvillei*, in coastal gillnetting in southern Brazil: 1994-1995. Rep. Int. Whal. Commn. 47, 653–658.

Secchi, E.R., Cremer, M.J., Danilewicz, D., Lailson-Brito, J., 2021. A synthesis of the ecology, human-related threats and conservation perspectives for the endangered franciscana dolphin. Frot. Mar. Sci. 8, 617956. http://doi.org/10.3389/fmars.2021.617956.

Shane, S.H., 1990. Behavior and ecology of the bottlenose dolphin at Sanibel Island, Florida. In: Leatherwood, S., Reeves, R.R. (Eds.). The Bottlenose Dolphin, 1. Academic Press, San Diego, CA, pp. 369–386.

Siciliano, S., 1994. Review of small cetaceans and fishery interactions in coastal waters of Brazil. Rep. Int. Whal. Commn. 15, 241–250.

Silva, D.F., Barbosa, Conversani, V.R.M., Botta, S., Hohn, A.A., Santos, M.C.O., 2020. Reproductive parameters of franciscana dolphins (*Pontoporia blainvillei*) of Southeastern Brazil. Mar. Mamm. Sci. 2020, 1–18. doi:10.1111/mms.12720.

Szymanski, M., Bain, D., Kiehl, K., Pennington, S., Wong, S., Henry, K.R., 1999. Killer whale (*Orcinus orca*) hearing: auditory brainstem response and behavioural audiograms. J. Acoust. Soc. Am. 106, 1134–1141.

Sucunza, F., Danilewicz, D., Cremer, M., Andriolo, A., Zerbini, A.N., 2018. Refining estimates of availability bias to improve assessments of the conservation status of an endangered dolphin. PLoS One 13, e0194213.

Sucunza, F., Danilewicz, D., Secchi, E., et al., 2021. Assessing bias in aerial surveys for threatened cetaceans: results from experiments conducted with the franciscana (*Pontoporia blainvillei*). SC/68B/ASI/04.

Todd, V.L.G., Pearse, W.D., Tregenza, N.C., Lepper, P.A., Todd, I.B., 2009. Diel echolocation activity of harbour porpoises (*Phocoena phocoena*) around North Sea offshore gas installations. ICES J. Mar. Sci. 66, 734–745.

Valsecchi, E., Zanelatto, R.C., 2003. Molecular analysis of the social and population structure of the franciscana (*Pontoporia blainvillei*): conservation implications. J. Cet. Res. Manag. 5, 69–75.

Wells, R.S., Boness, D.J., Rathbun, G.B., 1999. Behavior. In: Reynolds, J.E. III, Rommel, S.A. (Eds.), Biology of Marine Mammals. Smithsonian Institution Press, Washington, DC, pp. 324–422 578.

Wells, R.S., Bordino, P., Douglas, D.C., 2013. Patterns of social association in the franciscana, *Pontoporia blainvillei*. Mar. Mamm. Sci. 29, 520–528. http://doi.org/10.1111/mms.12010.

Wells, R.S., Cremer, M.J., Berninsone, L.G., Albareda, D., Wilkinson, K.A., Stamper, M.A., Paitach, R.L., Bordino, P., 2022. Tagging, ranging patterns and behavior of franciscana dolphins (*Pontoporia blainvillei*) off Argentina and Brazil: considerations for conservation. Mar. Mamm. Sci., 1–35. http://doi.org/10.1111/mms.1279.

Williamson, L., Brookes, K.L., Scott, B.E., Graham, I.M., Thompson, P.M., 2017. Diurnal variation in harbour porpoise detection—potential implications for management. Mar. Ecol. Prog. Series 570, 223–232.

Wisniewska, D.M., Johnson, M., Beedholm, K., Wahlberg, M., Teglberg, M.P., Madsen, P.T., 2012. Acoustic gaze adjustments during active target selection in echolocating porpoises. J. Exp. Biol. 215, 4358–4373.

Wisniewska, D.M., Johnson, M., Nachtigall, P., Madsen, P.T., 2014. Buzzing during biosonar-based interception of prey in the delphinids *Tursiops truncatus* and *Pseudorca crassidens*. J. Exp. Biol. 217, 4279–4282.

Wursig, B., Jefferson, G., 1990. Methods of photo-identification for small cetaceans. Rep. int. Whal. Commn 12 (Special Issue), 43–52.

CHAPTER 9

Acoustic repertoire and bioacoustics applications

Renan Lopes Paitach[a], Mariana Melcón[b], Artur Andriolo[c], Marta Jussara Cremer[a]

[a]Laboratório de Ecologia e Conservação de Tetrápodes Marinhos e Costeiros, Universidade da Região de Joinville, São Francisco do Sul, Santa Catarina, Brazil

[b]Fundación Cethus, Olivos, Buenos Aires, Argentina

[c]Laboratório de Ecologia Comportamental e Bioacústica, Universidade Federal de Juiz de Fora, Juiz de Fora, MG, Brazil

Introduction

The sounds produced by cetaceans can be generically divided into tonal and pulsed sounds (Tyack and Clark, 2000). Tonal sounds are signals in which the emission occurs with energy concentrated in a narrow band frequency (a few Hz) that undergoes modulation over its duration and can last for seconds or even minutes. The whistles produced by dolphins and the songs of whales, for example, are tonal sounds, that mainly fulfill the communication function (Janik, 2009; King et al., 2013). On the other hand, the pulsed sounds, commonly called "clicks," are exclusive to toothed whales (suborder Odontoceti) and are characterized by having dissipated energy in wide frequency bands (several kHz) with extremely short duration (microseconds). Clicks are produced in trains, that is, a sequence of pulses, and are mainly associated with echolocation (biosonar), but are also used for communication in some species (so-called burst pulsed sounds) (Blomqvist and Amundin, 2004; Clausen et al., 2010; Cremer et al., al, 2017).

The franciscana dolphin, *Pontoporia blainvillei* (Gervais and d'Orbigny, 1844), has a small body size, cryptic coloration, discrete behavior, and form small groups, imposing a special challenge to detect and visually monitor the species from boat surveys (Bordino et al., 2002). This difficulty in studying the species in the wild also applies to bioacoustics approaches that depend on visual tracking and approximation of animals, such as the use of active data loggers. Furthermore, the high frequency sounds that the species

The Franciscana Dolphin
DOI: https://doi.org/10.1016/B978-0-323-90974-7.00001-X

produce (above 120 kHz), make the detection range more restricted than that of other species (Melcón et al., 2012). Even so, some studies published in the last decade have provided a substantial advance in the knowledge about the franciscana's acoustic repertoire, in addition to other transversal aspects to bioacoustics, such as behavior, ontogenetic development, psychoacoustics, and evolution (e.g., Frainer et al., 2015; Melcón et al., 2016; Paitach et al., 2021).

Bioacoustics is a great opportunity to study the behavior and habitat use of the franciscana, while contributing to understanding the impacts to which the species is exposed. This chapter is divided into two parts. In the first session—Acoustic Repertoire Description—the main results and limitations of the studies on the acoustic behavior of the franciscanas, its ontogenetic and evolutionary aspects, are summarized. Recommendations for future acoustic studies of the species are also provided. The second session—Bioacoustics Applications for Conservation—presents the advances, potentials, and challenges of passive acoustic monitoring (PAM) for ecological studies of franciscanas, and conservation against the impacts of underwater anthropogenic noise. It is noteworthy that we seek to summarize the bioacoustics approaches that have already been practiced or are under development for the study of franciscanas, but still, there is a wide range of other possibilities that go beyond the topics covered here.

Acoustic repertoire description

Echolocation

The franciscanas produce mainly narrow-band high frequency clicks (NBHF). The first acoustic logs of franciscanas dolphins were obtained with very limited equipment (Busnel et al., 1974), and identified broad and narrow band clicks. Only in 1997, von Fersen et al. recorded the echolocation behavior of an individual in captivity using a single hydrophone. They observed that the main frequency of clicks was 130 kHz with a bandwidth of about 20 kHz, without further details of the parameters (Fersen et al. 1997). Similarly, Melcón et al. (2012) used a single hydrophone in the wild and were able to capture the whole spectral range of the echolocation behavior in the species. The authors observed a peak frequency at 139 kHz and a bandwidth of 19 kHz (at −10 dB, they ranged between 130 and 149 kHz, Fig. 9.1). Clicks were produced every 33 ± 4 ms, except in the case of buzzes (associated with foraging behavior), which were produced every 4–7 ms. Clicks within buzzes presented a similar spectral structure to

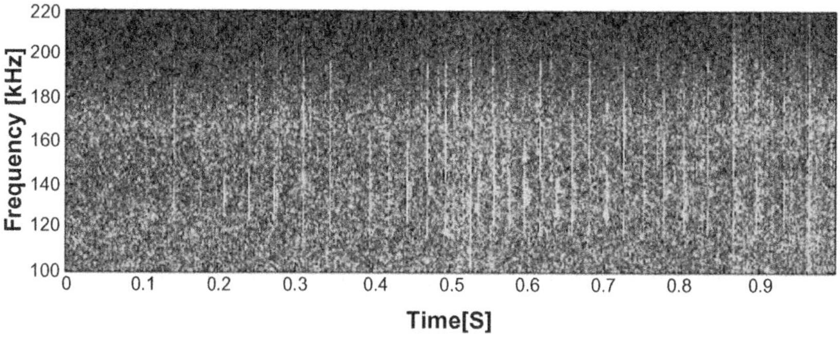

Fig. 9.1 ***Spectrogram of a representative click train of a franciscana (FFT = 1000, overlap = 0%).*** *Credit: Melcón et al. (2012).*

regular echolocation clicks (Fig. 9.2; Melcón et al. 2012). Note that these echolocation signals were recorded probably on and off-axis, and the purpose of this study was to characterize them to facilitate the identification of the species using PAM.

Tellechea et al. (2016) were able to acoustically detect franciscanas in the coast of Uruguay with a limitation of sensitivity of the hydrophone as well as the sampling rate, preventing researchers from observing frequencies above 96 kHz. Under these circumstances, and probably distorted with aliasing, the authors reported the average peak frequency of the high-frequency clicks of 86 kHz and 14 kHz in low-frequency clicks. Barcellos and Santos (2021) recorded franciscanas on the coast of São Paulo state, southern Brazil. Their equipment was able to capture more of the signal -up to ~144 kHz with a substantial decrease in amplitude after 130 kHz, and probably still showed aliasing. The authors of this work report a low

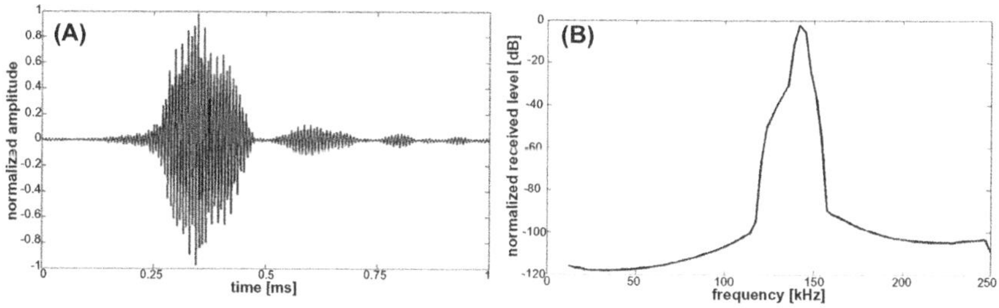

Fig. 9.2 (A) Normalized time series of a single echolocation signal analyzed. The onset of the click is represented at 200 ls. Note the presence of echoes proceeding from the surroundings after the click. (B) Normalized spectrum of the same echolocation signal emitted by a franciscana dolphin (FFT = 160, overlap = 95%, sampling frequency = 500 kHz). *Credit: Melcón et al. (2012).*

frequency of 84 kHz, peak frequency of 104 kHz, and center frequency of 107 kHz. While there could be geographical differences, most likely the different values reported are due to the challenges faced to get the right equipment to properly characterize the echolocation signals emitted by franciscanas.

Cetacean's acoustic repertoire can vary among habitats and the dolphins are able to modify acoustic parameters to increase the sensorial and communicational efficiency (Ey and Fischer, 2009). The only study specifically aimed at comparing the acoustic repertoire of franciscanas between different habitats was developed in the state of Santa Catarina, in southern Brazil, considering collections carried out with autonomous click data loggers (C-POD, Chelonia Limited) in two locations, approximately distant from 250 kilometers in between Babitonga Bay, an estuarine environment, and Itapirubá beach, an open sea environment (Paitach et al., 2021). The device used can record the entire frequency spectrum of the franciscana, but there is also no distinction regarding the directionality of the individuals detected. In addition, as the device has an automated processing system of the acoustic signals, the analysis of parameters is limited, but due to the large sample size (more than 17 thousand franciscana click trains), it was possible to consistently identify patterns and perform comparisons. The main acoustic difference in the click trains observed between the two sites was related to the click frequency spectrum and the emission rate. In general, franciscanas in the estuarine habitat emitted click trains with slightly lower modal frequency and with a frequency range almost twice that in the open sea, with average frequency ranges of 118–135 kHz and 124–134 kHz, respectively (Paitach et al., 2021). No difference was found in the train duration between habitats, which indicates that the animals in both populations have the same scanning behavior, but the average click rate was almost 20% higher in the estuarine environment (Paitach et al., 2021).

A higher click repetition rate may be associated with the dolphins locking on targets at closer ranges, which may be necessary for a more complex environment like the estuary (Monteiro-Filho and Monteiro, 2001). Also, the visibility is much lower in an estuarine environment, than in the open sea, due to the presence of murky waters (Oliveira et al., 2006). However, it is important to note that the study of Paitach et al., 2021 analyzed the general variability of the acoustic parameters between habitats, without distinction between regular echolocation clicks and feeding buzzes, but it would be important to analyze these parameters separately for each behavioral class of click trains for deeper knowledge.

Since data acquisition and analysis methods used in the different studies on the franciscanas' acoustic repertoire differ substantially from each other, any comparison must be made with caution. It is possible that the differences in the franciscanas' echolocation parameters are locality-specific. Based on genetic and bioecological characteristics of populations, the franciscana distribution range has been divided into four zones, known as "Franciscana Management Areas" (FMAs, for details, see Secchi et al., 2003). Perhaps the franciscana's acoustic repertoire has variations associated with these FMA's, but the limited bioacoustics information for most of the franciscana distribution prevents further elaboration on the possible reasons that may cause variations in the species' acoustic repertoire. Table 9.1 summarizes the main results and sampling features of the studies available in the literature on the pulsed sounds of franciscanas (except in Busnel et al., 1974 and Von Fersen et al., 1997, due to lack of information presented by the authors), both for regular echolocation click trains and burst pulsed sounds, which will be better presented below in the section about communication.

Communication (burst pulses and whistles)

Burst pulses and whistles play an important role in social communication (Janik, 2009), although they have also been reported in less-social species, such as *Inia geoffrensis* and *Lipotes vexillifer* (May-Collado and Wartzok, 2007; Wang et al., 2006). Burst pulses and whistles are not commonly used by the franciscanas and were recorded in Babitonga Bay, south Brazil, for free-ranging groups and for individuals during capture/tagging/release procedures (Cremer et al., 2017). The difficulty of studying the species in the wild is probably the main factor related to the absence of whistle recordings in other areas.

The whistles of franciscanas were produced between 0.7 and 94.6 kHz, and duration varied between 0.008 and 0.361 s. The whistle frequency amplitude can be considered high compared with many delphinids, and the frequency of the whistles recorded during capture/tagging/release procedures were the ones with the highest frequency (Fig. 9.3). Ten types of whistles were registered for franciscanas. Whistles recorded during capture/tagging/release procedures were also highly complex, with up to nine inflection points and the presence of harmonics. Cremer et al. (2017) suggest that they could be related to this stressful situation, as social signals for group cohesion, as proposed by May-Collado et al. (2007).

Burst pulses in Babitonga Bay were recorded in groups with calves, suggesting that these may have been produced by the calf, or perhaps by

Table 9.1 Summary of sampling features and parameters of pulsed sounds of franciscana dolphins available in the literature.

	Sampling features			Type of acoustic signals and analyzed parameters								
References	FMA	Habitat	Full frequency coverage	Train class	*n*	*N* clicks	Duration (ms)	ICI (ms)	Min F	Max F	Peak F	Center F
Melcón et al. (2012)	IV	Estuarine	Yes	Echolocation	61	–	212 ±56	33 ±4	130 ±3	149 ±3	139	–
Tellechea and Norbis (2014)	III	Captivity	No	Echolocation	13	17 ±13	200 ±100	29 ±1	–	–	81 ±3	–
Tellechea and Norbis (2014)	III	Captivity	No	Burst pulses	18	124 ±1	130 ±30	10 ±1	–	–	47 ±1	–
Melcón et al. (2016)	IV	Estuarine	Yes	No distinction	–	–	–	–	37 ±3	160 ±3	–	–
Tellechea et al. (2016)	III	Estuarine	No	Echolocation	26	38 ±4	210 ±230	35 ±2	–	–	86 ±2	–
Tellechea et al. (2016)	III	Estuarine	No	Burst pulses	5	52 ±6	150 ±7	10 ±1	–	–	41 ±5	–
Cremer et al. (2017)	II	Estuarine	No	Burst pulses	31	24 ±24	75 ±65	3 ±2	77 ± 4	–	–	–
Alves et al. (2016)	II	Captivity	No	Echolocation	52	62 ±45	540 ±410	9 ±3	–	–	74 ±29	–
Alves et al. (2016)	II	Captivity	No	Burst pulses	6	81 ±21	340 ±87	4 ±1	–	–	84 ±17	–
Barcellos and Santos (2021)	II	Open sea	No	Echolocation	182	27 ±24	668 ±624	30 ±43	107 ±63	144[a] ±0	104 ±4	108 ±3
Barcellos and Santos (2021)	II	Open sea	No	Burst pulses	60	33 ±40	153 ±215	6 ±20	98 ±7	144[a] ±0	99 ±7	104 ±5
Paitach et al. (2021)	II	Estuarine	Yes	No distinction	10,924	30 ±16	400 ±535	13 ±14	118 ±14	135 ±6	–	129 ±5
Paitach et al. (2021)	III	Open sea	Yes	No distinction	6,093	29 ±16	399 ±486	12 ±10	124 ±5	134 ±4	–	130 ±3

Parameter values represent mean ± standard deviation. FMA, franciscana management area; ICI, interclick interval; F, frequency (expressed in kHz). ([a] Limited by sampling rate).

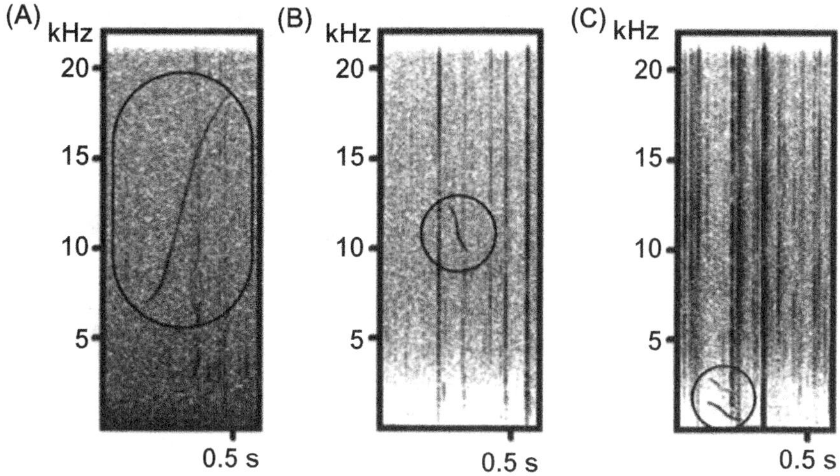

Fig. 9.3 ***Spectrograms show three examples of whistles produced by free-ranging franciscana dolphins.*** *Credit: Cremer et al. (2017).*

the mother, and could be useful for mother-calf communication (Cremer et al., 2017). Burst pulses were also recorded for neonatal franciscanas in captivity, during rehabilitation (Tellechea and Norbis, 2014), and for free-ranging franciscanas in Uruguay (Tellechea et al., 2016). In this case, the trains had a mean duration of 90 ± 0.40 s, and the maximum frequency was 63 kHz, with a minimum of 32 kHz. In Babitonga Bay, the burst pulses were shorter, varying between 0.019 and 0.276 s (0.075 ± 0.065), and the frequency was higher, with a minimum at 69.0 kHz and maximum at 82.1 kHz (Cremer et al., 2017).The minimum number of clicks was six and the maximum was 93 (mean 24.2 ± 23.9) (Fig. 9.4). Barcellos and Santos (2021) recorded burst pulses on the coast of São Paulo State, southern Brazil. Using a PAM system, they verified that there was a balanced production of burst pulses between day and night, and the lowest frequency was 79 kHz (97.49 ± 106 kHz).

It is worth noting, however, that most of the studies that recorded whistles and burst pulses in franciscanas were limited in relation to the maximum capacity of frequency recordings, which may have mainly influenced the results regarding pulsed sounds.

Table 9.1 summarizes all data available in the literature about the pulsed sounds produced by franciscanas, whether used for communication or echolocation.

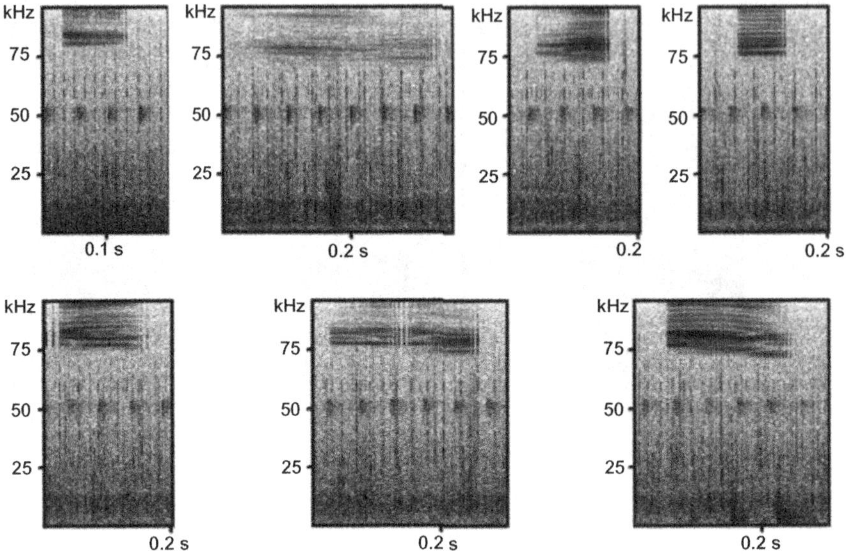

Fig. 9.4 ***Spectrograms show examples of burst pulses produced by franciscana dolphins.*** Note that higher frequencies are limited by the sampling rate. The 50 kHz pulses present in the figures constitute the frequency produced by the echosounder of the boat. *Credit: Cremer et al. (2017).*

Ontogenetic development

In recent years some authors were exposed to opportunistic encounters with franciscana dolphin calves, giving us a unique opportunity to start understanding the echolocation development in the species.

Tellechea and Norbis (2014) were the first ones to report sounds emitted by two calves found stranded in Uruguay. They recorded them in captivity during rehabilitation and observed that the sound most frequently emitted by these neonates was burst pulses. This work, together with another two studies available on echolocation in neonates, observe that the clicks emitted by calves have much higher bandwidth than adults, mostly due to the lower frequencies extending well below 100 kHz (Alves et al, 2016; Melcón et al, 2016; Tellechea and Norbis, 2014). From the spectrograms displayed in, we can observe very little consistency in the clicks emitted, which could be interpreted as a lack of skill in the control of sound production.

It is possible that the production of NBHF sounds is a learned characteristic that takes some time so that it is improved by the individual throughout their development. However, another theory raised in the literature concerns the subdevelopment of structures related to sound production in franciscanas calves that may be associated with differences in the bandwidth of their clicks. A comparative study performed with three neonates and two adults using digital imaging technology and macroscopic dissections reveals postnatal changes in allometry and shape of franciscana's biosonar relevant structures (Frainer et al., 2015). The authors also hypothesize that the underdevelopment of such structures may be an explanation for the high bycatch levels that franciscana suffers in the early stages of life since most animals found dead with incidental capture marks are neonates and juveniles with less than two years (e.g., Botta et al., 2010). However, there is still no conclusive evidence for this relationship, perhaps if the underdevelopment of structures related to biosonar impairs the ability to perceive the environment.

Evolutionary context

Pulsed sounds of odontocetes may differ between species and studies suggest that different types evolved convergently between different taxa (Au, 1993; Møhl et al., 2003; Zimmer et al., 2005). Among them, the NBHF clicks with a peak frequency at 130 kHz and no essential energy below 100 kHz are exclusively produced by the franciscanas, as mentioned above, porpoises of the family Phocoenidae, dwarf, and pygmy sperm whales (family Kogiidae), and at least six species of the genus *Cephalorhynchus* and *Lagenorhynchus* (family Delphinidae) (Kyhn et al., 2009; 2010; 2013; Madsen et al., 2005; Reyes et al., 2018; Tougaard and Kyhn, 2010; Von Fersen et al., 1997). NBFH pulsed sounds are stereotyped, showing similar time series and spectral features among the emitting species, even though they do not share a proximal phylogenetic lineage. Two main hypotheses were formulated in the literature and tried to answer why NBHF clicks evolved convergently in these groups.

The "acoustic crypsis" hypothesis postulates that predation pressure is the evolutionary drive of this characteristic, considering that NBHF clicks are above the auditory range of killer whales *Orcinus orca* (Szymanski et al., 1999), and thus allowing NBHF species to echolocate without being heard by this predator (Andersen and Amundin, 1976; Madsen et al., 2005). Galatius et al. (2019) revised this hypothesis, pointing out that the killer whale lineage originated more recently than that of the Phocoenidae and Pontoporiidae families. Thus, if the NBHF click in these families evolved in response to a predator it must have been another species than the killer whale, and they suggested extinct predators such as *Acrophyseter* sp., the raptorial sperm whale, which are believed to have had hearing like that of the killer whale. However, the NBHF clicks also evolved in the *Cephalorhynchus–Lagenorhynchus* clade, which originated after the killer whale. Thus, the killer whale may have provided selective pressure on these species.

The fact that the click power spectrum of all NBHF species peaks at ~130kHz points to another explanation, the "environmental hypothesis." The main natural noise source in the ocean is wind-generated waves, the noise of which decreases with increasing frequencies, to meet increasing thermal noise at 100–150 kHz (Richardson et al., 1995). This creates a window of minimum noise above 100 kHz which favors the echolocation and communication in NBHF species. Galatius et al. (2019) showed that wind speeds of 4–9 m/s were stable in the periods when NBHF species lineages evolved, suggesting this noise to have been a selective driver for the evolution of this particular frequency range in NBHF species. Possibly the combination of these hypotheses may be the best answer since the evolutionary process is dynamic and more than one factor may be responsible.

In relation to tonal sounds, Podos et al. (2002) suggested that whistles had evolved in the Delphinidae because until recently it was believed that these acoustic signals were absent from the repertoire of ancestral families, such as Iniidae, Lipotidae, Phocoenidae, Monodontidae, Platanistidae, and Pontoporiidae. Others have suggested that whistles evolved together with sociality (Herman and Tavolga, 1980), but May-Collado et al. (2007) stated that whistles are a synapomorphy of pandelphinids or a group including ziphiids, rejecting the hypothesis that they evolved for social communication in dolphins. In fact, whistles have also been reported in less-social species, such as *Inia geoffrensis* (May-Collado and Wartzok, 2007), the closest living relative of franciscanas (Cassens et al., 2000). May-Collado et al. (2007), analyzing the "delphinid-whistle" hypothesis, concluded that complexities of whistle structure appear to be related to different levels of species sociality.

Madsen et al. (2005) suggested a linkage between the convergent evolution of whistle loss and NBHF click production, but recently whistles were recorded for *Cephalorhynchus commersonii*, and for *C. heavisidii* was also observed that animals can produce both narrow and broadband echolocation signals, depending on the behavioral context and habitat (Martin et al., 2019; 2021; Reyes Reyes et al., 2016). All of these recent findings further challenge the evolutionary concept of NBHF signal production and evolution.

Considerations for future acoustic research

To properly record the echolocation clicks of franciscanas for characterization purposes, we need to keep some important things in mind.

Sensitivity of the hydrophone (so it can detect the whole range of spectrum of sounds emitted by these dolphins): knowing that the lowest frequency shown for NBHF clicks typically exceeds 100 kHz, it is crucial that the hydrophone can detect frequencies well above 100 kHz, preferably at least 150 kHz. Should the sensitivity of the hydrophone not cover the whole spectrum of the signal of interest, we may report only a part of the signal and potentially a distorted spectrum due to the device not being able to capture the whole signal or having much lower sensitivity at a still relevant frequency.

Sampling rate: Converting discrete waves into continuous waves is a challenge. The Nyquist theorem is a well-known law that states that the sampling interval needs to be at least $Dt = 1/2F_{max}$. In other words, the sampling rate needs to be equal to or larger than twice the maximum frequency of interest (Au, 1993). For example, if we want to record a signal of 150 kHz, we need a sampling rate of at least 300 kHz. Should there be higher frequencies present in the data than half of the sampling rate, the Fourier transform used to calculate the spectrum deals with ambiguity reflected in a distortion of the observed frequencies called "aliasing."

The angle of incidence to the hydrophone: it had been suggested by Au (1993) that echolocation signals are distorted as the observer moves out of the incidence point. The author hypothesized that these distortions may be due to the signal radiating from different parts of the head with a finite aperture and the internal reflection of the signal inside the head containing several structures (sacs, bones, soft tissue, etc.). Madsen et al. (2010) conducted an elegant experiment in which they observed that when Harbor porpoises (*Phocoena phocoena*) echolocate, the sound wave recorded is not symmetrical. The signal recorded in front of the animal (on-axis) is not the same as the one recorded on the left or right (off-axis). On the other hand, a set of several pulse paths outside the axis resulting from the internal pulse reflections can reveal the anatomy of the vocalizing animal, mainly through spectral peaks that are dependent on the morphology of the skull (Baumann-Pickering et al., 2010, Mello et al. 2021). An alternative to investigate which clicks were recorded on-axis is using a hydrophone array.

Calibration of the hydrophone: It is also important when trying to characterize sounds, to use a recording system that is calibrated. In other words, by measuring the sensitivity of the hydrophone at different frequencies, one can compensate for the different sensitivities observed at the time of processing the data to reconstruct the spectrum of the sound of interest. The calibration enables us to describe more realistically the underwater sounds right before being recorded.

Up to this date, no study was conducted on franciscana dolphins fulfilling the four requirements above. Nevertheless, it is important to stress that, given how cryptic this species can be, the acoustic research conducted (at least to detect the presence of animals, habitat use, seasonal behavior) is crucial for conservation purposes. One of the main applications of knowledge about the species' bioacoustics is for PAM.

Bioacoustics applications for conservation

Passive acoustic monitoring

Cetaceans PAM has gained more applicability and relevance on the investigation and management of coastal and oceanic ecosystems. It has enlarged the potential to investigate different ecological and behavioral aspects of free-living cetaceans, for long periods or large areas, of seldomly seen animals, like the franciscana (e.g., Andriolo et al., 2018a, Barcellos and Santos, 2021; Paitach, 2021). In conditions unsuitable for visual observations, such as poor visibility, darkness, and high sea states, acoustic monitoring has become an excellent opportunity to collect data (Andriolo et al., 2018a; Evans and Hammond, 2004; Mellinger et al., 2007; Paitach, 2021; Van Parijs et al., 2009).

The NBHF characteristic of the echolocation sounds emitted by the franciscana represents a great opportunity for its acoustic recognition in most of its distribution. The clicks of NBHF species are stereotyped and similar, making that these species can be differentiated and monitored through the same recording and acoustic recognition systems (Kyhn, 2010). Unlike Uruguay and Argentina, where two other NBHF species occur in potential sympatry with the franciscana, the Burmeister's porpoise, *Phocoena spinipinnis* (Reyes Reyes et al., 2018), and the Peale's dolphin, *Lagenorhynchus australis* (Kyhn et al., 2010) (only Argentina), in Brazil the franciscana is the only species that occur near the coast that produces NBHF clicks.

PAM is normally conducted from hydrophone devices that can be static, fixed in the environment, at structures of interest (e.g., fishing nets, marine wind farms, mineral extraction platforms) or even directly in the body of cetacean specimens (tags), launched adrift from the currents, or towed by vessels or submarines, manned or autonomous (e.g., Andriolo et al., 2018b; Bittencourt et al., 2018; Carstensen et al., 2006; Clay et al., 2018;

Kimura et al., 2014; Van Parijs et al., 2009). When the PAM device is towed, it provides data for large areas over short periods (days or weeks), while static devices can provide data for small areas over long periods (months or years) (Mellinger et al., 2007; Van Parijs et al., 2009). In this sense, the different approaches can be considered complementary and should be used to answer different issues. The following topics will present PAM initiatives already carried out with a focus on the study of franciscanas.

Static hydrophones

Since the creation of the first static acoustic monitoring devices for military purposes during the First World War until the methodology was appropriate in research with cetaceans, there have been great advances in terms of hardware and software and currently, there is a wide variety of models available on the market. Its specifications are the most varied, and the greater the storage capacity and rate, battery life, and range of sampled frequencies, the larger and more expensive the devices tend to be (Sousa-Lima et al., 2013). This trade-off is especially relevant when the objective is the study of franciscanas, which produce high-frequency sounds (>100 kHz), in which case the available device models suitable for recording them are strongly reduced.

Among the few studies that applied static arrays to study the acoustic repertoire of franciscanas, the pioneer was carried out on the coast of Uruguay (Tellechea et al., 2016), and the second one was close to Ilha Anchieta, the coast of São Paulo in Brazil (Barcellos and Santos, 2021), but in both studies the devices used were limited to log frequencies up to 96 and 144 kHz, respectively, making it impossible to fully describe the sounds of the species.

Among PAM devices, the C-POD (Chelonia Limited, UK) is a promising tool for long-term PAM of franciscanas (Paitach, 2021). It has a click detection system with associated software for automated species classification and data analysis, which saves time and reduces the researcher's possible subjectivity bias (Rayment et al., 2009). The nature of the data recorded by the C-PODs makes it particularly suitable for the study of NBHF species, as the sampling rate (> 150 kHz) requires a lot of storage capacity and makes long-term, full-bandwidth recordings a challenge. To be able to operate continuously for more than four months the C-POD has a limitation: the sounds captured are not recorded on the device, they are automatically processed and only information about a limited set of acoustic parameters is recorded.

C-PODs were first used in a study of franciscanas by Paitach (2021), where 60 PAM stations were implemented following a systematic sampling grid for 6 months of continuous data collection. The devices proved to be highly useful for studies of distribution, habitat use, and behavior patterns of franciscanas (Paitach, 2021). The C-PODs were mounted inside custom-made "cages," designed to protect them from being entangled in gillnets, and no buoys at the surface were used, to prevent theft (Fig. 9.5). Anchoring and signaling mode are important aspects in planning research with static acoustic monitoring devices (whatever the model).

The cages with the C-PODs were deployed close to the seafloor. This choice proved to be assertive for the study of franciscana foraging behavior since the species preys mainly on epibenthic species (Paitach, 2015), the positioning of the devices close to the seabed provided a high sampling rate of feeding buzzes (for details see, Paitach et al., 2021). In these cases, however, special attention should be paid to the topography and the presence of rocks and slabs that could increase the chance of acoustically shadowing the PAM device. Despite the protective cages, entanglement in nets became a problem throughout the study, and the authors strongly recommend

Fig. 9.5 ***Custom-made "cages," designed for anchoring and protecting static acoustic monitoring devices (Chelonia Limited, UK) from being entangled in gillnets.*** These cages had a steel foundation and two PVC tube arches over the logger, causing the least possible interference in the acoustic recording. *Credit: Paitach.*

extensive communication and involvement of the fishing communities in the study area so that such incidents can be avoided or at least minimized (Paitach, 2021).

In the analysis of Paitach (2021) was assumed a homogeneous probability of detection of franciscanas by C-PDOs over space and time. It is known, however, that sound propagation may be influenced by spatial and temporal variations in the behavior of the dolphins (Verfuß et al., 2009; Leeney et al., 2011), and also by environmental conditions, such as water temperature and salinity (Richardson et al., 1995). Understanding how different behaviors and environmental conditions can affect the detection probability of franciscanas by PAM devices can assist the accuracy of future studies.

Towed arrays

In the marine environment, acoustic data can be collected from mobile matrices such as gliders and towed arrays, suitable especially when the intention is to cover larger areas (Küsel et al., 2017; Marques et al., 2013; Nielsen and Møhl, 2006). Among different equipment and technics is the acoustic towed array, allowing to record underwater sounds continuously across a defined area of the survey on different species (e.g., Andriolo et al., 2018a, Amorim et al., 2019; Pagliani et al., 2021).

The towed array methods (Fig. 9.6) seem more adjustable to species whose vocalization presents a frequency above the noise range produced

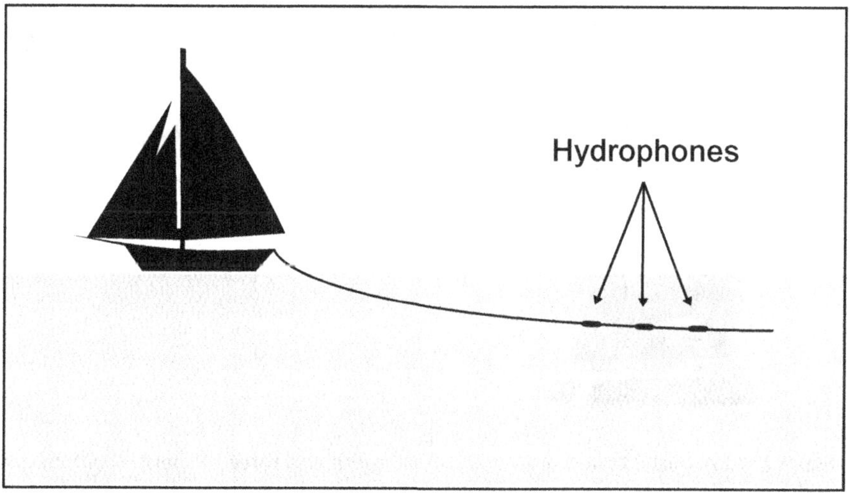

Fig. 9.6 ***Three-hydrophone array on a cable towed by a moving platform.*** *Credit: Gustavo Miranda.*

by the propeller cavitation and the water flow itself, which minimize the masking effect of the signals of interest received (Benda-Beckmann et al., 2010; Thode et al., 2010). The franciscana dolphin produces mainly NBHF echolocation clicks, but the range of frequency used by the species can be much broader (clicks—Melcón et al., 2016, whistles—Cremer et al., 2017). As they produce clicks whose frequency distribution extends above the dominant range of noise produced by the boat and water flow, they allow their clicks to be detected by passive towed acoustic systems (Andriolo et al., 2018b), making the association of this sound system viable.

The necessary equipment to investigate franciscana has some minimum requirements to correctly record high-frequency emissions produced by this species. The equipment arrangement implemented by Andriolo et al. (2018a) consisted of a three-elements omnidirectional hydrophones array (Fig. 9.7) on a cable towed between 30 meters (Auset) behind the sailboat coupled to an autonomous recording at 160 kHz/24 bits. The system was configured to give a variable frequency response from 499 Hz (High Pass Filter) to 160 kHz (Fig. 9.7).

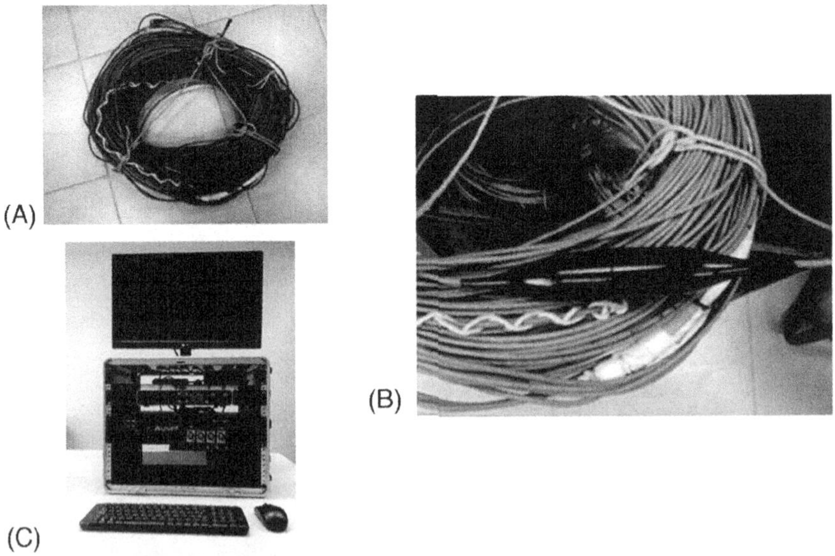

Fig. 9.7 (A) Acoustic towed array with 80 m length and three elements (hydrophone nodes) to investigate franciscana acoustic behavior. (B) Detail of one hydrophone node. (C) Acquisition system used to record franciscana produced by the Auset Company. *Credit: Andriolo.*

Taking into consideration the high sensitivity and strong avoidance to motorboats of the franciscana dolphin, an appropriate alternative is to use low noise impact boats as sailboats to tow the acoustic arrays (Andriolo et al., 2018b; RDDM, 2019). Results obtained by Andriolo et al. (2018b) suggest that acoustic surveys from sailboats are viable to assess the distribution of franciscanas and may, in the near future, provide an alternate, inexpensive, and effective methodology to estimate the abundance of this species. Weather characteristics are less impacting on data collecting but impose navigability issues. So, a stable and large sailboat is indicated for the surveys to establish a cost-effective good balance.

There is also an attempt to use C-PODs from towing arrangements for franciscana distribution studies (Teixeira, 2021). In this case, to deal with the buoyancy of the devices, a towing arrangement was created consisting of a three-meter-long concreted PVC bar (75 mm in diameter), connected by 4-meter cables at its two ends to the ends of a fiberglass kayak of equal size (Fig. 9.8). One or two C-PODs (one facing forward and one behind) were coupled to the bar by means of metal clamps (Fig. 9.8). Both the bar and the kayak were connected by cables to the towboat, in this case, an inflatable boat with a 200 Hp outboard motor. Teixeira (2021) carried out a series of tests for the ideation of this towing arrangement and considered its application to be promising for the detection of franciscanas, allowing its application in studies of species distribution. For this purpose, the author suggests a maximum displacement speed of 10km/h and a minimum distance of 100m from the tow rope, in order to reduce the capture of engine noise (Teixeira, 2021).

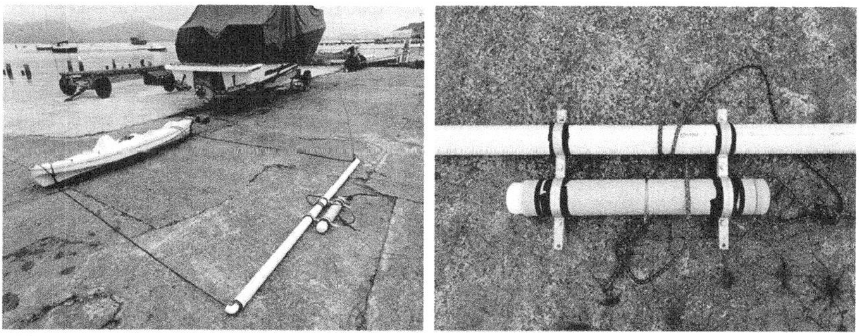

(A) (B)

Fig. 9.8 (A) An array made with caique, cables, and PVC bar with internal concrete for coupling C-POD (Chelonia Limited) in a horizontal position, used for towing by vessel. (B) Close-up view of monitoring device coupled with metal clamps. *Credit: Paitach.*

Prospect for abundance estimates

Different methods have been applied to estimate the abundance of franciscana where the aerial survey is the most prominent method (Crespo et al., 2010; Danilewicz et al., 2010; Secchi et al., 2003). The VIII Workshop for the Research and Conservation of the Franciscana (Anonymous, 2015) was defined as a priority evaluating the use of alternate, more economical, methods to assess trends in abundance, such as PAM.

Density estimates based on PAM were performed on a variety of cetacean species (Barlow and Taylor, 2005; Fais et al. 2016; Marques et al., 2009, 2011). Different derivations of the distance sampling methodological framework (Buckland et al., 2010) can be applied for abundance analysis from towed and static arrays. However, just detecting franciscana sounds with a PAM device does not allow obtaining information about animal density. Two basic pieces of information are needed: (1) the detection probability of an individual/group of franciscana as a function of the distance to the acoustic device (g(r)) and (2) the acoustic emission rate per individual/group.

For towed arrays, distance sampling is applied following linear transects (Buckland et al., 2001), and it is required to calculate the distance between the target species and the transect line, and the sounds produced by franciscana can be concomitantly used to detect and locate the sound source. The spacing between hydrophones considers the frequencies of vocalizations of the target species. In the case of franciscana, short distances are more adequate to perform the localization. Detected signal bearings can be estimated by the time difference of arrival of each signal to the pair of hydrophones (Hastie et al. 2003; Lewis et al. 2007), then displayed as bearings against time (Isojunno, 2014). Event locations can be estimated using target motion analysis, assuming a slow animal swim speed relative to the ship's speed (Barlow and Taylor, 2005; Hastie et al., 2003; Lewis et al., 2007). In this case, the estimated distances along the detection effort feed the model to define the g(r).

In the case of static arrays, density (and later abundance) estimates can be performed using the transect point sampling and the cue counting method (Marques et al., 2009; Marques et al., 2013). Unlike distance sampling in linear transects, in which detection probability distances are collected simultaneously with the population estimation effort, in the case of static devices, such an estimate must be obtained through a specific experiment for this purpose. In this case, the estimation of g(r) can be achieved in several ways, such as using theodolites from a fixed high-height platform to monitor the approach of animals to the PAM device (e.g., Kyhn, 2010), so that sampling acoustics and visuals must be combined (Mellinger et al., 2007). The increasing advancements in the use of unmanned aerial vehicles (like drones) and photogrammetry techniques can also be an interesting and promising alternative to estimate the g(r).

A second key required parameter necessary for both towed or static acoustic population estimate is the sound production rate, which in cetaceans can be influenced by factors like season, group size, gender, density, and possibly by anthropogenic noise. The most ideal way to obtain such information is through acoustic devices directly coupled to the animals (e.g., D-Tags). However, such methodology depends on the capture and physical containment of animals, a procedure considered extremely complex and of high risk for the franciscanas (Well et al., 2021). Therefore, alternative approaches have been employed. Andriolo et al. (2015) analyzed potential factors that could affect the production of high-frequency clicks in the resident population of franciscana dolphins in the Babitonga Bay, using acoustic and visual (from helicopter) sampling simultaneously. The group size was the only significant predictor and indicated that click frequency increases with the group's number. The results corroborate that using an average franciscana individual emission rate should be appropriate to obtain estimates of abundance using acoustic signals. Andriolo et al. (2014) calculated the median individual click rate at 0.290 clicks/s (10th = 0.036 and 90th = 1166 percentiles). The result is a general approximation of the individual echolocation signal emission rate.

As it is a methodology that is still under development, the use of static and towed arrays applied to abundance estimates of franciscanas has limitations that are expected to be overcome as efforts are made, seeking to improve equipment, methods of signal processing, and appropriate statistical correction factors. Further experiments are required to determine the probability of detecting franciscanas to perform the distance sampling methodology appropriately.

Acoustic recognition of calves

Considering the difference in the echolocation signals produced by the calves, mentioned in the ontogenetic development section, this character represents a promising possibility for their acoustic recognition through passive monitoring devices (Melcón et al., 2016). The first point that must be considered, however, is that all recorded calves that showed this characteristic were in a stress situation (isolated in tide pools or in captivity). Should the echolocation behavior be exhibited by calves irrespective of the behavioral state, the possibility of acoustic recognition through PAM would have important implications for franciscana research and management. In other words, if click trains produced by calves then can be identified and differentiated, it would make it possible to acoustically estimate the relative density of calves and, consequently, the birth rate of a studied population (Paitach et al., 2021). To make this feasible, however, it is necessary the prior knowledge of how long post-partum old a calf generates these broadband clicks, in addition to the other two baseline requirements to carry out population estimates mentioned above.

Conservation and anthropogenic underwater noise

A field in full development and extremely relevant is that of psychoacoustics, which seeks to assess the impacts of anthropogenic underwater noise on the behavior, hearing, and physiology of cetaceans (Guan et al., 2021). The strictly coastal distribution makes the franciscana especially vulnerable to anthropogenic impacts, which are more intense near the coast. Coastal occupation, vessel traffic, dredging, and rock removal work for the construction and expansion of port complexes, are some examples of human activities that represent a strong potential for intensifying underwater noise. In Babitonga Bay, it was observed that the franciscana population avoids the noisiest areas, close to the ports operating in the territory (Holz, 2014).

In Brazil, the National Action Plan for the Conservation of franciscanas is an instrument designed in a participatory manner with civil society and used by public management to identify and propose priorities for action and research for the species' conservation. This document contains a recommendation to "investigate potential noise impacts and propose mitigation measures in critical areas of franciscana occurrence." However, this is not an easy task. There is still a lack of methods and baselines to identify impacts, especially chronic ones, of this nature on franciscanas and cetaceans in general.

Definitions of underwater noise are not yet standardized, and this is an urgently needed demand to be able to assess potential noise impacts on species (Pires et al., 2021). In addition, the assessment of potential environmental impacts of human activities should consider an assessment of the acoustic landscape prior to the commencement of proposed activities (Jensen et al., 2009). The assessment of franciscana bioacoustics must also be considered before, during, and after implantation and potentially impacting human activities, making it possible to identify effects on the acoustic behavior of the species. Different bioacoustics approaches to the study of the franciscana, and its habitats must be involved in the environmental impact assessment process, and this chapter presents a compilation of useful information and recommendations that can contribute to this process.

References

Alves, A.K.M., Holz, A.C, Paitach, R.L., Maranho, A., Farah, R.F., Cremer, M.J., 2016. Pulsed sounds of a franciscana dolphin calf. In: Abstract book of the I Workshop LAMLA, Natal, Rio Grande do Norte.

Amorim, T.O.S., Rezende, D.E., Castro, F., Moron, J.R., Duque, B.R., Di Tullio, J.C., et al., 2019. Integrative bioacoustics discrimination of eight delphinid species in the western South Atlantic Ocean. PLoS One 14 (6), e0217977.

Andersen, S.H., Amundin, M., 1976. Possible predator-related adaption of sound production and hearing in the harbour porpoise (Phocoena phocoena). Aquat. Mamm. 4, 56–57.

Andriolo, A., Castro, F.R., Amorim, T.O.S., Miranda, G., Di Tullio, J., Moron, J., et al., 2018b. Marine mammal bioacoustics: using towed array systems in the Western South Atlantic Ocean. In: Rossi-Santos, M.R., Finkl, C.W. (Eds.), Advances in Marine Vertebrate Research in Latin America: Technological Innovation and Conservation. Springer International Publishing. Springer, Berlin, pp. 113–147.

Andriolo, A., Pizzorno, J.L., Sucunza, F., Zerbini, A.N., Danilewicz, D., Mura, J.P, 2018a. Acoustic sailboat surveys of franciscana dolphin (Pontoporia blainvillei) using towed arrays, Abstract book of the XII SOLAMAC Conference. Lima Peru.

Andriolo, A., Sucunza, F., Zerbini, A.N., Danilewicz, D., Cremer, M.J., Holz, A.C., 2014. Preliminary calculation of individual echolocation signal emission rate of Franciscana dolphins (Pontoporia blainvillei). J. Acoust. Soc. Am. 136, 2277.

Andriolo, A., Sucunza, F., Zerbini, A.N., Danilewicz, D., Cremer, M.J., Holz, A.C., et al., 2015. Investigating variables influencing click emission rate of franciscana dolphin (Pontoporia blainvillei), The 7th International DCLDE Workshop. La Jolla, CA.

Anonymous, 2015. Report of the VIII Workshop for the Research and Conservation of the Franciscana Dolphin (Pontoporia blainvillei) in the Soutwestern Atlantic. São Francisco do Sul, Brazil.

Au, W.W.L., 1993. The Sonar of Dolphins. Springer-Verlag, New York, NY, p. 277.

Barcellos, D.D., Santos, M.C.O., 2021. Echolocation characteristics of franciscana dolphins (Pontoporia blainvillei). Mar. Mam. Sci., 131, EL448–EL453.

Barlow, J., Taylor, B., 2005. Estimates of sperm whale abundance in the northeastern temperate Pacific from a combined acoustic and visual survey Mar. Mam. Sci. 2, 429–445.

Baumann-Pickering, S., Wiggins, S.M., Roth, E.H., Roch, M.A., Schnitzler, H.U., Hildebrand, J.A., 2010. Echolocation signals of a beaked whale at Palmyra Atoll. J. Acoust. Soc. Am. 127, 3790–3799.

Benda-Beckmann, A.M., Lam, F.P.A., Moretti, D.J., Fulkerson, K., Ainslie, M.A., Van IJsselmuide, S.P., et al., 2010. Detection of Blainville's beaked whales with towed arrays. Appl. Acoust. 71, 1027–1035.

Bittencourt, L., Soares-Filho, W., Lima, I.M.S., Pai, S., Lailson-Brito, J., Barreira, L.M., et al., 2018. Mapping cetacean sounds using a passive acoustic monitoring system towed by an autonomous Wave Glider in the Southwestern Atlantic Ocean. DRORE7 142, 58–68.

Blomqvist, C., Amundin, M., 2004. High-frequency burst-pulse sounds in agonistic/aggressive interactions in bottlenose dolphins, tursiops truncatus. In: Thomas, J.A., Moss, C.F.,

Vater, M. (Eds.), Echolocation in Bats and Dolphins. The University of Chicago Press, Chicago.

Bordino, P., Albareda, D., Palmerio, A., Mendez, M., Botta, S., 2002. Reducing incidental mortality of franciscana dolphin Pontoporia blainvillei with acoustic warning devices attached to fishing nets. Mar. Mam. Sci. 18, 833–842.

Botta, S., Secchi, E., Muelbert, M., Danilewicz, D., Negri, M., Cappozzo, H., Hohn, A., 2010. Age and growth of franciscana dolphins, Pontoporia blainvillei (Cetacea: Pontoporiidae) incidentally caught off southern Brazil and northern Argentina. J. Mar. Biolog. Assoc. 90 (8), 1493–1500.

Buckland, S.T., Anderson, D.R., Burnham, K.P., Laake, J.L., Borchers, D.L., Thomas, L., 2001. Introduction to Distance Sampling. Oxford University Press, Oxford.

Busnel, R., Dziedzic, G.A., Alcuri, G., 1974. Estudes preliminaries de signaux acoustiques de Pontoporia blainvillei Gervaiset D'Orbigny (Cetacea, Platanistidae). Mammalia 38, 449–459.

Carstensen, J., Henriksen, O., Teilmann, J., 2006. Impacts of offshore wind farm construction on harbour porpoises: acoustic monitoring of echolocation activity using porpoise detectors (T-PODs). Mar. Ecol. Prog. Ser. 321, 295–308.

Cassens, I., Vicario, S., Waddell, V.G., Balchowsky, H., Van Belle, D., Ding, W., et al., 2000. Independent adaptation to riverine habitats allowed survival of ancient cetacean lineages. Proc. Natl. Acad. Sci. USA 97, 11343–11347.

Clausen, K.T., Wahlberg, M., Beedholm, K., Deruiter, S., Madsen, P.T., 2010. Click communication in harbour porpoises (Phocoena phocoena). Bioacoustics 20, 1–28.

Clay, T.A., Mangel, J.C., Alfaro-Shigueto, J., Hodgson, D.J., Godley, B.J., 2018. Distribution and habitat use of a cryptic small cetacean, the burmeister's porpoise, monitored from a small-scale fishery platform. Front. Mar. Sci. 5, 220.

Cremer, M.J., Holz, A.C., Bordino, P., Wells, R.S., Simões-Lopes, P.C., 2017. Social sounds produced by franciscana dolphins, Pontoporia blainvillei (Cetartiodactyla, Pontoporiidae). J. Acoust. Soc. Am. 141, 2047–2054.

Crespo, E.A., Pedraza, S.N., Grandi, M.F., Dans, S.L., Graffo, G.V., 2010. Abundance and distribution of endangered Franciscana dolphins (Pontoporia blainvillei) in Argentine waters and conservation implications. Mar. Mammal Sci. 26, 17–35.

Danilewicz, D., Moreno, I.B., Ott, P.H., Tavares, M., Azevedo, A.F., Secchi, E.R., Andriolo, A., 2010. Abundance estimate for a threatened population of franciscana dolphins in southern coastal Brazil: uncertainties and management implications. J. Mar. Biolog. Assoc. 90 (8), 1659–1666.

Evans, P.G.H., Hammond, P.S., 2004. Monitoring cetaceans in European waters. Mammal Rev 34, 131–156.

Ey, E., Fischer, J., 2009. The "Acoustic Adaptation Hypothesis"—a review of the evidence from birds, anurans and mammals. Bioacoustics 19, 21–48.

Fais, A., Lewis, T.P., Zitterbart, D.P., Álvarez, O., Tejedor, A., Soto, N.A., 2016. Abundance and distribution of sperm whales in the Canary Islands: can sperm whales in the Archipelago sustain the current level of ship-strike mortalities? PLoS One 11, e0150660.

Frainer, G., Huggenberger, S., Moreno, I.B., 2015. Postnatal development of franciscana's (Pontoporia blainvillei) biosonar relevant structures with potential implications for function, life history, and bycatch. Mar. Mam. Sci. 31, 1193–1212.

Galatius, A., Olsen, M.T., Steeman, M.E., Racicot, R.A., Bradshaw, C.D., Kyhn, L.A., Miller, L.A., 2019. Raising your voice: evolution of narrow-band high frequency signals in toothed whales (Odontoceti). Biol. J. Linn. Soc. 126, 213–224.

Guan, S., Brookens, T., Vignola, J., 2021. Use of underwater acoustics in marine conservation and policy: previous advances, current status, and future needs. J. Mar. Sci. Eng. 9, 173.

Hastie, G.D., Swift, R.J., Gordon, J.C., Slesser, G., Turrell, W.R., 2003. Sperm whale distribution and seasonal density in the Faroe Shetland Channel. J. Cet. Res. Manag. 5, 247–252.

Herman, L.M., Tavolga, W.N., 1980. The communication systems of cetaceans. In: Herman, L.M. (Ed.), Cetacean Behavior: Mechanisms and Functions. Wiley, New York, NY, pp. 149–209.

Holz, A.C., 2014. Caracterização do ambiente acústico subaquático e uso de habitat pela toninha, Pontoporia blainvillei, na Baía da Babitonga, litoral norte de Santa Catarina, Brasil Master Thesis, Universidade da Região de Joinville – UNIVILLE, Joinville, 76.

Isojunno, S., 2014. Influence of natural fators and anthropogenic stressors on sperm whale foraging effort and success at high latitudes PhD Thesis, University of St. Andrews, St. Andrews.

Janik, V.M., 2009. Acoustic communication in delphinids. Adv. Study Behav. 40, 123–157.

Jensen, F.H., Bejder, L., Wahlberg, M., Soto, N.A., Johnson, M., Madsen, P.T., 2009. Vessel noise effects on delphinid communication. Mar. Ecol. Prog. Ser. 395, 161–175.

Kimura, S., Akamatsu, T., Dong, L., Wang, K., Wang, D., Shibata, Y., Araib, N., 2014. Acoustic capture-recapture method for towed acoustic surveys of echolocating porpoises. J. Acoust. Soc. Am. 135 (6), 3364–3370.

King, S., Sayigh, L., Wells, R., Fellner, W., Janik, V., 2013. Vocal copying of individually distinctive signature whistles in bottlenose dolphins. Proc. R. Soc. B 280, 1–9.

Küsel, E.T., Munoz, T., Siderius, M., Mellinger, D.K., Heimlich, S., 2017. Marine mammal tracks from two-hydrophone acoustic recordings made with a glider. Ocean Sci 13, 273–288.

Kyhn, L.A., 2010. Passive acoustic monitoring of toothed whales, with implications for mitigation, management and biology PhD Thesis, National Environmental Research Institute, Aarhus University, Denmark, 166.

Kyhn, L.A., Jensen, F.H., Beedholm, K., Tougaard, J., Hansen, M., Madsen, P.T., 2010. Echolocation in sympatric Peale's dolphins (Lagenorhynchus australis) and Commerson's dolphins (Cephalorhynchus commersonii) producing narrow-band high-frequency clicks. J. Exp. Biol. 213, 1940–1949.

Kyhn, L.A., Tougaard, J., Beedholm, K., Jensen, F.H., Ashe, E., Williams, R., Madsen, P.T., 2013. Clicking in a killer whale habitat: narrow-band, high frequency biosonar clicks of harbour porpoise (Phocoena phocoena) and Dall's porpoise (Phocoenoides dalli). PLoS One 8 (5), e63763.

Kyhn, L.A., Tougaard, J., Jensen, F., Wahlberg, M., Beedholm, K., Madsen, P.T., 2009. Feeding at a high pitch: source parameters of narrow band, high-frequency clicks from echolocating off-shore hourglass dolphins and coastal Hector's dolphins. J. Acoust. Soc. Am. 125, 1783–1791.

Leeney, R.H., Carslake, D., Elwen., S.H., 2011. Using static acoustic monitoring to describe echolocation behaviour of heaviside's dolphins (Cephalorhynchus heavisidii) in Namibia. Aquat. Mamm. 37 (2), 151–160.

Lewis, T., Gilliespie, D., Lacey, C., Matthews, J., Danbolt, M., Leaper, R., et al., 2007. Sperm whale abundance estimates from acoustic surveys of the Ionian Sea and Straits of Sicily in 2003. J. Mar. Biol. Assoc. UK 87, 353–357.

Madsen, P., Carder, D., Beedholm, K., Ridgway, S., 2005. Porpoise clicks from a sperm whale nose—convergent evolution of 130 kHz pulses in toothed whale sonars? Bioacoustic15, 195–206.

Madsen, P., Wisniewska, D., Beedholm, K., 2010. Single source sound production and dynamic beam formation in echolocating harbour porpoises (Phocoena phocoena). J. Exp. Biol. 213, 3105–3110.

Marques, T.A., Munger, L., Thomas, L., Wiggins, S.M., Hildebrand, J.A., 2011. Estimating North Pacific right whale Eubalaena japonica density using passive acoustic cue counting. Endang. Species Res. 13, 163–172.

Marques, T.A., Thomas, L., Martin, S.W., Mellinger, D.K., Ward, J.A., Moretti, D.J., et al., 2013. Estimating animal population density using passive acoustics. Biol. Rev. 88, 287–309.

Marques, T.A., Thomas, L., Ward, J., DiMarzio, N., Tyack, P.L., 2009. Estimating cetacean population density using fixed passive acoustic sensors: an example with Blainville's beaked whales. J. Acoust. Soc. Am. 125, 1982–1994.
Martin, M.J., Elwen, S.H., Kassanjee, R., Gridley, T., 2019. To buzz or burstpulse? The functional role of Heaviside's dolphin, Cephalorhynchus heavisidii, rapidly pulsed signals. Anim. Behav. 150, 273–284.
Martin, M.J., Torres Ortiz, S., Reyes Reyes, M.V., Marino, A., Iñíguez Bessega, M., Wahlberg, M., 2021. Commerson's dolphins (Cephalorhynchus commersonii) can relax acoustic crypsis. Behav. Ecol. 75 (6).
May-Collado, L.J., Agnarsson, I., Wartzok, D., 2007. Phylogenetic review of tonal sound production in whales in relation to sociality. BMC Evol. Biol. 7, 136.
May-Collado, L.J., Wartzok, D., 2007. The freshwater dolphin Inia geoffrensis produce high frequency whistles. J. Acoust. Soc. Am. 121, 1203–1212.
Melcón, M.L., Failla, M., Iñíguez, M.A., 2012. Echolocation behavior of franciscana dolphins (Pontoporia blainvillei) in the wild. J. Acoust. Soc. Am. 131, 448–453.
Melcón, M.L., Failla, M., Iñíguez, M.A., 2016. Towards understanding the ontogeny of echolocation in franciscana dolphins (Pontoporia blainvillei). Mar Mam Sci 32, 1516–1521.
Mellinger, D., Stafford, K.M., Moore, S.E., Dziak, R.P., Matsumoto, H., 2007. An overview of fixed passive acoustic observation methods for cetaceans. Oceanography 20, 36–45.
Melo, J.F., Amorim, T., Paschoalini, M., Andriolo, A., 2021. The biosonar of the boto: evidence of differences among species of river dolphins (Inia spp.) from the Amazon. PeerJ 9, e11105.
Møhl, B., Wahlberg, M., Madsen, P.T., Heerfordt, A., Lund, A., 2003. The monopulsed nature of sperm whale clicks. J. Acoust. Soc. Am. 114, 1143–1154.
Monteiro-Filho, E.L.A., Monteiro, K.D.K.A., 2001. Low frequency sounds emitted by Sotalia fluviatilis guianensis (Cetacea: Delphinidae) in an estuarine region in southeastern Brazil. Can. J. Zool. 79, 59–66.
Nielsen, B.K., Møhl, B., 2006. Hull-mounted hydrophones for passive acoustic detection and tracking of sperm whales (Physeter macrocephalus). Applied Acoust. 67, 1175–1186.
Oliveira, T.M.N., Tureck, C.R., Bassfeld, J., Faria, J., Brasil, K., Torrens, B.M.O., 2006. Integridade Ambiental da Baía da Babitonga: Características Físicoquímicas, Microbiológicas e Ecotoxicidade. In: Cremer, M.J., Morales, P.R.D., Oliveira, T.M.N. (Eds.), Diagnóstico Ambiental da Baía da Babitonga. UNIVILLE, Joinville, pp. 20–80.
Pagliani, B., Amorim, T.O.S., Castro, F.R., Andriolo, A., 2022. Intraspecific variation in short-beaked common dolphin's whistle repertoire. Bioacoustics. 31, 1–16. https://doi.org/10.1080/09524622.2020.1858449.
Paitach, R.L., 2015. Ecologia alimentar e espacial da toninha (Pontoporia blainvillei) e do boto-cinza (Sotalia guianensis) na Baía da Babitonga, Sul do Brasil [Feeding and spatial ecology of franciscana (Pontoporia blainvillei) and Guiana dolphin (Sotalia guianensis) in Babitonga Bay, Southern Brazil] [Unpublished Master thesis]. Federal University of Santa Catarina, Florianópolis.
Paitach, R.L., 2021. Padrões de uso de habitat e comportamento de toninhas: abordagens acústicas para monitoramento e conservação de uma espécie ameaçada [Habitat use and behavior patterns of franciscana dolphins: acoustic approaches for monitoring and conservation of a threatened species] [Unpublished PhD thesis]. Federal University of Santa Catarina, Florianópolis.
Paitach, R.L., Amundin, M., Teixeira, G., Cremer, M.J., 2021. Echolocation variability of franciscana dolphins (Pontoporia blainvillei) between estuarine and open-sea habitats, with insights into foraging patterns. J. Acoust. Soc. Am. 150, 3987–3998. https://doi.org/10.1121/10.0007277.
Podos, J., da Silva, V.M.F., Rossi-Santos, M.R., 2002. Vocalizations of Amazon River dolphins, Inia geoffrensis: insights into the evolutionary origins of delphinid whistles. Ethology 108, 601–612.

Rayment, W., Dawson, S., Slooten, L., 2009. Use of T-PODs for acoustic monitoring of Cephalorhynchus dolphins: a case study with Hector's dolphins in a marine protected area. Endanger. Species Res. 10, 333–339.

Reyes Reyes, M.V., Marino, A., Dellabianca, N.A., Hevia, M., Raya Rey, M.T.A., Melcón, M.L., 2018. Clicks of wild Burmeister's porpoises (Phocoena spinipinnis) in Tierra del Fuego. Argentina. Mar. Mam. Sci. 34, 1070–1081.

Reyes Reyes, M.V., Tossenberger, V.P., Iñiguez, M.A., Hildebrand, J.A., Melcón, M.L, 2016. Communication sounds of Commerson's dolphins (Cephalorhynchus commersonii) and contextual use of vocalizations. Mar. Mam. Sci. 32, 1219–1233.

Richardson, W.J., Green Jr., C.R., Malme, C.I., Thomson, D.H., 1995. Marine mammals and noise. Academic Press, San Diego, CA.

RRDM, 2019. Rede Rio Doce Mar. Relatório Anual – Anexo 6 Megafauna. RT-23, Programa de Monitoramento da Biodiversidade Aquática, Fundação Espírito-santense de Tecnologia. p. 531. http://www.ibama.gov.br/cif/notas-tecnicas/ct-bio/relatorios-da-rede-rio-doce-mar. Accessed date: January 20, 2022.

Secchi, E.R., Danilewicz, D., Ott, P.H., 2003. Applying the phylogeographic concept to identify franciscana dolphin stocks: implications to meet management objectives. J. Cetacean Res. Manage. 5, 61–68.

Sousa-Lima, R., Norris, T., Oswald, J., Fernandes, D., 2013. A review and inventory of fixed autonomous recorders for passive acoustic monitoring of marine mammals. Aquatic Mammals 39, 205–210.

Sucunza, F., Danilewicz, D., Andriolo, A., Azevedo, A.F., Secchi, E.R., Zerbini, A.N, 2019. Distribution, habitat use, and abundance of the endangered franciscana in southeastern and Southern Brazil. Mar. Mam. Sci. 36, 421–435.

Szymanski, M., Bain, D., Kiehl, K., Pennington, S., Wong, S., Henry, K.R., 1999. Killer whale (Orcinus orca) hearing: Auditory brainstem response and behavioural audiograms. J. Acoust. Soc. Am. 106, 1134–1141.

Teixeira, G., 2021. Elaboração e análise de detectabilidade de um conjunto de reboque para o monitoramento acústico passivo de toninhas, Pontoporia blainvillei, utilizando detectores de clicks (C-PODs) Unpublished Bachelor Thesis, Universidade da Região de Joinville. São Francisco do Sul, SC, 33.

Tellechea, J.S., Ferreira, M., Norbis, W., 2016. Echolocation and burst clicks from franciscana dolphins (Pontoporia blainvillei) on the coast of Uruguay. Mar. Mam. Sci. 33 (2), 687–694.

Tellechea, J.S., Norbis., W., 2014. Sound characteristics of two neonatal franciscana dolphins (Pontoporia blainvillei). Mar. Mam. Sci. 30, 1573–1580.

Thode, A., Skinner, J., Scott, P., Roswell, J., Straley, J., Folkert, K., 2010. Tracking sperm whales with a towed acoustic vector sensor. J. Acoust. Soc. Am. 128 (5), 2681–2694.

Tougaard, J., Kyhn, L.A., 2010. Echolocation sounds of hourglass dolphins (Lagenorhynchus cruciger) are similar to the narrow band high-frequency echolocation sounds of the dolphin genus Cephalorhynchus. Mar. Mam. Sci. 26, 239–245.

Tyack, P., Clark, C., 2000. Communication and acoustic behavior of dolphins and whales Eds. In: Au, W., Popper, A.N., Fay, R.R. (Eds.), Hearing by Whales and Dolphins. Springer-Verlag, New York, NY, pp. 156–224.

Van Parijs, S.M., Clark, C.W., Sousa-Lima, R.S., Parks, S.E., Rankin, S., Risch, D., Van Opzeeland, I.C., 2009. Management and research applications of real-time and archival passive acoustic sensors over varying temporal and spatial scales. Mar. Ecol. Prog. Ser. 395, 21–36.

Verfuß, U.K., Miller, LA., Pilz, P.K.D., Schnitzler, H.U., 2009. Echolocation by two foraging harbour porpoises (Phocoena phocoena). J. Exp. Biol. 212, 823834.

Von Fersen, L., Kamminga, C., Seidl, A., 1997. Estudios preliminares sobre el comportamiento de un ejemplar de franciscana (Pontoporia blainvillei) en Mundo Marino,

Argentina (DT10), Report of the third workshop for coordinated research and conservation of the franciscana dolphin (Pontoporia blainvillei) in the southwestern Atlantic, 26–28 November 1997. Buenos Aires, Argentina.

Wang, X., Wang, D., Akamatsu, T., Fujita, K., Shiraki, R., 2006. Estimated detection distance of a baiji's (Chinese river dolphin, Lipotes vexillifer) whistles using a passive acoustic survey method. J. Acoust. Soc. Am. 120, 1361–1365.

Wells, R.S., Cremer, M.J., Berninsone, L.G., Albareda, D., Wilkinson, K.A., Stamper, M.A., Paitach, R.L., Bordino, P., 2021. Tagging, ranging patterns and behavior of franciscana dolphins (Pontoporia blainvillei) off Argentina and Brazil: considerations for conservation. Mar. Mamm. Sci., 38, 571–605.

Zimmer, W.M.X., Johnson, M., Madsen, P.T., Tyack, P.L, 2005. Echolocation clicks of free-ranging Cuviers beaked whales (Ziphius cavirostris). J. Acoust. Soc. Am. 117, 3919–3927.

CHAPTER 10

Is the franciscana bycatch in gillnet fisheries sustainable?

Eduardo R. Secchi[a], Danielle Monteiro[a], Rodrigo Claudino[b]
[a]Laboratório de Ecologia e Conservação da Megafauna Marinha, Instituto de Oceanografia, Universidade Federal do Rio Grande, Rio Grande, Rio Grande do Sul, Brazil
[b]Programa de Pós-graduação em Oceanografia Biológica, Instituto de Oceanografia, Universidade Federal do Rio Grande, Rio Grande, Rio Grande do Sul, Brazil

"… one crucial role of conservation biologists and wildlife managers is to protect as many species as possible from extinction for as long as possible…" (Burgman et al., 1993)

Introduction

Bycatch as a global conservation concern

Bycatch, simply defined here as "the incidental catch of organisms that are not targeted by the fishery," is a global issue that has accelerated the collapse of several targets (Myers and Worm, 2003; Gilman et al., 2020) and increased the risk of extinction of non-target species (IUCN, 2019). Besides the ecological consequences, bycatch (and stock depletion) has negative economic and social welfare implications, especially in countries relying on fish as their main source of animal protein and income such as subsistence fishers in developing countries (FAO, 2021). Ultimately it represents a waste of high-quality food in a world of hunger. A recent study by the Food and Agriculture Organization of the United Nation (FAO/UN) indicates that the number of people affected by hunger globally has been slowly on the rise since 2014 (FAO, IFAD, UNICEF, WFP and WHO, 2020) and that it is not on the path to achieving the zero hunger target by 2030 according to the Agenda for Sustainable Development adopted at the UN Sustainable Development Summit held in New York in September 2015 (UN General Assembly, 2015). In Latin America and the Caribbean, for instance, food insecurity increased from 22.9% in 2014 to 31.7% in 2019 and it is expected that both food security and nutritional status of the most vulnerable groups will probably deteriorate further due to the impacts of the COVID-19 pandemic (FAO, IFAD, UNICEF, WFP and

The Franciscana Dolphin
DOI: https://doi.org/10.1016/B978-0-323-90974-7.00004-5

WHO, 2020). Therefore, mitigating or eliminating bycatch is an ecological, economical, and ethical need and should be a major priority for national, regional, and international fisheries management agencies or organizations.

Marine mammal bycatch occurs in most types of fishing gears such as gill and seine nets, hook and lines, traps (including pound nets, pots, and cages), and trawls (Read et al., 2006; Reeves et al., 2013; FAO 2021). For many species, however, gillnets pose the greatest risk (e.g., Reeves et al., 2013; Brownell et al., 2019). Bycatch of marine mammals and other large vertebrates in gillnet fisheries is a worldwide conservation issue that is far from being solved and has worsened in the last 40 years (e.g., Reeves et al., 2013, Brownell et al., 2019, FAO 2020). It has been estimated that more than half a million marine mammals are incidentally killed in fisheries every year, including several endangered and protected species (Read et al., 2006; Gray and Kennelly, 2018). After the Yangtze River dolphin or baiji (*Lipotes vexilifer*) was declared functionally extinct (Turvey et al., 2007), and it may be extinct now, the vaquita (*Phocoena sinus*) is considered the most critically endangered cetacean with only ca. 10 individuals remaining alive (Jaramillo-Legorreta et al., 2019; Sonne et al., 2021) and it may become extinct about 75 years after its description in the Upper Gulf of California, Mexico (Norris and McFarland, 1958). Vaquita's very critical situation is mostly due to the illegal gillnet fishery driven by the high value of totoaba (*Totoaba macdonaldi*) swim bladder in China. Despite the urgent need to eliminate gillnet fishing within a very small area as established in 2015, the new Mexican Government recently (December 2018) eliminated compensation to fishers without providing alternative leading to an accelerated illegal gillnetting for shrimp, totoaba, and other finfish (Brownell et al., 2019) and now (by July 2021), regrettably, the Government officially abandoned the policy of maintaining a fishing-free zone around the vaquita's refugia, replacing it with a sliding scale of punishments if more than 60 boats are seen in the area on multiple occasions (e.g., Sonne et al., 2021). From the conservation point of view, this decision is a setback that eliminates the only fishing management option that could save the vaquita.

Unfortunately, several other species are under threat. Among the 82 cetacean species in the world, 4 are listed as "Critically Endangered," 11 as "Endangered," 7 as "Vulnerable," 10 as "Near Threatened," 46 as "Least Concern," conservation-dependent, and 9 as "Data Deficient" on the IUCN Red List (IUCN 2019). The latter means that no adequate information is available for a proper assessment rather than a more favorable situation. Besides this, many subspecies or discrete populations of small cetaceans, that are in low numbers, are classified as critically endangered and face a high risk of collapse and eventual extinction. These demographically discrete units include the Atlantic humpback dolphin (*Sousa teuszii*), Maui dolphin (*Cephalorhynchus hectori maui*), Taiwanese humpback dolphin (*Sousa chinensis taiwanensis*), Yangtze finless porpoise (*Neophocaena asiaeorientalis asiaeorientalis*), Baltic Sea population of harbor porpoise (*Phocoena phocoena*), Cook Inlet population of beluga (*Delphinapterus leucas*), Fiordland population of bottlenose dolphin (*Tursiops truncatus*) and several populations of *Orcaella brevirostris* (see Brownell et al., 2019 and reference therein). Bycatch in gillnets is the main threat to most of them which have in common their restricted coastal distribution that greatly overlaps with fishing operations.

In South America, several fluvial species (the Tucuxi - *Sotalia fluviatilis* and the Amazon river dolphin, *Inia* spp.) and coastal marine small cetaceans including Burmeister porpoise (*Phocoena spinipinnis*), Guiana's (*Sotalia guianensis*), Commerson's (*Cephalorhynchus commersonii*), Chilean (*C. eutropia*), Lahille's bottlenose (*Tursiops truncatus gephyreus*) and franciscana (*Pontoporia blainvillei*) are also threatened by high levels of incidental catches in gillnet fisheries (e.g., Reyes and Oporto, 1994; Iñiguez et al., 2003; Iriarte and Marmontel, 2013; Clay et al., 2018; Crespo et al., 2010a; Silva et al., 2018; Fruet et al., 2021; Secchi et al., 2021). Some are either declining or present very small population sizes and are listed as Near Threatened (i.e., Chilean dolphin), or Vulnerable (i.e., Lahille's bottlenose and franciscana dolphins), in the IUCN Red List (Heinrich and Reeves, 2017; Zerbini et al., 2017; Vermeulen et al., 2019).

The franciscana bycatch

Franciscana is believed to be the most endangered cetacean in the western South Atlantic Ocean (Secchi et al., 2003b; 2021). Its low reproductive and survival rates make it vulnerable to high levels of non-natural mortality (e.g., Danilewicz et al., 2002; Secchi and Fletcher, 2004; Cáceres et al., 2020). The species is highly vulnerable to bycatch in fisheries. Although occasional incidental kills occur in trawl fisheries (e.g., Cappozzo et al., 2007; Montealegre-Quijano and Neves-Ferreira, 2010; Franco-Trecu et al., 2019), active gillnetting (Secchi et al., 1997) and small-scale driftnet (e.g., Bertozzi and Zerbini, 2002) mortality in the bottom set gillnets is currently, by far, the greatest threat to the species (Secchi et al., 2021) and occurs both in small-scale artisanal and large-scale industrial fisheries (Fig. 10.1) throughout its distribution range (e.g., Corcuera, 1994; Praderi, 1997;

Fig. 10.1 ***Small-scale artisanal (left) and large-scale industrial (right) gillnet fisheries in southeastern and southern Brazil, respectively.*** (photos by Projeto Biopesca and Ecomega/FURG).

Secchi et al., 1997; Di Beneditto et al., 1998; Bertozzi and Zerbini, 2002; Pinheiro and Cremer, 2006; Franco-Trecu et al., 2009; Frizzera et al., 2012; Marcondes et al., 2018). No reports of intentional kills exist. Carcasses of incidentally caught franciscanas are predominantly discarded overboard though they can occasionally be used for human consumption and other purposes. There are a few reports of blubber use as shark bait in the long-line fishery off Atafona, Rio de Janeiro, Brazil (Di Beneditto and Ramos, 2001). In Rio Grande do Sul and Paraná states there are a few records of using the meat for human consumption and to feed dogs (Secchi et al.,1997; Zanellato, 1997). In the former case, a few fishers reported to have extracted the oil from the blubber (Fig. 10.2) to waterproof boats.

Fig. 10.2 ***Blubber of a franciscana suspended in a fishing shed in Rio Grande, southern Brazil, in the mid-1980s.*** (photo: Salvatore Siciliano).

In Uruguay, informal trade of franciscana oil to treat horse's coats was reported (UNEP/CMS 2000). In Argentine, sun-dried and salted franciscana meat, locally known as "mushame," had been consumed by General Lavalle and San Clement del Tuyú community (Praderi et al.,1989; Carman and Carman, 2016).

Uruguay

Reports of incidental mortality in the shark gillnet fisheries of Punta del Diablo date back to the early 1940s (Van Erp, 1969). By the late 1960s, bycatch was estimated to be as high as 1500–2000 animals per year (Brownell and Ness, 1970; Pilleri, 1971). Between 1974 and 1994 at least 3683 franciscanas were estimated to be killed in gillnets, with annual estimates ranging from 66 to 418 dolphins (Praderi, 1997). Large mesh-sized nets set for sharks were responsible for about 70 to 90% of the kills (e.g., Praderi, 1997; 2000). The depletion of the targeted sharks led to a decrease of approximately 25% in the fishing effort using these nets by the early 1980s. In the mid-1990s, only 20% of the fishery targeted sharks (Praderi, 1997). It was expected that changes in the Uruguayan coastal fishery would benefit the recovery of the franciscana from the intense bycatch pressure of the past (Praderi, 1997). Nevertheless, an uncontrolled increase in fishing effort closer to shore using smaller mesh-sized nets to catch bony fishes (e.g., whitemouth croaker, *Micropogonias furnieri*, and striped weakfish, *Cynoscion guatucupa*) (Franco-Trecu et al., 2009), with an associated high rate of franscicana bycatch, in adjacent areas of southern Brazil (Prado et al., 2013, 2021), is likely to have offset or nullified any recovery at best. The most recent estimates, covering ca. 10% of the artisanal fishing vessels in the La Plata River estuary and ca. 50% in the Atlantic coast, were performed between 2006 and 2008 and resulted in mean annual mortality of 94 (CI: 85-118) and 195 (CI: 181-232) franciscanas in these ecosystems, respectively (Franco-Trecu et al., 2009). Bycatch in industrial trawl fishery is lower but not negligible. This fishery, composed of 33 fishing vessels, operates in the La Plata River estuary and the Atlantic coast, within the "Argentinian-Uruguayan Common Fishing Zone," up to the 50 m isobath, targeting mainly the whitemouth croaker. The mean number of franciscanas annually killed in this fishery, for the period 2009–2012, was estimated in 99 (IC: 49-210) individuals based on interviews with fishers (Franco-Trecu et al. 2019).

Brazil

Although gillnetting began around 1940 in southern Brazil (Haimovici et al., 1997), gillnet sets for bottom-fish were only recognized as a major threat to the franciscana in the 1980s (Pinedo, 1994). Bycatch estimates have been obtained from different approaches, including counting washed ashore carcasses found during beach surveys, indirect monitoring the fishing fleet through logbooks and interviews with fishers and, very occasionally, based on direct onboard observations. In Espírito Santo state, the extreme north of the species range, an average of 5.9 (maximum 14) carcasses/year have been recorded in beach monitoring between 2011 and 2017 (Marcondes et al., 2018). In Rio de Janeiro state, slightly over 100 franciscana were estimated to be annually killed in bottom set gillnet, based on interviews with fishers (Di Beneditto, 2003). South of Rio de Janeiro, from São Paulo to Santa Catarina states, the most complete estimates of mortality also came from near-daily beach monitoring carried out from 2016 to 2018. These surveys recorded 1628 carcasses that representing a mean of 543 individuals/year. Even though many animals were in an advanced state of decomposition, data indicated that most of them died because of incidental capture in fishing nets (Cremer et al., 2018). In Rio Grande do Sul state, where the most comprehensive fishing monitoring program to assess franciscana bycatch was taken place with the cooperation of fishers, estimates ranged from several hundred to more than one thousand individuals killed each year in commercial large-scale gillnetting (e.g., Secchi et al., 1997, 2004; Moreno et al., 1997). After the regulation of the gillnet fishery by the Federal Government (Brazil, 2012), which limits the effort throughout the states where franciscana occurs, provoked an aversive reaction by most fishers of Rio Grande do Sul who are not willing to cooperate with bycatch assessment research programs. Since then, most of the bycatch data come from beach surveys to account for washed ashore carcasses that are thrown back to sea by fishers (Fig. 10.3). During the last two years (2019 and 2020), 412 to 441 franciscanas were recorded dead along 340km on the central-south

Fig. 10.3 ***Franciscanas incidentally killed in gillnet fishery thrown back to the sea (left) are eventually washed ashore (right).*** (photos by Genoves/Ecomega and NEMA).

portion of the Rio Grande do Sul coast, from Lagoa do Peixe to Chuí river mouth/Santa Vitória do Palmar, in the border with Uruguay (ca. 55% of the state's coast). According to Prado et al. (2013), only around 10% of the incidentally killed franciscanas end washed ashore in this area. Therefore, applying a correction factor proposed by these authors, the number of carcasses found during beach surveys carried out in those two years would be as high as 3908 in 2019 and 4183 in 2020. These results show that the current magnitude of bycatch remains similar to periods prior to regulations to reduce fishing effort (see Fig. 10.4). Although it is not recommended to apply these correction factors to other areas due to regional differences in oceanographic conditions that influence the drift pattern and hence the chances of a carcass to wash ashore, it is likely that those figures obtained in other beach surveys programs are also underestimated.

Argentina

Similarly, to Brazil, gillnets also emerged in the 1940s, though no record of bycatch exists for that time (e.g., Praderi et al., 1989). Gillnet fisheries

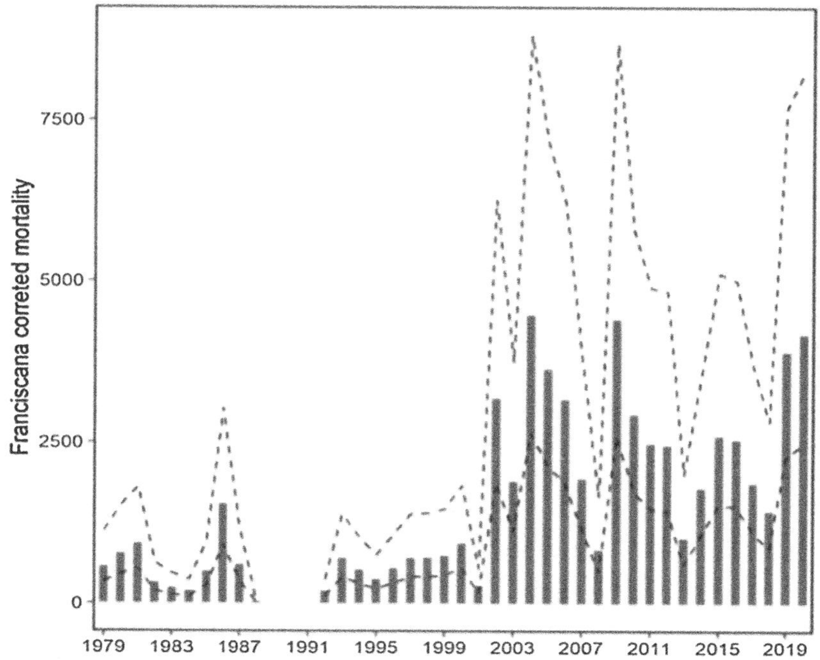

Fig. 10.4 ***Incidental mortality estimates for franciscanas obtained from stranded carcasses reported from 1979 to 2020.*** The dashed lines indicate 95% credible intervals. *(modified from Prado et al., 2013).*

targeting bottom-dwelling fish became the major conservation concern for franciscana in this country only in the 1980s (Crespo et al., 1986; Praderi et al., 1989; Corcuera et al., 1994). Most of the franciscana bycatch occurs in inshore gillnets targeting croakers and several species of sharks (*Galeorhinus galeus*, *Mustelus* spp., *Eugomphodus taurus* and *Squatina argentina*) in water less than 20 m deep (Corcuera et al.,1994; Crespo et al.,1994). Perez-Macri and Crespo (1989), based on a survey of several fishing communities conducted between 1984 and 1986, estimated that around 350 franciscanas were incidentally killed in Argentine gillnet fisheries per year. During the late 1980s, a marked decrease in franciscana bycatch in the Necochea region, which holds one of the largest gillnet fishing fleets in Argentina, was attributed to a movement of the fleet to fishing sites further offshore and a subsequent reduction in total fishing effort (Corcuera et al., 1994). Nevertheless, the coastal habitat of franciscana makes it vulnerable to small-scale inshore gillnetting. Most of the franciscana bycatch in Argentine waters occurs in fisheries based in small camps along with the Buenos Aires Province (Corcuera, 1994; Bordino and Albareda, 2004). Although most of the incidental mortality occurs in artisanal gillnetting, there seems to be a significant franciscana bycatch in trawl fishery targeting shrimps in the area of Ingeniero White and Puerto Rosales, southern Buenos Aires Province, and is an additional reason for concern (Cappozzo et al., 2007). The annual bycatch has been estimated in the range of 350 to 800 individuals/year, mostly in small-scale gillnet fisheries (Bordino and Albareda 2004; Capozzo et al., 2007; Negri et al., 2012). Since most of the available data on bycatch estimates were obtained from interviews, it is likely that the total annual bycatch for Argentina might be considerably underestimated.

Assessment of bycatch and conservation status

A proper assessment of the effects of bycatch and the conservation status of franciscana is a process that can be viewed as a puzzle, for which several pieces of key-information need to be put together (Fig. 10.5). Firstly, discreet units to be conserved need to be identified based on multiple sources of hierarchical evidence, including genetics, morphology, distribution, and populational parameters, with the former being the most relevant (Dizon et al., 1992). It is strongly recommended that the limits and substructuring of these discreet units be periodically reassessed as new data become available in order to minimize the risks of making type I or type II errors (e.g., Taylor, 2005). Type I error occurs if putative units are split when they

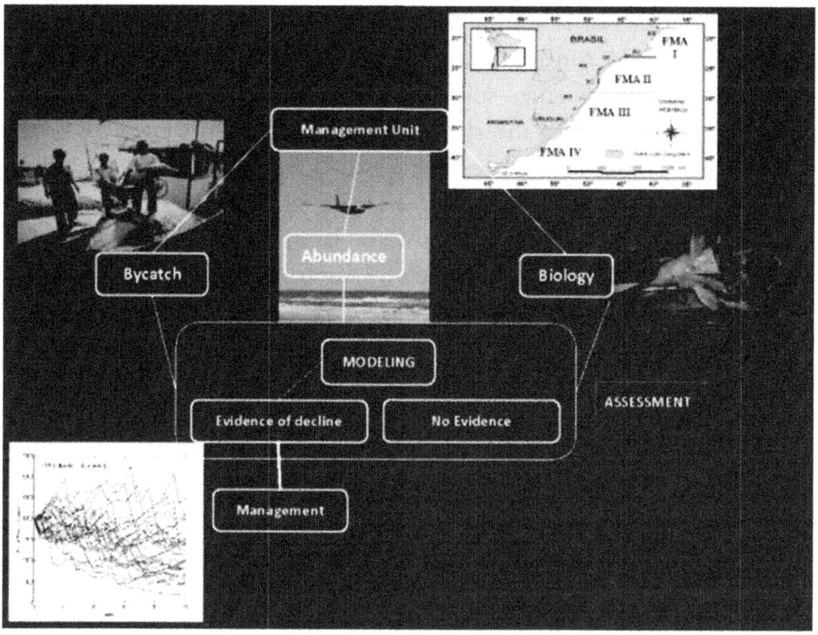

Fig. 10.5 ***The conservation process starts with the definition of discreet units to conserve (management units—MUs, sensu Moritz, 1994), for which, unit-specific biological parameters (Biology), such as age at first reproduction, fecundity, survival rates, and other demography-related parameters (i.e., abundance and non-natural removal—in this case, bycatch) are estimated.*** These data are used as input parameters and fed models (modeling) to assess the status of each unit. If evidence of decline exists, then management recommendations can be made to mitigate bycatch. *(adapted from Secchi et al., 2003a).*

should be clumped, leading to overprotection which is not desired as it represents a waste of resources and/or has socioeconomic consequences to vulnerable communities that depend upon fishing. Type II error, on the other hand, happens if potential units are clumped when they should be split, which may lead to under protection and eventual collapse of one or more units. Therefore, precisely identifying units to conserve is a crucial first step along the conservation process.

In the case of franciscana, four management units are recognized (Franciscana Management Areas, FMAs—sensu Secchi et al., 2003) with the following areas of occurrence: FMA I—coastal waters (up to ca. 30 m isobath) of Espírito Santo and Rio de Janeiro states, Brazil; FMA II—São Paulo, Paraná, and Santa Catarina states, Brazil; FMA III—coastal waters of Rio Grande do Sul state, southern Brazil, and Uruguay; and FMA

IV—coastal waters of Buenos Aires and Rio Negro Provinces (with occasional records as far south as Chubut), in Argentina. Updated genetic data have recommended further subdivisions within each FMA (e.g., Cunha et al., 2014). Given that the degree of differentiation within the FMAs was highly variable, that the evidence was based only on mitochondrial (i.e., female-transferred) DNA data, and that there is presently no agreement on threshold levels of gene flow, the latest refinements in franciscana subpopulation structure were restricted only to locations where the greatest levels of genetic differentiation were observed (Anonymous, 2015). Accordingly, FMA I was considered an Evolutionarily Significant Unit and was split into two distinct management units (FMA Ia and FMA Ib, Cunha et al., 2014). The other important change was the displacement of the boundary between FMA II and FMA III, which moved about 250 km northward, at the central coast of Santa Catarina state (Cunha et al., 2014: Ott et al., 2015) (Fig. 10.6).

After the management units were defined, unit-specific life-history traits (see Secchi et al., 2021 for a revision and references therein), such as age at first reproduction, fecundity, survival rates, and other demography-related parameters (i.e., abundance and non-natural removal—in this case, bycatch) need to be estimated for a proper assessment. For the purpose of assessing the current conservation status of franciscana, five management units are considered, FMA Ia, FMA Ib, and FMA II to IV (Secchi et al., 2003a; Cunha et al., 2014). Although other threats exist, such as habitat degradation that includes physical (noise) and chemical pollution, depletion of fish stocks, and climate change, incidental mortality in fisheries is currently the greatest threat to franciscanas and is the only source of non-natural removal that will be evaluated here. It is important to emphasize, however, that fishery-related cryptic mortality is not considered.

As stated earlier, the bycatch for each of the FMAs has been estimated from different approaches. A beach monitoring program in FMA Ia recorded an average of 5.9 carcasses/year in the period of 2011/2017, ranging from one to 14 franciscana' carcasses recorded each year (Marcondes et al., 2018). For the FMA Ia and Ib combined, bycatch was estimated at >110/year, based on interviews with fishers (Di Beneditto, 2003; Freitas-Neto and Barbosa, 2003). In FMA II, the most complete estimates of mortality also came from near-daily beach surveys carried out from 2016 to 2018. These surveys recorded 1628 carcasses that represent a mean of 543 individuals/year, most with evidence that incidental capture in fishing nets was the cause of death (Cremer et al., 2018). Since many carcasses do not

Fig. 10.6 ***Franciscana distribution in coastal waters of the western South Atlantic and the proposed limits of each Franciscana Management Areas (FMAs—sensu Secchi et al., 2003 and Cunha et al., 2014).*** ES: Espírito Santo state; RJ: Rio de Janeiro state; SP: São Paulo state; PR: Paraná state; SC: Santa Catarina state; RS: Rio Grande do Sul state. *(Adapted from Secchi et al., 2021).*

wash ashore (see Prado et al., 2013), these mortalities estimated from beach surveys are probably very low estimates. Bycatch estimates in FMA III and IV have been obtained through the monitoring of the fishing fleet using logbooks or interviews. Annual bycatch is higher in FMA III, with estimates ranging from several hundred to more than a few thousand animals/

year (Ott et al., 2002; Secchi et al. 2003b, 2004, Franco-Trecu et al., 2009) with much higher values since the early 2000s (Prado et al. 2013, 2016, see also Fig. 10.4). In FMA IV the annual bycatch was estimated to be 350-800 individuals/year (Bordino and Albareda 2004; Capozzo et al., 2007; Negri et al., 2012).

Once assessment-relevant information is obtained, the next step of the conservation process is to use these data as input parameters in assessment models (e.g., Population Viability Analysis (PVA)—Boyce, 1992; potential biological removal (PBR) estimates—Wade, 1998) to predict the current status or project the fate of each discreet unit (herein referred as population) into the future under different scenarios of removal rates. The two most comprehensive approaches that have been used to assess the conservation status of franciscana are the PVA (Secchi and Fletcher, 2004, Secchi, 2006) and the PBR (Secchi et al., 2021). The quality of information available determines the characteristics of the models and the level of uncertainty associated with the predictions or projections. Models can be either structured or unstructured. For example, if the age or stage-structure of the population is non-relevant to its dynamics, simple unstructured models can be as useful as structured models for calculating population growth and assessing population viability under a wide variety of modeling scenarios, including stochasticity, density dependence, parameter uncertainty and several management practices (Burgman et al., 1993). As with structured models, the outcomes depend on data quality. The main advantage of structured models (e.g., matrix models—Caswell, 2001) is that they are designed to capture the internal structure of a population, using a vector of numbers to represent the abundance for each of the components of the population. Although they often require considerably more data to parameterize, such models yield more detailed forecasts and may lead to different predictions about the future of the population. In cases where the age composition of the population has been disturbed (e.g., because of differential mortality among classes due to harvesting or, in this case, bycatch), structured models may be the only way to reasonably predict the consequence of the bycatch (e.g., Morris and Doak, 2003). Unstructured models have to assume that differences among individuals make no difference to the overall dynamics of the population. However, there are often important differences among individuals, both with respect to reproduction and survival rates as well as their vulnerability to bycatch. Individuals within an age class would be similar to one another, at least enough so that grouping them together results in a reasonable approximation. It is also recommended that parameter uncertainty be considered in any assessment process. Ignoring uncertainty in population modeling could result in misleading conclusions and poor management advice. PVA models ignoring parameter uncertainty (so-called point-estimate PVAs) develop estimates of risk based on single estimates for each model parameter and are of very limited practical use because parameter uncertainty is often very large in ecological data. Therefore, ignoring uncertainty in classifying populations according to risk or in estimating extinction probabilities can be misleading and inaccurate (e.g., Beissinger and Westphal, 1998; Ludwig, 1999; Taylor et al., 2002; Wade, 2002).

Considering that assessment-relevant parameters are uncertain, and bycatch may have different impacts on different age-classes and sexes in all franciscanas populations, Secchi and Fletcher (2004) used stage-structured matrix-based models to assess the effect of bycatch taking parameter uncertainty into account. They applied a stage-specific deterministic population model and suggested that the species has a low intrinsic annual population growth rate, with estimates from all populations ranging from 0.011 to 0.024. Using a stochastic version of the same model, that takes effects of environmental and demographic stochasticity (e.g., Caswell, 2001), it was estimated that population net growth lies between 0.003 and 0.016 (Fig. 10.7). The differences in growth rate among the populations are possibly due to a combination of varying population-specific reproductive potential and, in some populations, inaccuracy in parameter estimates.

These low population growth rates are insufficient to compensate for current levels of fishing-related mortality. According to the stochastic scenarios and estimated levels of bycatch in the study by Secchi and Fletcher (2004), populations presented negative (-0.008 in FMA IV, -0.005 in FMA II, -0.002 in FMA III) or nearly null (0.009 in FMA I) population growth rate that will cause their eventual collapse (Fig. 10.8). The population in FMA I would be classified as Vulnerable (VU) while all other populations would be listed as Endangered (EN) according to IUCN criteria (Table 10.1). Moreover, these authors emphasized that parameter uncertainty does not change the conclusion that immediate and important limitations of effort as well as modifications of fishing gear and operations are necessary to avoid the collapse of these populations and to reduce the species' risk of extinction. It is worthwhile to notice that current estimates indicate that abundances in FMA I to III (e.g., Danilewicz et al., 2010; Sucunza, 2020, Sucunza et al., 2020) were likely overestimated and that bycatch estimates were underestimated (e.g., Prado et al. 2013, 2016; Cremer et al., 2018) in the modeling exercises presented in Secchi and Fletcher (2004) and Secchi (2006).

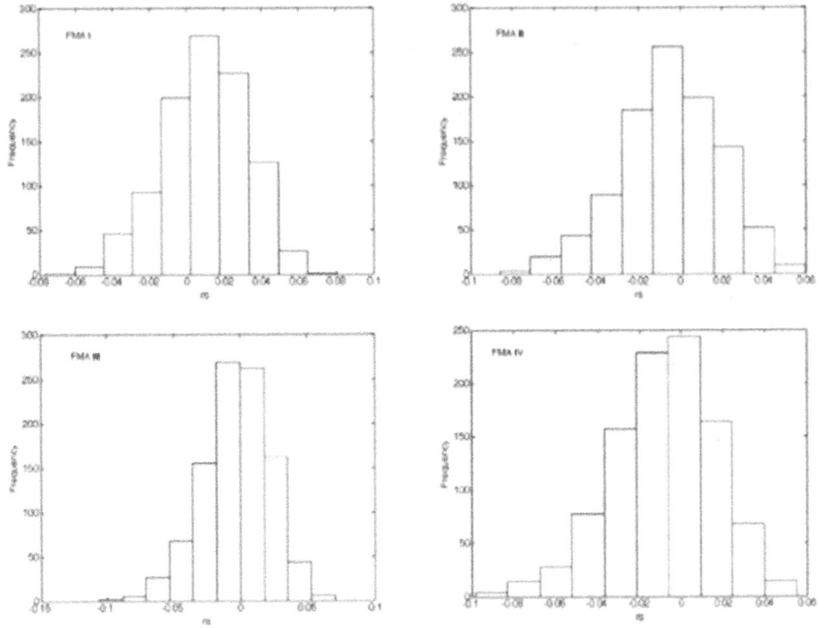

Fig. 10.7 ***Uncertainty in the estimation of the logarithm of the stochastic population growth rate (r_s) for all four franciscana populations underestimated levels of bycatch based on different sources of available evidence (n = 1000 runs).*** (Adapted from Secchi and Fletcher, 2004).

Nevertheless, at the time of the study by Secchi and Fletcher (2004), abundance used as input parameters for FMA I and II were mere approximations based on density estimated for FMA III after applying a correction factor based on estimates of bycatch per unit effort in similar fishing gillnets between compared areas assuming similar vulnerability. Furthermore, other assessment-relevant parameters, including life-history traits and bycatch estimates, were either highly uncertain or unavailable for some populations (Secchi, 2006). Furthermore, current abundance and bycatch estimates are, respectively, lower and higher than those available for the population viability analyses, hence, the actual risk of franciscana collapse is probably much higher. Therefore, new population projections are necessary for the light of up-to-date life history parameters (see Secchi et al., 2021 for a review and references therein) and abundance estimates (Crespo et al., 2010b; Danilewicz et al., 2010, 2012; Sucunza, 2020; Sucunza et al., 2020). Considering these new abundance estimates, Secchi et al. (2021) applied an unstructured PBR model to assess whether current levels of franciscana

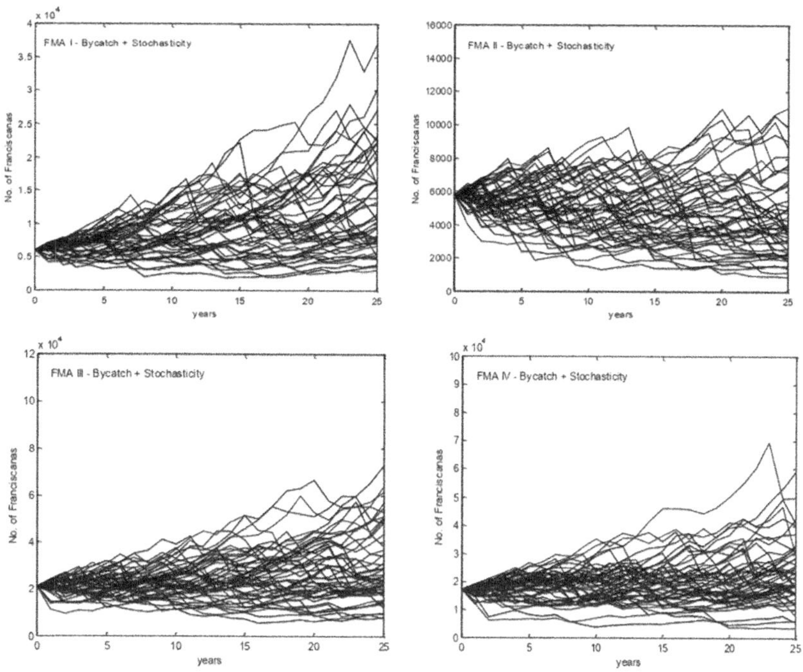

Fig. 10.8 ***Samples (n = 50) from 1000 projections of population size for a period of 25 years considering uncertainty in the estimates of stochastic population growth rate and with current levels of bycatch (i.e., current fishing effort assumed to be constant for the projection period).*** (Adapted from Secchi, 2006).

Table 10.1 Estimate (*r*) of population growth of franciscana under scenarios of available information of fishing effort, with lower (Low) and upper (Upp) confidence limits, probability (as a percentage of model runs) of population declining below initial size, below 70% (VU), 50% (EN) and 20% (CR) of initial size and of increasing 20% above initial size (values that reached the threshold[a] for listing in one of the IUCN's categories are in bold; modified from Secchi and Fletcher, 2004).

FMAs	*r*	Low	Upp	Decline (%)	VU (%)	EN (%)	CR (%)	Increase 0.2 (%)
FMA I	0.009	-0.040	0.053	33.8	**16.5**	7.2	0.3	56.1
FMA II	-0.005	-0.056	0.040	58.1	**32.8**	**15.8**	1.6	32.0
FMA III	-0.002	-0.056	0.045	49.9	**30.4**	**16.2**	1.5	37.4
FMA IV	-0.002	-0.050	0.043	51.8	**28.3**	**14.4**	0.9	36.5

[a]Considering parameter uncertainty and the very likely perspective of future increase in fishing effort, I arbitrarily adopted a risk-averse approach to recommending that if a 15% probability of decline below the VU, EN, and CR thresholds are projected, that population should be listed accordingly in that IUCN risk category.

bycatch are sustainable or not in each of the FMAs (Table 10.2). In this case, FMA Ia and Ib were treated as discrete units to conserve (Cunha et al., 2014). Bycatch ranged from approximately 100, in FMA I, to more than 1000 in FMA III, and exceed from near two in FMA IV to more than fivefold in FMA III the maximum allowed sustainable mortality rate based on PBR approach (Table 10.2). These estimates confirm that species cannot sustain the current levels of non-natural mortality, which are likely underestimated in some areas. Therefore, immediate and severe restrictions on fishing practice and effort are necessary to avoid the collapse of franciscana and to reduce its risk of extinction.

The PBR approach has a clear advantage for management programs related to bycatch as it does not rely on time-consuming and expensive surveys to detect negative population trends, but on simply estimating a mortality level that would cause depletion (Wade, 1998). However, this unstructured model does not capture neither population-specific differences in life-history traits (see Secchi et al., 2021) nor differential mortality between sexes and age classes that shape population dynamics and their potential to respond to impact (e.g., Secchi and Fletcher, 2004; Secchi, 2006). Furthermore, the PBR is a precautionary approach that uses half of the maximum net reproductive rate and minimum abundance estimated. This might be viewed as excessively cautious and sometimes resulting in negative responses by decision-makers in developing countries, where the ecosystem and social needs of economically and nutritionally vulnerable populations depend on reasonable and balanced decisions. Despite the

Table 10.2 Mean abundance estimate (N) from line transects and coefficient of variation (CV), minimum abundance (N_{min}), maximum net reproductive rate (R_{max}), recovery factor (*Rf*), approximate annual bycatch and maximum sustainable bycatch valued based on the potential biological removal approach (PBR, sensu Wade, 1998), for each franciscana management area (FMA—sensu Secchi et al., 2003a).

FMA	N	CV abundance	N_{min}	R_{max}	*Rf*	PBR	Annual bycatch ([a])
FMA Ia	893	0.30	697	0.4	0.1	14	ca. 6
FMA Ib	2,000	0.46	1,383	0.4	0.1	28	ca. 100
FMA II	6,827	0.26	5,504	0.4	0.1	110	ca. 500
FMA III ([b])	15,814	0.32	12,158	0.4	0.1	243	>1,000
FMA IV	14,175	0.42	10,096	0.4	0.1	202	500–800

[a]Recommended to be considered as a minimum value.
[b]It was assumed that the estimates provided by Danilewicz et al. (2010) for approximately 43% of the distribution area can be extrapolated to the whole FMA III.
Absolute values were rounded to the nearest integer (Adapted from Secchi et al., 2021).

unstructured and perhaps overly precautionary nature of PBR, the results converge with the structured matrix-based PVA models and indicate that levels of bycatch are not sustainable and that risks of decline are high for all FMAs. Furthermore, as stated earlier, it is very likely that estimates of bycatch and the effect of fisheries on the long-term survival of franciscana are downward biased to an unknown extent, primarily due to:

i. the occurrence of bycatch in other non-monitored categories of legal (e.g., active gillnetting and trawling—Secchi et al., 1997; Franco-Trecu et al., 2019) and illegal fisheries,
ii. entanglement in ghost nets and other cryptic mortality factors (Gilman et al., 2013, Bravington and Bisack, 1996),
iii. lack of data on the number of fishing boats carried out with gillnets,
iv. limited number of communities where landings are monitored (e.g., Bertozzi and Zerbini, 2002), and
v. under-reporting of incidental catches by fishers (Lien et al., 1994; Hall, 1999).

Therefore, the results of the modeling exercises presented here are likely optimistic and should be viewed as part of an adaptive conservation strategy and re-analyzed as new data become available. PBR should be reviewed in the light of new abundance estimates while PVA models should be rerun using the most up-to-date estimates of life-history traits, abundance, incidental mortality in fisheries, preferably taking sex ratio and age distribution of the bycatch into account.

Potential mitigation options

There are several potential mitigation options that could minimize the effect of bycatch on the fate of franciscana. Bycatch can be reduced either by reducing fishing effort, that is fishing less (e.g., fishing ban, spatial and/or temporal closures, limiting net dimensions and number of licenses), by lowering the bycatch per unit of effort, that is fishing better (e.g., through technological modifications of fishing gear, operational changes in fishing practice such as training fishers to avoid areas or periods of the higher likelihood of bycatch) (Hall et al., 2000), or a combination of both. Drastic alternatives such as large time-areas closures and other fishing effort restrictions are very unpopular among fishers and should be avoided whenever possible due to the negative socioeconomic outcomes that can affect fishing communities that are, in many cases, facing some degree of economic vulnerability. In such circumstances, operational, technological modifications or dynamic management (e.g., Dunn et al., 2016) are preferred. Often, the most successful mitigation strategies emerge from collaborations among fishers, cetacean researchers, fisheries administrators, and experts in gear technologies, each of which contributes with critical expertise that should form part of any project intending to evaluate bycatch reduction measures (e.g., Hall, et al., 2012; FAO 2018, 2020; Brownell et al., 2019; Gilman et al. 2019; Squires et al., 2021). Nevertheless, for the evaluation of their success, it is essential to appreciate the need for fishery-specific trials of the proposed modifications before implementation. For example, one technique may be ineffective in one area and reduce bycatch in another area or produce positive results after minor gear modification. This might happen due to variability that exists in fishing practices, marine mammal population characteristics, oceanographic conditions, local scientific and management capacities, and social, cultural, and economic circumstances (FAO, 2020). The efficiency of mitigation strategies can also vary in time and because of changes in environmental conditions and therefore their evaluation should be based on long-term observation. Moreover, effectiveness should be evaluated after scaling up from experimental trials to a real-fishery situation. In all cases, bycatch reduction measures and their success need robust scientific evidence before their adoption in order to avoid a false sense of security and a misperception that the problem is being solved. The adoption of management measures needs to consider not only the social and economic impacts mentioned above but the potential impact on other species (e.g., Gilman et al., 2019). Despite growing evidence that dynamic management (e.g., Dunn et al., 2016; Hazen et al., 2018) is likely more effective than static approaches, the former depends on data-rich fisheries and complex analytical procedures. Therefore, existing attempts to reduce franciscana, and all other cetaceans, bycatch are based predominantly on static approaches.

A list of potential mitigation strategies that can be used for reducing franciscana bycatch is presented, with their pros and cons identified and commented (Table 10.3). Other incipient mitigation methods less widely tested such as net illumination (e.g., Bielli et al., 2020) and glass bottles with a bolt inside that produce an alarming noise (e.g., Berggren et al., 2020) will not be considered in this review.

It is worthwhile noting, however, that despite the considerable effort that has been devoted to initiatives to seek for technological development and operational procedures to reduce cetacean bycatch during the last few decades, efficient implementation remains slow even in wealthy nations (e.g., Cox et al., 2007; Palka et al., 2008) not to mention in developing countries.

Table 10.3 List of potential mitigation measures to reduce franciscana bycatch in gillnet fishery, including the pros and cons, and comments on their likelihood of success from the author's perspective (adapted from FAO, 2020).

Mitigation Strategy	Pros	Cons	Likelihood of success	Comment
Time-area closures	1. Eliminate bycatch within the area (if enforcement is effective) 2. Likely to provide other ecosystem benefits such as fish stock recovery	1. Reliable information on franciscana and fisheries major overlap areas is required 2. Often unpopular with fishers excluded from fishing grounds 3. Effective enforcement and monitoring is required and may be logistically and economically expensive 4. Benefits are limited to the area(s) and fishing effort is displaced and may increase, potentially affecting franciscana and other vulnerable species outside the area(s) 5. Tend to be static instead of adaptive according to spatial and temporal changes in franciscana distribution and fishery dynamics	Medium	Likelihood of success is often high in wealthy nations where fishers' compliance exists, and enforcement is effective. That is not the case in the three countries where franciscana occurs. Compliance would be minimal if compensation or alternatives were not provided. In Brazil, for example, there is evidence that compliance is low and enforcement is nearly non-existent regarding the gillnet fishing exclusion zones established in the INI 12/2012 (Ecomega unpubl. data)
Reduction of fishing effort (net dimensions)	1. Reduce bycatch (if enforcement is effective) 2. May provide similar ecosystem benefits as time-area closures depending on the magnitude of the effort reduction	1. Unpopular with fishers 2. Enforcement and monitoring is required and may be logistically and economically expensive	Medium to Low	The reduction in fishing effort required to lower franciscana bycatch to sustainable levels would, very likely, turn the fishery economically unviable (e.g., Secchi et al., 2021). In southern Brazil, for example, there is evidence that compliance is low and enforcement is nearly non-existent regarding the reduction of nets length established in the INI 12/2012

(Continued)

Table 10.3 Cont'd

Mitigation Strategy	Pros	Cons	Likelihood of success	Comment
Acoustic deterrents	1. Have demonstrated reduction in bycatch in scientific trials and some fisheries 2. Tend to not affect target catch 3. Produced by a number of manufacturers with different models and accompanied by significant technological upgrades, lowering the price and logistic constraints 4. Help reduce depredation by pinnipeds with increased sound frequency	1. Do not work for all species 2. Effect may be nullified or reduced depending on where they are deployed 3. Can cause habituation and deterrent may become ineffective without adjustments (e.g., random emission of frequency) 4. Exclusion of species from critical habitats when used at a large scale and by overly insonifying the environment 5. Some models emit high power outputs that can cause hearing impairment and other adverse health effects 6. Bycatch reduction is likely insufficient to achieve sustainable level 7. Pingers increase the weight of the net and eventually can explode, posing risks to fishers 8. In some areas, and at certain frequencies, pingers can lead to increased depredation by South American sea lions, *Otaria flavescens* (dinner bell effect, e.g., Bordino et al., 2002).	Low	Despite technological advances, economical and logistical costs may remain prohibitive for many artisanal fishers. Also, large part of the bycatch occurs in large scale gillnetting (e.g., in Southern Brazil) where effective implementation would, very probably, lead to environmental insonification. Furthermore, scientific trial in a real fishery situation would require relatively large coverage for statistical robustness and would be difficult to carry on due to dimension of the fleet. The adoption in a real fishery without a proper monitoring program covering both small and large-scale fisheries would promote a misperception that the problem is being solved

Mitigation Strategy	Pros	Cons	Likelihood of success	Comment
Reflective and stiffnets	1. Reduction in bycatch observed in scientific trials and some fisheries 2. Tend to not affect target catch	1. Do not work for all species 2. Bycatch reduction may not be sufficient to achieve sustainable level (Bordino et al., 2013)	Low	Both stiff and reflective ($BaSO_4$, Fe_3O_4) gillnets are unlikely to reduce bycatch to desirable levels because animals do not echolocate all the time to detect the nets
Tie-downs and nets with lower profile	1. Reduces bycatch of some small cetaceans	1. May decrease target catch 2. Tie-downs can rise bycatch of other species such as sea turtles	Low	Although this strategy reduces the area of the nets, the amount of reduction would not be enough to reduce bycatch sustainable level
Gear switching (gillnet to other gears)	1. May provide comparable catch using different gear that is less risky to cetaceans in particular locations 2. Can provide catch of higher quality including other species of higher value to fishers and consumers 3. Demonstrated efficiency to reduce franciscana bycatch by switching gillnets to bottom longlines in scientific trials in small-scale fishery	1. Transitioning from successful trials to fishery implementation has rarely occurred and faces several challenges 2. Catches of target species may be slightly lower 3. May increase environmental damage (e.g., if switching to trawling) - see Cardoso et al., 2021).	Low to Medium	Fishers tend to be very conservative and resistant to changes from traditional to new fishing practices. It might be feasible in small-scale fisheries if catches of target species increase in quality and value hence fishers' profits (e.g., Berninsone et al., 2020)

Whatever the types of measure to be used it will involve a continued investment of monetary and human resources for scientific trials, implementation and monitoring in real fishery situations as well as assessment of effectiveness (e.g., Dawson et al., 2013; FAO, 2018). Often, effectiveness in trials evaluation is not transferred to the fishery. Moreover, statistical significance in scientific trials may not represent the necessary bycatch reduction to achieve the conservation goal of, for example, reducing incidental mortality to sustainable levels when the mitigation method is transferred to the fishery. Otherwise, it can erroneously be considered by fishers as "problem solved" and the bycatch remains at high levels. Drastic and extreme mitigation measures should be avoided whenever possible unless they are the only option to significantly reduce franciscana's bycatch to a sustainable level in all FMAs (see Secchi et al., 2021). Fishing effort is too high in some areas along with the species' range, such as in southern Brazil that even very strong restrictions of fishing effort would not be enough to achieve conservation goals (see Prado et al., 2021). In this area, the total amount of gillnet nets set in a fishing day can exceed 3500 km during the croaker season (Figs. 10.9 and 10.10). It is unlikely that technological mitigation measures can be implemented in fisheries with these characteristics. The bycatch in this area is nearly one order of magnitude above the maximum allowable fishing-related mortality according to the PBR estimate. As stated earlier, although PBR can be considered too precautionary

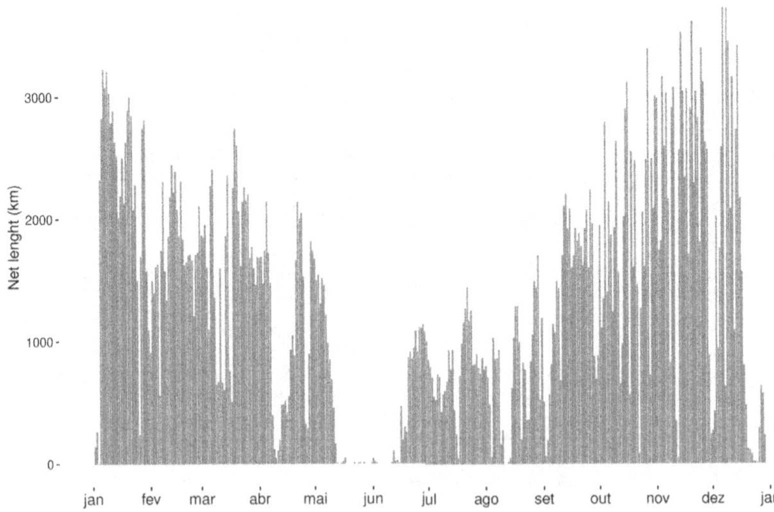

Fig. 10.9 ***The total daily amount of bottom gillnet nets along the year on the coastal waters (over the continental shelf) off Rio Grande do Sul state, Southern Brazil.***

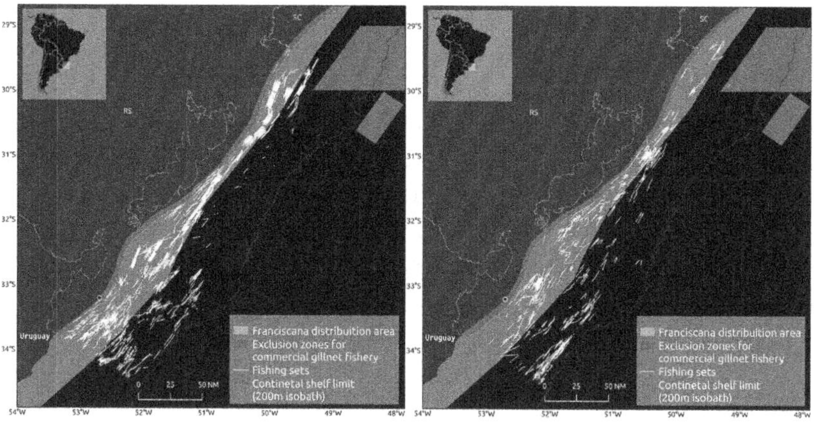

Fig. 10.10 ***Spatial distribution of gillnet nets set on cumulated 10-day period during the whitemouth croaker (left) and striped weakfish (right) fishing season on the Rio Grande do Sul state continental shelf, Brazil.***

and, perhaps, inappropriate for countries where conservation decisions should be balanced with socioeconomic aspects, convergent lines of evidence, using structured models, have demonstrated that bycatch of franciscana is most likely unsustainable and should be markedly reduced (Secchi and Fletcher, 2004; Secchi, 2006). It is reasonable that technological approaches continue to be tested and, if effective in a scientific trial, eventually implemented in fisheries where franciscana bycatch occurs in small-scale gillnetting such as in FMA I, IV, and part of FMA II. In FMA III, however, given the predominance of industrial fleet and the magnitude of the fishing effort, it is unlikely, considering the current knowledge, that any technological mitigation method, even if proved effective in experimental trials, could be voluntarily adopted by the entire fleet due to financial costs or logistic issues. If compulsory, implementation will depend on enforcement, which is often low or null.

Conservation perspectives

Despite the legal protection of the species in the three countries, high and probably unsustainable bycatch in fishing operations has occurred for several decades (Secchi et al. 2021). Other potential causes of human-related mortality, such as habitat degradation, may have a synergetic effect and increase the risk of franciscana collapse. However, the major danger to the species survival, in the short term, is incidental kills in gillnet fisheries (Secchi, 2010, Secchi et al., 2021).

In Brazil, where most of the franciscana bycatch occurs, gillnet fishery had been carried out unregulated for several decades and, despite the obvious negative effects of overfishing (e.g., Haimovici, 1998; Haimovici and Cardoso, 2017) and unsustainable bycatch of franciscana and other endangered species such as sea turtles, sharks, and rays, the fishing effort increased with no limits until recently. In 2012, a joint Interministerial Regulation of the Ministry of Environment and Ministry of Fisheries and Aquiculture (henceforth referred to as INI 12/2012) was issued to regulate the gillnet fishery in southern and southeastern regions of Brazil (Brasil, 2012), from Rio Grande do Sul to Espírito Santo states, which are the distribution limits of franciscana in the country. Among other restrictions, this regulation establishes that: i. the maximum permitted gillnet length was reduced, according to boat size; ii. gillnet fishing permits are no longer offered; and iii. no gillnet fishing is allowed between 15 May and 15 June, except for the artisanal fishery. Besides the restrictions above, a mosaic of no-fishing areas for gillnets with varying levels of restrictions was established (Fig. 10.11) as follows:

a. no powered boat can fish within the first nautical mile and rowing boats can operate but with nets no longer than 1000 m,

b. no powered gillnet fishing vessel can operate within the 5nm between the southern border of Brazil (with Uruguay) to Albardão lighthouse (located ca. 120 km to the north),

c. no gillnetting is allowed for commercial gillnetters within 4 nm between Albardão lighthouse and the northern limit of Rio Grande do Sul state and within 3nm from this point to the remaining area concerned in this Regulation, and

d. no gillnetting is allowed within the 15 nm off the Jurubatiba National Park, in Rio de Janeiro state.

Boats longer than 15 m must carry and turn on vessel monitoring system when at sea so their fishing location can be monitored. A fishing license is lost in case of an infraction to the Regulation.

Even before its publication, scientists had warned that the level of restriction of the INI 12/2012 was insufficient to reduce bycatch and lower than that recommended in the National Action Plan for Franciscana (ICMBio 2010). Nowadays, there is clear evidence that the Regulation is not efficient to reduce franciscana's bycatch to sustainable levels and fishing-related mortality, in some areas, remains in the same order of magnitude as observed in years that preceded the regulations (e.g., Prado et al., 2013, 2016; Cremer et al., 2018). This can be either due to lack of enforcement and, therefore, compliance, or insufficient level of reduction in

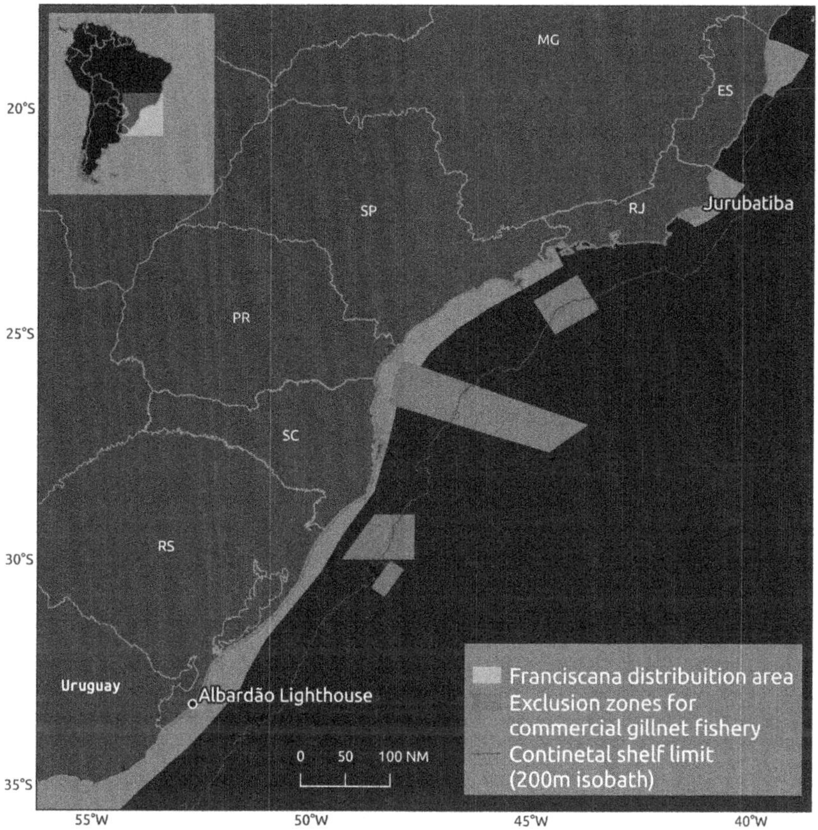

Fig. 10.11 ***Exclusion areas for commercial gillnet fishery in southern and southeastern states of Brazil: Rio Grande do Sul - RS, Santa Catarina – RS, Paraná – PR, São Paulo - SP, Rio de Janeiro – RJ and Espírito Santo – ES (obtained from Secchi et al.,*** 2021).

fishing effort. While lack of enforcement persists, no assessment regarding the effectiveness of the regulation can be made. Nevertheless, a recent study showed that even with full compliance, the current fishing regulations cannot reduce franciscana bycatch to sustainable levels in southern Brazil (Prado et al., 2021). This is most likely the scenario throughout the species distribution in Brazilian waters.

Concluding remarks and recommendations

It is substantial the amount of information regarding franciscana's biology, ecology, and conservation issues that have been generated since the first workshop on river dolphins held in China in 1986 (Perrin et al., 1989).

Research and conservation priorities were identified for all Platanistoid dolphins, including franciscana (Pinedo et al. 1989; Praderi et al., 1989) and several specific workshops were carried out ever since. These workshops fostered collaboration and articulated research among scientists and institutions, leading to a fast increase in the knowledge about franciscana's ecology and conservation needs. This species is perhaps one of the best-studied small cetaceans in the world (see Secchi et al., 2021 for a review). High numbers of franciscana bycatch in gillnets have been reported since the late 1960s in Uruguay (van Erp 1969; Brownell and Praderi, 1974) and focused the attention of scientists and conservation organizations on the potential threat to the species survival (e.g., Praderi et al. 1989; Perrin, 1994). Despite the persistent warnings of the problem and the testing of many technological and management solutions for other small cetaceans, that could be applied to franciscana, at least in some areas where small-scale artisanal gillnetting predominates, the fact is that the bycatch still exceeds by far our ability to deal with it. This situation is not due to a lack of understanding of bycatch rates or insufficient data on the demographic response to bycatch at least for FMA III (e.g., Secchi and Fletcher, 2004; Secchi et al., 2021), but to lack of fisheries management policies and political will, that would be required to implement the changes needed. These changes comprise effective and enforced management actions together with the testing and implementation of modifications in gear and operations to produce sustainable fisheries in both artisanal and industrial scales. Also, the lack of support for research aiming to collect quality data to understand the causes of bycatch that can be useful to design, test, and implement mitigation measures. Such measures should consider ecosystem-based and dynamic management approaches (e.g., Dunn et al., 2016; Hazen et al., 2018; Gilman et al., 2019). These approaches should pursue, for example, higher resolution closure areas (i.e., to reduce total time–area required to achieve the desired bycatch reduction), lower bycatch/catch ratios (i.e., to reduce relative damage to the ecosystem). Baseline information is necessary and, therefore, some priority research should include but not be restricted to: determine spatio-temporal patterns of bycatch in FMA I, II and IV; study franciscana ecology and behavior that clearly impact bycatch (e.g., movement patterns, preferential habitat for feeding and reproduction); identify environmental and operational variables that increase the probability of bycatch; design and test mitigation measures to address the bycatch; design postimplementation and monitoring schemes; collect data on their efficacy in regular fishing conditions; study the social and economic impact of mitigation measures;

and understand the complex legal framework of the three countries that are necessary to harmonize regional, federal and international legislation.

The seemingly ideal dynamic management depends on data-rich and sophisticated analytical tools and might not be feasible in the three countries where franciscana occurs. Nevertheless, information generated from these studies can be useful for designing effective bycatch mitigation alternatives through a multi and transdisciplinary participatory approach involving multiple stakeholders of the public and private sectors (e.g., Hall et al., 2007; Gilman et al., 2019; Squires et al., 2021). Mitigation solutions, however, should be relatively easy to implement that are not logistically complex and economically costly.

Acknowledgments

We are indebted to research team ECOMEGA at FURG for decades of dedication to collect data to support franciscana conservation. M. Hall reviewed the chapter and provided useful advice for its improvement. This research was partially supported by an environmental offset measure established through a Consent Decree/Conduct Adjustment Agreement between Petrorio and the Brazilian Ministry for the Environment, with the Brazilian Biodiversity Fund—FUNBIO as implementer. We also thank the Organization for the Conservation of South American Aquatic Mammals—YAQU PACHA e.V, Nuremberg Zoo for continuously funding the franciscana research. Financial support was also provided by the National Council for Research and Development (CNPq) and Coordination for the Improvement of Higher Education Personnel (CAPES). CNPq provided a research fellowship to ES (PQ 310597/2018-8) and funding (grant # 426503/2018-0). CAPES provided free access to many relevant journals through the portal "Periódicos CAPES." This study is a contribution of the Research Group Ecologia e Conservação da Megafauna Marinha – ECOMEGA-FURG/CNPq.

References

Beissinger, S.R., Westphal, M.I, 1998. On the use of demographic models of population viability in endangered species management. J. Wildl. Manage. 62, 821–841.

Berggren, P., Temple, A.J., Alfaro-Shigueto, J., Mangel, J.C, 2020. Low-cost solutions to cetacean bycatch in global gillnet fisheries, Scientific Committee Technical Document SC/68B/HIM11 Rev1. Virtual Meeting.

Berninsone, L.G., Bordino, P., Gnecco, M., Foutel, M., Mackay, A.I., Werner, T.B, 2020. Switching gillnets to longlines: an alternative to mitigate the bycatch of Franciscana dolphins (*Pontoporia blainvillei*) in Argentina. Front. Mar. Sci. 7, 699. http://doi.org/10.3389/fmars.2020.00699.

Bertozzi, C.P., Zerbini, A.N, 2002. Incidental mortality of franciscana (*Pontoporia blainvillei*) in the artisanal fishery of Praia Grande, São Paulo State, Brazil Lat. Am. J. Aq. Mamm. 1, 153–160.

Bielli, A., Alfaro-Shigeto, J., Doherty, P.D., Godley, B.J., Ortiz, C., Pasara, A., Wang, J.H., Mangel, J.C, 2020. An illuminating idea to reduce bycatch in the Peruvian small-scale gillnet fishery. Biol. Cons. 241, 108277.

Bordino, P., Albareda, D., 2004. Incidental mortality of Franciscana dolphin Pontoporia blainvillei in coastal gillnet fisheries in northern Buenos Aires, Argentina.

Bordino, P., Kraus, S., Albareda, D., Fazio, A., Palmerio, A., Mendez, M., et al., 2002. Reducing incidental mortality of franciscana dolphin *Pontoporia blainvillei* with acoustic warning devices attached to fishing nets. Mar. Mamm. Sci. 18, 833–842.

Boyce, M.S., 1992. Population viability analysis. Annual Rev. Ecol. Systemat. 23, 481–506.

Brasil. 2012. Instrução Normativa Interministerial INI 12/2021. 22 de agosto de 2012.

Bravington, M.V., Bisack, K.D., 1996. Estimates of harbour porpoise bycatch in the Gulf of Maine sink gillnet fishery, 1990-1993. Rep. Int. Whal. Commn. 46, 567–574.

Brownell Jr., R.L., Praderi, R., 1974. Present research and conservation problems with the franciscana, *Pontoporia blainvillei*, in Uruguayan waters. FAO Advisory Commitee on Marine Resources Research, La Jolla.

Brownell, R.L., Reeves, R.R., Read, A.J., et al., 2019. Bycatch in gillnet fisheries threatens critically endangered small cetaceans and many others. End. Sp. Res. 40, 285–296.

Burgman, M.A., Ferson, S., Akçakaya, H.R., 1993. Risk Assessment in Conservation Biology. Chapman & Hall, New York, NY.

Cáceres, M.O., Cáceres-Saez, I., Secchi, E.R., Negri, M.F., Panebianco, M.V., Cappozzo, H.L, 2020. Assessing the growth rate of endangered Franciscana dolphin in Argentina. South America. Reg. Stud. Mar. Sci. 40, 101479.

Cappozzo, H.L., Negri, M.F., Pérez, F.H., Albareda, D., Monzón, F., Corcuera, J.F, 2007. Incidental mortality of franciscana dolphin, (Pontoporia blainvillei) in Argentina. Lat. Am. J. Aq. Mamm. 6, 127–137.

Cardoso, L.G., Haimovici, M., Abdallah, P.R., Secchi, E.R., Kinas, P.G., 2021. Prevent bottom trawling in southern Brazil. Science. doi:10.1126/science.abh0279.

Carman, M., Carman, V.G, 2016. La fragilidad de las especies: tensiones entre biólogos y pescadores artesanales en torno a la conservación marina. Etnográfica 20, 411–438.

Caswell, H., 2001. Matrix Population Models: Construction, Analysis and Interpretation, 2nd ed. Sinauer Associates, Inc, Sunderland.

Clay, T.A., Mangel, J.C., Alfaro-Shigueto, J., Hodgson, D.J., Godley, B.J, 2018. Distribution and habitat use of a cryptic small cetacean, the Burmeister's porpoise, monitored from a small-scale fishery platform. Front. Mar. Sci. 5, 220.

Corcuera, J., 1994. Mortality of *Pontoporia blainvillei* in Northern Buenos Aires Province: the threat of small fishing camps. Rep. Int. Whal. Commn (special issue) 1, 291–294.

Corcuera, J., Monzón, F., Crespo, E.A., Aguilar, A., Raga, J.A, 1994. Interactions between marine mammals and the coastal fisheries of Necochea and Claromecó (Buenos Aires, Argentina). Rep. Int. Whal. Commn(special issue) 15, 283–290.

Cremer, M.J., Barreto, A.S., Castilho, P.V., Dick, J., Domit, C., Barbosa, C.B., Groch, K.R., et al., 2018. Franciscana dolphin mortality in Franciscana Management Area II (FMA II). Southern Brazil SM WP03.

Crespo, E.A., Alarcon, D., Alonso, M., Bazzalo, M., Borobia, M., Cremer, M., et al., 2010a. Report of the working group on major threats and conservation. Lat. Am. J. Aq. Mamm 8, 47–56.

Crespo, E.A., Perez-Macri, G., Praderi, R., 1986. Estado atual de la población de franciscana (*Pontoporia blainvillei*) en las costas uruguayas. Actas I Reunión de Trabajo de Especialistas en Mamíferos Acuáticos de America del Sur. Buenos Aires, pp. 92–105.

Crespo, E.A., Pedraza, S.N., Grandi, M.F., Dans, S.L., Garaffo, G.V, 2010b. Abundance and distribution of endangered Franciscana dolphins in Argentine waters and conservation implications. Mar. Mam. Sci. 26, 17–35.

Cunha, H.A., Medeiros, B.V., Barbosa, L.A., Cremer, M.J., Marigo, J., Lailson-Brito, J., Azevedo, A.F, Solé-Cava, A.M., 2014. Population structure of the endangered franciscana dolphin (*Pontoporia blainvillei*): reassessing management units. PLoS One 9, e85633. http://doi.org/10.1371/journal.pone.0085633.

Danilewicz, D., Rosas, F., Bastida, R., Marigo, J., Muelbert, M., Rodriguez, D., et al., 2002. Special Issue on the Biology and Conservation of Franciscana. Lat. Am. J. Aq. Mamm. 1 (1), 25–42.

Danilewicz, D.S., Moreno, I.B., Ott, P.H, Tavares, M., Azevedo, A.F., Secchi, E.R., Andriolo, A., 2010. Abundance estimate for a threatened population of franciscana dolphins in southern coastal Brazil: uncertainties and management implications. J. Mar. Biol. Assoc. UK. 90, 1649–1657.

Danilewicz, D., Zerbini, A.N., Andriolo, A., Secchi, E.R., Sucunza, F., Ferreira, E., et al., 2012. Abundance and distribution of an isolated population of franciscana dolphins (*Pontoporia blainvillei*) in southeastern Brazil: red alert for FMA I?, Paper SC/64/SM17 presented to the IWC Scientific Committee. Panama City.

Di Beneditto, A.P.M., 2003. Interactions between gillnet fisheries and small cetaceans in northern Rio de Janeiro, Brazil: 2001-2002. Lat. Am. J. Aq. Mamm. 2, 79–86.

Di Beneditto, A.P.M., Ramos, R.M.A., 2001. Biology and conservation of the franciscana (*Pontoporia blainvillei*) in the north of Rio de Janeiro State, Brazil. J. Cet. Res. Manage. 3, 185–192.

Di Beneditto, A.P.M., Ramos, R.M.A., Lima, N.R.W., 1998. Fishing activity in northern Rio de Janeiro State (Brazil) and its relation with small cetaceans. Braz. Arch. Biol. Tech. 41 (3) 0-0.

Dizon, A.E., Lockyer, C., Perrin, W.F., Demaster, D.P., Seisson, J., 1992. Rethinking the stock concept: a phylogeographic approach. Cons. Biol. 6, 24–36.

Dunn, D.C., Maxwell, S.M., Boustany, A.M., Halpin, P.N, 2016. Dynamic ocean management increases the efficiency and efficacy of fisheries management. Proc. Natl. Acad. Sci. U.S.A. 113, 668–673.

FAO, IFAD, UNICEF, WFP, WHO, 2020. The State of Food Security and Nutrition in the World 2020. Transforming food systems for affordable healthy diets. FAO, Rome. https://doi.org/10.4060/ca9692en.

FAO. 2020. Report of the Expert Meeting to Develop Technical Guidelines to Reduce Bycatch of Marine Mammals in Capture Fisheries. Rome, Italy, 17–19 September 2019. FAO Fisheries and Aquaculture Report No. 1289, Rome. https://doi.org/10.4060/CA7620EN

FAO. 2021. *Fishing operations. Guidelines to prevent and reduce bycatch of marine mammals in capture fisheries.* FAO Technical Guidelines for Responsible Fisheries No. 1, Suppl. 4. Rome. https://doi.org/10.4060/cb2887en

Franco-Trecu, V., Costa, P., Abud, C., Dimitriadis, C., Laporta, P., Passadore, C., Szephegyi, M.N, 2009. By-catch of franciscana *Pontoporia blainvillei* in Uruguayan artisanal gillnet fisheries: an evaluation after a twelve-year gap in data collection. Lat. Am. J. Aquat. Mamm. 7, 11–22.

Franco-Trecu, V., Szephegyi, M.N., Doño, F., Forselledo, R., Reyes, F., Passadore, C., et al., 2019. Marine mammal bycatch by the industrial bottom trawl fishery at the Río de la Plata Estuary and the adjacent Atlantic Ocean. Lat. Am. J. Aquat. Res. 47, 89–101.

Frizzera, F.C., Tosi, C., Pinheiro, H., Marcondes, M., 2012. Captura acidental de toninha (*Pontoporia blainvillei*) na costa norte do Espírito Santo, Brasil. Bol do Mus Biol Mello Leitão 29, 81–86.

Fruet, P.F., Möller, L.M., Secchi, E.R, 2021. Dynamics and viability of a small, estuarine-resident population of Lahille's bottlenose dolphins from southern Brazil. Front. Mar. Sci. 7, 593474. http://doi.org/10.3389/fmars.2020.593474.

Gilman, E., Suuronen, P., Hall, M., Kennelly, S., 2013. Causes and methods to estimate cryptic sources of fishing mortality. J. Fish Biol. doi:10.1111/jfb.12148.

Gilman, E., Chaloupka, M., Dagorn, L., Hall, M., Hobday, A., Musyl, M., et al., 2019. Robbing Peter to pay Paul: replacing unintended cross-taxa conflicts with intentional tradeoffs by moving from piecemeal to integrated fisheries bycatch management. Rev. Fish. Biol. and Fish. 29, 93–123. https://doi.org/10.1007/s11160-019-09547-1.

Gilman, E., Pérez Roda, M.A., Huntington, T., Kennelly, S.J., Suuronen, P., Chaloupka, M., Medley, P.A.H., 2020. Benchmarking global fisheries discards. Nat. Sci. Rep. 10, 14017. https://doi.org/10.1038/s41598-020-71021-x.

Gray, C.A., Kennelly, S.J, 2018. Bycatches of endangered, threatened and protected species in marine fisheries. Rev. Fish Biol. and Fish. 28, 521–541.

Haimovici, M., 1998. Present state and perspectives for the southern Brazil shelf demersal fisheries. Fish. Manag. Ecol. 5, 277–289.

Haimovici, M., Cardoso, L.G, 2017. Long-term changes in the fisheries in the Patos Lagoon estuary and adjacent coastal waters in Southern Brazil. Mar. Biol. Res. 13, 135−150.

Hall, M.A., 1999. Estimating the ecological impacts of fisheries: what data are needed to estimate bycatches? In: Nolan, C.P. (Ed.), Proceedings of the International Conference on Integrated Fisheries Monitoring. Sydney, Rome. FAO, pp. 175–184.

Hall, M., Alverson, D., Metuzal, K., 2000. By-catch: problems and solutions. Mar. Poll. Bull. 41, 204–219.

Hall, M., Nakano, H., Clarke, S., Thomas, S., Molloy, J., Peckham, S.H., et al., 2007. Working with fishers to reduce bycatches. In: Kennelly, S. (Ed.), Bycatch Reduction in the World's Fisheries. Springer, Dordrecht.

Hall, M., Swimmer, Y., Parga, M., 2012. No "silver bullets" but plenty of options. Working with artisanal fishers in the Eastern Pacific to reduce incidental sea turtle mortality in longline fisheries. In: Seminoff, J., Wallace, B. (Eds.), Sea turtles of the Eastern Pacific: advances in research and conservation. The University of Arizona Press, Tucson, pp. 136–153.

Hazen, E.L., Scales, K.L., Maxwell, S.M., Briscoe, D.K., Welch, H., Bograd, S.J., et al., 2018. A dynamic ocean management tool to reduce bycatch and support sustainable fisheries. Sci. Adv. 4, eaar3001.

Heinrich, S., Reeves, R. 2017. *Cephalorhynchus eutropia*. The IUCN Red List of Threatened Species 2017: e.T4160A50351955. https://dx.doi.org/10.2305/IUCN.UK.2017-3.RLTS.T4160A50351955.en. accessed July 17, 2021.

Iriarte, V., Marmontel, M., 2013. River dolphin (*Inia geoffrensis, Sotalia fluviatilis*) mortality events attributed to artisanal fisheries in the western Brazilian Amazon. Aquatic Mammals 39, 10–18.

IUCN. 2019. The IUCN Red List of Threatened Species. Version 2019-2. http://www.iucnredlist.org. accessed July 18, 2019

Jaramillo-Legorreta, A.M., Cardenas-Hinojosa, G., Nieto-Garcia, E., Rojas-Bracho, L., et al., 2019. Decline towards extinction of Mexico's vaquita porpoise (*Phocoena sinus*). R. Soc. Open Sci. 6, 190598.

Ludwig, D., 1999. Is it meaningful to estimate a probability of extinction? Ecology 80, 298–310.

Marcondes, M.C.C., Angeli, M., Fontes, F., Palazzo Júnior, J.T., Campos, R., Dapper, C., Cremer, M.J, 2018. Report on franciscana fisheries interaction in FMA Ia, Paper submitted to the 68 IWC Scientific Meeting, SC/67B/SM/03. Brazil, p. 17.

Montealegre-Quijano, S., Neves-Ferreira, C., 2010. Incidental catch of a franciscana dolphin Pontoporia blainvillei (Cetacea, Pontoporiidae) in a pair trawl off southern Brazil. Pan-Am. J. Aq. Sci. 5, 465–468.

Moreno, I.B., Ott, P.H., Danilewicz, D.S., 1997. Análise preliminar do impacto da pesca artesanal costeira sobre *Pontoporia blainvillei* no litoral norte do Rio Grande do Sul, sul do Brasil. In: Pinedo, M.C., Barreto, A.S. (Eds.), Anais do 2o Encontro sobre Coordenação de Pesquisa e Manejo da Franciscana. FURG, Rio Grande, pp. 31−41.

Moritz, C., 1994. Defining 'Evolutionarily Significant Units' for conservation. Tree 9 (10), 373–375.

Morris, W.F., Doak, D.F, 2003. Quantitative Conservation Biology: Theory and Practice of Population Viability Analysis. Sinauer Associates, Sunderland.

Myers, R.A., Worm, B., 2003. Rapid worldwide depletion of predatory fish communities. Nature 423, 280–283.

Negri, M.F., Denuncio, P., Panebianco, M.V., Capozzo, H.L, 2012. Bycatch of franciscana dolphins *Pontoporia blainvillei* and the dynamic of artisanal fisheries in the species' southernmost area of distribution. Braz. J. Ocean. 60, 151–160.

Norris, K.S., Mcfarland, W.N, 1958. A new harbor porpoise of the genus Phocoena from the Gulf of California. J. Mamm. 39, 22–39.

Ott, P.H., Domit, C., Siciliano, S., Flores, P.A.C., 2015Memórias do VII Workshop para a coordenação de pesquisa e conservação de *Pontoporia blainvillei* 1844. Gervais & d'Orbigny. http://www.pontoporia.org (Accessed 12 March 2020).

Ott, P.H., Secchi, E.R., Moreno, I.B., Danilewicz, D., Crespo, E.A., Bordino, P., et al., 2002. Report of the working group on fishery interactions. Lat. Am. J. Aq. Mamm 1 (special issue), 55–64.

Pages, P.R.Wade, 2002. Bayesian population viability analysis. In: Beissinger, S.R., McCullough, D.R. (Eds.), Population Viability Analysis. The University of Chicago Press, Chicago, pp. 213–238.

Perez-Macri, G., Crespo, E.A, 1989. Survey of the franciscana, *Pontoporia blainvillei*, along the Argentine coast, with a preliminary evaluation of mortality in coastal fisheries. In: Perrin, W.F., Brownell, R.L, Kaiya, Z., Jiankang, L. (Eds.). Biology and Conservation of the River Dolphins, 3. Occasional Papers. IUCN SSC, Gland, pp. 57–63.

Perrin, W.F., Brownell, R.L., Zhou, K., Liu, J., 1989. Biology and conservation of the river dolphins. World Conservation Union, Gland.

Pinedo, M.C., 1994. Review of small cetacean fishery interactions in southern Brazil with special reference to the franciscana. Pontoporia blainvillei. Rep. Int. Whal. Commn 15, 251–259.

Praderi, R., 1997. Análisis comparativo de estadísticas de captura y mortalidad incidental de Pontoporia blainvillei en Uruguay durante 20 años. In: Pinedo, M.C., Barreto, A.S. (Eds.), Anais do 2o Encontro sobre Coordenação de Pesquisa e Manejo da Franciscana. FURG, Rio Grande, pp. 42–53.

Pinedo, M.C., Praderi, R., Brownell, R.L, 1989. Review of the biology and status of the franciscana, *Pontoporia blainvillei* eds. In: Perrin, W.F., Brownell, R.L., Zhou, K., Liu, J. (Eds.), Biology and Conservation of the River Dolphins. Occas. Pap. IUCN SSC3, Gland, pp. 46–51 173.

Praderi, R., Pinedo, M.C., Crespo, E.A, 1989. Conservation and management of *Pontoporia blainvillei* in Uruguay, Brazil and Argentina. In: Perrin, W.F., Brownell, R.L., Kaiya, Z., Jiankang, L. (Eds.). Biology and Conservation of the River Dolphins, 3. Occasional Papers. IUCN SSC, Gland, pp. 52–56 173.

Prado, J.H.F., Secchi, E.R., Kinas, P.G, 2013. Mark-recapture of the endangered franciscana dolphin (*Pontoporia blainvillei*) killed in gillnet fisheries to estimate past bycatch from time series of stranded carcasses in southern Brazil. Ecol. Indicators. 32, 35–41.

Prado, J.H.F., Mattos, P.H., Silva, K.G., Secchi, E.R, 2016. Long-term seasonal and interannual patterns of marine mammal strandings in subtropical western South Atlantic. PLoS One 11, 1–23.

Prado, J.H.F., Kinas, P.G., Pennino, M.G., Seyboth, E., Silveira, F.R.G., Ferreira, E.C., Secchi, E.R., 2021. Definition of no-fishing zones and fishing effort limits to reduce franciscana bycatch to sustainable levels in southern Brazil. Animal Cons. doi:10.1111/acv.12679 Doi.

Read, A.J., Drinker, P., Northridge, S., 2006. Bycatch of marine mammals in the United States of America and global fisheries. Cons. Biol. 20, 163–169.

Reeves, R.R., McClellan, K., Werner, T.B., 2013. Marine mammal bycatch in gillnet and other entangling net fisheries, 1990 to 2011. End. Sp. Res. 20, 71–97.

Reyes, J.C., Oporto, J.A, 1994. Gillnet fisheries and cetaceans in the southeast Pacific. Rep. Int. Whal. Commn. 15, 467–474.

Secchi, E.R., 2006. Modeling the population dynamics and viability analysis of franciscana (*Pontoporia blainvillei*) and Hector's dolphins (*Cephalorhynchus hectori*) under the effects of bycatch in fisheries, parameter uncertainty and stochasticity. University of Otago, Dunedin Ph.D. Thesis.

Secchi, E.R., 2010. Review on the threats and conservation Status of Franciscana, *Pontoporia blainvillei* (Cetacea, Pontoporiidae). In: Shostell, J.M., Ruiz-Garcia, M (Eds.). Biology, Evolution and Conservation of River Dolphins within South America and Asia, 1. Nova Science Publishers Inc., Hauppauge, pp. 323–339.

Secchi, E.R., Danilewicz, D., Ott, P.H, 2003a. Applying the phylogeographic concept to identify franciscana dolphin stocks: implications to meet management objectives. J. Cetacean Res. Manage. 5, 61–68.

Secchi, E.R., Ott, P.H., Danilewicz, D.S., 2003. Effects of fishing bycatch and conservation status of the franciscana dolphin, *Pontoporia blainvillei*. In: Gales, N., Hindell, M., Kirkwood, R. (Eds.), Marine mammals: fisheries, tourism and management issues. CSIRO Publishing, Collingwood, pp. 174–191.

Secchi, E.R., Fletcher, D., 2004. Modelling population growth and viability analysis for four franciscana stocks: effects of stock-specific differences in life traits, fishing bycatch, parameter uncertainty and stochasticity.

Secchi, E.R., Kinas, P.G., Muelbert, M., 2004. Incidental catches of franciscana in coastal gillnet fisheries in the Franciscana Management Area III: period 1999-2000. Lat. Am. J. Aq. Mamm. 3, 61–68.

Secchi, E.R., Zerbini, A.N., Bassoi, M., Dalla Rosa, L., Möller, L.M., Rocha-Campos, C.C., 1997. Mortality of franciscanas, *Pontoporia blainvillei*, in coastal gillnetting in southern Brazil: 1994-1995. Rep. Int. Whal. Commn. 47, 653–658.

Secchi, E.R., Cremer, M.J., Danilewicz, D., Lailson-Brito, J., 2021. A Synthesis of the Ecology, Human-Related Threats and Conservation Perspectives for the Endangered Franciscana Dolphin. Front. Mar. Sci. 8, 617956. http://doi.org/10.3389/fmars.2021.617956.

Silva, Da, V.M.F., F., E.C., C., Dias, R.L., Martin, A.R, 2018. Both cetaceans in the Brazilian amazon show sustained, profound population declines over two decades. PLoS One 13, e0191304.

Sonne, C., Diaz-Jaimne, P., Adams, D.H., 2021. Mexico's final death blow to the vaquita. Science 373, 863–864.

Squires, D., Balance, L.T., Dagorn, L., Dutton, P.H., Lent, R., 2021. Mitigating bycatch: novel insights to multidisciplinary approaches. Front. Mar. Sci. 8, 613285. http://doi.org/10.3389/fmars.2021.61328.

Sucunza, F., 2020. Estimativas de densidade e abundância da toninha (Pontoporia blainvillei) a partir de levantamentos aéreos: produção e aplicação de fatores de correção Doctoral Thesis. Universidade Federal de Juiz de Fora.

Sucunza, F., Danilewicz, D., Andriolo, A., Azevedo, A.F., Secchi, E.R., Zerbini, A.N., 2020. Distribution, habitat use, and abundance of the endangered franciscana in southeastern and southern Brazil. Mar. Mamm. Sci. 36, 421–435.

Taylor, B.L, 2005. Identifying Units to Conserve. In: Reynolds, J.E. III, Perrin, W.F., Reeves, R.R., Montgomery, S., Ragen, T.J. (Eds.), Marine Mammal Research: Conservation Beyond Crisis. The John Hopkins University Press, Baltimore, MD, pp. 149–164.

Taylor, B.L., Wade, P.R., Ramakrishnan, U.M., Gilpin, M.E., Akçakaya, R., 2002. Incorporating uncertainty in population viability analysis for the purpose of classifying species by risk. In: Beissinger, S.R., McCullough, D.R. (Eds.), Population Viability Analysis. The University of Chicago Press, Chicago, pp. 239–252.

Turvey, S.T., Pitman, R.L, Taylor, B.L., Barlow, J., et al., 2007. First human-caused extinction of a cetacean species? Biol. Lett. 3, 537−540.

UNEP/CMS, 2000. Report of the Third Workshop for Coordinated Research and Conservation of the Franciscana Dolphin (*Pontoporia blainvillei*) in the Soutwestern Atlantic. UNEP/CMS, Bonn.

UN General Assembly. 2015. Transforming our world: The 2030 Agenda for Sustainable Development, October 21, 2015, A/RES/70/1. www.refworld.org/docid/57b6e3e44.html.

Van Erp, I., 1969. In quest of the La Plata dolphin. Pacif. Discov. 22, 18–24.

Vermeulen, E., Fruet, P., Costa, A., Coscarella, M., Laporta, P. 2019. *Tursiops truncatus* ssp. *gephyreus. The IUCN Red List of Threatened Species* 2019: e.T134822416A135190824. https://dx.doi.org/10.2305/IUCN.UK.2019-3.RLTS.T134822416A135190824.en. (accessed 17 July 2021).

Wade, P.R., 1998. Calculating limits to the allowable human-caused mortality of cetaceans and pinnipeds. Mar. Mam. Sci. 14, 1–37.

Zanelatto, R.C., 1997. Captura acidental de toninha, *Pontoporia blainvillei*, (Gervais & D'Orbigni, 1844) (Cetacea, Pontoporiidae) no litoral do estado do Paraná, Brasil. In: Pinedo, M.C., Barreto, A.S. (Eds.), Anais do 2o Encontro sobre Coordenação de Pesquisa e Manejo da Franciscana. FURG, Rio Grande, pp. 22–29.

Zerbini, A.N., Secchi, E., Crespo, E., Danilewicz, D., Reeves, R. 2017. *Pontoporia blainvillei* (errata version published in 2018). The IUCN Red List of Threatened Species 2017: e.T17978A123792204. https://dx.doi.org/10.2305/IUCN.UK.20173.RLTS.T17978A50371075.en. (accessed April 17, 2020).

CHAPTER 11

Chemical pollution and franciscana—a review

José Lailson-Brito Júnior, Nara Oliveira-Ferreira, Bárbara Moura Reis Manhães, Tatiana Lemos Bisi, Elitieri Santos-Neto
Laboratório de Mamíferos Aquáticos e Bioindicadores, Faculdade de Oceanografia, Universidade do Estado do Rio de Janeiro, Rio de Janeiro, Brazil

Chemical pollution can be an important threat to marine life. Some contaminants are of particular concern due to specific features such as how they are easily transported in currents and in the atmosphere, their long half-life, and their potential to bioaccumulate in marine organisms over time and biomagnify along with trophic webs (Gray, 2002).

Persistent organic pollutants (POPs), essential and nonessential elements, and contaminants of emergent concern have been released into the environment through natural and anthropogenic processes (Agency for Toxic Substances and Disease Registry ATSDR, 1999; UNEP, 2017). While certain classes of contaminants have had their manufacture and use prohibited worldwide many years ago, they are extremely persistent and ubiquitous (Jepson and Law, 2016), and areas close to their sources can be heavily impacted by their presence.

The franciscana dolphin, *Pontoporia blainvillei*, is a small cetacean and the only extant species of the Pontoporiidae family, occurring from the Chubut province, in Argentina, until the north of Espírito Santo state, in Brazil. Across its distribution, genetic and ecological differences were observed among individuals inhabiting different localities, hence being suggested a division of the area of occurrence in franciscana management areas (FMAs) (Secchi et al., 2003). At the moment, four FMAs are described (FMAs I, II, III, and IV) and FMAs I, II, and III are subdivided into a and b, while FMA IV into a, b and c (Cunha et al., 2014).

Franciscana dolphin is a sentinel species of the environment due to high trophic level, and the low metabolization/excretion processes of contaminants in its tissues/organs (Lailson-Brito et al., 2002; Seixas et al., 2007). Because of the coastal/estuarine habits, the species is chronically exposed to anthropogenic contaminants and susceptible to their toxic effects. These

The Franciscana Dolphin
DOI: https://doi.org/10.1016/B978-0-323-90974-7.00017-3

contaminants can play an important role as immunosuppressants and endocrine disruptors that can negatively affect the survival and replacement rates of vulnerable populations. While information regarding the current contamination status, contaminants' temporal trends, and toxic effects are still scarce, this chapter aims to summarize the knowledge of the contamination on franciscana dolphins across its distribution range.

Organic compounds

Contamination by legacy organic pollutants and those of emergent concern is one of the main threats for coastal cetaceans inhabiting areas close to point sources and occupying high trophic levels, once the diet is the main via of exposure for these animals (Jepson and Law, 2016). This is particularly true for franciscana dolphins that present fidelity to sites known as hotspots for chemical pollution. POPs are a group of synthetic organic compounds with wide use in industrial activities, agriculture, and disease vector control in recent decades. POPs have had their production and applicability restricted and/or prohibited for many decades now (UNEP, 2017), safeguarding exceptional events when the use is still allowed (e.g., vectors controls in outbreaks, such as for malaria). Organic compounds are lipophilic and stable, presenting high resistance to environmental and biological degradations. These compounds bioaccumulate and biomagnify through marine food webs, reaching high levels in top predators. Pollutants' concentrations in these apex predators are of concern due to their potential effects on population viability. They are known for their effects on the immune, endocrine and reproductive systems (Das et al., 2006; Desforges et al., 2016; Hall et al., 2003) and could impair growth rates in populations already heavily impacted by elevated removal due to bycatch.

Organochlorine contaminants

Organic pollutants containing chlorine in their composition were widely applied during the urbanization process. The polychlorinated biphenyls (PCBs) were mostly used as fluids in capacitors and transformers due to their high thermal stability, hence, associated with industrialized and urbanized areas. On the other hand, organochlorine pesticides (OCPs, e.g., dichlorodiphenyltrichloroethane (DDT), hexachlorocyclohexane (HCH), mirex) were used in agriculture, wood conservation processes, and as vector control in tropical areas. These compounds were globally applied indiscriminately in the 20th century (UNEP, 2017). Since their prohibition in the southern hemisphere around the 1980s, organochlorine compounds (OCs) are still detected in elevated concentrations in franciscana dolphins. OCs physical-chemical properties ensure their persistence in the environment, bioaccumulation in lipid-rich tissues, and biomagnification along with trophic webs (Gray, 2002).

The first study regarding organochlorine contaminants in franciscana dolphins was performed in individuals from FMA IIIa (Uruguay) in the early 1970s (O'Shea et al., 1980). The OCPs DDT and its metabolites, were the predominant compounds found in the blubber but were not detected in other tissues. Other OCPs (hexachlorobenzene (HCB), dieldrin, and *trans*-Nonachlor) were also observed in the blubber. PCBs were detected not only in the blubber of male and female franciscana dolphins but also in tissues containing lower lipidic content (e.g., brain and muscle). The lipophilic character of these compounds and their corporal distribution point toward a preferred accumulation in the blubber, but their detection in other organs and tissues show their efficient transport in the bodies of franciscana dolphins (Fig. 11.1; Barbosa et al., 2018; O'Shea et al., 1980). Recently, Barbosa et al. (2018) reported the presence of PCBs and DDTs in the blubber, liver, kidney, and muscle of fetuses, reflecting the placental transference of such compounds from mother to the offspring at a critical developmental stage. As observed in adults, the blubber was the main storage tissue for fetuses (Barbosa et al., 2018). Some studies also reported other OCs (dioxin-like PCBs, mirex, and dechloranes, the latter used as flame retardants) in the liver of juvenile and adult specimens across FMAs (De La Torre et al., 2012; Dorneles et al., 2013), as the hepatic tissue shows detox function for most compounds.

Most studies concerning OCs were performed in the FMA II area, specifically in São Paulo state. The contamination status in specimens from FMA Ia, FMA IIa, and FMA IV are still widely scarce, some of which (i.e., FMA Ia and FMA IIa) are only now starting to be elucidated

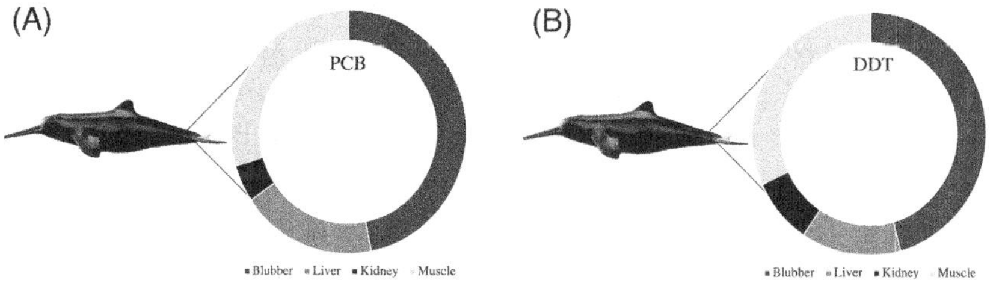

Fig. 11.1 Distribution of PCBs (A) and DDTs (B) in the blubber, liver, kidney and muscle of franciscana dolphins, *Pontoporia blainvillei*, from São Paulo state coast (FMA II). Based on Barbosa et al. (2018).

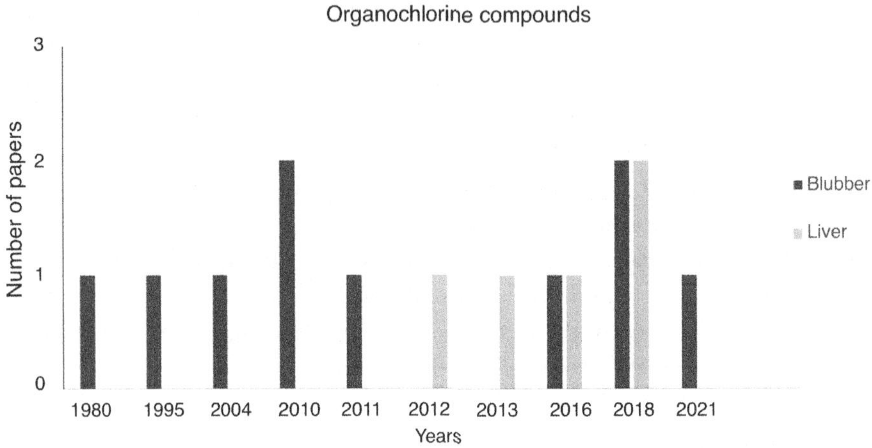

Fig. 11.2 ***Overview containing the number of papers available in the literature concerning organochlorine compounds in blubber and liver of franciscana dolphin,*** **Pontoporia blainvillei.**

(Oliveira-Ferreira et al., (*in prep.*). Furthermore, within the FMA IIb range, Santa Catarina state is still underexplored regarding OCs contamination. Considering the number of papers per year, the studies started to be published in 1980, being only three studies until 2000. In 2010, two studies were published, as well as in 2018 (Fig. 11.2). The most analyzed tissue was blubber, followed by the liver.

PCB concentrations in the blubber of franciscana dolphin are mostly similar across the different FMAs in Brazilian waters (Table 11.1). The most recent estimates are of mean 2.4 μg/g lipid weight (lw) in individuals from FMA Ia, 6.5 μg/g lw in FMA IIb and 4.5 μg/g lw in FMA IIIa (Leonel et al., 2010; Oliveira-Ferreira et al., *in prep*). Lower PCB concentration of 0.6 μg/g lw was found in mature females and fetuses from São Paulo state (FMA IIb) while the higher mean of 20.6 μg/g lw in males and females from southern Rio de Janeiro (FMA IIa). Comparatively, the lower values found in females and fetuses from FMA IIb may be associated with the biological parameters of the data set due to depuration processes in pregnant and lactating females (Aguilar et al., 1999). The southeastern and southern regions of Brazil are important urban centers, with industrial and agricultural activities. The southeastern concentrates over 40% of the entire Brazilian population and it is an area of high anthropogenic impact. In this region, Rio de Janeiro and São Paulo states are the most industrialized areas and with higher demographic density. Similarities between PCB concentrations within FMAs are of particular alarm for populations inhabiting areas considered to have a less anthropogenic

Table 11.1 Review of organochlorine compounds reported in the blubber and liver of franciscana dolphins, *Pontoporia blainvillei*, along with their distribution.

n	ΣPCB	ΣDDT	ΣHCH	HCB	Mirex	ΣChlordanes	Sampling site	FMA	References
Blubber									
8	6.4 ± 4.8[a]	19.7 ± 7.3[a]	NA	0.1 ± 0.03[a]	NA	ND	Uruguay	FMA IIIa	O'Shea et al. (1980)
74	1.93 ± 1.02[a]	1.55 ± 1.58[a]	NA	NA	NA	NA	Argentina	FMA IVc	Borrell et al. (1995)
26	3.0 ± 1.6	3.9 ± 4.0	3.4 ± 1.0	0.01 ± 0.001	NA	0.05 ± 0.01	São Paulo/ Paraná	FMA IIb	Kajiwara et al. (2004)
26	4.5 ± 2.4	1.1 ± 0.8	NA	0.04 ± 0.02	0.06 ± 0.02	0.1 ± 0.07	Rio Grande do Sul	FMA IIIa	Leonel et al. (2010)
8	4.0 ± 3.0	2.1 ± 2.1	0.01 ± 0.002	0.03 ± 0.03	0.05 ± 0.03	0.005 ± 0.002	São Paulo	FMA IIb	Yogui et al. (2010)
1	3.7	1.6	NA	0.05	NA	NA	São Paulo	FMA IIa	Lailson-Brito et al. (2011)
9	2.9 ± 1.8	1.5 ± 1.7	NA	0.03 ± 0.02	NA	NA	Paraná	FMA IIb	Lailson-Brito et al. (2011)
41	0.4 ± 0.8	NA	NA	NA	NA	NA	Buenos Aires	FMA IVc	Arias et al. (2016)
8	0.6 ± 0.5	0.3 ± 0.2	0.008	0.009 ± 0.006	0.009 ± 0.005	0.003 ± 0.001	São Paulo	FMA IIb	Barbosa et al. (2018)
10	NA	0.1 ± 0.1	NA	NA	NA	NA	Buenos Aires	FMA IVc	Romero et al. (2018)
10	2.4 ± 3.7	1.5 ± 1.5	ND	0.03 ± 0.02	0.2 ± 0.3	NA	Espírito Santo	FMA Ia	Oliveira-Ferreira et al. (*in prep.*)

(Continued)

Table 11.1 Cont'd

n	ΣPCB	ΣDDT	ΣHCH	HCB	Mirex	ΣChlordanes	Sampling site	FMA	References
10	20.6 ± 30.4	9.7 ± 20.0	ND	0.04 ± 0.02	0.2 ± 0.4	NA	Rio de Janeiro	FMA IIa	Oliveira-Ferreira et al. (*in prep.*)
25	6.5 ± 4.6	2.3 ± 1.7	0.1 ± 0.02	0.04 ± 0.02	0.05 ± 0.02	NA	São Paulo	FMA IIb	Oliveira-Ferreira et al. (*in prep.*)
Liver									
3	NA	NA	NA	NA	0.2 ± 0.04	NA	Espírito Santo	FMA Ia	de La Torre et al. (2012)
12	NA	NA	NA	NA	0.02 ± 0.007	NA	São Paulo	FMA II	de La Torre et al. (2012)
2	NA	NA	NA	NA	0.04 ± 0.01	NA	Santa Catarina	FMA IIb	de La Torre et al. (2012)
3	NA	NA	NA	NA	0.03 ± 0.01	NA	Rio Grande do Sul	FMA IIIa	de La Torre et al. (2012)
19	0.3 ± 0.2	NA	NA	NA	NA	NA	Espírito Santo, São Paulo, Santa Catarina, Rio Grande do Sul	FMA I, FMA II e FMA III	Dorneles et al. (2013)
9	12.1 ± 6.4	NA	NA	NA	NA	NA	Rio de Janeiro	FMA Ib	Lavandier et al. (2016)
10	NA	4.3 ± 3.4	NA	NA	NA	NA	Buenos Aires	FMA IVc	Romero et al. (2018)
6	0.2 ± 0.03	0.09 ± 0.08	0.05 ± 0.04	0.02	0.007 ± 0.004	ND	São Paulo	FMA IIb	Barbosa et al. (2018)

Legend: ([a]) μg/g wet weight; NA, not analyzed; ND, not detected.
Concentrations are expressed in μg/g lipid weight (Mean ± Standard Deviation) unless stated otherwise.

impact (Espírito Santo, Paraná, Santa Catarina, and Rio Grande do Sul states). Otherwise, PCB concentrations are reported as one order of magnitude lower in the Argentinian population from FMA IVc, with a mean of 0.4 μg/g lw (Table 11.1; Fig. 11.3).

As reported for PCBs, the concentrations of DDTs found in franciscana dolphins from FMAs I, II, and III are similar amongst them, and lower in specimens from Argentina (FMA IV) (Table 11.1). DDT is one of the most used pesticides in the world, and was the main OCP detected in franciscana dolphin blubber. Recent studies indicate that *p.p'*-DDE, a metabolite of the aerobic biotransformation of DDT and correlated to the historical contamination by this ubiquitous pesticide, is the main isomer detected in cetaceans, including franciscana dolphins. Other OCPs, such as chlordane, mirex, HCB, HCHs, heptachlor-epoxide, were also reported in the blubber of this species, as well as pesticides metabolites like TCPMOH and TCPMe. However, the levels were lower than those observed for DDTs (Table 11.1; Fig. 11.3).

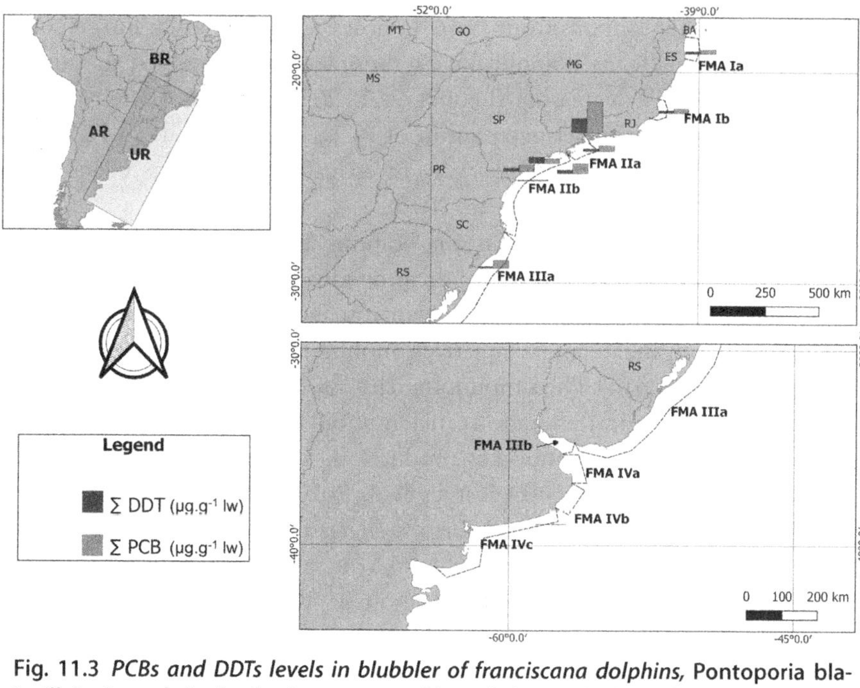

Fig. 11.3 ***PCBs and DDTs levels in blubbler of franciscana dolphins,*** **Pontoporia blainvillei*****, along their distribution, expressed in μg/g lw.*** For ΣPCB and ΣDDT values, see Table 11.1. References: Kajiwara et al. (2004), Leonel et al. (2010), Yogui et al. (2010), Lailson-Brito et al. (2011), Arias et al. (2016), Barbosa et al. (2018), Romero et al. (2018), Oliveira-Ferreira et al. (*in prep.*).

The contamination profile of OCs varies according to their distribution. Since franciscana dolphin populations present strong evidence of site fidelity, their profile shows a specific contamination scenario from their small-ranging foraging habitat. In general, PCBs were the dominant compounds followed by DDT, except in the Uruguayan population (while caution is demanded since these samples date from the 1970s) and in the FMA IIb (Kajiwara et al., 2004; Lailson-Brito et al., 2011). This is a common profile found in coastal odontocetes due to the highly industrialized feature of the study areas and to the recalcitrant character of these compounds (UNEP, 2017). However, a change in compounds predominance was observed in the FMA Ia population after the collapse of the Fundão dam, in Minas Gerais state. The disaster led to the leakage of millions of cubic meters of mud rich in trace elements into the habitat of the most isolated population of franciscana dolphins. The drag of the mud through the margins of Doce River triggered the upsurge of DDT availability and its transformation products in FMA Ia, which were previously immobilized in other compartments (Oliveira-Ferreira et al., *in prep.*). This increase in DDTs levels in specimens from FMA Ia poses a major setback in the conservation of franciscana dolphin once temporal trends suggested that these OCPs were slowly decreasing (Leonel et al., 2010; Oliveira-Ferreira et al., *in prep.*). Conversely, the concentration of PCBs over the years in FMAs Ia and IIIa, where temporal trends were investigated, show signs of stability since the mid-1990s (Leonel et al., 2010).

The effects of OCs in franciscana dolphins are mainly associated with the impairment these could pose to the individual's homeostasis. OCs found in elevated concentrations in marine mammals were already linked to a decrease in both innate and adaptative immune responses at different levels (Desforges et al., 2016). The immunotoxicity associated with such pollutants leads to morphological alterations in lymphoid tissue, decreased lymphocytes proliferation and antibodies production, impaired NK cells activity, phagocytosis, among others (Desforges et al., 2017, 2016). The contaminants profile can also influence the deleterious effects caused to the immune system as heavier compounds like highly chlorinated PCBs, tend to be more cytotoxic than lighter profiles (Desforges et al., 2017). The reduction in the individuals' immune function against pathogens weakens their fitness and is ultimately associated with increased mortality by infectious diseases (Jepson et al., 2005). Although concentrations of PCBs in franciscana dolphins were not so high as reported for other coastal cetacean species, their profile was dominated mainly by heavier congeners (Lailson-Brito et al., 2011), associated with increased toxicity (Desforges et al., 2017). Nevertheless, OCs also play an important role as endocrine disruptors while are associated with deregulation of stress and sex hormones, like testosterone and cortisol in male and female cetaceans (Galligan et al., 2019). This disturbance can pose a negative impact on the reproductive process of franciscana dolphins by delaying maturation cycles, for example. Exposure to OCs via maternal transference is also related to a decrease in the survival probability of the offspring in other cetacean species (Hall et al., 2005; Schwacke et al., 2002). Populations across FMAs are exposed to extremely high removal rates, in particular due to bycatch (Bertozzi and Zerbini, 2002). The negative effects of OCs on the immune endocrine system and reproductive function may increase these rates and impair the replacement of individuals in the population.

Organobromine contaminants

Organobromine compounds (OBs), as the polybrominated diphenyl ethers (PBDEs), pentabromoethylbenzene (PBEB), and hexabromobenzene (HBBZ), were used as brominated flame retardants in a diverse range of products. In the southern hemisphere, the records of their production/importation are highly unspecific, but these were already reported in elevated concentrations in franciscana dolphin blubber (Barbosa et al., 2018; Lailson-Brito et al., 2011; Leonel et al., 2014;Yogui et al., 2011) and liver (Alonso et al., 2012a; De La Torre et al., 2012; Lavandier et al., 2016). Additionally, the methoxylated brominated diphenyl ethers (MeO-BDEs) natural compounds synthesized by algae, sponges, and associated organisms like cyanobacteria, have an analog structure to PBDEs and were also described in blubber and liver of the species (Alonso et al., 2012a; Oliveira-Ferreira et al., *in prep.*).

Fewer studies comprising contamination by OBs in franciscana dolphins were published up to this moment in comparison to OCs. Considering the number of papers per year, recently, studies regarding this class of contaminants started to be published, in 2011 (Fig. 11.4). In 2012, two papers were published and most of them used blubber and liver as the analyzed matrix.

A comprehensive overview of the presence of these contaminants in the liver of franciscana dolphin from FMA Ia, FMA Ib, FMA IIb, and FMA IIIa was performed by Alonso et al. (2012a), but the information is still scarce. There are no data available regarding OBs in the Uruguayan and Argentinian populations. PBDEs trended to increase between the early

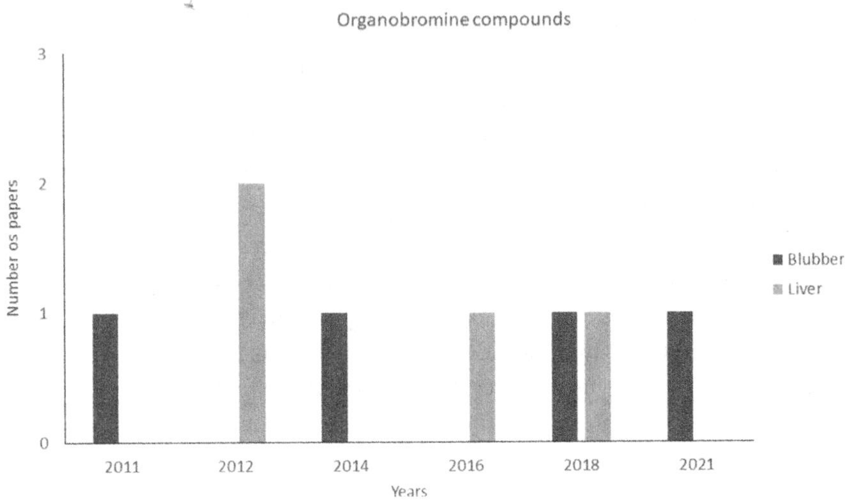

Fig. 11.4 ***Overview containing the number of papers available in the literature concerning organobromine compounds in blubber and liver of franciscana dolphins,*** **Pontoporia blainvillei.**

1990s and 2000s (Leonel et al., 2014). However, with the prohibition and recent replacement in the use of PBDEs for other flame retardants, new assessments are necessary to determine the current trends as in the northern hemispheres PBDEs contamination already started to decline (Law et al., 2012).

The concentrations of OBs ranged across the Brazilian distribution of franciscana dolphins, with a mean blubber concentration of 0.2 μg/g lw in FMA IIa and 0.02 μg/g lw in FMA Ia, IIb and III. Lower concentrations were obtained in FMA IIb, of 0.01 μg/g lw since pregnant females and their fetuses were assessed. Mature females tend to present lower concentrations than juveniles and mature males due to the transference of organic compounds via placenta and lactation (Kajiwara et al., 2008). As for OCs, there is evidence that the blubber tends to accumulate higher levels of OBs than the liver (Barbosa et al., 2018). Brominated compounds were not observed in the kidney or muscle of franciscana dolphin (Barbosa et al., 2018). The emergent contaminant PBEB was reported in both liver and blubber of franciscana dolphin while HBBZ was not found in the blubber (Alonso et al., 2012a; Oliveira-Ferreira et al., *in prep.*). MeO-BDEs were analyzed in fewer studies than PBDEs, but individuals from FMA Ia showed the highest concentrations probably because of their proximity to the Abrolhos Bank, a hotspot for biodiversity that influences the presence of the natural compounds (Alonso et al., 2012a; Oliveira-Ferreira et al., *in prep.*) (Table 11.2; Fig. 11.5).

PBDEs are endocrine disruptors that may alter the histological composition of the thyroid and the circulation of its hormones, crucial to maintaining the physiological homeostasis as it controls the metabolism and development of the individuals (Das et al., 2006; Hall et al., 2003). Their effects are alarming especially when PBDEs are combined with other compounds (Das et al., 2006). Ultimately, hormonal alterations triggered by OBs could influence reproductive parameters, adding to the threat of other organohalogen contaminants found in franciscana dolphins. Although MeO-BDEs concentrations are higher than the OBs of anthropogenic sources, no negative effects were to date described in cetaceans (Vetter et al., 2002). However, their structural similarities to PBDEs raise concerns about their potential as endocrine disruptors. More studies are necessary to assess the impact that these biosynthesized compounds may represent on franciscana dolphins, particularly those exposed to elevated concentrations.

Other organic pollutants

In the last decade, studies regarding emerging contaminants have demonstrated their presence in franciscana dolphins. For instance, pyrethroids (PYR) started to be used as an alternative to prohibiting persistent pesticides (Alonso et al., 2015). Literature suggests an efficient metabolization of pyrethroids in mammals, but the bioaccumulation of these pesticides in franciscana dolphins indicates that their metabolization capacity is slower than their bioaccumulation rate (Alonso et al., 2012b). Studies have suggested toxic effects of these compounds, such as carcinogenic, neurotoxin, and impairment of reproductive and immune systems (Jin et al., 2012). PYR were reported in the liver, milk, and placenta of franciscana dolphins, evidencing their bioaccumulation and maternal transference (Alonso et al., 2015, 2012b). In the liver, PYR values were similar in franciscana dolphin from FMA II and FMA IIIa, with the highest mean values observed in calves (0.03 and 0.04 μg/g lw for FMA II and IIIa, respectively) (Alonso et al., 2012b).

UV-filters (sunscreen agents) are also emerging contaminants, and there is a lack of knowledge regarding their effects. Octocrylene is of concern in marine organisms due to its high liposolubility, and it was the most frequent UV-filters compound detected in franciscana dolphins' liver. Octocrylene concentrations ranged from not detect values in FMA Ib to 0.4 μg/g lw in FMA IIb (Gago-Ferrero et al., 2013).

Table 11.2 Review of organobromine compounds reported in the blubber and liver of franciscana dolphins, *Pontoporia blainvillei*, along with their distribution.

n	ΣPBDE	PBEB	HBBZ	ΣMeO-BDE	Sampling site	FMA	References
Blubber							
8	0.1 ± 0.1	NA	NA	NA	São Paulo	FMA IIb	Yogui et al. (2011)
41	0.2 ± 0.1	NA	NA	NA	São Paulo	FMA II	Leonel et al. (2014)
72	0.02 ± 0.01	NA	NA	NA	Rio Grande do Sul	FMA III	Leonel et al. (2014)
8	0.01 ± 0.02	NA	NA	NA	São Paulo	FMA IIb	Barbosa et al. (2018)
10	0.02 ± 0.03	0.002 ± 0.001	ND	1.5 ± 0.6	Espírito Santo	FMA Ia	Oliveira-Ferreira et al. (*in prep.*)
10	0.2 ± 0.2	ND	ND	0.4 ± 0.4	Rio de Janeiro	FMA IIa	Oliveira-Ferreira et al. (*in prep*)
25	0.04 ± 0.03	ND	ND	0.04 ± 0.02	São Paulo	FMA IIb	Oliveira-Ferreira et al. (*in prep.*)
Liver							
13	0.1 ± 0.1	0.01 ± 0.005	0.01 ± 0.01	2.4 ± 1.7	Espírito Santo	FMA Ia	Alonso et al. (2012a)
1	0.04	0.1	ND	0.3	Rio de Janeiro	FMA Ib	Alonso et al. (2012a)
26	0.2 ± 0.2	0.01 ± 0.004	0.002 ± 0.002	0.3 ± 0.2	São Paulo	FMA II	Alonso et al. (2012a)
13	0.03 ± 0.01	0.01 ± 0.003	0.001 ± 0.0002	0.3 ± 0.1	Rio Grande do Sul	FMA IIIa	Alonso et al. (2012a)
3	0.02 ± 0.01	NA	NA	NA	Espírito Santo	FMA Ia	de la Torre et al. (2012)
14	0.05 ± 0.04	NA	NA	NA	São Paulo	FMA II	de la Torre et al. (2012)
3	0.01± 0.002	NA	NA	NA	Rio Grande do Sul	FMA IIIa	de la Torre et al. (2012)
9	0.1 ± 0.04	NA	NA	NA	Rio de Janeiro	FMA Ib	Lavandier et al. (2016)
6	0.002	NA	NA	NA	São Paulo	FMA IIb	Barbosa et al. (2018)

Legend: NA, not analyzed; ND, not detected.
Concentrations are expressed in μg/g lipid weight (Mean ± Standard Deviation).

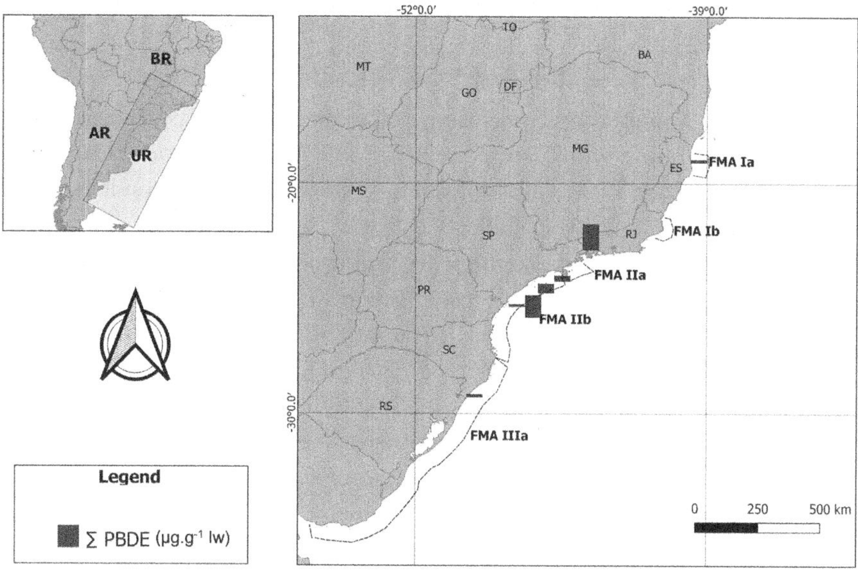

Fig. 11.5 ***Revisional map of PBDEs concentrations (µg/g lw) reported in the blubber of franciscana dolphin,*** **Pontoporia blainvillei*****, across their home range.*** For ΣPBDEs values, see Table 11.2. References: Yogui et al., (2011), Leonel et al., (2014), Barbosa et al., (2018), and Oliveira-ferreira et al., (*in prep.*).

Although there are few studies on contamination by organic pollutants, bioaccumulation by organochlorine, organobromine, and emerging contaminants compounds has been reported in franciscana dolphins along with their distribution. It is imperative to elucidate the effects of these contaminants in the species. Usually, the reported level of organic pollutants for franciscana dolphins was lower than in coastal delphinid species. But caution is required when addressing such comparisons due to their small size and weight. Hence, the dosage available for the specimens may be similar to those that delphinids are exposed to, leading ultimately to the deleterious effects that organic contaminants can cause to this threatened species.

Trace elements

Trace elements can be found in the environment by natural and anthropogenic sources, such as volcanic emissions and mining activities (Agency for Toxic Substances and Disease Registry ATSDR, 1999). They can be essential to biological functions in the organisms and/or non-essencial, being toxic even at low concentrations (Romero et al., 2016b; Zhang et al.,

2018). For instance, low levels of mercury (Hg) and cadmium (Cd), both non-essential elements, were already associated with immunosuppression in marine mammals (Desforges et al., 2016; Marsili et al., 2019).

There are some studies regarding trace elements bioaccumulation in franciscana dolphins, which are present in Table 11.3. It is worth mentioning that there is not a consensus to weight basis to express the results, which can hinder the comparison (Table 11.3).

The non-essential elements mercury and cadmium, and copper and the essential copper and zinc are the trace elements most frequently reported in franciscana dolphin dolphins in the literature (e.g., De Carvalho et al., 2008; Lailson-Brito et al., 2002; Polizzi et al., 2014). The liver is, commonly, the most analyzed organ and usually has the highest concentrations, followed by kidney and muscle (Das et al., 2003; Romero et al., 2016a). The exception is cadmium, which presents higher concentrations in kidneys due to the affinity with metallothioneins (Das et al., 2000; Dorneles et al., 2007b). Most studies of contamination by trace elements in franciscana dolphins regard mercury concentrations in all FMAs (I, II, III, and IV), especially FMAs Ib and IVc, and the overview is shown in Figs. 6 and 7. The papers concerning this element started to be published in 1990. Since then, there was a gap, and in the early 2000's more studies were published. Within the published studies, liver was the most assessed organ (Table 11.3; Figs. 6 and 7).

The trace element concentrations reported for the franciscana dolphin along its distribution varied according to the tissue analyzed and study site. The differences among FMAs can be related to geographical differences in bioavailability in trace elements in each area, feeding habits, temporal trends, and changes in the dynamics in the environment (Gui et al., 2017; Campos et al., 2020; Rodríguez et al., 2002; Manhães et al., 2022).

For franciscana dolphins' muscular tissue, THg concentrations were similar between FMA Ia and FMA Ib (Baptista et al., 2016; De Carvalho et al., 2008; Kehrig et al., 2016) and lower than franciscana dolphins from Argentina (FMA IVc) (Marcovecchio et al., 1990). For Cd, FMA Ia individuals presented lower concentrations than the ones reported by previous studies on FMA Ib and FMA IVc (De Carvalho et al., 2008; Marcovecchio et al., 1990). Among the essential trace elements, Cu and Zn muscular concentrations, franciscana dolphins from FMA Ia also showed lower concentrations than the previous study from FMA IVc (Marcovecchio et al., 1990). However, in a recent study performed in FMA IVc, Zn concentrations in FMA Ia were in the same order of magnitude (Panebianco et al., 2013), showing that temporal trends can

Table 11.3 Review of elements concentrations (total mercury—THg, arsenic—As, cadmium—Cd, copper—Cu, iron—Fe, manganese—Mn, and zinc—Zn) reported in muscle, liver, and kidney of franciscana dolphin, *Pontoporia blainvillei*, along with their distribution.

N	Weight basis	THg	As	Cd	Cu	Fe	Mn	Zn	Sampling site	FMA	References
2	Wet	3.0 ± 1.2	NA	0.1 ± 0.1[#]	2.5 ± 1.5	NA	NA	49.3 ± 4.8	Buenos Aires	FMA IVc	Marcovecchio et al. (1990)
7	Wet	0.19	NA	0.21	NA	NA	NA	25.8	Rio de Janeiro	FMA Ib	De Carvalho et al. (2008)
36	Wet	NA	NA	NA	NA	NA	NA	12.1 ± 3.8	Buenos Aires	FMA IVc	Panebianco et al. (2013)
16	Dry	1.92 ± 0.96	NA	NA	NA	NA	NA	NA	Rio de Janeiro	FMA Ib	Baptista et al. (2016)
7[a]	Wet	0.3 ± 0.04	NA	NA	NA	NA	NA	NA	Rio de Janeiro	FMA Ib	Kehrig et al. (2016)
4[b]	Wet	0.5 ± 0.02	NA	NA	NA	NA	NA	NA	Rio de Janeiro	FMA Ib	Kehrig et al. (2016)
9[c]	Dry	1.03 ± 0.2	NA	NA	NA	NA	NA	NA	Buenos Aires	FMA IIIb/IV	Romero et al. (2016a)
22[d]	Dry	1.49 ± 0.72	NA	NA	NA	NA	NA	NA	Buenos Aires	FMA IIIb/IV	Romero et al. (2016a)
9[e]	Dry	3.10 ± 1.96	NA	NA	NA	NA	NA	NA	Buenos Aires	FMA IIIb/IV	Romero et al. (2016a)
32	Dry	2.27 ± 1.5	1.28 ± 0.53	<0.03	3.54 ± 1.85	266 ± 88	1.54 ± 2.57	54 ± 14	Espirito Santo	FMA Ia	Manhães et al. (2022) - BTM
19	Dry	2.24 ± 1.57	0.51 ± 0.06	<0.03	2.99 ± 1.51	241 ± 83	2.44 ± 2.40	67 ± 15	Espirito Santo	FMA Ia	Manhães et al. (2022)- ATM

(Continued)

Table 11.3 Cont'd

N	Weight basis	THg	As	Cd	Cu	Fe	Mn	Zn	Sampling site	FMA	References
10	Dry	1.97 ± 1.08	<1.11	0.02 ± 0.03	3.7 ± 2.4	197 ± 90	0.8 ± 0.7	49 ± 22	Rio de Janeiro	FMA IIa	Vannuci-Silva et al., *in prep*
25	Dry	1.53 ± 1.15	<1.11	0.04 ± 0.09	3.5 ± 3.0	243 ± 111	1.0 ± 3.7	58 ± 26	São Paulo	FMA IIb	Vannuci-Silva et al., *in prep*
2	Wet	3.8 ± 1.6	NA	3.3 ± 1.4	16.0 ± 3.3	NA	NA	83.4 ± 40.0	Buenos Aires	FMA IVc	Marcovecchio et al. (1990)
17	Wet	1.24 ± 1.02	NA	0.43 ± 0.43	8.21 ± 4.45	216 ± 114	3.34 ± 1.76	36 ± 14	Rio de Janeiro	FMA Ib	Lailson-Brito et al. (2002)
23	Dry	3.5 ±2.1	1.2 ± 0.4	0.41 ± 0.78	44.5 ± 89.0	816 ± 280	14.8 ± 2.3	152 ± 82	São Paulo-Paraná	FMA II	Kunito et al. (2004)
18	Wet	2.6	1.1	0.6	NA	NA	NA	NA	Southeastern BR	FMA Ib, IIa	Seixas et al. (2007)
13	Wet	10.7	1.7	1.5	NA	NA	NA	NA	Southern, BR	FMA IIIa	Seixas et al. (2007)
18	Dry	0.7 – 9.7#	NA	NA	NA	NA	NA	NA	Southeastern BR	FMA Ib, IIa	Seixas et al. (2008)
13	Dry	1.2 – 52#	NA	NA	NA	NA	NA	NA	Southern, BR	FMA IIIa	Seixas et al. (2008)
18	Dry	NA	0.79	0.41	26.48	NA	NA	NA	Rio de Janeiro	FMA Ib	Seixas et al. (2009)
7	Wet	0.85	NA	0.08	NA	NA	NA	37.8	Rio de Janeiro	FMA Ib	De Carvalho et al. (2008)
31	Wet	5.98	NA	NA	NA	NA	NA	NA	Southeastern/South, BR	–	Moreira et al. (2009)

N	Weight basis	THg	As	Cd	Cu	Fe	Mn	Zn	Sampling site	FMA	References
24	Wet	NA	NA	2.01	5.03	NA	NA	29.3	Buenos Aires	FMAVIc	Panebianco et al. (2011)
1	Dry	1.1*	NA	0.53*	5.8*	NA	1.73*	100.1*	Rio de Janeiro	FMA Ib	Lemos et al. (2013)
14[f]	Wet	NA	NA	1.81	NA	NA	NA	NA	Buenos Aires	FMA IVa/IVb	Polizzi et al. (2014)
13[g]	Wet	NA	NA	0.19	NA	NA	NA	NA	Buenos Aires	FMA IVa/ IVb)	Polizzi et al. (2014)
7/9[c]	Dry	1.29 ± 0.79	NA	NA	NA	1289[#]	17.21[#]	NA	Buenos Aires	FMA IIIb/IV	Romero et al. (2016a,b)
8/22[d]	Dry	3.62 ± 3.46	NA	NA	NA	1232[#]	13.15[#]	NA	Buenos Aires	FMA IIIb/IV	Romero et al. (2016a,b)
5/9[e]	Dry	26.9 ± 22.3	NA	NA	NA	1397[#]	12.70[#]	NA	Buenos Aires	FMA IIIb/IV	Romero et al. (2016a,b)
7[a]	Wet	0.5 ± 0.2	NA	NA	NA	NA	NA	NA	Rio de Janeiro	FMA Ib	Kehrig et al. (2016)
4[b]	Wet	1.6 ± 1.0	NA	NA	NA	NA	NA	NA	Rio de Janeiro	FMA Ib	Kehrig et al. (2016)
44	Wet	NA	NA	0.35	NA	NA	NA	NA	Rio Grande do Sul	FMA IIIa	Dorneles et al. (2007b)
26	Dry	5.58 ± 8.36	1.80 ± 0.65	0.24 ± 0.23	24.3 ± 26.6	583 ± 192	7.32 ± 4.11	133 ± 62	Espirito Santo	FMA Ia	Manhães et al. (2022)- BTM
19	Dry	2.79 ± 0.40	0.60 ± 0.08	0.12 ± 0.04	32.1 ± 18.8	689 ± 83	17.05 ± 2.14	149 ± 25	Espirito Santo	FMA Ia	Manhães et al. (2022)-ATM

(Continued)

Table 11.3 Cont'd

N	Weight basis	THg	As	Cd	Cu	Fe	Mn	Zn	Sampling site	FMA	References
8	Dry	9.80 ± 10.0	<1.11	5.68 ± 5.67	138.6 ± 209.1	1781 ± 1075	12.1 ± 5.7	221 ± 234	Rio de Janeiro	FMA IIa	Vannuci-Silva et al., *in prep*
25	Dry	6.91 ± 17.0	0.53 ± 0.12	0.86 ± 1.34	36.5 ± 64.4	745 ± 418	13.4 ± 5.2	146 ± 161	São Paulo	FMA IIb	Vannuci-Silva et al., *in prep*
2	Wet	1.9 ± 0.7	NA	9.9 ± 3.9	14.0 ± 4.9	NA	NA	79 ± 21	Buenos Aires	FMA IVc	Marcovecchio et al. (1990)
15	Wet	5.37 ± 10.9	NA	0.30 ± 0.32	4.91 ± 4.06	156 ± 50	0.89 ± 0.32	23 ± 9.5	Rio de Janeiro	FMA Ib	Lailson-Brito et al. (2002)
18	Wet	1.4	0.7	3.4	NA	NA	NA	NA	Southeastern BR	FMA Ib, IIa	Seixas et al. (2007)
13	Wet	1.7	1.4	5.5	NA	NA	NA	NA	Southern BR	FMA IIIa	Seixas et al. (2007)
18	Dry	0.5 – 5.1[#]	NA	NA	NA	NA	NA	NA	Southeastern, BR	FMA Ib, IIa	Seixas et al. (2008)
13	Dry	0.6 – 4.7[#]	NA	NA	NA	NA	NA	NA	Southern, BR	FMA IIIa)	Seixas et al. (2008)
37	Wet	NA	NA	2.99	3.31	NA	NA	20.9	Buenos Aires	FMA IIIb/IV	Panebianco et al. (2011)
14[f]	Wet	NA	NA	4.22	NA	NA	NA	NA	Buenos Aires	FMA IVa/IVb	Polizzi et al. (2014)
13[g]	Wet	NA	NA	0.43	NA	NA	NA	NA	Buenos Aires	FMA IVa/IVb	Polizzi et al. (2014)

N	Weight basis	THg	As	Cd	Cu	Fe	Mn	Zn	Sampling site	FMA	References
7/9[c]	Dry	0.76 ± 0.72	NA	NA	NA	653[#]	6.43[#]	NA	Buenos Aires	FMA IIIb/IV	Romero et al. (2016a,b)
8/22[d]	Dry	1.35 ± 1.11	NA	NA	NA	816[#]	3.65[#]	NA	Buenos Aires	FMA IIIb/IV	Romero et al. (2016a,b)
5/9[e]	Dry	3.78 ± 4.35	NA	NA	NA	973[#]	3.85[#]	NA	Buenos Aires	FMA IIIb/IV	Romero et al. (2016a,b)
23	Dry	2.65 ± 2.30	1.72 ± 0.47	0.97 ± 0.57	12.18 ± 3.24	488 ± 170	3.03 ± 1.37	96 ± 30	Espirito Santo	FMA Ia	Manhães et al. (2022)-BTM
11	Dry	2.36 ± 1.71	0.71 ± 0.30	0.98 ± 1.23	12.47 ± 2.80	377 ± 87	3.95 ± 1.72	96 ± 28	Espirito Santo	FMA Ia	Manhães et al. (2022)-ATM
6	Dry	1.94 ± 1.11	<1.11	20.1 ± 19.1	14.7 ± 3.0	477 ± 166	3.2 ± 1.4	85 ± 16	Rio de Janeiro	FMA IIa	Vannuci-Silva et al., *in prep*
25	Dry	1.10 ± 0.68	0.55 ± 0.11	2.42 ± 1.94	9.5 ± 2.9	401 ± 97	2.8 ± 0.5	59 ± 10	São Paulo	FMA IIb	Vannuci-Silva et al., *in prep*

[#]min–max and median: when mean values were not reported in the study; *absolute value=when only one individual was analyzed; NA = not analyzed/available, AR, Argentina; BR, Brazil; FMA, franciscana management area; BTM, Franciscana dolphins collected before the tailing mud reached the Espírito Santo coast; ATM, franciscana dolphins collected after the tailing mud reached the Espírito Santo coast; determined based on map sampling area provide for the study; a = immature; b = mature; c = calf; d = juvenile; e = adult; f = marine; g = estuarine.

THg concentrations in muscle, liver, and kidney of franciscana dolphins (*Pontoporia blainvillei*) along with their distribution are expressed in μg/g dry and wet weight (Mean ± Standard Deviation).

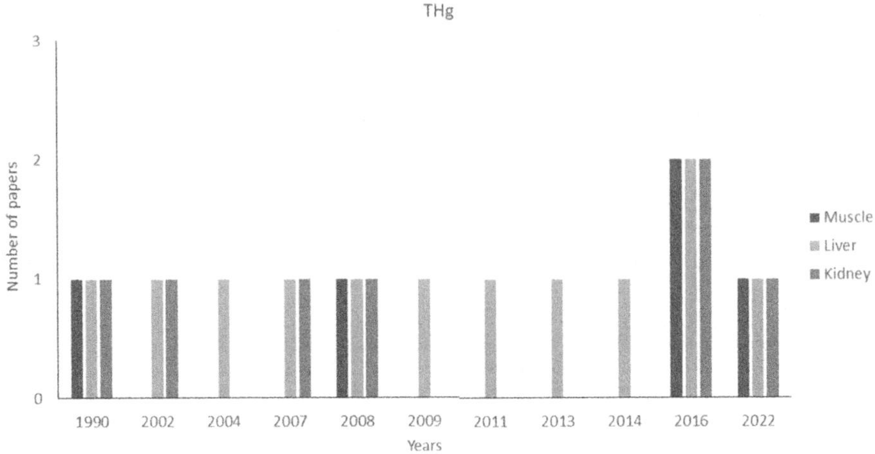

Fig. 11.6 ***Overview containing the number of papers available in literature concerning THg in liver, muscle and kidney of franciscana dolphins,*** **Pontoporia blainvillei.**

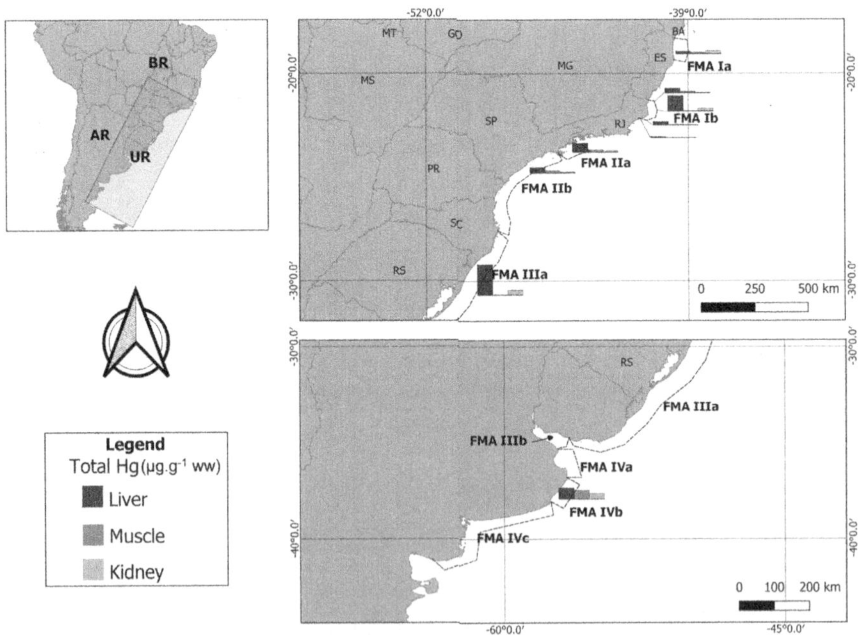

Fig. 11.7 ***THg levels in muscle, liver and kidney of franciscana dolphin dolphins*** **(Pontoporia blainvillei*)*** ***along with their distribution, expressed in μg/g, wet weight.*** For THg values, see Table 11.3. References: Marccovecchio et al. (1990), De Carvalho et al. (2008), Lailson-Brito et al. (2002), Lemos et al. (2013), Kehrig et al. (2016), Seixas et al. (2007), Manhães et al. (2022), Vanucci-Silva et al. (*in prep.*).

influence these results in the same studied area (Marcovecchio et al., 1990; Panebianco et al., 2013) (Table 11.3).

More studies are available for comparison regarding hepatic trace element concentrations. Considering studies on a dry basis, concentrations of THg in the liver of franciscana dolphins from FMA Ia were in the same order of magnitude than franciscana dolphins from FMA Ib and II (Kunito et al., 2004; Lemos et al., 2013). But they were lower than other FMAs, such as FMA IIIb/IV, which considered only adult individuals (Romero et al., 2016b, 2016a) (Table 11.3). Dry weight Cd concentration in franciscana dolphins from FMA IIa was higher than FMA Ia, II, and IVc (Kunito et al., 2004; Panebianco et al., 2013). In FMA IVc, Cd concentrations were higher than FMA Ib, considering the wet basis (Lailson-Brito et al., 2002; Marcovecchio et al., 1990). These results are probably related to differences in foraging habits since Cd accumulation is higher cephalopods consumers (Bustamante et al., 1998; Dorneles et al., 2007a). For As, which is a very toxic trace element, concentrations in livers were in the same order of magnitude for FMA II and FMA Ia (collected before the tailing mud reached the Espírito Santo coast). However, it was lower in franciscana dolphins from FMA Ia (collected after the tailing mud reached the Espírito Santo coast) (Kunito et al., 2004; Manhães et al., 2022). These differences can be due to the sampling period of the studies and the impact of the dam failure in the environment, especially because of the changes in trace elements bioavailability (Kunito et al., 2004; Manhães et al., 2022). Among the essential elements, Fe, for instance, were lower in franciscana dolphins from FMA Ia, FMA IIa, and IIb (Lemos et al., 2013) when compared to FMAs IIb, IVa and IVb (Kunito et al., 2004; Romero et al., 2016b).

And for kidneys, considering studies on a dry basis, THg concentrations were higher in FMA Ia and IIIb/IV than other FMAs, such as FMA IIa and IIb (Manhães et al., 2022; Romero et al., 2016a) (Table 11.3). When considering studies on a wet basis, franciscana dolphins from FMA Ib were higher than IVc (Lailson-Brito et al., 2002; Marcovecchio et al., 1990). It is important to highlight that the studies available for this organ present a high range in sampling year, which can also play an important role in the results. Dry weight Cd concentration in franciscana dolphins from FMA IIa was higher than FMA Ia (Manhães et al., 2022). In FMA IVc, wet weight Cd concentrations were also higher than FMA Ib (Lailson-Brito et al., 2002; Marcovecchio et al., 1990). Among the essential elements (Fe, Mn, and Zn) concentrations in franciscana dolphins were in the same order of magnitude across their distribution (Manhães et al., 2022; Romero et al., 2016a).

Despite trace elements toxicity in organisms, there are detoxification processes were already reported in literature in odontocete species. The bind between Hg and selenium (Se) was described in Guiana dolphins, including the important function of Kuppfer cells in the detoxification of Hg (Lailson-Brito et al., 2012). Furthermore, metallothioneins, which are proteins found mainly in kidneys and livers also protect the organism from toxic trace elements (Das et al., 2000). Romero et al. (2016a) investigated these two mechanisms in franciscana dolphins and also detected Hg-Se binding as an important pathway for Hg detoxification in the species.

Recently, the presence of trace elements of emerging concern, such as titanium (Ti) was evaluated in franciscana dolphins. The interest in this trace element has increased due to its use in daily use products, such as toothpaste. The study also identified the detoxification process to which Ti is subjected to, via its binding to metalothioneins (Monteiro et al., 2020).

Franciscana dolphins have lower trace element concentrations in their tissues when compared to other cetacean species (e.g., Chen et al., 2020; García-Alvarez et al., 2015; Lemos et al., 2013), probably due to their feeding ecology. The morphological features of their oral and digestive apparatus enable it to feed on small size prey (Di Beneditto et al., 2011), such as fishes and cephalopods (Campos et al., 2020). Additionally, franciscana dolphins are among the smallest cetacean species and have a shorter life span (Ramos et al., 2000). Thereby, relatively lower concentrations cannot be disregarded. The toxic effects of trace elements in cetaceans are described in the literature, especially related to chronic low-level exposure, resulting in immune suppression and less resistance to diseases (Marsili et al., 2019; Pelissó et al., 2008; Rawson et al., 1993).

Long-term effects to franciscana dolphin dolphins' conservation in FMA Ia

The Fundão dam failure was the largest environmental disaster that occurred in Brazil, and resulted in the release of millions of m^3 of tailing mud from mining activities (Hatje et al., 2017). The sediment plume from tailing mud spreaded into the area of the Rio Doce basin, reaching the coastal area of Espírito Santo state in a region that have the highest density of franciscana dolphins individuals in FMA Ia (Danilewicz et al., 2012; Magris et al., 2019; Sucunza et al., 2020). As a consequence, the high trace element concentrations found in the water column and sediment during and after the dam failure has the potential to induce toxic effects in organisms at different trophic levels (Sá et al., 2021; Vergilio et al., 2020). Concentrations of As, Cd, Cu, Fe, Mn, and Zn found in the water column were above Brazilian and International standards for safe levels in water quality (Sá et al., 2021). Despite that, several impacts of the Fundão dam failure and its environmental consequences are still unknown. However, the consequences for franciscana dolphins are beginning to be elucidated. A recent study showed changes in trace elements bioavailability in the environment and the bioaccumulation in the species, such as variation in temporal trends for Hg, Zn, As, Cu, and Fe in franciscana dolphins tissues/organs (Manhães et al., 2022). The disaster also increased the bioavailability of organochlorine pesticides for franciscana dolphins, that could lead to endocrine disruption (Oliveira-Ferreira, *in prep.*).

FMA Ia represents one of the least known stocks, considering many ecological features, including ecotoxicological studies. Recent studies showed lower population' estimative in the area (Danilewicz et al., 2012; Sucunza et al., 2019). Additionally, a reduced area of use by franciscana dolphins was reported in FMA Ia, and it was related to a large shipyard construction (Jurong Aracruz) (Danilewicz et al., 2012; do Amaral et al., 2018). All these issues can represent a concern for the species.

Conservation implications

Studies of contamination have evidenced bioaccumulation of different elements and organic pollutants in muscle, liver, kidney, blubber, and other tissues of species (e.g., Lailson-Brito et al., 2002; Dorneles et al., 2007b; Alonso et al., 2012a; Lavandier et al., 2016; Baptista et al., 2016). However, in general, the concentrations were lower than those reported for other species of cetaceans on the Brazilian, Uruguayan, and Argentinean coast. These results have been associated, mainly, with the dietary habits and trophic position of the franciscana dolphin. It is necessary to be careful and take some aspects into account when comparing the concentration of contaminants between the franciscana dolphins and other coastal species. The species is from a distinct taxonomic group, and it has different biological characteristics from delphinids. The franciscana dolphin is one of the smallest cetaceans, with earlier reproduction and shorter life expectancy than most delphinids. The smaller size makes the dose that the franciscana dolphin is exposed to be similar to that of delphinids. And the maternal transfer of contaminants to a female calf is soon retransferred to a new offspring, once the species reaches sexual maturity at just two years old. Still, it is necessary to highlight the synergic effect of pollutants already reported for other cetaceans and the coastal distribution of franciscana dolphins, which causes the species to be continuously exposed to high levels of contaminants. The Fundão dam failure, with a consequent temporal increase in bioavailability and bioaccumulation of mercury and pesticides by franciscana dolphin from FMA Ia, is an example of how the species is vulnerable to the impact of anthropogenic actions in the coastal zone. Contamination by trace elements and organic pollutants is an effective threat to franciscana dolphins' conservation that cannot be disregarded. Systematic studies throughout its distribution are still needed to understand the levels of contamination and the effects of pollutants on franciscana dolphin survival and reproduction rate.

References

Agency for Toxic Substances and Disease Registry (ATSDR), 1999. Toxicological profile for mercury. Department of Health and Human Services, Public Health services, Atlanta, p. 617.

Aguilar, A., Borrell, A., Pastor, T., 1999. Biological factors affecting variability of persistent pollutant levels in cetaceans. J. Cetacean Res. Manag. 1, 83–116.

Alonso, M.B., Eljarrat, E., Gorga, M., Secchi, E.R., Bassoi, M., Barbosa, L., Bertozzi, C.P., Marigo, J., Cremer, M., Domit, C., Azevedo, A.F., Dorneles, P.R., Torres, J.P.M., Lailson-Brito, J., Malm, O., Barceló, D., 2012a. Natural and anthropogenically-produced brominated compounds in endemic dolphins from Western South Atlantic: another risk to a vulnerable species. Environ. Pollut. 170, 152–160.

Alonso, M.B., Feo, M.L., Corcellas, C., Gago-Ferrero, P., Bertozzi, C.P., Marigo, J., Flach, L., Meirelles, A.C.O., Carvalho, V.L., Azevedo, A.F., Torres, J.P.M., Lailson-Brito, J., Malm, O., Diaz-Cruz, M.S., Eljarrat, E., Barceló, D., 2015. Toxic heritage: maternal transfer of pyrethroid insecticides and sunscreen agents in dolphins from Brazil. Environ. Pollut. 207, 391–402.

Alonso, M.B., Feo, M.L., Corcellas, C., Vidal, L.G., Bertozzi, C.P., Marigo, J., Secchi, E.R., Bassoi, M., Azevedo, A.F., Dorneles, P.R., Torres, J.P.M., Lailson-Brito, J., Malm, O., Eljarrat, E., Barceló, D., 2012b. Pyrethroids: a new threat to marine mammals? Environ. Int. 47, 99–106.

Arias, A.H., Panebianco, M.V., Net, S., Dumoulin, D., Souissi, S., Ouddane, B., Cappozzo, H.L., Marcovecchio, J.E., 2016. Franciscana dolphins as PCBs marine biomonitors in Argentina, south-west Atlantic Ocean. J. Mar. Biolog. Assoc. U.K. 96 (4), 979–992.

Baptista, G., Kehrig, H.A., Di Beneditto, A.P.M., Hauser-Davis, R.A., Almeida, M.G., Rezende, C.E., Siciliano, S., de Moura, J.F., Moreira, I., 2016. Mercury, selenium and stable isotopes in four small cetaceans from the southeastern brazilian coast: influence of feeding strategy. Environ. Pollut. 218, 1298–1307.

Barbosa, A.P.M., Méndez-Fernandez, P., Dias, P.S., Santos, M.C.O., Taniguchi, S., Montone, R.C., 2018. Transplacental transfer of persistent organic pollutants in La Plata dolphins (Pontoporia blainvillei; Cetartiodactyla, Pontoporiidae). Sci. Total Environ 631–632, 239–245.

Bertozzi, C.P., Zerbini, A.N., 2002. Incidental mortality of franciscana (Pontoporia blainvillei) in the artisanal fishery of Praia Grande, São Paulo State, Brazil. Lat. Am. J. Aquat. Mamm. 1, 153–160.

Borrell, A., Pastor, T., Aguilar, A., Corcuera, J., Monzón, F., 1995. DDTs and PCBs in La Plata dolphins (*Pontoporia blainvillei*) from Argentina: age and sex trends. Europ. Res. Cetac. 9, 273–276.

Bustamante, P., Caurant, F., Fowler, S.W., Miramand, P., 1998. Cephalopods as a vector for the transfer of cadmium to top marine predators in the north-east Atlantic Ocean. Sci. Total Environ. 220, 71–80.

Campos, L.B., Lopes, X.M., Silva, Da, E., S.M.C.D.O, 2020. Feeding habits of the franciscana dolphin (Pontoporia blainvillei) in south-eastern Brazil. J. Mar. Biol. Assoc. United Kingdom 100, 301–313.

Chen, M.H., Lin, Y.T., Lai, C.C., Chou, L.S., Chen, C.Y., 2020. Tissue concentrations of Fe, Zn, Cu and Mn of four taiwanese toothed cetaceans. Mar. Pollut. Bull. 158, 1–8.

Cunha, H.A., Medeiros, B.V., Barbosa, L.A., Cremer, M.J., Marigo, J., Lailson-Brito, J., Azevedo, A.F., Solé-Cava, A.M., 2014. Population structure of the endangered Franciscana dolphin (Pontoporia blainvillei): Reassessing management units. PLoS One 9, 1–8.

Danilewicz, D., Zerbini, A.N., Andriolo, A., Secchi, E.R., Sucunza, F., Ferreira, E., Denuncio, P., Flores, P.A.C., 2012. Estimating Abundance of an Isolated Population of the Threatened Franciscana: Moving Towards Conservation Actions. International Whaling Commission (IWC) Final Report, Ilhéus, September, 16p.

Das, K., Debacker, V., Bouquegneau, J.M., 2000. Metallothioneins in marine mammals. Cell Mol. Bio. 46, 283–294.

Das, K., Debacker, V., Pillet, S., Bouquegneau, J., 2003. Heavy metals in marine mammals. Toxicology of Marine Mammals, Chapter 7, 1st Edition, CRC Press, London, p. 33.

Das, K., Vossen, A., Tolley, K., Víkingsson, G., Thron, K., Müller, G., Baumgärtner, W., Siebert, U., 2006. Interfollicular fibrosis in the thyroid of the harbour porpoise: an endocrine disruption? Arch. Environ. Contam. Toxicol. 51, 720–729.

De Carvalho, C.E.V., Di Beneditto, A.P.M., Souza, C.M.M., Ramos, R.M.A., Rezende, C.E., 2008. Heavy metal distribution in two cetacean species from Rio de Janeiro State, south-eastern Brazil. J. Mar. Biol. Assoc. United Kingdom 88, 1117–1120.

De La Torre, A., Alonso, M.B., Martínez, M.A., Sanz, P., Shen, L., Reiner, E.J., Lailson-Brito, J., Torres, J.P.M., Bertozzi, C., Marigo, J., Barbosa, L., Cremer, M., Secchi, E., Malm, O., Eljarrat, E., Barceló, D., 2012. Dechlorane-related compounds in franciscana dolphin (Pontoporia blainvillei) from southeastern and southern coast of Brazil. Environ. Sci. Technol. 46, 12364–12372.

Desforges, J.P., Levin, M., Jasperse, L., De Guise, S., Eulaers, I., Letcher, R.J., Acquarone, M., Nordøy, E., Folkow, L.P., Hammer Jensen, T., Grøndahl, C., Bertelsen, M.F., St Leger, J., Almunia, J., Sonne, C., Dietz, R., 2017. Effects of polar bear and killer whale derived contaminant cocktails on marine mammal immunity. Environ. Sci. Technol. 51, 11431–11439.

Desforges, J.P.W., Sonne, C., Levin, M., Siebert, U., De Guise, S., Dietz, R., 2016. Immunotoxic effects of environmental pollutants in marine mammals. Environ. Int. 86, 126–139.

Di Beneditto, A.P.M., de Souza, C.M.M, Kehrig, H.A., Rezende, C.E., 2011. Use of multiple tools to assess the feeding preference of coastal dolphins. Mar. Biol. 158, 2209–2217.

do Amaral, K.B., Danilewicz, D., Zerbini, A., Di Beneditto, A.P., Andriolo, A., Alvares, D.J., Secchi, E., Ferreira, E., Sucunza, F., Borges-Martins, M., de Oliveira Santos, M.C., Cremer, M., Denuncio, P., Ott, P.H., Moreno, I.B., 2018. Reassessment of the franciscana Pontoporia blainvillei (Gervais & d'Orbigny, 1844) distribution and niche characteristics in Brazil. J. Exp. Mar. Bio. Ecol. 508, 1–12.

Dorneles, P.R., Lailson-Brito, J., dos Santos, R.A., Silva da Costa, P.A., Malm, O., Azevedo, A.F., Machado Torres, J.P., 2007a. Cephalopods and cetaceans as indicators of offshore bioavailability of cadmium off Central South Brazil Bight. Environ. Pollut. 148, 352–359.

Dorneles, P.R., Lailson-Brito, J., Secchi, E.R., Bassoi, M., Lozinsky, C.P.C., Torres, J.P.M., Malm, O., 2007b. Cadmium concentrations in franciscana dolphin (Pontoporia blainvillei) from south brazilian coast. Brazilian J. Oceanogr. 55, 179–186.

Dorneles, P.R., Sanz, P., Eppe, G., Azevedo, A.F., Bertozzi, C.P., Martínez, M.A., Secchi, E.R., Barbosa, L.A., Cremer, M., Alonso, M.B., Torres, J.P.M., Lailson-Brito, J., Malm, O., Eljarrat, E., Barceló, D., Das, K., 2013. High accumulation of PCDD, PCDF, and PCB congeners in marine mammals from Brazil: A serious PCB problem. Sci. Total Environ. 463–464, 309–318.

Gago-Ferrero, P., Alonso, M.B., Bertozzi, C.P., Marigo, J., Barbosa, L., Cremer, M., Secchi, E.R., Domit, C., Azevedo, A., Lailson-Brito, J., Torres, J.P.M., Malm, O., Eljarrat, E., Díaz-Cruz, M.S., Barceló, D., 2013. Erratum: first determination of UV filters in marine mammals. Octocrylene levels in franciscana dolphins. Environ. Sci. Technol. 47 (11), 5619–5625. http://doi.org/10.1021/es400675y.

Galligan, T.M., Balmer, B.C., Schwacke, L.H., Bolton, J.L., Quigley, B.M., Rosel, P.E., Ylitalo, G.M., Boggs, A.S.P., 2019. Examining the relationships between blubber steroid hormones and persistent organic pollutants in common bottlenose dolphins. Environ. Pollut. 249, 982–991.

García-Alvaréz, N., Fernández, A., Boada, L.D., Zumbado, M., Zaccaroni, A., Arbelo, M., Alumunia, J., Luzardo, O.P., 2015. Mercury and selenium status of bottlenose dolphins (*Tursiops truncatus*): a study in stranded animals on the Canary Islands. Sci. Total Environ. 536, 489–498.

Gray, J.S., 2002. Biomagnification in marine systems: the perspective of an ecologist. Mar. Pollut. Bull. 45, 46–52.

Gui, D., Yu, R.Q., Karczmarski, L., Ding, Y., Zhang, H., Sun, Y., Zhang, M., Wu, Y., 2017. Spatiotemporal trends of heavy metals in Indo-Pacific Humpback Dolphins (Sousa chinensis) from the Western Pearl River Estuary. China. Environ. Sci. Technol. 51, 1848–1858.

Hall, A.J., Kalantzi, O.I., Thomas, G.O., 2003. Polybrominated diphenyl ethers (PBDEs) in grey seals during their first year of life—are they thyroid hormone endocrine disrupters? Environ. Pollut. 126, 29–37.

Hall, A.J., McConnell, B.J., Rowles, T.K., Aguilar, A., Borrell, A., Schwacke, L., Reijnders, P.J.H., Wells, R.S., 2005. Individual-based model framework to assess population consequences of polychlorinated biphenyl exposure in bottlenose dolphins. Environ. Health Perspect. 114, 60–64.

Hatje, V., Pedreira, R.M.A., De Rezende, C.E., Schettini, C.A.F., De Souza, G.C., Marin, D.C., Hackspacher, P.C., 2017. The environmental impacts of one of the largest tailing dam failures worldwide. Sci. Rep. 7, 1–13.

Jepson, B.P.D., Law, R.J., 2016. Persistent pollutants, persistent threats: polychlorinated biphenyls remain a major threat to marine apex predators such as orcas. Science 352 (80-), 1–2.

Jepson, P.D., Bennett, P.M., Deaville, R., Allchin, C.R., Baker, J.R., Law, R.J., 2005. Relationships between polychlorinated biphenyls and health status in harbor porpoises (Phocoena phocoena) stranded in the United Kingdom. Environ. Toxicol. Chem. 24, 238–248.

Jin, Y., Liu, J., Wang, L., Chen, R., Zhou, C., Yang, Y., Liu, W., Fu, Z., 2012. Permethrin exposure during puberty has the potential to enantioselectively induce reproductive toxicity in mice. Environ. Int. 42, 144–151.

Kajiwara, N., Kamikawa, S., Amano, M., Hayano, A., Yamada, T.K., Miyazaki, N., Tanabe, S., 2008. Polybrominated diphenyl ethers (PBDEs) and organochlorines in melon-headed whales, Peponocephala electra, mass stranded along the Japanese coasts: Maternal transfer and temporal trend. Environ. Pollut. 156, 106–114.

Kajiwara, N., Matsuoka, S., Iwata, H., Tanabe, S., Rosas, F.C.W., Fillmann, G., Readman, J.W., 2004. Contamination by persistent organochlorines in cetaceans incidentally caught along Brazilian coastal waters. Arch. Environ. Contam. Toxicol. 46, 124–134.

Kehrig, H.A., Hauser-Davis, R.A., Seixas, T.G., Pinheiro, A.B., Di Beneditto, A.P.M., 2016. Mercury species, selenium, metallothioneins and glutathione in two dolphins from the southeastern Brazilian coast: mercury detoxification and physiological differences in diving capacity. Environ. Pollut. 213, 785–792.

Kunito, T., Nakamura, S., Ikemoto, T., Anan, Y., Kubota, R., Tanabe, S., Rosas, F.C.W., Fillmann, G., Readman, J.W., 2004. Concentration and subcellular distribution of trace elements in liver of small cetaceans incidentally caught along the Brazilian coast. Mar. Pollut. Bull. 49, 574–587.

Lailson-Brito, J., Azeredo, M.A.A., Malm, O., Ramos, R.A., Di Beneditto, A.P.M., Saldanha, M.F.C., 2002. Trace metals in liver and kidney of the franciscana (Pontoporia blainvillei) from the northern coast of Rio de Janeiro State. Brazil. Lat. Am. J. Aquat. Mamm. 1, 107–114.

Lailson-Brito, J., Cruz, R., Dorneles, P.R., Andrade, L., Azevedo, A.F., Fragoso, A.B., Vidal, L.G., Costa, M.B., Bisi, T.L., Almeida, R., Carvalho, D.P., Bastos, W.R., Malm, O., 2012. Mercury-selenium relationships in liver of Guiana Dolphin: the possible role of kupffer cells in the detoxification process by Tiemannite formation. PLoS One 7, 1–10.

Lailson-Brito, J., Dorneles, P.R., Azevedo-Silva, C.E., Azevedo, A.F., Vidal, L.G., Marigo, J., Bertozzi, C., Zanelatto, R.C., Bisi, T.L., Malm, O., Torres, J.P.M., 2011. Organochlorine concentrations in franciscana dolphins, Pontoporia blainvillei, from Brazilian waters. Chemosphere 84, 882–887.

Lavandier, R., Arêas, J., Quinete, N., De Moura, J.F., Taniguchi, S., Montone, R., Siciliano, S., Moreira, I., 2016. PCB and PBDE levels in a highly threatened dolphin species from the Southeastern Brazilian coast. Environ. Pollut. 208, 442–449.

Law, R.J., Barry, J., Barber, J.L., Bersuder, P., Deaville, R., Reid, R.J., Brownlow, A., Penrose, R., Barnett, J., Loveridge, J., Smith, B., Jepson, P.D., 2012. Contaminants in cetaceans from UK waters: Status as assessed within the Cetacean Strandings Investigation Programme from 1990 to 2008. Mar. Pollut. Bull 64, 1485–1494.

Lemos, L.S., de Moura, J.F., Hauser-Davis, R.A., de Campos, R.C., Siciliano, S., 2013. Small cetaceans found stranded or accidentally captured in southeastern Brazil: Bioindicators of essential and non-essential trace elements in the environment. Ecotoxicol. Environ. Saf. 97, 166–175.

Leonel, J., Sericano, J.L., Fillmann, G., Secchi, E., Montone, R.C., 2010. Long-term trends of polychlorinated biphenyls and chlorinated pesticides in franciscana dolphin (Pontoporia blainvillei) from Southern Brazil. Mar. Pollut. Bull. 60, 412–418.

Leonel, J., Sericano, J.L., Secchi, E.R., Bertozzi, C., Fillmann, G., Montone, R.C., 2014. PBDE levels in franciscana dolphin (Pontoporia blainvillei): Temporal trend and geographical comparison. Sci. Total Environ. 493, 405–410.

Magris, R.A., Marta-Almeida, M., Monteiro, J.A.F., Ban, N.C., 2019. A modelling approach to assess the impact of land mining on marine biodiversity: Assessment in coastal catch ments experiencing catastrophic events (SW Brazil). Sci. Total Environ. 659, 828–840.

Manhães, B.M.R., Vannuci-Silva, M., Brião, J.A., Guari, E.B., Botta, S., Colosio, A.C., Ramos, H.G.C., Barbosa, L.A., Cunha, I.A.G., Azevedo, A.F., Cunha, H.A., Bisi, T.L., Lailson-Brito, J., 2022. Temporal trends of trace elements bioaccumulation by a vulnerable cetacean (Pontoporia blainvillei) before and after one of the largest mining disasters worldwide. Sci. Total Environ. 804, 1–12.

Marcovecchio, J.E., Moreno, V.J., Bastida, R.O., Gerpe, M.S., Rodríguez, D.H., 1990. Tissue distribution of heavy metals in small cetaceans from the Southwestern Atlantic Ocean. Mar. Pollut. Bull. 21, 299–304.

Marsili, L., Di Guardo, G., Mazzariol, S., Casini, S., 2019. Insights into cetacean immunology: do ecological and biological factors make the difference? Front. Immunol. 10, 1–11.

Monteiro, F., Lemos, L.S., de Moura, J.F., Rocha, R.C.C., Moreira, I., Di Beneditto, A.P.M., Kehrig, H.A., Bordon, I.C., Siciliano, S., Saint'Pierre, T.D., Hauser-Davis, R.A., 2020. Total and subcellular Ti distribution and detoxification processes in Pontoporia blainvillei and Steno bredanensis dolphins from Southeastern Brazil. Mar. Pollut. Bull. 153, 1–6.

Moreira, I., Seixas, T.G., Kehrig, H.A., Fillmann, G., Di Beneditto, A.P.M., Souza, C.M., Malm, O., 2009. Selenium and mercury (Total and Organic) in tissues of a coastal small cetacean, *Pontoporia blainvillei*. J. Coast. Res. I, 866–870.

O'Shea, T.J., Brownell, R.L., Clark, D.R., Walker, W.A., Gay, M.L., Lamont, T.G., 1980. Organochlorine pollutants in small cetaceans from the Pacific and South Atlantic Oceans, November 1968-June 1976. Pestic. Monit. J. 14, 35–46.

Panebianco, M.V., Negri, M.F., Botté, S.E., Marcovecchio, J.E., Cappozzo, H.L., 2011. Metales pesados en el riñón del delfín franciscana, *Pontoporia blainvillei* (Cetacea:Pontoporiidae) y su relación con parámetros biológicos. Lat. Am. J. Aqua. Res. 39, 526–533.

Panebianco, M.V., Negri, M.F., Botte, S.E., Marcovecchio, J.E., Cappozzo, H.L., 2013. Essential and non-essential heavy metals in skin and muscle tissues of franciscana dolphins (Pontoporia blainvillei) from the southern Argentina coast. Chem. Ecol. 29, 511–518.

Pelissó, S.C., Munoz, M.J., Carballo, M., Sanchez-Vizcaino, J.M., 2008. Determination of the immunotoxic potential of heavy metals on the functional activity of bottlenose dolphin leucocytes in vitro. Vet. Immunol. Immunopathol. 121, 189–198.

Polizzi, P.S., Romero, M.B., Chiodi Boudet, L.N., Das, K., Denuncio, P.E., Rodríguez, D.H., Gerpe, M.S., 2014. Metallothioneins pattern during ontogeny of coastal dolphin, Pontoporia blainvillei, from Argentina. Mar. Pollut. Bull. 80, 275–281.

Ramos, R.M.A., Di Beneditto, A.P.M., Lima, N.R.W., 2000. Growth parameters of Pontoporia blainvillei and Sotalia fluviatilis (Cetacea) in northern Rio de Janeiro. Brazil. Aquat. Mamm. 261, 65–75.

Rawson, A.J., Patton, G.W., Hofmann, S., Pietra, G.G., Johns, L., 1993. Liver abnormalities associated with chronic mercury accumulation in stranded Atlantic Bottlenosed Dolphins. Ecotoxicol. Environ. Saf. 25, 41–47.

Rodríguez, D., Rivero, L., Bastida, R., 2002. Feeding ecology of the franciscana (Pontoporia blainvillei) in marine and estuarine waters of Argentina. Lat. Am. J. Aquat. Mamm. 1, 77–94.

Romero, M.B., Polizzi, P., Chiodi, L., Das, K., Gerpe, M., 2016a. The role of metallothioneins, selenium and transfer to offspring in mercury detoxification in Franciscana dolphins (Pontoporia blainvillei). Mar. Pollut. Bull. 109, 650–654.

Romero, M.B., Polizzi, P., Chiodi, L., Robles, A., Das, K., Gerpe, M., 2016b. Metals as chemical tracers to discriminate ecological populations of threatened Franciscana dolphins (Pontoporia blainvillei) from Argentina. Environ. Sci. Pollut. Res. 24, 3940–3950.

Romero, M.B., Polizzi, P., Chiodi, L., Medici, S., Blando, M., Gerpe, M., 2018. Preliminary Assessment of Legacy and Current-Use Pesticides in Franciscana Dolphins from Argentina. Bull. Environ. Contam. Toxicol. 101, 14–19.

Sá, F., Menolli, C., Schettini, E., Alexandro, C., Caiado, R., Eduardo, L., Gomes, D.O., Teresa, A., Fraga, A., Rodrigues, R., 2021. Time-sequence development of metal(loid)s following the 2015 dam failure in the Doce river estuary. Brazil. Sci. Total Environ. 769, 1–12.

Schwacke, L.H., Voit, E.O., Hansen, L.J., Wells, R.S., Mitchum, G.B., Hohn, A.A., Fair, P.A., 2002. Probabilistic risk assessment of reproductive effects of polychlorinated biphenyls on bottlenose dolphins (Tursiops truncatus) from the Southeast United States coast. Environ. Toxicol. Chem. 21, 2752–2764.

Secchi, E.R., Danilewicz, D., Ott, P.H., 2003. Applying the phylogeographic concept to identify franciscana dolphin stocks: implications to meet management objectives. J. Cetacean Res. Manag. 5, 61–68.

Seixas, T.G., Kehrig, H.A., Costa, M., Fillmann, G., Di Beneditto, A.P.M., Secchi, E.R., Souza, C.M.M., Malm, O., Moreira, I., 2008. Total mercury, organic mercury and selenium in liver and kidney of a South American coastal dolphin. Environ. Pollut. 154, 98–106.

Seixas, T.G., Kehrig, H.A., Di Beneditto, A.P.M., Souza, C.M.M., Malm, O., Moreira, I., 2009. Trace elements in different species of cetacean from Rio de Janeiro coast. J. Braz. Chem. Soc. 20, 243–251.

Seixas, T.G., Kehrig, H.A., Fillmann, G., Di Beneditto, A.P.M., Souza, C.M.M., Secchi, E.R., Moreira, I., Malm, O., 2007. Ecological and biological determinants of trace elements accumulation in liver and kidney of Pontoporia blainvillei. Sci. Total Environ. 385, 208–220.

Sucunza, F., Danilewicz, D., Ott, P.H., Neves, M., Berchieri, N., Farro, A.P., Martins, A., Zerbini, A.N., 2019. Population size and IUCN Red Listing of the isolated northern population of the franciscana (Pontoporia blainvillei). Int. Whal. Comm.

Sucunza, F., Danilewicz, D., Ott, P.H., Neves, M., Berchieri, N., Farro, A.P., Martins, A., Zerbini, A.N., 2020. Populational size and IUCN Red Listing of the isolated northern population of the franciscana (Pontoporia blainvillei). IWC Sci. Committee SC/68B/ASI, 24.

UNEP, 2017. Stockholm Convention on Persistent Organic Pollutants (POPs). Secr. Stock. Conv. Geneva 76.

Vergilio, C.S., Lacerda, D., Oliveira, B.C.V., Sartori, E., Campos, G.M., Pereira, A.L.S., Aguiar, D.B., Souza, T.S., Almeida, M.G., Thompson, F., Rezende, C.E., 2020. Metal concentrations and biological effects from one of the largest mining disasters in the world (Brumadinho, Minas Gerais, Brazil). Sci. Rep. 10, 1–12.

Vetter, W., Stoll, E., Garson, M.J., Fahey, S.J., Gaus, C., Müller, J.F., 2002. Sponge halogenated natural products found at parts-per-million levels in marine mammals. Environ. Toxicol. Chem. 21, 2014–2019.

Yogui, G.T., Santos, M.C.O., Bertozzi, C.P., Montone, R.C., 2010. Levels of persistent organic pollutants and residual pattern of DDTs in smallcetaceans from the coast of São Paulo, Brazil. Mar. Pollut. Bull. 60, 1862–1867.

Yogui, G.T., Santos, M.C.O., Bertozzi, C.P., Sericano, J.L., Montone, R.C., 2011. PBDEs in the blubber of marine mammals from coastal areas of São Paulo, Brazil, southwestern Atlantic. Mar. Pollut. Bull. 62, 2666–2670.

Zhang, W., Guo, Z., Song, D., Du, S., Zhang, L., 2018. Arsenic speciation in wild marine organisms and a health risk assessment in a subtropical bay of China. Sci. Total Environ. 626, 621–629.

CHAPTER 12

Coastal development and habitat loss: understanding and resolving associated threats to the franciscana, *Pontoporia blainvillei*

Camila Domit[a], Tailise Hoppe Trevizani[b], Ana Paula Cazerta Farro[c], Angela Zaccaron Silva[d,e], Tara Van Belleghem[d], Danniele Firme Herbst[f], Rodrigo Hipólito Tardin[g], Tatiana Lemos Bisi[h], Paula Laporta[i], Lara Gama Vidal[d], Henrique Chupil[j], Daiane Santana Marcondes[d], Gabriel Fraga da Fonseca[d], Rodrigo del Rio do Valle[k], Leonardo G. Berninsone[l], Carla Beatriz Barbosa[m], Marta Jussara Cremer[n], Matt K. Broadhurst[o]

[a]Laboratório de Ecologia e Conservação, Universidade Federal do Paraná, Pontal do Sul, Paraná, Brazil
[b]Instituto Oceanográfico, Universidade de São Paulo, São Paulo, Brazil
[c]Laboratório de Genética e Conservação Animal, Universidade Federal do Espírito Santo, São Mateus, Espírito Santo, Brazil
[d]Laboratório de Ecologia e Conservação, Centro de Estudos do Mar, Universidade Federal do Paraná, Pontal do Paraná, PR, Brazil
[e]Associação MarBrasil, Pontal do Paraná, PR, Brazil
[f]Institute of Environmental Science and Technology and Department of Prehistory, Universitat Autònoma de Barcelona, Bellaterra, Spain
[g]Laboratório de Bioacústica e Ecologia de Cetáceos, Universidade Federal do Rio de Janeiro, Rio de Janeiro, Brazil
[h]Laboratório de Mamíferos Aquáticos e Bioindicadores, Faculdade de Oceanografia, Universidade do Estado do Rio de Janeiro, Rio de Janeiro, Brazil
[i]Yaqu Pacha Uruguay, Punta del Diablo, Rocha, Uruguay
[j]Instituto de Pesquisas Cananéia, Cananéia, São Paulo, Brazil
[k]Instituto Biopesca, Praia Grande, São Paulo, Brazil
[l]Fundación AquaMarina, Centro de Estudios en Ciencias Marinas, Pinamar, Buenos Aires, Argentina
[m]Instituto Argonauta para Conservação Costeira e Marinha, Ubatuba, São Paulo, Brasil
[n]Laboratório de Ecologia e Conservação de Tetrápodes Marinhos e Costeiros, Universidade da Região de Joinville, São Francisco do Sul, Santa Catarina, Brazil
[o]NSW Department of Primary Industries, Fisheries Conservation Technology Unit, National Marine Science Centre, Southern Cross University, Coffs Harbour, NSW, Australia

Introduction

Habitat loss is one of five globally identified ecological pressures (along with fishing pressure, climate change, water pollution, and the introduction of alien species or genotypes) affecting oceans and biodiversity (Ocean Health Index, Halpern et al., 2017 - accessed in march 2021). Globally, the greatest predicted cumulative impacts to habitats occur across the continental shelf and slope

The Franciscana Dolphin
DOI: https://doi.org/10.1016/B978-0-323-90974-7.00010-0

areas subjected to land- and ocean-based anthropogenic activities (Halpern et al., 2008). The expansion and intensity of anthropogenic activities generate many environmental stressors and changes, which directly and indirectly impact biodiversity (Maxwell et al., 2013). Multistressors across coastal and marine habitats include, but are not limited to, fishing, aquaculture, port activities, marine traffic, mining, hydropower, and industrial, agricultural, and urban development. The consequences of these activities often act in synergy and are cumulative, generating coastal syndromes and even local and global changes (Newton et al., 2012).

Coastal infrastructure (such as ports, jetties, and other structures) can change environmental stability and hydrodynamics, contributing to erosion, eutrophication, and shifts in ecological dynamics. Moreover, unplanned urban and industrial infrastructure development increase sewage and marine debris across estuarine and coastal areas, while deforestation and erosion across adjacent continental areas facilitate the runoff of chemical pollutants from multiple activities. Some of these activities exacerbate noise pollution, but all collectively increase changes in environmental processes and dynamics, with cascading ecosystem effects, including habitat loss (Hawkins et al., 2017; Alonso et al., 2015; Magris et al., 2020a; Nelms et al., 2021). If sustained, such anthropogenic stressors evoke behavioral and physiological variation among some species, altering their prey and habitats, which can manifest as long-term population effects (Cubero-Pardo et al., 2011).

Many species of marine mammals occur across coastal areas and are highly susceptible to anthropogenic activities that, with sufficient intensity, can cause health disorders and population declines (Avila et al., 2018). Specifically, the distribution of the franciscana, *Pontoporia blainvillei* (a small delphinid also known as the La Plata dolphin or "toninha") is restricted to a coastal area with some of the most intense anthropogenic activities in South America, comprising southern and southeastern Brazil, and marginal areas along the Río de la Plata estuary (Secchi et al., 2021). The species lives in shallow waters (30–50 m) and has suffered from the actions of multiple stressors and pressures mostly encompassing anthropogenic sources (e.g., fishing and port activities), but also natural perturbations (Bracarense and Domiciano, 2022, this volume; Lailson-Brito et al., 2022, this volume; Secchi et al., 2021, 2022, this volume; Fig. 12.1).

Recently, a review of the current global risk to marine mammals considered the franciscana one of the most threatened species, mainly due to its endemism and restricted coastal distribution (Avila et al., 2018). Owing to its exposure to coastal and maritime impacts, the franciscana is globally

Fig. 12.1 The franciscana, *Pontoporia blainvillei* is exposed to multiple anthropogenic threats throughout their distribution in coastal areas off South America. *(Credit: Marta J. Cremer, Babitonga bay.)*

classified as "Vulnerable" under the IUCN Red List (Zerbini et al., 2017), but is jurisdictionally listed as "Critically Endangered" in Brazil (MMA, 2018), "Endangered" in Argentina (Denuncio et al., 2019) and among "Priority Conservation Species" in Uruguay (González et al., 2013).

Fishery bycatch and associated mortalities (especially by gillnets) is the greatest perceived threat to franciscana populations and evoked by both industrial and small-scale fleets (Ott et al., 2002; Zerbini et al., 2017; Secchi et al., 2021). However, the species is also exposed to other potentially severe threats that increase their mortality risk and hinder health (Alonso et al., 2012; Avila et al., 2018; Denuncio et al., 2011; Trevizani 2018; Secchi et al., 2021; Lailson-Brito et al., 2022, this volume; Manhães et al., 2022). These threats include pollutants, diseases, overfishing, and other consequences of coastal development and habitat degradation that can independently or cumulatively affect an individual's immunological system and, ultimately, population health and dynamics (Bossart, 2011; Gall et al., 2015; Carbery et al., 2018).

Based on genetic, morphological, distribution, and population parameters, the species' distribution range has been divided into four "franciscana management areas" (FMAs), which were recently partitioned into other subareas, facilitating management and conservation actions (Secchi et al., 2003; Cunha et al., 2014). During the past decade, coastal and maritime developments have increased throughout the designated FMAs (Amaral et al., 2018; World Bank, 2018); but with an absence of rigorous environmental regulations and planning, which means sustainability levels have been surpassed. The effects of such development, along with sustained historic fishing effort, have systematically affected the different regional/jurisdictional populations of franciscana and their habitats.

To minimize the extinction risk for franciscana, it is necessary to evaluate the current key threats faced and prioritize these for mitigation via available conservation mechanisms (Nelms et al., 2021). Integrating socioeconomic development and conservation needs is a considerable challenge, but crucial to guarantee healthy habitats and a robust ecosystem, within which megafauna, like franciscana (and indeed all marine mammals), are pivotal. Considering the above, here we sought to (1) summarize the key coastal impacts to franciscana and their habitats; (2) assess coherent methodological and analytical processes to determine and manage anthropogenic threats; and then (3) identify knowledge gaps and future steps for conservation actions. Ultimately, we sought to provide guidance to improve the future outcome for the species throughout its distribution.

Franciscana management areas (FMAs) and coastal impacts

The four FMAs encompass the entire known distribution of the species, and while all are characterized by anthropogenic pressures, there are some spatial distinctions that warrant overview prior to assessing risks (Table 12.1; Figs. 12.2 and 12.3). Each of the FMAs is described below.

Franciscana management area I

Franciscana management area I (FMAI) encompasses an area from northern Espírito Santo state (18°25´S) to Rio de Janeiro state (22°44´S), but with a spatial separation in species distribution delimitated by two genetically isolated populations (FMAIa and FMAIb), considered evolutionary significant units (Amaral et al., 2018; Cunha et al., 2014; Secchi et al., 2021). Both populations are particularly vulnerable to anthropogenic stressors owing to their geographical isolation, low genetic diversity and gene flow, and very small population sizes (Cunha et al., 2014; Danilewicz et al., 2010; Oliveira et al., 2020; Zerbini et al., 2022) (Table 12.1; Figs. 12.2 and 12.3A and B).

The coastal area bound by FMAI includes densely urbanized areas north of Rio de Janeiro state, and is influenced by the Paraíba do Sul River, which incurs substantial fishing effort, passes through densely populated areas with industrial parks, and receives effluent from domestic sewage, industrial activities, and agricultural runoff (Di Beneditto, 2003). In addition, the region has one of the largest ports in Brazil: Porto do Açu in São João da Barra. The port is regularly dredged and represents an operational support base for offshore oil and gas exploration activities in the Campos Basin, as well as industries in the petrochemical sector and fishing activities.

Further, the coast of Espírito Santo state has been affected by intensive small-scale fisheries, port facilities, and contamination from pulp/paper and mining industries. But most concerningly, in the last six years, the area suffered the impacts of two major environmental disasters: the arrival of the ore tailings mud from the Fundão tailings dam failure in Mariana, Minas Gerais state in 2015 (Carmo et al., 2017; Pinheiro et al., 2019); and a substantial oil spill that reached the northeastern and northern coasts of Espírito Santo in 2019 (Soares et al., 2020).

The Fundão tailings dam failure released 60 million cubic meters of mining waste into the Doce River which then deposited along the coast of Espírito Santo (Frainer, 2016; Marta-Almeida et al., 2016), creating substantial environmental impacts in an area that overlaps the northern section of FMAI. Water, sediment and biota were contaminated with metals and a large sediment load (Costa et al., 2021; Manhães et al., 2022). An aquatic biodiversity monitoring program (PMBA), executed by the Rede Rio Doce Mar (RRDM) since 2018, showed, among other effects, temporal increases in metal concentrations among aquatic biota, including Cu, Fe, and Pb in marine teleosts; Fe in marine phytoplankton, zooplankton and shrimps; and Hg in phytoplankton (Costa et al., 2021).

The affected area and species encompass key habitat for franciscana, and the bioaccumulation of pollutants along the food chain (Fossi et al., 2018) and/or changes in the availability or quality of prey (Frainer, 2016) are serious concerns for population health (Manhães et al., 2022). Such deleterious effects have exacerbated the sustained population impacts evoked through fishery bycatch, which remains the regional main threat to the species in this region (Di Beneditto et al., 1998; Di Beneditto, 2003).

Franciscana management area II

The FMAII encompasses southern Rio de Janeiro, São Paulo, Paraná, and Santa Catarina states (22°59′S–27°35′S) (Amaral et al., 2018; Cunha et al.,

2014) (Fig. 12.2). Estimates of the entire franciscana population size are also limited, but as for FMAI, there is evidence supporting separate subpopulations (FMAIIa and FMAIIb), although additional data are required to confirm any boundaries (Cunha et al., 2014; Sucunza et al., 2020).

The entire coast within FMAII is highly urbanized with approximately 3.2 million inhabitants (IBGE, 2010), and has suffered considerable environmental changes and habitat loss (Magris et al., 2020b). Important shipping ports are distributed along the coasts of all states, including Angra dos Reis in Rio de Janeiro, São Sebastião and Santos in São Paulo, Paranaguá and Antonina in Paraná, Itapoá, São Francisco do Sul and Itajaí in Santa Catarina (ANTAQ, 2021). All ports have intense, regular activities, such as dredging, heavy marine traffic, and infrastructure for port expansion (Marcondes et al., 2020). Combined with petrochemical storage, industrial areas, small- and large-scale fisheries, inadequate waste disposal, and effluent discharge, these various activities threaten regional marine biodiversity (Magris et al., 2020a and 2020b; Fuentes et al., 2020) and particularly the franciscana (Bertozzi and Zerbini, 2002; Alonso et al., 2012; Trevizani, 2018; Lailson-Brito et al., 2022, this volume).

The cumulative impacts caused by intense coastal development in FMAII expose the franciscana to other potential threats that increase their risks (Table. 12.1; Figs. 12.2 and 12.3). Specifically, the threats induced by coastal development are even greater in the central and southern areas of FMAII. These areas are also characterized by large abundances of franciscana prey (teleosts) that are concomitantly targeted in small- and large-scale fisheries, and mostly gillnets. Consequently, despite evidence of contamination and disease throughout regional cetaceans (Domiciano et al., 2016), fishing mortality among franciscanas in FMAII is assumed to be a key cause of their ongoing strandings and mortality (Cremer et al., 2022; Secchi et al., 2022).

Notwithstanding anthropogenic influences throughout FMAII, franciscana frequently use bays and estuarine areas, such as Ilha Grande Bay in Rio de Janeiro, the Cananéia-Iguape estuarine lagoon complex (CIELC) in São Paulo, the Paranaguá estuarine complex (PEC) in Paraná, and Babitonga Bay in Santa Catarina state (Cremer and Simões Lopes, 2008; Santos et al., 2009). Some of these areas have been internationally identified as conservation hotspots, including the PEC and CIELC; both of which are categorized as World Heritage sites and Biosphere Reserves by (UNESCO - United Nations Educational, Scientific and Cultural, 1999), and were recently included in the RAMSAR site listings (i.e., wetlands of international importance; MMA, 2018). In addition, FMAII comprises other important ecological areas (e.g., Babitonga Bay) and marine protected areas (e.g., PARNA Tamoios), which collectively are relevant conservation tools for reducing anthropogenic impacts and protecting franciscana (Ott et al., 2022).

Franciscana management area III

The third management area is the largest, extending from southern Santa Catarina state, southern Brazil (Palhoça—around 27°36′S) to Uruguay, including the waters of the Río de la Plata estuary (34°00′—36°10′ S, 55°00′—58°10′W) (i.e., La Plata river) (Figs. 12.2 and 12.3; Amaral et al., 2018; Cunha et al., 2014). While the Santa Catarina coast is highly urbanized, further south into Uruguay remains relatively undeveloped and encompasses two RAMSAR sites (MMA, 2018)—although development increases throughout the Río de la Plata estuary (34°00′–36°10′S, 55°00′–58°10′W). There are no estimates of franciscana numbers throughout FMAIII, but like in all other FMAs, the species is resident or regionally transient across inshore coastal waters and estuaries, and especially in the Rio de la Plata, which may have a different sub-population to those found in adjacent areas (Costa-Urrutia et al., 2012; Mendez et al., 2008).

The FMAIII covers the Rio Grande and Río de la Plata marine ecoregions (MEOW—Spalding et al., 2007), which are described as important marine productivity areas; however, they also incur intense cumulative impacts along the continental shelf (Magris et al., 2020b). The Rio de la Plata is the second-largest estuary basin in South America (~38,800 km), comprising a large and highly dynamic mix of salt and freshwater (Acha et al., 2004; Guerrero et al., 1997). The latter mostly comes from the Uruguay and Paraná rivers, which drain water from a large and complex basin >3,170,000 km^2 extending across five countries and home to 65 million inhabitants and several megacities, and with concomitant pollution issues raising concerns over impacts to aquatic flora and fauna (Menafra et al., 2009). Along the coast, the river discharge and the latitudinal shifts of the Brazil-Malvinas Confluence generate seasonal variability in primary productivity, with cold and nutrient-rich sub-Antarctic waters dominating in winter, and warm, nutrient-poor subtropical waters during summer (Ortega and Martínez, 2007). The confluence of these two currents produces one of the most energetic and productive regions in the world (Guerrero et al., 1997; Mianzan et al., 2001; Ortega and Martinez, 2007), supporting large fisheries (involving various gears, but especially gillnets) for commercially important aquatic species (Macchi and Acha, 2000; Macchi et al., 2003).

The extent of fisheries within and adjacent to the Rio de la Plata basin and known associated widespread collateral mortalities among franciscana are of most concern for the sustainability of the regional population. Reports of incidental catches of franciscana in the shark-gillnet fisheries of Punta del Diablo, Uruguay date back to the early 1940s, and have maintained since (Van Erp, 1969). Bottom-set gillnetting in southern Brazil started around the same time as in Uruguay (Haimovici et al., 1997), although these fisheries were only documented as a major threat to franciscana during the 1980s. The available data on collateral mortalities across other coastal regions in FMAIII are provided in Chapter 10, this volume (Secchi et al., 2022).

The FMAIII includes a vast estuarine and river-mouth ecosystem that transfers freshwater to the continental shelf, but also adds effluents and multiple pollutants. Thus, fisheries, land-based activities, and coastal infrastructure have been the most severe threats to biodiversity and habitat loss in FMAIII. The land and coastal-based activities include coal mining, different industries, agriculture crops, and harbor activities along the Brazilian and Uruguayan coasts. Among the latter, the Rio de la Plata estuary coast has extensive maritime traffic and pollution from the estuary (García-Rodríguez et al., 2010; Lozoya et al., 2015; Muniz et al., 2019). These impacts evoke considerable threats for franciscana and other marine species.

Franciscana management area IV

Extending to the southern extent of the franciscanas distribution, FMAIV encompasses the coastal waters of Argentina, including the provinces of Buenos Aires (35°16′S) and Rio Negro (41°10′S) (Figs. 12.2 and 12.3; Crespo, 2018). As for FMAIII, there is no population estimate of franciscana for the entire management area, but individuals occur throughout, and especially in estuarine areas like Bahía Samborombón, Bahía Blanca, Bahia San Blas, and the mouth of the Rio Negro, where individuals move in and out of bays following tidal patterns (Wells et al., 2021).

Among the key estuaries in FMAIV, Bahia Blanca (southern Buenos Aires) includes several ports for exporting grains and general merchandise. The maintenance of this area requires regular dredging activities that suspend sediments that inhibit phytoplankton blooms, resulting in habitat degradation and impacting the trophic chain, which culminates in impacts to franciscana (Berninsone et al., 2018; Perillo and Cuadrado, 1991).

Table 12.1 Critical threats to franciscana, *Pontoporia blainvillei* and potential effects with corresponding references, throughout the species' distribution off Brazilian (Espirito Santo, ES; Rio de Janeiro, RJ; São Paulo, SP; Paraná, PR; and Santa Catarina , SC), Uruguayan (UY), and Argentinean (ARG) coastal areas, separated into four management areas (FMAI, II, III, and IV).

Anthropogenic stressors/ activities/pressures	Potential effects		FMA (state/province)	References
	Direct impacts	Indirect impacts		
Fishery interactions (small-scale and industrial fisheries)	Incidental capture (bycatch)	Overfishing or depletion of fish stocks	FMAI (ES, RJ) FMAII (SP, PR, SC) FMAIII (RS, UY) FMAIV (ARG)	Ott et al. (2002), Rosas et al. (2002), Praderi et al. (1989), Franco-Trecu et al. (2009, 2019), Negri et al. (2012), Bertozzi and Zerbini (2002), Di Beneditto (2003), Freitas-Netto and Di Beneditto (2008), Barbosa-Filho et al. (2020), Secchi et al. (2021), Secchi et al. (2022)
Sewage (wastewater discharge)	Pathogen introduction	Increase the source of emerging pollutants (synthetic and natural compounds)	FMAI (ES, RJ) FMAII (SP, PR, SC)	Secchi et al. (2021), Bracarense and Domiciano (2022) and Lailson-Brito et al. (2022)
Chemical contaminants	Immunosuppression, endocrine disruptions, neurological and other types of diseases, such as allergies and cancer	Habitat loss, and reduced reproductive success	FMAI (ES, RJ) FMAII (SP, PR, SC) FMAIII (RS, UY) FMAIV (ARG)	Gerpe et al. (2002), Kajiwara et al. (2004), Lailson-Brito et al. (2002), 2011), Kunito et al. (2004), Seixas et al. (2007, 2008), Dorneles et al. (2007, 2013), Alonso et al. (2012, 2015), Gago-Ferrero et al. (2013), Foglia (2013), Leonel et al. (2014), Romero et al. (2017), Carmo et al. (2017), Trevizani (2018), Soares et al. (2020), Costa et al. (2021), Trevizani et al. (2021), Manhães et al. (2022), Lailson-Brito et al. (2022)

(Continued)

Table 12.1 (Cont'd)

Anthropogenic stressors/activities/pressures	Potential effects: Direct impacts	Potential effects: Indirect impacts	FMA (state/province)	References
Marine debris	Ingestion of plastic waste (implications are uncertain) and entanglement	Ingestion of plastic via prey	FMAI (ES, RJ) FMAII (SP, PR, SC) FMAIII (RS, UY) FMAIV (ARG)	Pinedo (1982), Bassoi (1997), Bassoi and Secchi (1997), Danilewicz et al. (2002), Denuncio et al. (2011), Lopes (2012), Di Beneditto and Awabdi (2014), Di Beneditto and Ramos (2014), Seabra et al. (2018), Di Beneditto and Oliveira (2019), Kuhn and Franeker (2020)
Noise pollution	Masking, shifting and impoverishment of acoustic repertoire	Area abandonment	FMAII	Cremer and Simões-Lopes (2008), Holtz (2014)
Maritime and coastal infrastructure	Collisions, morphological and physiological harm, changes in behavior, and area abandonment	Habitat degradation and loss, trophic changes (cascading effects), immunosuppression	FMAI (ES, RJ) FMAII (SP, PR, SC)	Pinheiro et al. (2019), Cremer et al. (2018), Sechi et al. (2021)
Climate change	Changes in the behavior and distribution of dolphins and their prey	Prey availability, habitat loss, and emergent diseases	FMAI (ES, RJ) FMAII (SP, PR, SC) FMAIII (RS, UY) FMAIV (ARG)	No studies available

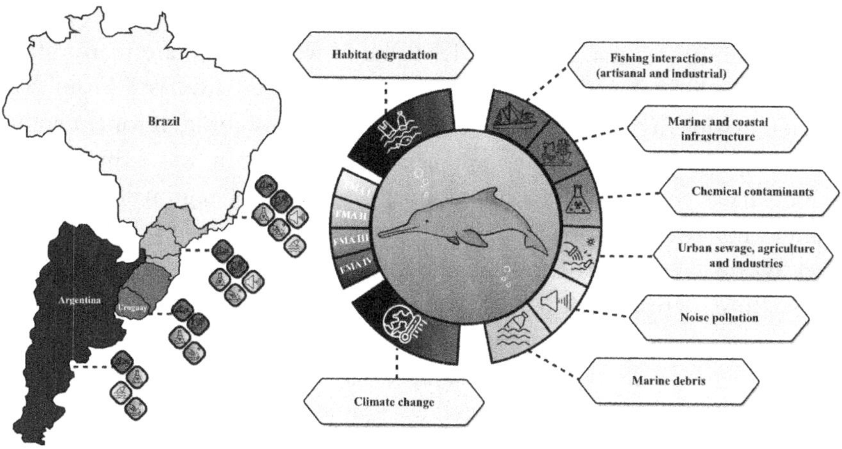

Fig. 12.2 Distribution of anthropogenic impacts affecting franciscana, *Pontoporia blainvillei* in each franciscana management area (FMA) as described in the literature. Anthropogenic impacts are categorized from most intense (red) to least intense (yellow). Habitat degradation and climate change are considered ecosystemic and global impacts. (*Credit: Gabriel Fonseca.*)

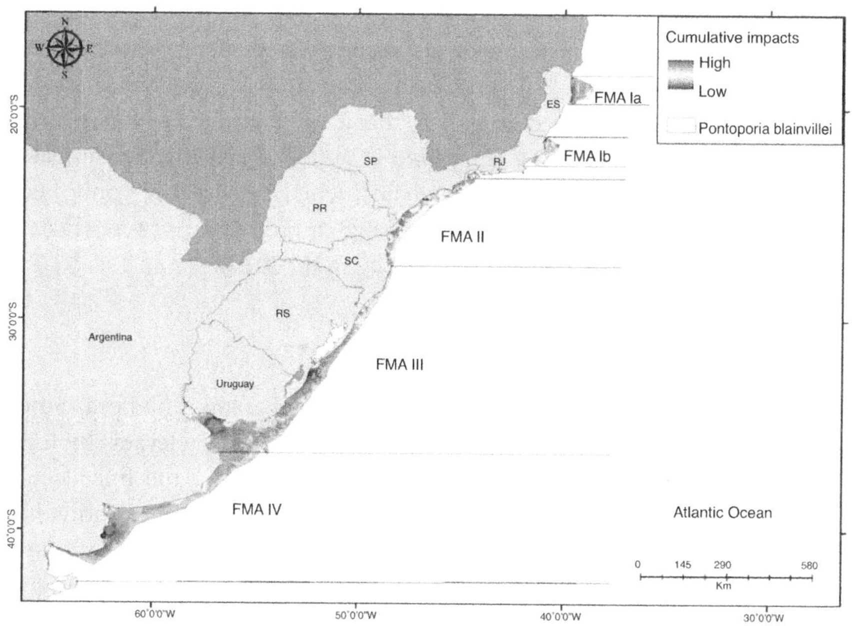

Fig. 12.3 Cumulative impact intensity obtained from Halpern et al. (2015) throughout the franciscanas', *Pontoporia blainvillei*, entire distribution. (*Credit: Rodrigo Tardin.*)

Within the region, the main developed urban areas are the cities of Bahía Blanca, Viedma, and Mar del Plata; however, there are many other areas including the coast which is popular among tourists (from San Clemente del Tuyú up to Necochea). A combination of high tourist activity and poor beach management has increased urban waste and debris, which have been found in the stomach contents of regionally stranded franciscana (Becherucchi et al., 2017; Denuncio et al., 2011; Denuncio and Bastida, 2014). Further, in the province of Buenos Aires, intense agriculture, petrochemical poles, and untreated sewage disposal produce high contaminant and trace metal concentrations in biota (Arias et al., 2016; Gerpe et al., 2002; Panebianco et al., 2013; Romero et al., 2018). For franciscana, this has manifested as significantly greater concentrations of nonessential trace elements among estuarine than marine conspecifics (Romero et al., 2016).

Beyond environmental contamination and poor health among resident franciscana are the ever-present concerns regarding fishing mortalities. Small-scale fishing is common in the province of Buenos Aires; predominantly in estuarine areas such as Bahía Samborombón and Bahía Blanca, but including coastal regions like Cabo San Antonio, Necochea, Quequén, Monte Hermoso and Pehuen Co. By comparison, industrial fishing trawlers operate mainly from the ports at General Lavalle (Bahía Samborombón), Mar del Plata and Bahía Blanca. As for all FMAs, bycatch in artisanal gillnets is considered a key threat for franciscana off Argentina (Bordino and Albareda, 2004, Cappozzo et al., 2007). Mortalities probably extend to industrial trawlers, although incidental catches have only been reported from Uruguayan vessels operating in Argentine waters (Franco Trecu et al., 2019; see Chapter 10 - Secchi et al., 2022).

General threats of coastal development

In terms of managing cumulative impacts to aquatic fauna and flora caused by coastal and infrastructure development, the most relevant biological organization is the ecosystem response; but to detect the finer trends of changes, the most sensitive outcomes are the molecular or individual responses (Pirotta et al., 2018). Therefore, to mitigate impacts and improve conservation gains, it is imperative to first identify and map the anthropogenic drivers and then assess impacts from individuals to ecosystems and their risk levels. Doing so can facilitate identifying the best regional conservation tools to support management actions (Nelms et al., 2021).

Most coastal anthropogenic activities are vital to the socioeconomic development of their surrounding areas; however, virtually all cause at least some negative environmental impacts. These impacts can be divided into two subcategories; problems caused to the species by the activity itself (individual drivers), and problems caused to the food chain or habitats (ecological drivers). In general, impacts to franciscana occur via fisheries, coastal and maritime infrastructure, land and maritime-based exploitation, and multiple sources of pollutants. More specific threats, described in detail below and in other chapters, this volume (see Secchi et al., 2022, this volume; Lailson-Brito et al., 2022, this volume; Bracarense and Domiciano et al., 2022, this volume), include depleted fish stocks and food-chain disturbances, noise and chemical pollutants, marine debris and sewage, induced pathogens, and habitat degradation in general, which can independently affect an individuals' immune system and, more likely, cumulatively affect environmental dynamics and ultimately, population health and resilience.

Depletion of prey by regional fisheries

Synonymy in species that are overfished, but important prey for franciscana, increases their vulnerability through less-available food and greater exposure to fisheries interactions. It is well established that fishing activities are responsible for critical ecological drivers affecting multiple components of ecosystems, including dolphin species.

As one example of the likely effect of commercial fishing on prey, in southern Brazil (FMAIII) gillnet and trawl fisheries have declined regional stocks of the sciaenids *Micropogonias furnieri* (Whitemouth croaker) and *Macrodon ancylodon* (King weakfish) (Haimovici et al., 1997; Haimovici, 1998), with concomitant changes in the diets of regional franciscana (Bassoi and Secchi, 2000; Secchi et al., 2003). While historical food sources might be replaced with other species, consequent changes in habitat use can expose dolphins to more impacted areas, or even reduce their energy intake and health condition.

Another example of the effects of overfishing on the traditional prey of franciscana in FMAIII is off Uruguay. The two regionally most important commercial species (*M. furnieri* and *Cynoscion guatucupa* - Striped weakfish) are fully or over-exploited (Defeo et al., 2011; Gianelli and Defeo, 2017). These species are important components of the diets of franciscana in both the Río de la Plata estuary and adjacent oceanic coast (Fitch and Brownell, 1971; Paso-Viola et al., 2014; Tellechea et al., 2017). A decline in the availability of these species owing to fishing pressure has resulted in franciscana sourcing other juvenile demersal fishes and overlapping the diets of the

South American fur seal (*Arctocephalus australis*) and South American sea lion (*Otaria flavescens*) (Drago et al., 2018; Secchi et al., 2021). Such shifts in resource partitioning via forced competition demonstrates the impact of fishing on marine mammals extends beyond a simple reduction in prey biomass and dietary changes. The energetic implications of such changes might be of some concern (Drago et al., 2018).

No studies have assessed the potential effect of overfishing franciscana's key prey in FMAI, FMAII, or FMAIV. But, based on the available information concerning the diet of this species, many important species (e.g., *Pellona harroweri*, *Isopisthus parvipinnis*, *Cetengraulis edentulus*, *Ophistonema oglinum*, *Lycengraulis grossidens*, *Anchoa filifera*, and *M. furnieri*) are heavily exploited by regional fisheries (Di Benetido et al., 2001; Henning et al., 2018; Campos et al., 2020). Consequently, at least some shifts in diet have probably occurred across space and time.

Noise pollution

Anthropogenic noise contributions to underwater soundscapes have substantially increased in coastal waters (Hildebrand, 2009; Duarte et al., 2021). Human noise can have different influences on the natural underwater soundscape according to the type, distribution, and persistence of the source (Richardson et al., 1995; Hildebrand, 2009; Miksis-Olds and Nichols, 2016; Fredianelli et al., 2021). Particularly for marine mammals, noise pollution can cause behavioral, perceptual, physical, chronic, and indirect disturbances (e.g., avoidance and displacement behaviors; acoustic masking; hearing impairment; stress; and reduced in prey availability: Richardson et al., 1995; Clark et al., 2009; Erbe, 2016; Pirotta et al., 2013; Bittencourt et al., 2017).

Any biological effects of noise pollution are especially concerning for small, endangered resident coastal species (Forney et al., 2017, Marcondes et al., 2020), such as the franciscana, which according to hearing capability predictions could be included in the "functional hearing group" of high-frequency (HF) cetaceans. Specifically, franciscana produce HF echolocation clicks at around 139 kHz (Melcón et al., 2012), but the total frequency range used can be much broader (clicks: Busnel et al., 1974; Von Fersen et al., 1997; Melcón et al., 2016; burst clicks: Tellechea et al., 2017; and whistles: Cremer et al., 2017). Their classification as HF cetaceans implies the lowest onset hearing thresholds for behavioral and physical disturbance/injuries (NMFS, 2018, Guan and Brookens, 2021). Consequently, this species is among the most sensitive to temporary and permanent threshold shifts when exposed to both impulsive and nonimpulsive noises.

However, despite the intensification of coastal and maritime infrastructure throughout the species' distribution which generates an array of underwater noise, the potential any effects of noise pollution on franciscana remain unknown for all FMAs—although there is some consistency among observed behaviors. Franciscana apparently exhibits similar habitat-use patterns of avoiding port surroundings, which seems to be related to the high ambient noise in these areas (Holz, 2014). For example, a 10-year study in Babitonga Bay (a port area in Santa Catarina state) demonstrated that franciscana concentrated in the inner bay area where noise intensity was lowest (Cremer and Simões-Lopes, 2008). Around the port area, the underwater ambient noise reached 116.9 dB between a range of 0 to 12 kHz (Cremer et al., 2018; Holz, 2014). In another study in FMAII, which includes the Paranaguá port (the third largest in Brazil), franciscana have been documented to more frequently use areas around designated marine protected areas and well away from the port (Santos et al., 2009; Zappes et al., 2016). Underwater noise at this port has been recorded at 0 to 8 kHz, reaching 206 dB (Marcondes, 2022), and at a relatively greater intensity than Babitonga Bay.

The other FMAs are also characterized by various areas of excessive underwater noise, and especially FMAIII, which encompasses the bioceanic Atlantic–Pacific corridor and includes the port of Montevideo, and one of the main centers of the cargo movement in South America (Lozoya et al., 2015; Puig et al., 2010). Because of the constant traffic of cargo and fishing vessels, some focal areas of high underwater noise levels are expected.

Considering the cumulative effects of anthropogenic noise on the coastal environment, we propose this is an important factor that could threaten the dynamics and resilience of franciscana populations. Based on the known hearing capabilities of franciscana, and the likelihood for at least some behavioral impacts, relevant work should be given priority.

Vessel collisions

Vessel collisions with small cetaceans are rarely reported (Schoeman et al., 2020; Van Waerebeek et al., 2007) and, to our knowledge, there are no records of these types of interactions with franciscana in the literature. Based on data from a beach monitoring project (https://simba.petrobras.com.br/simba/web/) in southern and southeastern Brazil, injuries among stranded franciscana reported due to vessels collisions were scarce, recorded in fewer than 1% of the stranded animals (n = 2860 individuals). However, accurately identifying such impacts can be very difficult and easily underestimated, especially considering the small size of the franciscana and the fact that most carcasses arrive in an advanced state of decomposition on the beach.

Chemical contaminants

Pollutant discharges into the marine environment are heavy ecological drivers and collectively expose fauna to cumulative stressors, including sewage effluents and runoff from urban areas (including litter, oils, pathogens, industrial chemicals, heavy metals, and sediments) and agriculture (including fertilizers, pesticides, and other agrochemicals; Islam and Tanaka, 2004). Several studies assessing chemical contamination have been done for franciscana with a focus on two classes of chemical pollutants: (1) trace elements (metals and metalloids); and (2) organic compounds (including organochlorines, dechlorane-related compounds, octocrylene, brominated compounds, and pyrethroids) (Alonso et al., 2012; Gago-Ferrero et al., 2013; Lavandier et al., 2015; Lailson-Brito et al., 2022, this volume; Polizzi et al., 2014; Torre et al., 2012). Most studies indicated multiple intraspecific contaminations and chronic contaminations by both classes of pollutants, with high levels in all ages (including neonates) when compared with the global literature. Maternal transfer for some compounds has also been reported. The main results of chemical contaminants are presented in Chapter 11, this volume (see Lailson-Brito et al., 2022, this volume).

Land-based activities are the main sources of pollutants in the ocean. However, ocean-based activities, such as marine traffic and oil and gas exploration, contribute to impacts across coastal zones (Alonso et al., 2012; Baptista et al., 2016; Carvalho Aguiar et al., 2016; Dorneles et al., 2013; Gago-Ferrero et al., 2013; Lailson-Brito et al., 2011; Lavandier et al., 2015; Leonel et al., 2014; Manhães et al., 2022; Negri et al., 2012; Trevizani et al., 2021; Torre et al., 2012). In addition, such regions are also exposed to environmental accidents (e.g., oil spills). Among the FMAs, the more developed regions, such as the state of Rio de Janeiro and the province of Buenos Aires, present chronic chemical contamination, which affects environmental quality and, consequently, is of considerable concern for impacts among franciscana (Aguiar et al., 2016; Arias et al., 2016; Gerpe et al., 2002; Panebianco et al., 2013; Romero et al., 2018; Trevizani et al., 2021).

The threats of different compounds and trace elements can be individual and cumulative among regional populations of franciscana. Many contaminants cause endocrine changes, cancerous cells, neurotoxic disorders, allergies, influence immunosuppression, and affect reproductive parameters (Bossart, 2011; O'shea and Odell, 2008). If unique in the animals' tissues, such compounds should rarely cause death; but the synergistic and cumulative effects could represent significant contributors to lethality among franciscana. Further, any sublethal impacts to endocrine and reproductive parameters might not only affect individuals, but the entire population (National Academies of Sciences, 2017), manifesting as weakened ability to respond to other direct impacts, including interactions with fisheries.

Marine debris

Among cetaceans, interactions with anthropogenic marine debris, especially from fishing (such as nets, lines, and rope), but also general human pollution are well documented and encompass both entanglement and ingestion (Moore et al., 2009; Stelfox et al., 2016). Interactions with various fishing gears and materials that are either secured or derelict (ghost fishing) can cause entangling around animals' bodies, leading to immobility, strangulation of limbs and consequent amputation, suffocation, and certainly death. Many stranded individuals (across all FMAs) have been reported entangled in fishing material and other marine debris and, while the cause of death is not always definitive, the implication of causality is robust in many cases.

The effects of marine debris ingestion by marine fauna are not well defined because these are frequently limited to small amounts, or in reduced sizes, including for franciscana (Denuncio et al., 2011; Walker and Coe, 1990). However, even small amounts in an animal's digestive tract can result in long-term effects or even cause partial or total blockages (Laist, 1987).

Several relevant studies have been done for franciscana. For example, Bassoi (1997) found plastic fragments in 17% of the digestive tracts of analyzed franciscana stranded on beaches in Rio Grande do Sul (FMAIII). In Argentina (FMAIV), Denuncio et al. (2011) recorded residue in 14% of all animals analyzed, concluding that estuarine individuals had more than their coastal conspecifics (34.6 and 19.2%, respectively). In São Paulo (FMAII), Seabra et al. (2018) found plastics in 14% of the animals analyzed, and the same results were recorded by Lopes and Santos (2012) in São Paulo and Paraná states (FMAII). Di Beneditto and Ramos (2014) and Di Beneditto and Oliveira (2019) recorded residuals in 15% of the individuals collected on the northern coast of Rio de Janeiro (FMAI). More recently, data from the beach monitoring projects in southern Brazil (SIMBA, 2021), including FMAI and FMAII, showed 4.6% of the stranded franciscana had some kind of debris in their gastrointestinal tracts.

No studies on marine debris ingestion in franciscana have been carried out in Uruguayan waters (FMAIII); however, Uruguayan aquatic systems (freshwater, coastal marine, and benthic) are affected by plastic pollution, both from land- and marine-based activities (Lozoya et al., 2015). In the coastal zone, a smaller fraction of plastic debris (i.e., microplastics < 5 mm) is widely deposited on beaches (Weinstein et al., 2014). This phenomenon is due to the great availability of debris in the Río de la Plata estuary, especially in the turbid mouth (Lozoya et al., 2015).

Compared with other megafauna, such as sea turtles or some species of sea birds, the number of franciscana ingesting marine debris may seem trivial. But, a preliminary assessment of plastic effects on marine fauna revealed some nutritional changes according to the ingestion of hard or soft (sheet-like) plastics (Machovsky-Capuska et al., 2019). Franciscana consumes many benthic teleosts, crustaceans, and cephalopods, and uses a collection of behavior strategies to inspect the sandy bottom (Cremer et al., 2012; Di Beneditto and Ramos, 2014; Henning et al., 2018). Such behavior can lead to the accidental ingestion of small plastic particles with the food (Di Beneditto and Ramos, 2014), and so the bioaccumulation of microplastics is likely to be high.

Evaluating the impacts from debris ingestion will be challenging because these residues may have a cumulative effect, and the animal may still suffer from the toxicity of absorbed pollutants over time. Thus, notwithstanding uncertainty regarding the lethality of marine debris and plastics toward franciscana, this remains an anthropogenic impact that needs to be included in coastal management and conservation planning for franciscana.

Climate change

Climate change has emerged as the most important challenge to conservation in the 21st century. Various predicted and unknown changes to the ecosystems of coastal and oceanic regions are expected, and with increasing rates that will ultimately modify functions and services and in many cases to the detriment of existing systems (Bellard et al., 2012). As high trophic-level predators that have evolved to their current habitats, cetaceans will be affected through modifying sea temperatures, sea levels, ice extent, seawater chemistry (e.g., ocean acidity and salinity), seawater mixing, precipitation, and even ocean waves (Buchholz et al., 2019).

Several studies have described potential cascading effects from the impacts of climate change on cetaceans, related to changes in their distributional patterns and habitat use, mainly due to variations in the abundance and distribution of prey (Learmonth et al., 2006; Simmonds and Eliott, 2009; Simmonds and Isaac, 2007; Whitehead et al., 2008). Changes have also been reported in the duration and timing of migrations (Ramp et al., 2015) and in reproductive success (Leaper et al., 2015), among other effects to their immunological system and exposure to emerging diseases. As stated by Nelms et al. (2021), the most vulnerable species will be those with restricted ranges and small populations, such as the franciscana. But, despite these entirely foreseeable impacts, there are no specific empirical studies being done to focus on climate-change effects in franciscana throughout any of the FMAs.

Nevertheless, based on known biology and ecology, some climate-change impacts can be postulated. Specifically, the franciscana will likely be indirectly affected by climate change, particularly with regard to food resource availability. In Uruguay, studies conducted on commercial fishes showed that populations of *M. furnieri* and *C. guatucupa* (both important prey for franciscana) have been affected by ocean warming (Gianelli et al., 2019) and other climatic indexes since 1997 (Santana, 2020). As for other anthropogenic disturbances affecting the availability and abundance of prey (e.g., overfishing), there will most likely be shifts in distributions of franciscana and/or to other sources of food, with implications for other habitats and interacting/competing species.

Future studies assessing the effects of climate change as an environmental stressor for the franciscana and its habitats and prey are necessary to better understand the potential cumulative effects and predict specific impacts. Further, monitoring shifts in the dolphins' behavior related to climate change are crucial to guide conservation and management actions, such as designing protected areas (Avila et al., 2018).

Pathogens and diseases

Diseases are frequently the result of changes in the ecology of the host or pathogen and are often driven by anthropogenic environmental modifications, including habitat degradation and the effects of global climate change (Nelms et al., 2021). Diseases may trigger massive mortalities, constrain the growth of wild animal populations, combine with other factors to increase the risk of extinction of small populations, and ultimately provoke loss of biodiversity (Daszak et al., 2000). Of concern here is that an increase in disease is correlated with the urbanization of coastal areas, and most commonly the associated sewage which increases terrestrial pathogens in marine life and several emerging diseases in marine mammals.

Health assessments are not often applied to marine mammals in South America, although recent initiatives have contributed toward addressing the historical deficit of knowledge. Approximately 1500 franciscana have been necropsied and investigated for pathological findings and mortality; mostly during 2015–2020, and within FMAII, as part of beach monitoring (PMP-BS). In general, drowning was the most common cause of death (~70%); however, 17% of fatalities were related to infections (bacterial or viral infection agents) (PMP-BS). In addition, emaciation (potentially disease-related) was noted in ~5% of individuals examined (Domiciano et al., 2016).

Histopathological analyses were performed on different tissues of 288 of the above necropsied franciscana (data available at http://simba.petrobras.com.br), and of these, 75.4% had histopathological findings in the lungs (n = 217), 55.2% in the kidneys (n = 159), 36.1% in the liver (n = 104), 22.6% in the heart (n = 65) and 20.1% in the lymph nodes (n = 58). Further, 102 individuals from FMAII and FMAIII were analyzed for *Toxoplasma gondii* by immunohistochemistry—although none was IHC-positive (Costa-Silva et al., 2019). Brucella has been detected in 2.5% of the franciscana specimens analyzed in southern Brazil (Sánchez-Sarmiento et al., 2019). Additional information on health parameters is provided in Chapters 3 and 4, this volume (Bracarense and Domiciano, 2022, this volume; Rocha et al., 2022).

The derived results are incipient and new techniques are being developed to support a better understanding of dolphin health conditions and the potential effects of environmental disturbance on their immunological system. In addition, the franciscana was identified by the scientific committee of the International Whaling Commission as a priority species to be evaluated in terms of cumulative impacts during future efforts.

Cumulative effects, risk assessment, and management in coastal areas

The exposure of franciscana to all described anthropogenic activities and stressors (intrinsic and extrinsic; Table 12.1, Figs. 12.3 and Fig. 12.4) leads to various direct and indirect impacts (Hawkins et al., 2017; Nelms et al., 2021). The cumulative effect of these impacts may result in habitat loss (a major factor in species declines/extinctions worldwide), and further consequences at individual or population levels, including long-term shifts in habitat suitability, decreased reproductive success, hindered health parameters, increased mortality, and population declines (Sih et al., 2000). These

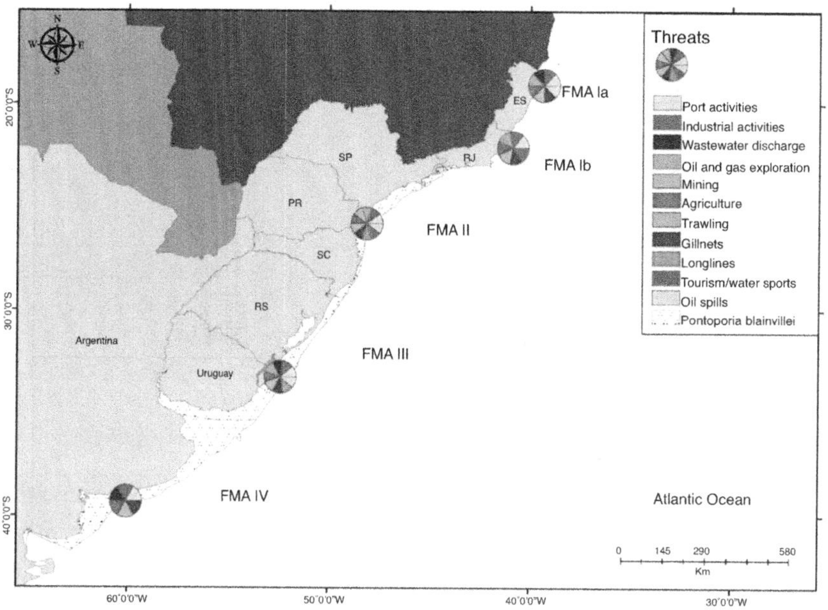

Fig. 12.4 Map of anthropogenic activities potentially impacting fransciana, *Pontoporia blainvillei* identified in the literature for each franciscana management area (FMA). *(Credit: Rodrigo Tardin.)*

possible ecological, evolutionary and genetic consequences, induced by the intense exposure of franciscana to the cumulative effects of multiple coastal threats off South America, triggered the IWC Scientific Committee to endorse a Conservation Management Plan to improve government actions and guide scientific research for conserving both the species and its habitat (see more about the CMP in Iñiguiz, 2022).

In general, management and conservation strategies and actions, which are established to minimize, mitigate and regulate anthropogenic activities impacting biodiversity, require identifying important areas for key life-cycle activities (e.g., breeding and feeding). This information is integral to impact assessments, mainly in a cumulative-effect framework. However, many different factors will determine the most effective approaches for developing management strategies to detect and monitor existing and potential impacts of coastal and maritime development on marine mammals (Table 12.2). Selecting the best approach will invariably be constrained by available time and resources, ecological factors, and should be adapted to local biodiversity distribution and abundance (Hawkins et al., 2017). Currently, there is little consensus on the most appropriate approach to

Table 12.2 Established methodologies that could be applied to franciscana, *Pontoporia blainvillei* to assess systemic anthropogenic impacts with benefits (pros) and downsides (cons) of each.

Methods/approach	Pros	Cons	References
Habitat/species risk assessment (HRA)	Assesses spatial variation in cumulative risk from multiple human activities for using an area of a species and/or critical habitat/ecosystem	The risk value is relative to the total of risks considered for an area	Arkema et al. (2014), InVest - Natural Capital (2020), Herbst et al. (2020)
Population consequences of disturbance (PCoD)	Assesses non-lethal effects of disturbance and their population-level consequences	Extensive baseline information concerning behavior, life-history and demography	Pirotta et al. (2018)
Mapping of cumulative impacts	Provides a spatial assessment of cumulative impacts by identifying areas that are strongly impacted by human activities and where these impacts overlap with the presence of species occurrence hotspots	Need for a broad spatial database and requires many assumptions due to data limitations	Halpern et al. (2008), Maxwell et al. (2013)
Environmental impact assessment (EIA)	Quantifies the potential adverse consequences of coastal development on biodiversity and identifies measures to avoid, mitigate or offset impacts (Vlachos, 1985), and enables adaptive management measures	Requires substantial baseline data	Elliott (2014), Hawkins et al. (2017)

evaluate these potential impacts and monitor cumulative effects at different organizational levels (from molecular to ecosystem). This chapter does not intend to bring a comprehensive review of methods and approaches; rather we highlight some examples applied in the marine ecosystem, their benefits, and downsides, focusing on endangered species (Table 12.2).

Overall, mapping the cumulative impacts of anthropogenic threats potentially affecting franciscana might allow for spatially identifying areas that are heavily impacted by human activities, where these impacts overlap with the presence of species hotspots, which threats majorly affect ecological components and how they evolve over time (e.g., following the methods applied by Avila et al., 2018; Halpern et al., 2008b; Korpinen and Andersen, 2016; Magris et al, 2020b; Maxwell et al., 2013). Further, cumulative impact spatial assessments can provide tools that help reduce conflicts, guide allocating resources for conservation and implementing ecosystem-based management, and, in addition, provide crucial support to spatial-conservation strategies (Ehler and Douvere, 2009; Halpern et al., 2008a; National Academies of Sciences, 2017) (Table 12.2).

Ecosystem-based management and ecosystem impact assessments could also contribute to identifying the most critical areas to implement conservation actions, supporting spatial planning, which includes determining the best strategies, interventions, and activities to be implemented for species conservation (i.e., strategic planning) (Table 12.2). Strategic-spatial planning can facilitate a systematic process toward supporting decision making, termed the "conservation planning approach" (Redford et al., 2003; Pressey et al., 2007). Systematic-conservation planning enables integrating predictive and simulation models supporting territory governance and addressing conservation actions for the short-, mid-, and long-term. The use of this management tool is especially important in coastal areas, where decision makers face the challenges of weighing the trade-offs of coastal development projects and subsequent activities considering economic, social, and environmental impacts (Halpern et al., 2008a). Complexities may also arise due to cross-jurisdictional boundaries and the need for multi-agency involvement, which can increase tensions due to political, economic, and social pressures, in addition to the different goals, agendas, and interests among agencies (Braulik et al., 2015; Grech et al., 2013).

Unfortunately, the combination of management obstacles and the inadequate or complete lack of assessment of coastal-development impacts on franciscana, aggravated by impacts of fisheries interactions, has led to catastrophic declines in some populations (Zerbini et al., 2017). Because effective conservation actions are a result of both rigorous ecological assessments and adaptive management frameworks, prior to and during coastal development, it is fundamental that decision makers and society are informed by rigorous empirical assessments that robustly predict the potential effects of human activities on marine biota and ecosystems (Hawkins et al., 2017; Higham and Higham, 2009). Further, this process must be conducted in an integrated manner with all stakeholders, based on the best scientific knowledge available to support developing more effective conservation strategies and management actions to balance development demands with the need for marine biodiversity protection.

Emerging technologies and approaches are helping to improve ecological knowledge and predict and monitor risks for threatened species and habitats. These methods harness computational and statistical language with greater power to integrate complex data, address ecological questions, as well as run simulations, and detect intrinsic and extrinsic changes/effects of impacts at multiple organizational levels (Valle and Berdanier, 2012). Some examples of emerging approaches include molecular, biochemical (biomarkers), and ecological markers (stable isotope), epigenetic approaches, the use of biologging devices and techniques, image analysis, predictive and simulation models, and individual assessment (Valle and Berdanier, 2012).

Notwithstanding the available techniques and approaches, few efforts to develop models to predict impacts, assess risks, and understand cumulative effects have been done focusing on the franciscana and its habitats (Amaral et al., 2018; Avila et al., 2018; Manhães et al., 2022; Trevizani, 2018). But, considering the impacts described in this chapter (and information in Chapters 3, 4, 10, 11, and 13, this volume) and the high priority assessment for cumulative effects on franciscana recommended by IWC, we stress that ongoing efforts should integrate emerging technologies to assess past, present and future regional and broader anthropogenic risks and impacts. The results of such work should form the basis of conservation policies and government decisions in the near future.

Conclusion and future actions

Human occupation patterns and increased demand for resources, place considerable pressure on coastal biodiversity and ecosystems (Lotze et al., 2006). In the absence of effective strategic planning and mitigation, detrimental impacts on wildlife from human activities can cause permanent and irreversible change (Hawkins et al., 2017).

While fisheries bycatch and associated sublethal and lethal effects are the most critical threats to the conservation of franciscana, other threats are reported. Nevertheless, large knowledge gaps must still be filled to better assess impacts and risks to the species, both regionally and across their entire distribution. Limited studies on habitat degradation and exposure to threats have been carried out within the FMAs; however, coastal and maritime activities are increasing, and emerging pathogens and chemicals have the potential to negatively affect reproduction, population resilience, and cause mortality—and not only among franciscana but also other regional cetaceans (Bossart, 2011). In general, many cause-and-effect relationships between anthropogenic activities and franciscana health have not been determined, but the potential for long-term synergistic and cumulative impacts on population viability should not be overlooked.

A coordinated and systematic monitoring program is crucial to assess the effects of anthropogenic threats, particularly integrating multiple-anthropogenic activities in marine and coastal areas, such as fisheries interactions and other environmental stressors that affect franciscana and the spatial areas of influence. Such studies would also further support actions and decision-making concerning the conservation of this species within the context of the IWC-CMP for franciscana (https://iwc.int/franciscana-dolphin), and other national conservation planning (e.g., the Brazilian Action Plan for the conservation of the franciscana—https://www.icmbio.gov.br/portal/faunabrasileira/plano-de-acao-nacional-lista/147-plano-de-acao-nacional-para-conservacao-da-toninha).

Considering the reviewed information here, we postulate eight actions to better place natural resource managers with the required information to conserve the species into the future.

1. Strengthen the development and implementation of effective environmental impact assessments, marine spatial planning, conservation-planning approaches, and other coastal management tools, to address the conservation needs of franciscana throughout the FMAs. Conservation and management plans and actions should be supported by the best available scientific information and technology.
2. Through legislation, reinforce the conservation needs and critical habitats of franciscana in governmental development plans, and require environmental impact assessments prior to and during coastal and maritime infrastructure development.
3. Reinforce the importance of decision making on coastal development to be an integrative and transdisciplinary process between society, government, science, and decision makers.

4. Develop systematic impact-assessment monitoring throughout the FMAs, integrating conservation actions, regulations, and mitigation efforts.
5. Improve scientific knowledge concerning the critical habitats of franciscana, including risk assessments and mapping hotspot areas to support coastal management, impact mitigation, and conservation actions, using emerging and improved methodologies, especially technologies such as drones, biologging devices, image analyses, and multiple predictive models.
6. Develop and apply tools able to monitor or predict synergistic and cumulative effects of multiple impacts on franciscana and improve knowledge including PCoD models, genomics, epigenomics, transcriptomics, proteomics, and metabolomics biomarkers as novel tools to investigate the overall health of populations.
7. Encourage implementing "national action plans" in Brazil, Uruguay, and Argentina to reduce the risks of coastal and maritime development on franciscana populations.
8. Foster an integrative strategy to increase multilateral discussions to protect franciscana with regard to the challenges and desired outcomes of the Ocean Decade (2021-2030 COI/UNESCO) and Agenda 2030 (ONU).

Perhaps most importantly, an implicit requirement within all eight proposed actions is the need to engage communities and raise awareness of the plight of endemic species like the franciscana. Such work should encourage regional identification of the species' (and other megafauna) value in terms of environmental economics and facilitate broader recognition of the need to nurture and protect the mostly unseen marine environment. Doing so, and avoiding megafauna extinctions, will ultimately realize substantial socioeconomic benefits more broadly—a key message that should be repeatedly conveyed to promote action.

References

Acha, E.M., Mianzan, H., Guerrero, R., Favero, M., Bava, J., 2004. Marine fronts at the continental shelves of austral South America. Physical and ecological processes. J. Mar. Systems 44, 83–105.

Alonso, M.B, Feo, M.L., Corcellas, C., Gago-Ferrero, P., Bertozzi, C.P., Marigo, J., Flach, L., Meirelles, A.C.O., Carvalho, V.L., Azevedo, A.F., Torres, J.P.M., Lailson-Brito, J., Malm, O., Diaz-Cruz, M.S., Eljarrat, E., Barceló, D., 2015. Toxic heritage: maternal transfer of pyrethroid insecticides and sunscreen agents in dolphins from Brazil. Environ. Pollut. 207, 391–402. http://doi.org/10.1016/j.envpol.2015.09.039.

Alonso, M.B., Eljarrat, E., Gorga, M., Secchi, E.R., Bassoi, M., Barbosa, L., Bertozzi, C.P., Marigo, J., Cremer, M., Domit, C., Azevedo, A.F., Dorneles, P.R., Torres, J.P.M., Lailson-Brito, J., Malm, O., Barceló, D., 2012. Natural and anthropogenically-produced brominated compounds in endemic dolphins from Western South Atlantic: another risk to a vulnerable species. Environ. Pollut. 170, 152–160.

Amaral, K.B., Danilewicz, D., Zerbini, A., Di Beneditto, A.P., Andriolo, A., Alvares, D.J., Secchi, E., Ferreira, E., Sucunza, F., Borges-Martins, M., de Oliveira Santos, M.C., Cremer, M., Denuncio, P., Ott, P.H., Moreno, I.B., 2018. Reassessment of the franciscana Pontoporia blainvillei (Gervais & d'Orbigny, 1844) distribution and niche characteristics in Brazil. J. Exp. Mar. Biol. Ecol. 508, 1–12. https://doi.org/10.1016/j.jembe.2018.07.010.

Arias, A., Panebianco, M., Net, S., Dumoulin, D., Souissi, S., Ouddane, B., Marcovecchio, J., 2016. Franciscana dolphins as PCBs marine biomonitors in Argentina, southwest Atlantic Ocean. J. Mar. Biolog. Assoc. U.K. 96 (4), 979–992. doi:10.1017/S0025315415000776.

Arkema, K.K., Verutes, G., Bernhardt, J.R., Clarke, C., Rosado, S., Canto, M., Wood, S.A., Ruckelshaus, M., Rosenthal, A., McField, M., De Zegher, J., 2014. Assessing habitat risk from human activities to inform coastal and marine spatial planning: a demonstration in Belize. Environ. Res. Lett. doi:10.1088/1748-9326/9/11/114016.

Avila, I.C., Kaschner, K., Dormann, C.F., 2018. Current global risks to marine mammals: taking stock of the threats. Biol. Conserv. 221, 44–58. https://doi.org/10.1016/j.biocon.2018.02.021.

Baptista, G., Kehrig, H.A., Di Beneditto, A.P.M., Hauser-Davis, R.A., Almeida, M.G., Rezende, C.E., Siciliano, S., de Moura, J.F., Moreira, I., 2016. Mercury, selenium and stable isotopes in four small cetaceans from the Southeastern Brazilian coast: influence of feeding strategy. Environ. Pollut. 218, 1298–1307. http://doi.org/10.1016/j.envpol.2016.08.088.

Barbosa-Filho, M.L.V., Seminara, C.I., Tavares, D.C., Siciliano, S., Hauser-Davis, R.A., Mourão, J.S., 2020. Artisanal fisher perceptions on ghost nets in a tropical South Atlantic marine biodiversity hotspot: challenges to traditional fishing culture and implications for conservation strategies. Ocean Coast. Manage., 192. doi:10.1016/j.ocecoaman.2020.105189.

Bassoi, M., 1997. Avaliação da dieta alimentar de toninha, Pontoporia blainvillei (Gervais and D' Orbigny, 1844), capturadas acidentalmente na pesca costeira de emalhe no sul do Rio Grande do Sul. Bachelor Thesis, Universidade Federal do Rio Grande, 68.

Bassoi, M., Secchi, E.R., 1997. Feeding of franciscana, Pontoporia blainvillei, accidentallly caught in southern Rio Grande do Sul State, Brazil, Abstracts of the 3o Workshop sobre Pesquisa e Conservação da Franciscana, Pontoporia blainvillei, no Atlântico Sul Ocidental. Porto Alegre, DT33.

Bassoi, M., Secchi, E.R., 2000. Temporal variation in the diet of franciscana Pontoporia blainvillei (Cetacea, Pontoporiidae) as a consequence of fish stocks depletion off southern Brazil. In: Abstracts of the 4o. Workshop sobre Pesquisa e Conservação da Franciscana, Pontoporia blainvillei, no Atlântico Sul Ocidental. Porto Alegre, DT9.

Becherucci, M.E., Rosenthal, A.F., Pon, J.P.S., 2017. Marine debris in beaches of the Southwestern Atlantic: an assessment of their abundance and mass at different spatial scales in northern coastal Argentina. Mar. Pollut. Bull. 119, 299–306.

Bellard, C., Bertelsmeier, C., Leadley, P., Thuiller, W., Courchamp, F., 2012. Impacts of climate change on the future of biodiversity. Ecol. Lett. 15, 365–377.

Berninsone, L.G., Newton, A., Icely, J., 2018. A co-designed, transdisciplinary adaptive management framework for artisanal fisheries of Pehuen Co and Monte Hermoso (Argentina). Ocean Coast. Manage. 152, 37–47. https://doi.org/10.1016/j.ocecoaman.2017.11.002.

Bertozzi, C.P., Zerbini, A.N., 2002. Incidental mortality of franciscana (Pontoporia blainvillei) in the artisanal fishery of Praia Grande, São Paulo State, Brazil. LAJAM - The Latin American Journal of Aquatic Mammals (special issue) 1 1 (1), 153–160. doi:10.5597/lajam00019.

Bittencourt, L., Lima, I.M.S., Andrade, L.G., Carvalho, R., Bisi, T., Lailson-Brito, J., Azevedo, A.F., 2017. Underwater noise in an impacted environment can affect Guiana dolphin communication. Mar. Pollut. Bull. 114, 1130–1134. http://doi.org/10.1016/j.marpolbul.2016.10.037.

Bordino, P., Albareda, D., 2004. Incidental mortality of franciscana dolphin Pontoporia blainvillei in coastal gillnet fisheries in Northern Buenos Aires, Argentina, Abstracts of the International Whaling Commission Meeting. Sorrento. Paper SC/56/SM1.

Bossart, G.D., 2011. Marine mammals as sentinel species for oceans and human health. Veter. Pathol. 48, 676–690. http://doi.org/10.1177/0300985810388525.

Bracarense, A.P.F.R.L., Domiciano, I.G., 2022. Pathologies and diseases. Chapter 4. In: Simões-Lopes, P.C., Cremer, M.J. (Eds.), The Franciscana Dolphin: on the edge of survival. Academic Press, Elsevier, London.

Braulik, G.T., Noureen, U., Arshad, M., Reeves, R.R., 2015. Review of status, threats, and conservation management options for the endangered Indus River blind dolphin. Biolog. Conserv. 192, 30–41. https://doi.org/10.1016/j.biocon.2015.09.008.

Buchholz, R., Banusiewicz, J.D., Burgess, S., Crocker-Buta, S., Eveland, L., Fuller, L., 2019. Behavioural research priorities for the study of animal response to climate change. An. Behav. 150, 127–137. https://doi.org/10.1016/j.anbehav.2019.02.005.

Busnel, R., Dziedzic, G.A., Alcuri, G., 1974. Études préliminaries de signaux acoustiques de Pontoporia blainvillei Gervaiset D'Orbigny (Cetacea, Platanistidae). Mammalia 38, 449–459.

Campos, L.B., Lopes, X.M., da Silva, E., Santos, M.C.D.O., 2020. Feeding habits of the franciscana dolphin (Pontoporia blainvillei) in south-eastern Brazil. J. Mar. Biol. Assoc. UK 100, 301–313 doi.org/10.1017/S0025315420000120.

Cappozzo, H.L., Negri, M.F., Pérez, F.H., Albareda, D., Monzón, F., Corcuera, J.F., 2007. Incidental mortality of franciscana dolphin, (Pontoporia blainvillei) in Argentina. Lajam 6, 127–137.

Carbery, M., O'Connor, W., Palanisami, T., 2018. Trophic transfer of microplastics and mixed contaminants in the marine food web and implications for human health. Environ. Internat. 115, 400–409. https://doi.org/10.1016/j.envint.2018.03.007.

Clark, C.W., Ellison, W.T., Southall, B.L., Hatch, L., Van Parijs, S.M., Frankel, A., Ponirakis, D., 2009. Acoustic masking in marine ecosystems: intuitions, analysis, and implication. Mar. Ecol. Prog. Ser. 395, 201–222. doi:10.3354/meps08402.

do Carmo, F.F., Kamino, L.H.Y., Junior, R.T., de Campos, I.C., do Carmo, F.F., Silvino, G., Castro, K.J., de da, S.X., Mauro, M.L., Rodrigues, N.U.A., Miranda, M.P.S., Pinto, C.E.F., 2017. Fundão tailings dam failures: the environment tragedy of the largest technological disaster of Brazilian mining in global context. Perspectives Ecol. Conserv. 15, 145–151. doi.org/10.1016/j.pecon.2017.06.002.

Carvalho Aguiar, V.M., de Lima, M.N., Abuchacra, R.C., Abuchacra, P.F.F., Neto, J.A.B., Borges, H.V., de Oliveira, V.C., 2016. Ecological risks of trace metals in Guanabara Bay, Rio de Janeiro, Brazil: An index analysis approach. Ecotox. Environ. Safety 133, 306–315. doi.org/10.1016/j.ecoenv.2016.07.012.

Costa, P.G., Marube, L.C., Artifon, V., Escarrone, A.L., Hernandes, J.C., Zebral, Y.D., Bianchini, A., 2021. Temporal and spatial variations in metals and arsenic contamination in water, sediment and biota of freshwater, marine and coastal environments after the Fundão dam failure. Science Total Environ. 151340. doi:10.1016/J.SCITOTENV.2021.151340.

Costa-Silva, S., Sacristán, C., Gonzales-Viera, O., Díaz-Delgado, J., Sánchez-Sarmiento, A.M., Marigo, J., Groch, K.R., Carvalho, V.L., Ewbank, A.C., Colosio, A.C., Marcondes, M.C.C., Meirelles, A.C.O., Bertozzi, C.P., Lailson-Brito, J., Azevedo, A.D.F., Ruoppolo, V., Oliveira, L., Ott, P.H., Catão-Dias, J.L., 2019. Toxoplasma gondii in cetaceans of Brazil: a histopathological and immunohistochemical survey. Revta. Bras. Parasit. Veter. 28, 395–402. http://doi.org/10.1590/s1984-29612019051.

Costa-Urrutia, P., Abud, C., Secchi, E.R., Lessa, E.P., 2012. Population genetic structure and social kin associations of franciscana dolphin, Pontoporia blainvillei. J. Hered. 103, 92–102. http://doi.org/10.1093/jhered/esr103.

Cremer, M.J., Prado, J.H., Chupil, H., Kolesnikovas, C.K.M., Tavares, M., Valle, R.R., Denuncio, P., Rosenthal, A., Rodriguez-Heredia, S., Zapata, M.F., Ott, P.H., Barbosa, C.B., Castilho, P.V., Domit, C., Estima, S., Colosio, A.C., Maranho, A., Fallabrino, A., Vélez-Rubio, G., Laporta, P., Valdivia, M., Jiménez, S., Barreto, A.S., 2022. Long-term patterns of franciscana strandings throughout its distribution. Chapter 13. In: Simões-Lopes, P.C., Cremer, M.J. (Eds.), The Franciscana Dolphin: on the edge of survival. Academic Press, Elsevier, London.

Cremer, M.J., Simões-Lopes, P.C., 2008. Distribution, abundance and density estimates of franciscanas, Pontoporia blainvillei (Cetacea: Pontoporiidae), in Babitonga bay, southern Brazil. Revta. Bras. Zool. 25, 397–402. http://doi.org/10.1590/S0101-81752008000300003.

Cremer, M.J., Holz, A.C., Sartori, C.M., Schulze, B., Paitach, R.L., Simões-Lopes, P.C., 2018. Behavior and ecology of endangered species living together: long-term monitoring of resident sympatric dolphin populations. In: Rossi-Santos, M.R., Charles, W. (Eds.), Advances in Marine Vertebrate Research in Latin America. Springer, pp. 477–508. http://doi.org/10.1007/978-3-319-56985-7_17.

Cremer, M.J., Pinheiro, P.C., Simões-Lopes, P.C., 2012. Prey consumed by Guiana dolphin Sotalia guianensis (Cetacea, Delphinidae) and franciscana dolphin Pontoporia blainvillei (Cetacea, Pontoporiidae) in an estuarine environment in southern Brazil. Iheringia. Série Zoologia 102, 131–137. http://doi.org/10.1590/s0073-47212012000200003.

Cremer, M.J., Holz, A.C., Bordino, P., Wells, R.S., Simões Lopes, P.C., 2017. Social sounds produced by franciscana dolphins, Pontoporia blainvillei (Cetartiodactyla, Pontoporiidae). J. Acoust. Soc. Am. 141 (3), 2047–2054.

Crespo, E.A., 2018. Franciscana Dolphin. Encyclopedia of Marine Mammals, Second edition. Academic Press, pp. 388–392.

Cubero-Pardo, P., Donovan, C., Urbán-Ramírez, J., 2011. A proposal to define vulnerability of cetacean areas to human development: variables and analysis procedures applied to the Gulf of California. Aq. Conserv. Mar. Fresh. Ecosyst. 21, 433–447. http://doi.org/10.1002/aqc.1205.

Cunha, H.A., Medeiros, B.V., Barbosa, L.A., Cremer, M.J., Marigo, J., Lailson-Brito, J., Azevedo, A.F., Solé-Cava, A.M., 2014. Population structure of the endangered Franciscana dolphin (Pontoporia blainvillei): reassessing management units. PLoS One 9. doi:10.1371/journal.pone.0085633.

Danilewicz, D., Rosas, F., Bastida, R., Marigo, J., Muelbert, M., Rodríguez, D., Lailson-Brito, J., Ruoppolo, V., Ruoppolo, R., Bassoi, M., Ott, P.H., Caon, G., Rocha, A.M., Catão-Dias, J.L., Secchi, E.R., 2002. Report of the Working Group on Biology and Ecology. Lat. Am. J. Aquat. Mamm. 1, 25–42. doi:10.5597/lajam00005.

Danilewicz, D., Moreno, I.B., Ott, P.H., Tavares, M., Azevedo, A.F., Secchi, E.R., Andriolo, A., 2010. Abundance estimate for a threatened population of franciscana dolphins in southern coastal Brazil: uncertainties and management implications. J. Mar. Biol. Assoc. UK 90, 1649–1657. http://doi.org/10.1017/S0025315409991482.

Daszak, P., Cunningham, A.A., Hyatt, A.D., 2000. Emerging infectious diseases of wildlife-threats to biodiversity and human health. Science 287, 1756.

Defeo, O., McLachlan, A., 2011. Coupling between macrofauna community structure and beach type: a deconstructive meta-analysis. Mar. Ecol. Progr. Series 433, 29–41. https://doi.org/10.3354/meps09206.

Denuncio, P.E., Paso Viola, N., Cáceres-Saez, I., Cappozzo, H.L., Rodríguez, D., Mandiola, A., 2019. Pontoporia blainvillei. In: SAyDS–SAREM. (Ed.), Categorización 2019 de los mamíferos de Argentina según su riesgo de extinción. Lista Roja de los mamíferos de Argentina, Buenos Aires.

Denuncio, P., Bastida, R., 2014. Composition, distribution and waste management of Playa Grande, the most important touristic beach of Mar del Plata city. Argentina. Waste Manage. 34, 839–841.

Denuncio, P., Bastida, R., Dassis, M., Giardino, G., Gerpe, M., Rodríguez, D., 2011. Plastic ingestion in Franciscana dolphins, Pontoporia blainvillei (Gervais and d'Orbigny, 1844), from Argentina. Mar. Pollut. Bull. 62, 1836–1841. http://doi.org/10.1016/j.marpolbul.2011.05.003.

Di Benedito, A.P.M., Awadbi, D.R., 2014. How marine debris ingestion differs among megafauna species in a tropical coastal area. Mar. Pollut. Bull. 88, 86–90. http://dx.doi.org/10.1016/j.marpolbul.2014.09.020.

Di Beneditto, A.P.M., 2003. Interactions between gillnet fisheries and small cetaceans in northern Rio de Janeiro, Brazil: 2001-2002. Lajam 2. doi:10.5597/lajam00035.

Di Beneditto, A.P.M., Oliveira, A.S., 2019. Debris ingestion by carnivorous consumers: does the position in the water column truly matter? Mar. Pollut. Bull. 144, 134–139. http://doi.org/10.1016/j.marpolbul.2019.04.074.

Di Beneditto, A.P.M., Ramos, R.M.A., 2014. Marine debris ingestion by coastal dolphins: What drives differences between sympatric species? Mar. Pollut. Bull. 83, 298–301. http://doi.org/10.1016/j.marpolbul.2014.03.057.

Di Beneditto, A.P.M., Ramos, R.M.A., Lima, N.R.W., 1998. Fishing activity in Northern Rio de Janeiro State (Brazil) and its relation with small cetaceans. Braz. Archiv. Biol. Technol. 41, 296–302. http://doi.org/10.1590/s1516-89131998000300004.

Di Beneditto, A.P.M., Ramos, R.M.A., 2001. Biology and conservation of the franciscana (Pontoporia blainvillei) in the north of Rio de Janeiro State. Brazil. J. Cetac. Res. Manage. 3, 185–192.

Domiciano, I.G., Domit, C., Broadhurst, M.K., Koch, M.S., Bracarense, A.P.F.R.L., 2016. Assessing disease and mortality among small cetaceans stranded at a world heritage site in Southern Brazil. PLoS One 11. doi:10.1371/journal.pone.0149295.

Dorneles, P.R., Lailson-Brito, J., Secchi, E.R., Bassoi, M., Pereira, C., Lozinsky, C., Torres, P.M., Malm, O., 2007. Cadmium concentrations in franciscana dolphin (Pontoporia blainvillei) from south Brazilian coast. Braz. J. Oceanog. 55, 179–186. http://doi.org/10.1590/S1679-75922007000300002.

Dorneles, P.R., Sanz, P., Eppe, G., Azevedo, A.F., Bertozzi, C.P., Martínez, M.A., Secchi, E.R., Barbosa, L.A., Cremer, M., Alonso, M.B., Torres, J.P.M., Lailson-Brito, J., Malm, O., Eljarrat, E., Barceló, D., Das, K., 2013. High accumulation of PCDD, PCDF, and PCB congeners in marine mammals from Brazil: A serious PCB problem. Sci. Total Environ. 463–464, 309–318. http://doi.org/10.1016/j.scitotenv.2013.06.015.

Drago, M., Franco-Trecu, V., Segura, A.M., Valdivia, M., González, E.M., Aguilar, A., Cardona, L., 2018. Mouth gape determines the response of marine top predators to long-term fishery-induced changes in food web structure. Scient. Rep. 8. doi:10.1038/s41598-018-34100-8.

Duarte, C.M., Chapuis, L., Collin, S.P., Costa, D.P., Devassy, R.P., et al., 2021. The soundscape of the Anthropocene ocean. Science 371, 1–10. doi:10.1126/science.aba4658.

Ehler, C., Douvere, F., 2009. Marine spatial planning: a step-by-step approach toward ecosystem-based management, Intergovernmental Oceanographic Commission and Man and the Biosphere Programme. IOC Manual and Guides n° 53, ICaM Dossier n° 6. Paris: UNESCO.

Elliott, M., 2014. Environmental Impact Assessment in Australia: Theory and Practice, 6th edn. Federation Press, Leichhardt, NSW.

Erbe, C., Reichmuth, C., Cunningham, K., Lucke, K., Dooling, R., 2016. Communication masking in marine mammals: a review and research strategy. Mar. Pollut. Bull. 103, 15–38. http://doi.org/10.1016/j.marpolbul.2015.12.007.

Fitch, J.E., Brownell, R.L., 1971. Food habits of the Franciscana Pontoporia blainvillei (Cetacea: Platanistidae) from South America. Bull. Mar. Sci. 21, 626–636.

Foglia, M.C., 2013. Evaluación de Pontoporia blainvillei como potencial especie bio-indicadora de metales en el estuario del Río de la Plata y su frente marítimo. Thesis Universidad de la República (Uruguay).

Forney, K.A., Southall, B.L., Slooten, E., Dawson, S., Read, A.J., Baird, R.W., Brownell, R.L., 2017. Nowhere to go: noise impact assessments for marine mammal populations with high site fidelity. Endang. Species Res. 32, 391–413. http://doi.org/10.3354/esr00820.

Fossi, M.C., Pedà, C., Compa, M., Tsangaris, C., Alomar, C., Claro, F., Ioakeimidis, C., Galgani, F., Hema, T., Deudero, S., Romeo, T., Battaglia, P., Andaloro, F., Caliani, I., Casini,

S., Panti, C., Baini, M., 2018. Bioindicators for monitoring marine litter ingestion and its impacts on Mediterranean biodiversity. Environ. Pollut. 237, 1023–1040. https://doi.org/10.1016/j.envpol.2017.11.019.

Frainer, G., 2016. Franciscana calls for help: the short and long-term effects of Mariana's disaster on small cetaceans of Southeastern Brazil. In: International Whaling Commission Meeting, SC/66b/SM/04.

Franco-Trecu, V., Costa, P., Abud, C., Dimitriadis, C., Laporta, P., Passadore, C., Szephegyi, M.N., 2009. By-catch of franciscana Pontoporia blainvillei in Uruguayan artisanal gillnet fisheries: an evaluation after a twelve-year gap in data collection. Lajam 7, 11–22.

Franco-Trecu, V., Szephegi, M.N., Doño, F., Forselledo, R., Reyes, F., Passadore, C., Crespo, E.A., Inchausti, P., 2019. Marine mammal bycatch by the industrial bottom trawl fishery at the Rio de la Plata estuary and adjacent Atlantic Ocean. Lajam 47, 89–101.

Fredianelli, L., Bolognese, M., Fidecaro, F., Licitra, G., 2021. Classification of noise sources for port area noise mapping. Environments 8, 1–16. http://doi.org/10.3390/environments8020012.

Freitas-Netto, R., Di Beneditto, A.P.M., 2008. Interactions between fisheries and cetaceans in Espírito Santo State coast, southeastern Brazil. Revista Brasileira de Zoociências 10 (1).

Fuentes, M.M.P.B., Wildermann, N., Gandra, T.B.R., Domit, C., 2020. Cumulative threats to juvenile green turtles in the coastal waters of southern and southeastern Brazil. Biodiv. Conserv. 29, 1783–1803. http://doi.org/10.1007/s10531-020-01964-0.

Gago-Ferrero, P., Alonso, M.B., Bertozzi, C.P., Marigo, J., Barbosa, L., Cremer, M., Secchi, E.R., Azevedo, A., Lailson-Brito, J., Torres, J.P.M., Malm, O., Eljarrat, E., Díaz-Cruz, M.S., Barceló, D., 2013. First determination of UV filters in marine mammals. octocrylene levels in Franciscana dolphins. Environ. Science Technol. 47, 5619–5625. http://doi.org/10.1021/es400675y.

Gall, S.C., Thompson, R.C., 2015. The impact of debris on marine life. Mar. Pollut. Bull. 92 (1–2), 170–179. doi:10.1016/j.marpolbul.2014.12.041.

García-Rodríguez, Hutton, M., Brugnoli, E., Venturini, N., Del Puerto, L., Inda, H., Bracco, R., Burone, L., Muniz, P., 2010. Assessing the effect of natural variability and human impacts on the environmental quality of a coastal metropolitan area (Montevideo Bay, Uruguay). Pan-Am. J. Aquat. Sci. 5, 91–100.

Gerpe, M., Rodríguez, D., Moreno, V.J., Bastida, R.O., Moreno, J.E., 2002. Accumulation of heavy metals in the franciscana (Pontoporia blainvillei) from Buenos Aires Province, Argentina. Lajam 1. doi:10.5597/lajam00013.

Gianelli, I., Defeo, O., 2017. Uruguayan fisheries under an increasingly globalized scenario: Long-term landings and bioeconomic trends. Fish. Res. 190, 53–60. http://doi.org/10.1016/j.fishres.2017.02.002.

Gianelli, I., Ortega, L., Marín, Y., Piola, A.R., Defeo, O., 2019. Evidence of ocean warming in Uruguay's fisheries landings: the mean temperature of the CAtch approach. Mar. Ecol. Progr. Series 625, 115–125. http://doi.org/10.3354/meps13035.

González, E.M., Martínez-Lanfranco, J.A., Juri, E., Rodales, A.L., Botto, G., Soutullo, A., 2013. Mamíferos. en: Soutullo, A., Clavijo, C., Martinez-Lanfranco, J.A. (Eds.), Especies prioritarias para la conservación en Uruguay. pp. 175–207. Vertebrados, moluscos continentales y plantas vasculares. snap/dinama/mvotma y dicyt/mec. Montevideo, pp. 222.

Grech, A., Bos, M., Brodie, J., Coles, R., Dale, A., Gilbert, R., Hamann, M., Marsh, H., Neil, K., Pressey, R.L., Rasheed, M.A., Sheaves, M., Smith, A., 2013. Guiding principles for the improved governance of port and shipping impacts in the Great Barrier Reef. Mar. Pollut. Bull. 75, 8–20. https://doi.org/10.1016/j.marpolbul.2013.07.013.

Guan, S., Brookens, T., 2021. The use of psychoacoustics in marine mammal conservation in the United States: from science to management and policy. J. Mar. Sci. Engin. 9. doi:10.3390/jmse9050507.

Guerrero, R.A., Acha, E.M., Frami, M.B., Framinam, M.B., Lasta, C.A., 1997. Physical oceanography of the Rio de la Plata Estuary. Argentina 1. Contin. Shelf Res. 17, 7.

Haimovici, M., 1997. Recursos Pesqueiros Demersais da Região Sul. Aval. do Potencial Sustentável de Recursos Vivos da ZEE. 81.

Haimovici, M., 1998. Present state and perspectives for the southern Brazil shelf demersal fisheries. Fish. Manage. Ecol. 5, 277–289.

Halpern, B.S., McLeod, K.L., Rosenberg, A.A., Crowder, L.B., 2008a. Managing for cumulative impacts in ecosystem-based management through ocean zoning. Ocean Coast. Manage. 51, 203–211. http://doi.org/10.1016/j.ocecoaman.2007.08.002.

Halpern, B., Walbridge, S., Selkoe, K.A., Kappel, C.V., Micheli, F., D'Agrosa, C., Bruno, J.F., Casey, K.S., Ebert, C., Fox, H.E., Fujita, R., Heinemann, D., Lenihan, H.S., Madin, E.M.P., Perry, M.T., Selig, E.R., Spalding, M., Steneck, R., Watson, R., 2008b. A global map of human impact on marine ecosystems. Science 319, 948–952.

Halpern, B., Frazier, M., Potapenko, J. et al. 2015. Spatial and temporal changes in cumulative human impacts on the world's ocean. Nat. Commun. 6, 7615. https://doi.org/10.1038/ncomms8615.

Halpern, B.S., Frazier, M., Afflerbach, J., O'Hara, C., Katona, S., Stewart Lowndes, J.S., et al. (2017) Drivers and implications of change in global ocean health over the past five years. PLoS ONE 12(7): e0178267. https://doi.org/10.1371/journal.pone.0178267.

Hawkins, S.J., Evans, A.J., Mieszkowska, N., Adams, L.C., Bray, S., Burrows, M.T., Firth, L.B., Genner, M.J., Leung, K.M.Y., Moore, P.J., Pack, K., Schuster, H., Sims, D.W., Whittington, M., Southward, E.C., 2017. Distinguishing globally driven changes from regional- and local-scale impacts: the case for long-term and broad-scale studies of recovery from pollution. Mar. Pollut. Bull. 124, 573–586. http://doi.org/10.1016/j.marpolbul.2017.01.068.

Herbst, D., Gerhardinger, L.C., Hanazaki, N., 2020. Linking user-perception diversity on ecosystems services to the inception of coastal governance regime transformation. Frontiers Mar. Sci. 7, 83. http://doi.org/10.3389/fmars.2020.00083.

Henning, B., de Sá Carvalho, B., Pires, M.M., Bassoi, M., Marigo, J., Bertozzi, C., Araújo, M.S., 2018. Geographical and intrapopulation variation in the diet of a threatened marine predator, Pontoporia blainvillei (Cetacea). Biotropica 50, 157–168. http://doi.org/10.1111/btp.12503.

Higham, C., Higham, T., 2009. A new chronological framework for prehistoric Southeast Asia, based on a Bayesian model from Ban Non Wat. Antiquity 83, 125–144.

Hildebrand, J.A., 2009. Anthropogenic and natural sources of ambient noise in the ocean. Mar. Ecol. Progr. Series 395, 5–20 doi.org/10.3354/meps08353.

Holz, A.C., 2014. Caracterização do ambiente acústico subaquático e uso de habitat pela toninha, Pontoporia blainvillei, na Baía da Babitonga, litoral norte de Santa Catarina, Brasil. Master Thesis, Universidade da Região de Joinville, Joinville, 76.

InVest - Natural Capital, 2020. Stanford University, Stanford, CA 94305.

Islam, M.S., Tanaka, M., 2004. Impacts of pollution on coastal and marine ecosystems including coastal and marine fisheries and approach for management: a review and synthesis. Mar. Poll. Bull. 48, 624–649. https://doi.org/10.1016/j.marpolbul.2003.12.004.

Kajiwara, N., Matsuoka, S., Iwata, H., Tanabe, S., Rosas, F.C.W., Fillmann, G., Readman, J.W., 2004. Contamination by persistent organochlorines in cetaceans incidentally caught along Brazilian coastal waters. Archiv. Environ. Contamin. Toxicol. 46, 124–134.

Korpinen, S., Andersen, J.H., 2016. A global review of cumulative pressure and impact assessments in marine environments. Frontiers Mar. Sci. 3 (Issue AUG). doi:10.3389/fmars.2016.00153.

Kühn, S., Franeker, J.A.V., 2020. Quantitative overview of marine debris ingested by marine megafauna. Mar. Pollut. Bull. 151. doi:10.1016/j.marpolbul.2019.110858.

Kunito, T., Nakamura, S., Ikemoto, T., Anan, Y., Kubota, R., Tanabe, S., Rosas, F.C.W., Fillmann, G., Readman, J.W., 2004. Concentration and subcellular distribution of trace elements in liver of small cetaceans incidentally caught along the Brazilian coast. Mar. Pollut. Bull. 49, 574–587. http://doi.org/10.1016/j.marpolbul.2004.03.009.

Lailson-Brito, J.Jr., Azeredo, M.A.A., Malm, O., Ramos, R.A., Dibeneditto, A.P.M., Saldanha, M.F.C., 2002. Trace metal concentrations in liver and kidney of franciscana, Pontoporia blainvillei, of the North coast of the Rio de Janeiro State, Brazil. LAJAM, 1, (1), 107–114.

Lailson-Brito, J., Oliveira-Ferreira, N., Manhães, B.M.R., Bisi, T.L., Santos-Neto, E., 2022. Chemical pollution and franciscana—a review. Chapter 11. In: Simões-Lopes, P.C., Cremer, M.J. (Eds.), The Franciscana Dolphin: on the edge of survival. Academic Press, Elsevier, London.

Lailson-Brito, J., Dorneles, P.R., Azevedo-Silva, C.E., Azevedo, A.D., Vidal, L.G., et al., 2011. Organochlorine concentrations in franciscana dolphins, Pontoporia blainvillei, from Brazilian waters. Chemosphere 84, 882–887.

Laist, D.W., 1987. Overview of the biological effects of lost and discarded plastic debris in the marine environment. Mar. Pollut. Bull. 18 (6, Supplement B), 319–326. https://doi.org/10.1016/S0025-326X(87)80019-X.

Lavandier, R., Arêas, J., Quinete, N., de Moura, J.F., Taniguchi, S., Montone, R., Siciliano, S., Moreira, I., 2015. PCB and PBDE levels in a highly threatened dolphin species from the Southeastern Brazilian coast. Environ. Pollut. 208, 442–449. http://doi.org/10.1016/j.envpol.2015.10.013.

Leaper, D., Assadian, O., Edmiston, C.E., 2015. Approach to chronic wound infections. British J. Dermat. 173, 351–358. http://doi.org/10.1111/bjd.13677.

Learmonth, J.A., Macleod, C.D., Santos, M.B., Pierce, G.J., Crick, H.Q.P., Robinson, R.A., 2006. Potential effects of climate change on marine mammals. Oceanog. Mar. Biol. 44, 431–464. http://doi.org/10.1201/9781420006391.ch8.

Leonel, J., Sericano, J.L., Secchi, E.R., Bertozzi, C., Fillmann, G., Montone, R.C., 2014. PBDE levels in franciscana dolphin (Pontoporia blainvillei): temporal trend and geographical comparison. Sci. Total Environ. 493, 405–410. http://doi.org/10.1016/j.scitotenv.2014.06.003.

Lopes, X.M., Santos, O., 2012. Hábitos alimentares de toninha, Pontoporia blainvillei (Gervais D'Orbigny, 1844) (Mammalia, Cetacea), no sul do estado de São Paulo e norte do Paraná, Brasil. Dissertação - (mestrado). Universidade Estadual Paulista, Instituto de Biociências de Rio Claro, São Paulo, p. 92. . http://hdl.handle.net/11449/99539.

Lotze, H.K., Lenihan, H.S., Bourque, B.J., Bradbury, R.H., Cooke, R.G., Kay, M.C., Kidwell, S.M., Kirby, M.X., Peterson, C.H., Jackson, J.B.C., 2006. Depletion, degradation, and recovery potential of estuaries and coastal seas. Science 312, 1806–1809.

Lozoya, J.P., Carranza, A., Lenzi, J., Machín, E., de Mello, F.T., González, S., Hernández, D., Lacerot, G., Martínez, G., Scarabino, F., Sciandro, J., Vélez-Rubio, G., Burgues, F., Carrizo, D., Cedrés, F., Chocca, J., de Álava, D., Jiménez, S., Leoni, V., Weinstein, F., 2015. Management and research on plastic debris in Uruguayan Aquatic Systems: update and perspectives. J. Integ. Coastal Zone Manage. 15, 377–393. http://doi.org/10.5894/rgci583.

Macchi, G.J., Acha, E.M., 2000. Spawning frequency and batch fecundity of Brazilian menhaden, Brevoortia aurea, in the Río de la Plata estuary off Argentina and Uruguay. Fish. Bull. 98, 283–289.

Macchi, G.J., Acha, E.M., Militelli, M.I., 2003. Seasonal egg production of whitemouth croaker (Micropogonias furnieri) in the Río de la Plata Estuary, Argentina-Uruguay. Fish. Bull. 101, 332–342.

Machovsky-Capuska, G.E., Amiot, C., Denuncio, P., Grainger, R., Raubenheimer, D., 2019. A nutritional perspective on plastic ingestion in wildlife. Sci. Total Environ. 656, 789–796.

Magris, R.A., Giarrizzo, T., 2020a. Mysterious oil spill in the Atlantic Ocean threatens marine biodiversity and local people in Brazil. Mar. Pollut. Bull. 153, 110961. http://doi.org/10.1016/j.marpolbul.2020.110961.

Magris, R.A., Costa, M.D.P., Ferreira, C.E.L., Vilar, C.C., Joyeux, J.C., Creed, J.C., Copertino, M.S., Horta, P.A., Sumida, P.Y.G., Francini-Filho, R.B., Floeter, S.R., 2020b. A blueprint for securing Brazil's marine biodiversity and supporting the achievement of global conservation goals. Divers. Distrib. doi:10.1111/ddi.13183.

Manhães, B.M.R., Vannuci-Silva, M., Brião, J.A., Guari, E.B., Botta, S., Colosio, A.C., Ramos, H.G.C., Barbosa, L.A., Cunha, I.A.G., Azevedo, A.F., Cunha, H.A., Bisi, T.L., Lailson-Brito, J., 2022. Temporal trends of trace elements bioaccumulation by a vulnerable cetacean (Pontoporia blainvillei) before and after one of the largest mining disasters worldwide. Sci. Total Environ. 804. doi:10.1016/j.scitotenv.2021.150196.

Marcondes, D.S., 2022. Comportamento acústico de botos-cinza (Sotalia guianensis) em áreas portuária e de conservação ambiental. Master dissertation, Universidade Federal do Paraná, Curitiba.

Marcondes, D.S., Nunes, T.Y., Fonseca, G.F., Moura, S.P.G., Belleghem, T.V., Domit, C., 2020. Atividades de derrocagem subaquática e potenciais impactos em golfinhos costeiros: avaliação, monitoramento e medidas de mitigação. Bol. Soc. Bras. Mastozoologia 89 135–145.

Marta-Almeida, M., Mendes, R., Amorim, F.N., Cirano, M., Dias, J.M., 2016. Fundão Dam collapse: oceanic dispersion of River Doce after the greatest Brazilian environmental accident. Mar. Pollut. Bull. 112, 359–364. http://doi.org/10.1016/j.marpolbul.2016.07.039.

Maxwell, S.M., Hazen, E.L., Bograd, S.J., Halpern, B.S., Breed, G.A., Nickel, B., Teutschel, N.M., Crowder, L.B., Benson, S., Dutton, P.H., Bailey, H., Kappes, M.A., Kuhn, C.E., Weise, M.J., Mate, B., Shaffer, S.A., Hassrick, J.L., Henry, R.W., Irvine, L., Costa, D.P., 2013. Cumulative human impacts on marine predators. Nature Communic 4. doi:10.1038/ncomms3688.

Melcón, M.L., Failla, M., Iñíguez, M.A., 2012. Echolocation behavior of franciscana dolphins (Pontoporia blainvillei) in the wild. J. Acoust. Soc. Am. 131, EL448–EL453. http://doi.org/10.1121/1.4710837.

Melcón, M.L., Failla, M., Iñíguez, M.A., 2016. Towards understanding the ontogeny of echolocation in franciscana dolphins (Pontoporia blainvillei). Mar. Mamm. Sci. 32, 1516–1521. http://doi.org/10.1111/mms.12336.

Menafra, R., Conde, D., Roche, I., Gorfinkiel, D., Píriz, C., Baliero, W., Biasco, E., Fossati, M., Lorenzo, E., Cortazzo, R., Fournier, R., 2009. Challenges and perspectives for integrated coastal management in Uruguay. Ocean Yearbook 23, 403–432.

Mendez, M., Rosembaum, H.C., Bordino, P., 2008. Conservation genetics of the franciscana dolphin in Northern Argentina: population structure, bycatch impacts and management implications. Conserv. Genetics 9, 419–435. http://10.1007/s10592-007-9354-7.

Mianzan, H., Lasta, C., Acha, E., Guerrero, R., Macchi, G., Bremec, C., 2001. The Río de la Plata Estuary, Argentina-Uruguay. In: Seeliger, U., Kjerfve, B. (Eds.), Coastal Marine Ecosystems of Latin America. Ecological Studies (Analysis and Synthesis), 144. Springer, Berlin, Heidelberg.

Miksis-Olds, J.L., Nichols, S.M., 2016. Is low frequency ocean sound increasing globally? J. Acoust. Soc. Am. 139, 501–511. http://doi.org/10.1121/1.4938237.

Moore, J.E., Wallace, B.P., Lewison, R.L., Žydelis, R., Cox, T.M., Crowder, L.B., 2009. A review of marine mammal, sea turtle and seabird bycatch in USA fisheries and the role of policy in shaping management. Mar. Policy 33, 435–451. http://doi.org/10.1016/j.marpol.2008.09.003.

Muniz, P., Marrero, A., Brugnoli, E., Kandratavicius, N., Rodríguez, M., Bueno, C., Venturini, N., Figueira, R.C., 2019. Heavy metals and As in surface sediments of the north coast of the Río de la Plata estuary: spatial variations in pollution status and adverse biological risk. Reg. Studies in Mar. Sci. 28, 100625.

National Academies of Sciences, Engineering, and Medicine, 2017. Approaches to Understanding the Cumulative Effects of Stressors on Marine Mammals. The National Academies Press, Washington, DC. https://doi.org/10.17226/23479.

Negri, M.F., Denuncio, P., Panebianco, M.V., Cappozzo, H.L., 2012. Bycatch of franciscana dolphins Pontoporia blainvillei and the dynamic of artisanal fisheries in the species' southernmost area of distribution. Braz. J. Oceanog. 60, 149–158. http://doi.org/10.1590/S1679-87592012000200005.

Nelms, S.E., Alfaro-Shigueto, J., Arnould, J.P.Y., Avila, I.C., Nash, S.B., Campbell, E., Carter, M.I.D., Collins, T., Currey, R.J.C., Domit, C., Franco-Trecu, V., Fuentes, M.M.P.B., Gilman, E., Harcourt, R.G., Hines, E.M., Hoelze, A.R., Hooker, S.K., Kelkar, N., Kiszka, J.J., Godley, B.J., 2021. Marine mammal conservation: over the horizon. Endang. Species Res. 44, 291–325. http://doi.org/10.3354/ESR01115.

Newton, A., Carruthers, T.J.B., Icely, J., 2012. The coastal syndromes and hotspots on the coast. Estuar. Coast. Shelf Sci. 96, 39–47. http://doi.org/10.1016/j.ecss.2011.07.012.

NMFS. Revisions to: technical Guidance for assessing the effects of anthropogenic sound on marine mammal hearing (Version 2.0): underwater thresholds for onset of permanent and temporary threshold shifts (Issue April). Data access on 1st October 2021. https://www.fisheries.noaa.gov/action/2018-revision-technical-guidance-assessing-effects-anthropogenic-sound-marine-mammal-hearing

O'shea, T.J., Odell, D.K., 2008. Large-scale marine ecosystem change and the conservation of marine mammals. J. Mamm. 89, 529–533.

Oliveira, V.K.M., Faria, D.M., Cunha, H.A., dos Santos, T.E.C., Colosio, A.C., Barbosa, L.A., Freire, M.C.C., Farro, A.P.C., 2020. Low genetic diversity of the endangered franciscana (Pontoporia blainvillei) in Its Northernmost, Isolated Population (FMAIa, Espírito Santo, Brazil). Frontiers Mar. Sci. 7. doi:10.3389/fmars.2020.608276.

Ortega, L., Martínez, A., 2007. Multiannual and seasonal variability of water masses and fronts over the Uruguayan shelf. J. Coastal Res. 23 (3), 618–629. doi:10.2112/040221.1.

Ott, P.H., Di Tullio, J., Prado, J., Danilewicz, D., Passadore, C., Failla, M., Iñíguez, M.B, Rodriguez, R., 2022. Marine protected areas: their potential for the conservation of the franciscana. Chapter 18. In: Simões-Lopes, P.C., Cremer, M.J. (Eds.), The Franciscana Dolphin: on the edge of survival. Academic Press, Elsevier, London.

Ott, P.H., Secchi, E.R., Moreno, I.B., Danilewicz, D., Crespo, E.A., Bordino, P., Ramos, R., di Beneditto, A.P., Bertozzi, C., Bastida, R., Zanelatto, R., Perez, J.E., Kinas, P.G., 2002. Report of the working group on fishery interactions. Lajam 1, 55–64. http://doi.org/10.5597/lajam00008.

Panebianco, M.V., Negri, M.F., Botte, S.E., Marcovecchio, J.E., Cappozzo, H.L., 2013. Essential and nonessential heavy metals determination in skin and muscle tissues in franciscana dolphins (Pontoporia blainvillei) from Argentina southern coast. Chem. Ecol. 29, 511–518.

Paso-Viola, U., Natalia, M., Fernanda, M., Luis, H., 2014. Diet composition of franciscana dolphin Pontoporia blainvillei from southern Buenos Aires, Argentina and its interaction with fisheries. Revta. Biol. Mar. Oceanog. 49, 393–400.

Perillo, G.M.E, Cuadrado, D.G, 1991. Geomorphologic evolution of El Toro Channel, Bahia Blanca Estuary (Argentina) prior to dredging. Mar. Geol. 97 (3–4), 405–412. doi:10.1016/0025-3227(91)90129-R.

Pinheiro, F.C.F., Pinheiro, H.T., Teixeira, J.B., Martins, A.S., Cremer, M.J., 2019. Opportunistic development and environmental disaster threat franciscana dolphins in the southeast of Brazil. Tropical Conserv. Sci. 12, 1–7. http://doi.org/10.1177/1940082919847886.

Pinedo, M.C., 1982. Análise dos conteúdos estomacais de Pontoporia blainvillei (Gervais e D'Orbigny, 1844) e Tursiops gephyreus (Lahille, 1908) (Cetacea, Platanistidae e Delphinidae) na zona estuarial e costeira de Rio Grande, RS, Brasil. Master Thesis, Fundação Universidade Federal do Rio Grande, Rio Grande.

Pirotta, E., Booth, C.G., Costa, D.P., Fleishman, E., Kraus, S.D., Lusseau, D., Moretti, D., New, L.F., Schick, R.S., Schwarz, L.K., Simmons, S.E., Thomas, L., Tyack, P.L., Weise, M.J., Wells, R.S., Harwood, J., 2018. Understanding the population consequences of disturbance. Ecology and Evolution 8, 9934–9946. doi:10.1002/ece3.4458.

Pirotta, E., Laesser, B.E., Hardaker, A., Riddoch, N., Marcoux, M., Lusseau, D., 2013. Dredging displaces bottlenose dolphins from an urbanised foraging patch. Mar. Pollut. Bull. 74, 396–402. http://doi.org/10.1016/j.marpolbul.2013.06.020.

Polizzi, P.S., Romero, M.B., Chiodi Boudet, L.N., Das, K., Denuncio, P.E., Rodríguez, D.H., Gerpe, M.S., 2014. Metallothioneins pattern during ontogeny of coastal dolphin,

Pontoporia blainvillei, from Argentina. Mar. Pollut. Bull. 80, 275–281. https://doi.org/10.1016/j.marpolbul.2013.10.037.

Praderi, R., Pinedo, M.C., Crespo, E.A., 1989. Conservation and management of Pontoporia blainvillei in Uruguay, Brazil and Argentina. In: Perrin, W.F., Brownell, Kaya, Z., Jiankang, L. (Eds.), Biology and Conservation of the River Dolphins. IUCN/Species Survival Commission, Gland, Switzerland, pp. 52–56.

Pressey, R.L., Cabeza, M., Watts, M.E., Cowling, R.M., Wilson, K.A., 2007. Conservation planning in a changing world. Trends Ecol. Evol. 22, 583–592. http://doi.org/10.1016/j.tree.2007.10.001.

Puig, P., Grunwaldt, P., Gonzalez, S., 2010. Pesquería artesanal de corvina en Uruguay. Frente Maritimo 21, 23–35.

Ramp, C., Delarue, J., Palsbøll, P.J., Sears, R., Hammond, P.S., 2015. Adapting to a warmer ocean—seasonal shift of baleen whale movements over three decades. PLoS One, 10. doi:10.1371/journal.pone.0121374.

Redford, K.H., Coppolillo, P., Sanderson, E.W., da Fonseca, G.A.B., Dinerstein, E., Groves, C., Mace, G., Maginnis, S., Mittermeier, R.A., Noss, R., Olson, D., Robinson, J.G., Vedder, A., Wright, M., 2003. Mapping the conservation landscape. Conserv. Biol. 17, 116–131. http://doi.org/10.1046/j.1523-1739.2003.01467.x.

Richardson, W.J., Greene, C.R., Malme, C.I., Thomson, D.H., Moore, S.E., Wiirsig, B., 1995. Marine Mammals and Noise, 1st ed. Academic Press, San Diego, CA.

Rocha, A., Valente, A.L., Marigo, J., 2022. Macroparasites of franciscana: what can their coexistence say to conservationists? Chapter 3. In: Simões-Lopes, P.C., Cremer, M.J. (Eds.), The Franciscana Dolphin: on the edge of survival. Academic Press, Elsevier, London.

Romero, A.F., Oliveira, M., Abessa, D.M.S., 2018. A simple Bird Sensitivity to Oil Index as a management tool in coastal and marine areas subject to oil spills when few biological information is available. Mar. Pollut. Bull. 128, 460–465. https://doi.org/10.1016/j.marpolbul.2017.12.008.

Romero, M.B., Polizzi, P., Chiodi, L., Robles, A., Das, K., Gerpe, M., 2017. Metals as chemical tracers to discriminate ecological populations of threatened Franciscana dolphins (Pontoporia blainvillei) from Argentina. Environ. Science Pollut. Res. 24. doi:10.1007/s11356-016-7970-9.

Romero-Romero, S., Molina-Ramírez, A., Höfer, J., Duineveld, G., Rumín-Caparrós, A., Sanchez-Vidal, A., Canals, M., Acuña, J.L., 2016. Seasonal pathways of organic matter within the Avilés submarine canyon: food web implications. Deep Sea Res. Part I: Oceanog. Res. Papers 117, 1–10. https://doi.org/10.1016/j.dsr.2016.09.003.

Rosas, F.C.W., Monteiro-Filho, E.L.A., Oliveira, M.O., 2002. Incidental catches of franciscana (Pontoporia blainvillei) on the southern coast of São Paulo State and the coast of Paraná State, Brazil. Lat. Am. J. Aquat. Mamm. 1 (1), 161–167. doi:10.5597/lajam00020.

Sánchez-Sarmiento, A.M., Carvalho, V.L., Díaz-Delgado, J., Ressio, R.A., Fernandes, N.C.C.A., Guerra, J.M., Sacristán, C., Groch, K.R., Silvestre-Perez, N., Ferreira-Machado, E., Costa-Silva, S., Navas-Suárez, P., Meirelles, A.C.O., Favero, C., Marigo, J., Bertozzi, C.P., Colosio, A.C., Marcondes, M.C.C., Cremer, M.J., Catão-Dias, J.L., 2019. Molecular, serological, pathological, immunohistochemical and microbiological investigation of Brucella spp. in marine mammals of Brazil reveals new cetacean hosts. Transb. Emerg. Diseas. 66, 1674–1692. http://doi.org/10.1111/tbed.13203.

Santana, F.R., 2020. Climate Variability Effects on Uruguayan Whitemouth Croaker and Striped Weakfish Fisheries. Thesis.

Santos, M.C.O., Oshima, J.E.F., Silva., E., 2009. Sightings of franciscana dolphins (Pontoporia blainvillei): the discovery of a population in the Paranaguá estuarine complex, southern Brazil. Braz. J. Oceanog. 57, 57–63.

Schoeman, R.P., Patterson-Abrolat, C., Plön, S., 2020. A global review of vessel collisions with marine animals. Frontiers Mar. Sci. 7. doi:10.3389/fmars.2020.00292.

Seabra, C.B., Ribeiro, V.L., Bertozzi, C.P., Junior, I.R.B., Luiz, V.S., Valle, R.R., 2018. Ingestão de resíduos sólidos por toninhas (Pontoporia blainvillei) do litoral centro-sul de São

Paulo, Brasil. In: abstracts of the 4° Congresso Latino Americano de Reabilitação de Fauna Marinha, 50–52.

Secchi, E.R., Monteiro, D., Claudino, R., 2022. Is the franciscana bycatch in gillnet fisheries sustainable? Chapter 10. In: Simões-Lopes, P.C., Cremer, M.J. (Eds.), The Franciscana Dolphin: on the edge of survival. Academic Press, Elsevier, London.

Secchi, E.R., Cremer, M.J., Danilewicz, D., Lailson-Brito, J., 2021. A synthesis of the ecology, human-related threats and conservation perspectives for the endangered franciscana dolphin. Frontiers Mar. Sci. 8, 1–18. http://doi.org/10.3389/fmars.2021.617956.

Secchi, E.R., Danilewicz, D., Ott, P.H., 2003. Applying the phylogeographic concept to identify franciscana dolphin stocks: implications to meet management objectives. J. Cetacean Res. Manage. 5, 61–68.

Seixas, T.G., Kehrig, H.A., Costa, M., Fillmann, G., Di Beneditto, A.P.M., Secchi, E.R., Souza, C.M.M., Malm, O., Moreira, I., 2008. Total mercury, organic mercury and selenium in liver and kidney of a South American coastal dolphin. Environ. Pollut. 154, 98.

Seixas, T.G., Kehrig, H.A., Fillmann, G., Di Beneditto, A.P.M., Souza, C.M.M., Secchi, E.R., Moreira, I., Malm, O., 2007. Ecological and biological determinants of trace elements accumulation in liver and kidney of Pontoporia blainvillei. Science Total Environ. 385, 208–220.

Sih, A., Jonsson, B.G., Luikart, G., 2000. Habitat loss: ecological, evolutionary and genetic consequences. Tree 15, 132–133.

Simmonds, M.P., Eliott, W.J., 2009. Climate change and cetaceans: concerns and recent developments. J. Mar. Biol. Assoc. UK 89, 203–210. http://doi.org/10.1017/S0025315408003196.

Simmonds, M.P., Isaac, S.J., 2007. The impacts of climate change on marine mammals: early signs of significant problems. Orynx 41, 19–260. http://doi.org/10.1017/S0030605307001524.

Soares, M.O., Teixeira, C.E.P., Bezerra, L.E.A., Paiva, S.V., Tavares, T.C.L., Garcia, T.M., de Araújo, J.T., Campos, C.C., Ferreira, S.M.C., Matthews-Cascon, H., Frota, A., Mont'Alverne, T.C.F., Silva, S.T., Rabelo, E.F., Barroso, C.X., de Freitas, J.E.P., de Melo Júnior, M., Campelo, R.P.S., Santana, C.S., Cavalcante, R.M., 2020. Oil spill in South Atlantic (Brazil): Environmental and governmental disaster. Mar. Policy, 115. doi:10.1016/j.marpol.2020.103879.

Spalding, M.D., Fox, H.E., Allen, G.R., Davidson, N., Ferdaña, Z.A., Finlayson, M., Halpern, B.S., Jorge, M.A., Lombana, A., Lourie, S.A., Martin, K.D., McManus, E., Molnar, J., Recchia, C.A., Robertson, J., 2007. Marine ecoregions of the world: a bioregionalization of coastal and shelf areas. Bioscience 57, 573–583.

Stelfox, M., Hudgins, J., Sweet, M., 2016. A review of ghost gear entanglement amongst marine mammals, reptiles and elasmobranchs. Mar. Pollut. Bull. 111, 6–17. https://doi.org/10.1016/j.marpolbul.2016.06.034.

Sucunza, F., Danilewicz, D., Andriolo, A., Azevedo, A.F., Secchi, E.R., Zerbini, A.N., 2020. Distribution, habitat use, and abundance of the endangered franciscana in southeastern and southern Brazil. Mar. Mamm. Sci. 36, 421–435. http://doi.org/10.1111/mms.12650.

Tellechea, J.S., Ferreira, M., Norbis, W., 2017. Echolocation and burst clicks from franciscana dolphins (Pontoporia blainvillei) on the coast of Uruguay. Mar. Mamm. Sci. 33, 687–694. http://doi.org/10.1111/mms.12383.

Torre, A.La, Alonso, M.B., Martínez, M.A., Sanz, P., Shen, L., Reiner, E.J., Lailson-Brito, J., Torres, J.P.M., Bertozzi, C., Marigo, J., Barbosa, L., Cremer, M., Secchi, E., Malm, O., Eljarrat, E., Barceló, D., 2012. Dechlorane-related compounds in franciscana dolphin (Pontoporia blainvillei) from southeastern and southern coast of Brazil. Environ. Sci. Technol. 46, 12364–12372. http://doi.org/10.1021/es302934p.

Trevizani, T.H., 2018. Bioacumulação e biomagnificação de metais pesados em teias tróficas de estuários do sul-sudeste do Brasil. PhD thesis, Instituto Oceanográfico de São Paulo. Universidade de São Paulo, São Paulo/SP.

Trevizani, T.H., Figueira, R.C.L., Santos, M.C.O., Domit, C., 2021. Mercury in trophic webs of estuaries in the southwest Atlantic Ocean. Mar. Pollut. Bull. 167. doi:10.1016/j.marpolbul.2021.112370.

UNESCO - United Nations Educational, Scientific and Cultural, 1999. Convention concerning the protection of the world cultural and natural heritage: Atlantic Forest South/East Brazil. World Heritage Centre.

Von Fersen, L., Kamminga, C., Seidl, A., 1997. Preliminary studies on the behavior of one individual of franciscana dolphin (Pontoporia blainvillei) in Mundo Marino, Argentina. Report of the Third Workshop for Coordinated Research and Conservation of the Franciscana Dolphin. Buenos Aires, pp. 30–33.

Valle, D., Berdanier, A., 2012. Computer programming skills for environmental sciences. Bull. Ecol. Soc. Am. 93, 373–389. http://doi.org/10.1890/0012-9623-93.4.373.

Van Erp, I., 1969. In quest of the La Plata dolphin. Pacific Discovery 22, 18–24.

Van Waerebeek, K., Baker, A.N., Félix, F., Gedamke, J., Iñiguez, M., Sanino, G.P., Secchi, E., Sutaria, D., van Helden, A., Wang, Y., 2007. Vessel collisions with small cetaceans worldwide and with large whales in the Southern Hemisphere, an initial assessment. Lajam 6. doi:10.5597/lajam00109.

Walker, W.A., Coe, J.M., 1990. Survey of marine debris ingestion by odontocete cetaceans. NOAA Technical Memorandum NMFS-SWFSC 154, 747–774.

Weinstein, F., Olivera, Y., Cedrés, F., Pereira, M., Teixeira de Mello, F., Carrizo, D., Lozoya, J.P., 2014. Evaluación y análisis de microplásticos en playas de Punta del Este: caracterización de una potencial contaminación para la fauna marina, Proceedings of the Tercer Congreso Uruguayo de Zoología "Prof. Dr. Raúl Vaz-Ferreira [Resúmenes]. Montevideo, Facultad de Ciencias, Universidad de la República, pp. 46. http://cuz.szu.org.uy/Libro%20resumenes%20IIICUZ.pdf.

Wells, R.S., Cremer, M.J., Berninsone, L.G., Albareda, D., Wilkinson, K.A., Stamper, M.A., Paitach, R.L., Bordino, P., 2021. Tagging, ranging patterns, and behavior of franciscana dolphins (Pontoporia blainvillei) off Argentina and Brazil: considerations for conservation. Mar. Mamm. Sci. 1–35. http://doi.org/10.1111/mms.12879.

Whitehead, P.J., Purdon, P., Russell-Smith, J., Cooke, P.M., Sutton, S., 2008. The management of climate change through prescribed Savanna burning: Emerging contributions of indigenous people in Northern Australia. Public Administ. Develop. 28, 374–385. http://doi.org/10.1002/pad.512.

World Bank, 2018. The World Bank Annual Report 2018. © World Bank, Washington, DC. https://openknowledge.worldbank.org/handle/10986/30326.

Zappes, C.A., Gama, R.M., Domit, C., Gatts, C.E.N., Di Beneditto, A.P.M., 2016. Artisanal fishing and the Franciscana (Pontoporia blainvillei) in Southern Brazil: ethnoecology from the fishing practice. J. Mar. Biolog. Assoc. UK 98, 867–877. http://10.1017/S0025315416001788.

Zerbini, A.N., Secchi, E., Crespo, E., Danilewicz, D., Reeves, R., 2017. Pontoporia blainvillei (errata version published in 2018). The IUCN Red List of Threatened Species 2017, 8235 e.T17978A123792204.

Zerbini, A.N., Andriolo, A., Cremer, M.J., Crespo, E.A., Danilewicz, D., Domit, C., Sucunza, F., 2022. Abundance estimates of franciscana dolphins (Pontoporia blainvillei): a review and future recommendations. Chapter 7. In: Simões-Lopes, P.C., Cremer, M.J. (Eds.), The Franciscana Dolphin: on the edge of survival. Academic Press, Elsevier, London.

SIMBA (Sistema Integrado de Monitoramento da Biota Aquática), 2021. Available at: https://simba.petrobras.com.br/simba. (Accessed April 18, 2021).

IBGE - Instituto Brasileiro de Geografia e Estatística, 2010. Available on https://www.ibge.gov.br/. (Accessed 10 November 2021).

ANTAQ - Agência Nacional de Transportes Aquaviários, 2021. Available on https://www.gov.br/antaq/pt-br. (Accessed 11 November 2021).

MMA - Ministério do Meio Ambiente, 2018. Available on https://www.gov.br/mma/pt-br (Accessed 10 November 2021).

CHAPTER 13

Long-term patterns of franciscana strandings throughout its distribution

Marta Jussara Cremer[a], Jonatas Henrique Prado[b], Henrique Chupil[c], Cristiane Kiyomi Miyagi Kolesnikovas[d], Maurício Tavares[e], Rodrigo del Rio do Valle[f], Pablo Denuncio[g], Alan Rosenthal[h], Sergio Rodriguez-Heredia[i], Maria Fernanda Zapata[j], Paulo Henrique Ott[k], Carla Beatriz Barbosa[l], Pedro Volkmer de Castilho[m], Camila Domit[n], Sergio Estima[o], Adriana Castaldo Colosio[p], Andrea Maranho[q], Alejandro Fallabrino[r], Gabriela Vélez-Rubio[s], Paula Laporta[t], Meca Valdivia[u], Sebástian Jiménez[v], André Silva Barreto[w]

[a]Laboratório de Ecologia e Conservação de Tetrápodes Marinhos e Costeiros, Universidade da Região de Joinville, São Francisco do Sul, Santa Catarina, Brazil
[b]Cooperativa para Conservação e Proteção dos Recursos Naturais, Florianópolis, Santa Catarina, Brazil
[c]Instituto de Pesquisas Cananéia, Cananéia, São Paulo, Brazil
[d]Associação R3 Animal, Florianópolis, Santa Catarina, Brazil
[e]Centro de Estudos Costeiros, Limnológicos e Marinhos, Campus Litoral Norte, Universidade Federal do Rio Grande do Sul, Imbé, Rio Grande do Sul, Brazil
[f]Instituto Biopesca, Praia Grande, São Paulo, Brazil
[g]Instituto de Investigaciones Marinas y Costeras, Mar del Plata, Buenos Aires, Argentina
[h]Asociación de Naturalistas Geselinos, Villa Gesell, Argentina
[i]Fundación Mundo Marino, San Clemente del Tuyú, Argentina
[j]Fundación AquaMarina, Centro de Estudios en Ciencias Marinas, Buenos Aires, Argentina
[k]Universidade Estadual do Rio Grande do Sul (Uergs), Osório, Rio Grande do Sul, Brazil
[l]Instituto Argonauta para Conservação Costeira e Marinha, Ubatuba, São Paulo, Brasil
[m]Laboratório de Zoologia, Universidade do Estado de Santa Catarina (UDESC), Laguna, Santa Catarina, Brazil
[n]Laboratório de Ecologia e Conservação, Universidade Federal do Paraná, Pontal do Sul, Paraná, Brazil
[o]Núcleo de Educação e Monitoramento Ambiental, Rio Grande, Rio Grande do Sul, Brazil
[p]Instituto Baleia Jubarte, Caravelas, Bahia, Brazil
[q]Instituto Gremar Pesquisa, Educação e Gestão de Fauna, Guarujá, SP, Brazil
[r]Karumbé NGO, Montevideo, Uruguay
[s]Sección de Oceanografía y Ecología Marina, Instituto de Ecología y Ciencias Ambientales, Facultad de Ciencias, Universidad de la República, Montevideo, Uruguay
[t]Yaqu Pacha Uruguay, Punta del Diablo, Rocha, Uruguay
[u]Sección Mamíferos, Museo Nacional de Historia Natural, Montevideo, Uruguay
[v]Laboratorio de Recursos Pelágicos, Dirección Nacional de Recursos Acuáticos, Constituyente, Montevideo, Uruguay
[w]Laboratório de Informática da Biodiversidade e Geomática, Universidade do Vale do Itajaí, Itajaí, SC, Brazil

The Franciscana Dolphin
DOI: https://doi.org/10.1016/B978-0-323-90974-7.00019-7

Introduction

The franciscana dolphin is the most endangered cetacean species in the Southwestern Atlantic Ocean (Secchi et al., 2001; Crespo, 2002). Its distribution extends from Espírito Santo State, Brazil (18°S), to the province of Chubut, Argentina (42°S) (Bastida et al., 2007). The incidental capture in fishing nets is considered the main threat, putting at risk the survival of its populations (Secchi et al. 1997; Kinas, 2002; Secchi et al., 2003). For this reason, the species is listed as "Vulnerable" on a global scale (Zerbini et al., 2017).

The precise number of animals that die in the fishing activity is unknown and only local estimates have been obtained (e.g., Praderi et al., 1989; Bertozzi and Zerbini, 2002; Di Beneditto et al., 1998; Di Beneditto and Ramos, 2001; Franco-Trecu et al., 2009; Frizzera et al., 2012; Negri et al., 2012; Rosas et al., 2002; Secchi et al., 2004). Some authors have made broader bycatch estimates, indicating mortality of about 2,900 animals per year (e.g., Ott et al., 2002; Secchi et al., 2003b), but it is understood that they are underestimated, to an unknown extent (Secchi et al., 2021). Despite that, the best available information indicates that the annual mortality of franciscanas is unsustainable (Ott et al., 2002; Prado et al., 2013; Secchi et al., 2021).

Due to its coastal habitat, the franciscana is also vulnerable to impacts caused by various other human activities (Pinheiro et al., 2019; Secchi et al., 2021). Contamination by chemical pollutants (Lailson-Brito et al., 2011; Panebianco et al., 2013; Polizzi et al., 2014), ingestion of plastic debris (Denuncio et al., 2011; Seabra et al., 2018), and the displacement from areas with intense port activity (Cremer et al., 2018) are some of the impacts reported so far. However, how much each of these impacts contributes to the mortality of the species can be even more difficult to assess. In addition, some of these factors, or even a combination of them, may be responsible for the diseases already reported for the species (Bracarense and Domiciano, 2022).

The franciscana range was divided into four management stocks, known as franciscana management areas (FMAs) (Secchi et al., 2003), that were reassessed based on genetic differentiation (Cunha et al., 2014). The FMA I correspond to the coastal waters of Espírito Santo and Rio de Janeiro States, Brazil. FMA II comprises part of the southeast and south coast of Brazil, from Piraquara, in the south of Rio de Janeiro State, to the central region of Santa Catarina Island, in Santa Catarina State. FMA III is the only one shared between Brazil and Uruguay, and extends from Santa Catarina Island to Uruguay; and FMA IV includes the species' distribution in Argentina, extending from the Río de la Plata estuary up to the southern limit of occurrence of the species, in Golfo Nuevo (Crespo et al., 1998). The existence of different stocks throughout the species' distribution is an important factor to be considered, as well as population size and in impact intensity, which may vary in different regions.

Information about the number of franciscanas that incidentally die in fishing nets should be the best data to evaluate fishery mortality. However, this information is not easy to be obtained systematically, especially covering a broad temporal-space scale. In addition, fishermen tend to under-report bycatch, for fear of reprisal or due to constraints (Negri et al., 2012; Pinheiro and Cremer, 2003; Secchi et al., 2003). This in part seems to be related to the historical absence of integrating fishermen knowledge into the research fisheries and/or the absence of involvement of fishermen in decisions relating to fishery management, making it difficult to obtain reliable information. In this context, carcasses can provide valuable knowledge about cetacean mortality, as well as about biological parameters, feeding ecology, relative abundance, diseases, and contamination, among others, regarding individuals and populations (e.g., Méndez-Fernandez et al., 2014; Peltier et al., 2012; Quirós et al., 2018; Yochum et al., 2011). However, several factors influence the number of carcasses that strand, including the spatial distribution of populations, oceanographic conditions, trophic web relationships, carcasses' buoyancy, the type of beach survey effort to record the carcasses (systematically or opportunistically), among others (e.g., Evans et al., 2005; Jepson et al., 2003; Kirschvink, 1983; Peltier et al., 2012; Saavedra et al., 2017). Therefore, despite its importance, strandings need to be interpreted with caution.

This chapter presents an unprecedented compilation of franciscana stranding data throughout the entire distribution of the species along Brazil, Uruguay, and Argentina, including live and dead animals. The aim of this work was to describe the temporal and spatial distribution of the franciscana strandings and evaluate which portion of the population is dying, considering total length (TL) and sex. We expect that gathering stranding information for the entire range of the species will contribute to providing a comprehensive picture to assess the mortality of franciscanas.

Material and methods

Study area

The study area comprises the whole franciscana distribution, from the north of the Espírito Santo State, Brazil, to the province of Chubut,

Argentina (Bastida et al., 2007). This region has a wide continental shelf in the south (~300 km) (Lucas et al., 2005), but became progressively narrow toward the north (~50 km) (Palma et al., 2008). The tidal amplitude is relatively small, and the wind regime is seasonally variable mainly in the southern portion of the study area (Palma et al., 2008). The shelf circulation is influenced by two western boundary currents: the Brazilian Current (BC) and the Malvinas/Falkland Current. The latter carries the cooler and fresher subantarctic water (SAW) equatorward along the continental shelf, and the former carries tropical waters (TW) and the South Atlantic Central Water (mixing between the SAW and TW) poleward along the edge of South America. The encounter of both currents, known as the subtropical convergence, occurs approximately between the latitude of 36°S and 39°S (Reid et al., 1977).

Wind-driven and shelf-break upwelling induced by BC eddies play an important role in pumping up nutrient-rich waters to the euphotic zone in this region. During austral summer, along the coasts of Uruguay and southern Brazil (e.g., Cabo Frio and Cabo de Santa Marta; Punta del Este), there is a development of nearshore upwelling driven by the prevailing northeasterly winds (upwelling-favorable) (Palma et al., 2008, Trinchin et al., 2019). This region is also influenced by shelf fronts that also increase productivity in some areas, for example, in southern Brazil and the north portion of the Argentina shelf (Guinder et al., 2018).

The two major freshwater inflow into the region are associated with the Río de la Plata estuary (38°S) and the Patos Lagoon estuary system (32°S) discharges. During austral winter the southwesterly winds displace the Plata/Patos-Mirim plume far from the Cabo de Santa Marta Grande (28°S), while in summer it retracts to approximately 32°S (Möller et al., 2008; Guerrero et al., 2014; Matano et al., 2014).

The influence of subantarctic waters, upwelling and frontal systems, and freshwater input, provides the region with high nutrient and chlorophyll concentrations, that sustain high diversity of top predators, such as sea turtles, seabirds, and marine mammals (Ortega and Martinez, 2007; Monteiro et al., 2016; Di Tulio et al., 2016).

Data collection

In the present study, stranding was defined as individuals washed ashore alive or dead with each one considered as a separeted event. Along the study area, different levels of effort were put into data collection, with stranding data being collected both opportunistically (when records are obtained through reports and telephone calls from local people, government officials, and tourists) and systematically (when records derived from active monitoring of coastline). We used unpublished data from the authors' institutions and data available at public databases, which include data published in technical literature. For the latter, three sources were used: Global Biodiversity Information System (GBIF), Marine Mammal Monitoring Support System (SIMMAM) (*Sistema de Apoio ao Monitoramento de Mamíferos Marinhos*) and Aquatic Biota Monitoring Information System (SIMBA) (*Sistema de Informação de Monitoramento da Biota Aquática*) (SIMBA, 2021). From GBIF (http://www.gbif.org), 458 records with date information were considered (1842–2020, with 70% of them recorded from 1970 to 1975) from 26 datasets (GBIF, 2021). SIMMAM (http://simmam.acad.univali.br) is a repository for data on aquatic mammal occurrence in Brazil that includes unpublished data from researchers and data from the literature. Only public records were included in this study, or where the data owners are authors of this chapter. SIMMAM dataset consisted of 2,714 records from 1970 to 2015, with 88% of the data being from 2000 to 2015. SIMBA (http://simba.petrobras.com.br) holds stranding data collected systematically from the *Projeto de Monitoramento de Praias da Bacia de Santos* (PMP-BS) and *Projeto de Monitoramento de Praias da Bacia de Campos* (PMP-BC) for a requirement of the Brazilian Environmental Federal Agency (IBAMA) for environmental licensing of the oil and natural gas production and transport by Petrobras at the presalt province. According to the logistics and difficulties in the access to the beaches, some stretches were monitores by these projects daily or weekly. Additionally, a network was established with local communities contributing to the indirect monitoring of some stretches. The SIMBA dataset includes 2,749 records, from 2015 to 2020, collected from Itaúnas (18°25′S), Espírito Santo State, to Laguna (28°29′S), in Santa Catarina State. Table 13.1 contains the list of institutions that collaborated with the input of data for this work including their respective area and monitored period.

Data collection followed different protocols between the institutions, with different information being recorded. For this work, only information referring to the date, geographic location, carcass condition, total length (TL), and sex of individuals were considered. In the absence of information about specific coordinates, but with the name of beach or municipality, the geographic position was estimated using the central coordinates of the location. Despite the different protocols, we assumed that the TL was measured following Norris (1961) and the carcass condition follows the classification proposed by Geraci and Lounsbury (2005).

Table 13.1 List of institutions that carried out the registration of stranding records of live or dead franciscanas, *Pontoporia blainvillei*, and collaborated with this study, respective areas monitored, and data collection periods.

Country	Franciscana management area	Institution/initials	Area monitored	Period
Brazil	FMA I	Instituto Baleia Jubarte/IBJ	18°20′S–19°56′S	1996–2020[a]
	FMA II	Instituto Argonauta	23°0′S–23°45′S	2000–2020[b]
		Instituto Biopesca	23°22′S–25°2′S	1998–2020[b]
		Instituto Gremar	23°45′S–24°20′S	2002–2020[b]
		Instituto de Pesquisas Cananéia/IPeC	24°41′S–25°37′S	1997–2020[b]
		Universidade Federal do Paraná/UFPR	25°12′S–25°58′S	1992–1997 and 2007–2020[b]
		Universidade da Região de Joinville/UNIVILLE	25°58′S–26°41′S	2001–2020[b]
		Associação R3 Animal	27°31′S–27°31′S	2002–2020[b]
		Universidade do Estado de Santa Catarina/UDESC	28°08′S–28°53′S	2011–2020[b]
	FMA III	Grupo de Estudos de Mamíferos Aquáticos do Rio Grande do Sul/GEMARS	29°21′S–31°31′S	1991–2011
		Centro de Estudos Costeiros, Limnológicos e Marinhos – Universidade Federal do Rio Grande do Sul/CECLIMAR-UFRGS	29°21′S–30°23′S	2012–2020
		Núcleo de Educação Monitoramento Ambiental—NEMA	31°26′S–33°45′S	1993–2020

Country	Franciscana management area	Institution/initials	Area monitored	Period
Uruguay	FMA III	Karumbé/Yaqu Pacha Uruguay	Uruguayan coast	2003–2020
		Museo Nacional de Historia Natural/MNHN	Uruguayan coast	1961–2020
		Dirección Nacional de Recursos Acuáticos–Ministerio de Ganadería, Agricultura y Pesca/DINARA-MGAP	Uruguayan coast	2019–2020
Argentina	FMA IV	Fundación Aquamarina	37°04′S–37°09′S	2002–2008, 2018–2020
		Fundación Mundo Marino (Marine Fauna Rescue and Rehabilitation Center)	36°17′S–37°01′S	1991–2020
		Asociación Naturalistas Geselinos/ANG and Universidad Nacional de Mar del Plata/UNMDP	37°01′S–37°06′S	2011–2020

[a]This institution participates in the Campos and Espírito Santo Basin Beach Monitoring Project (PMP-BC/ES) and from October 2010 onward, stranding data were included in the SIMBA database.

[b]These institutions participate in the Santos Basin Beach Monitoring Project (PMP-BS) and from August 2015 onward, stranding data were included in the SIMBA database.

All databases mentioned were pooled together in order to analyze all data jointly and multiple records of the same stranding event were unified (some institutions can survey the same stretch of beach) and the information was complemented among all institutions. As there was the possibility of a record being included in two or more databases, all duplicated records were eliminated. We considered duplicated records as (1) carcasses with equal values for date, location (coordinates or beach name), sex, and TL; and (2) individuals without the two last information, but with equal date and coordinates. Records without information on sex, TL, and coordinates, but recorded on the same date and beach, were maintained as we could not rule out that these could be different records.

Brazil

Franciscana stranding records at the north of FMA I were obtained opportunistically from 1978 to 2009 by the *Instituto Baleia Jubarte* (1996–2009) between Itaúnas (18°25′S) and Santa Cruz (19°57′S), Espírito do Santo State. Since October 2010, both the north and the south regions (between Barra de Itabapoana—21°18′S, and Armação de Búzios—22°44′S) began to be systematically monitored by the PMP-BC.

In the FMA II stranding records were obtained opportunistically from 1979 to 2015. Afterwards, with the beginning of the PMP-BS project in August 2015, the area from Saquarema (23°21′S), Rio de Janeiro State, to Laguna (27°34′S), Santa Catarina State, were sistematically surveyed.

In the north part of FMA III (Brazil), systematic beach surveys have been conducted since 1976 in the southern portion of Rio Grande Sul State (from Chuí—33°45′S, to Lagoa do Peixe—31°26′S), by the *Universidade Federal do Rio Grande*—FURG and *Núcleo de Educação e Monitoramento Ambiental*—NEMA, and in the northern portion (between Lagoa do Peixe—31°26′S, to Torres—29°19′S) from 1991 to 2011 by the *Grupo de Estudos de Mamíferos Aquáticos do Rio Grande do Sul*—GEMARS. Afterwards (2012–2020), a smaller portion of this area (from Dunas Altas/Palmares do Sul—30°25′S, to Torres—29°19′S) was systematically surveyed by *Centro de Estudos Costeiros, Limnológicos e Marinhos*—CECLIMAR-UFRGS. In the north portion of FMA III (from Torres—29°19′S, to Florianópolis—27°23′S), the franciscana records were mainly opportunistic and were made by the *Laboratório de Zoologia*—UDESC. Only in August 2015, systematic beach surveys started from Laguna (28°29′S) to north Espírito Santo State by the PMP-BS.

Uruguay

In the south part of FMA III (Uruguay), franciscana strandings were opportunistically recorded since 1961 along all the Uruguayan coast by four different institutions: *Karumbé* and *Yaqu Pacha Uruguay*; *Museo Nacional de Historia Natural de Montevideo* (MNHN); and *Dirección Nacional de Recursos Acuáticos* (DINARA). Systematic surveys were carried out since 2003, from October to April, along the Rocha Department coast by *Karumbé* and *Yaqu Pacha Uruguay* (Vélez-Rubio et al., 2013). The area from La Coronilla (33°54′S) to Barra del Chuy (33°44′S) and from La Coronilla to Punta del Diablo (34°02′S) were covered weekly between 2003 to 2021, while the area from La Coronilla to Cabo Polonio (34°25′S), and from Cabo Polonio to La Pedrera (34°35′S) were covered biweekly between 2003 and 2020 (Table 13.1).

Argentina

Beach monitoring programs were exclusively carried out in the north coast of the Buenos Aires province on an opportunistic basis, between Punta Rasa (South Bahía Samborombón—36°17′S) and the northern limit of Mar Chiquita MAB-UNESCO Protected Area (37°60′S), covering 160 kilometers of sandy beaches (Table 13.1). Monitoring has been carried out by the *Fundación Mundo Marino* (FMM) since 1991, as it is part of the Marine Fauna Rescue Network of the Provincial Agency for Sustainable Development (OPDS) of the Buenos Aires Province; and by *Fundación Aquamarina*. Since 2017 the franciscana carcasses record in the Buenos Aires province has to be made under the framework of a provincial law, as the franciscana has been declared a Natural Monument of the Province of Buenos Aires (Law Number 14.992). South of this region, monitoring of stranded marine megafauna has been conducted since 2011 on a biweekly basis by the cooperation between *Asociación de Naturalistas Geselinos* (ANG) and *Universidad Nacional de Mar del Plata* (UNMDP), and includes an extension of 60 km.

Analysis

The stranding data analyzed here extends from 1961 to 2020. We follow the FMA division proposed by Secchi et al. (2003), but with the boundary between FMA II and FMA III at Florianópolis, Santa Catarina State, as proposed by Ott et al. (2002) and Cunha et al. (2014), rounding to the nearest degree.

To evaluate which portion of the population is dying, six categories based on TL were defined: (1) 70–90 cm; (2) 91–110 cm; (3) 111–130 cm; (4) 131–150 cm; (5) 151–170 cm; (6) individuals > 170 cm. Individuals

Table 13.2 The mean length at franciscana, *Pontoporia blainvillei*, sexual maturity for each franciscana management area (FMA), according to the specific literature for each region.

	Mature		
	Male	**Female**	**References**
FMA I	>115 cm	>130 cm	Ramos et al. (2000) and Di Beneditto and Ramos (2001)
FMA II	>116 cm	>130 cm	Silva et al. (2020) and Cordeiro (2021)
FMA III	>128 cm	>139 cm	Kasuya and Brownell (1979) and Danilewicz (2003); Danilewicz et al. (2004)
FMA IV	>126 cm	>133.4 cm	Panebianco et al. (2012, 2016)

with a TL lower than 70 cm were excluded from this analysis ($n = 400$), as they were probably fetuses (defined as the cutting line considering the birth size estimated for different regions along with the species distribution—Danilewicz, 2002; Denuncio et al., 2013; Rosas and Monteiro-Filho, 2002). Moreover, to verify the proportion of mature and immature individuals, we used the estimated mean length at sexual maturity, available in the literature (Table 13.2). Diferences in sex, TL, and maturity were analyzed for each FMA separately.

All records of franciscanas (with and without biological information, and individuals with TL < 70 cm) were used to analyze the spatial and temporal pattern of strandings. To visualize the seasonal variability in space, monthly values were compared within each latitude degree. The seasons were defined as spring (October, November, and December), summer (January, February, and March), autumn (April, May, and June), and winter (July, August, and September).

Because of the heterogeneity of beach surveys conducted (opportunistically or systematically) along the study area and years, the strandings could not be standardized by km of beach survey. Therefore, all temporal and spatial patterns were presented in terms of absolute numbers. To allow direct comparisons between FMAs and sex, we obtained the proportion of stranding numbers.

Results and discussion

Distribution of the strandings

A total of 8,163 franciscana strandings in the period from 1961 to 2020 were recorded. From the total records in which location was available, 3,541 were recorded in FMA III, 2,985 in FMA II, 474 in FMA IV, and 160

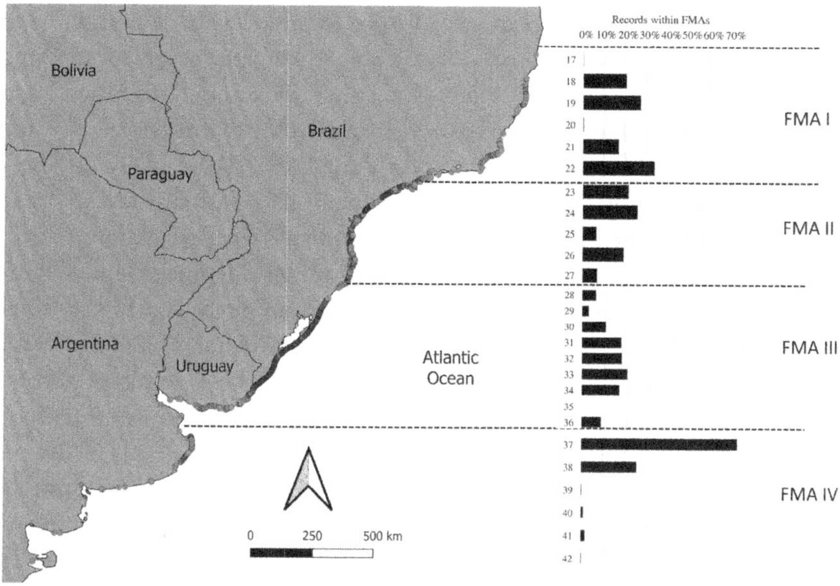

Fig. 13.1 ***Records of franciscanas,*** **Pontoporia blainvillei,** ***along with its distribution with bars indicating the percent of records within each degree of latitude, for each franciscana management area (*****FMA*****) from 1961 to 2020.*** Limits of the different FMAs follow Amaral et al. (2018), rounding to the nearest degree. The geographic position was estimated when records did not have a location but included beach or municipality names; in this way, one point may include more than one record. The coordinates of the graph (on the right) represent an approximation to the coordinates in the figure (on the left).

in FMA I. The annual mean stranding number follow a same pattern, with highest value observed in FMA III (93; sd = 107), followed by FMA II (78; sd = 162), FMA IV (12; sd = 18), and FMA I (4; sd = 6). The strandings were more evenly distributed in FMA I, FMA II, and FMA III, compared with FMA IV, and the largest concentration was observed in the central portion of FMA III (Fig. 13.1). The distribution of strandings probably reflects variations in the abundance of the species throughout its distribution (Crespo et al., 2010; Danilewicz et al., 2009; Sucunza et al., 2020), the level and consistency of mortality caused by anthropogenic factors, mainly incidental mortality in gillnet fishery (Secchi et al., 2021), oceanographic conditions and differences in beach survey effort, or a combination of these factors. Based on estimates carried out along the franciscana distribution in the Brazilian coast, FMA III has the highest values of abundance (Danilewicz et al., 2010). This, coupled with the high levels of incidental mortality in gillnet fisheries off Rio Grande do Sul State (Prado et al. 2013,

2016; Secchi et al., 2004), can explain the highest levels of strandings in FMA III. In addition, this area is also where systematic beach monitoring has been carried out for the longest time, since the late 1970s (Pinedo and Polacheck, 1999, Prado et al., 2016). It is important to emphasize that the stranding numbers presented in this study for FMA III are likely underestimated, since we did not include all dataset available for this FMA. Prado et al. (2016) analyzing temporal stranding pattern of marine mammals over 38 years (1976–2013) in Rio Grande do Sul coast, reported 4,574 franciscanas, which represents 120 individuals stranded per year. On the other hand, the high number of strandings observed in FMA II, which has lower abundance estimates (Sucunza et al., 2020), could be a result of a daily beach survey effort started in 2015 by the PMP-BS program (see a temporal distribution of the strandings section), and of the high levels of fishery mortality in FMA II (Cremer et al., 2018). The more regular distribution of records in FMA I, FMA II and FMA III is probably a consequence of regular monitoring carried out in these regions, different from what is observed in FMA IV, where greater effort has been made to the north.

The presence of one record in the central portion of FMA I and the absence of records in the southern portion of it is due to the two gaps where franciscana does not occur (Amaral et al., 2018).

Temporal distribution of the strandings

Generally, all FMAs had an increase in stranding number at the beginning of the 2000s, except in FMA II, which started in 2015 (Fig. 13.2). These patterns are likely related to the increase in public awareness and/or logistical improvement of the institutions in reporting stranding data. For FMA II, it is visible that the systematic monitoring started in 2015 by the PMP-BS program profoundly changed the existing knowledge regarding the mortality of the species in this region, demonstrating the importance of regular monitoring as a tool for assessing the mortality of the species (Fig. 13.2). The number of records, which was no more than 50 animals per year until 2014, jumped to an average of more than 500 individuals per year from 2015 to 2016, remaining at this high level since then.

In all FMAs autumn and spring were the seasons with the lowest and highest number of franciscana records respectively, with the exception of FMA IV, whereas the summer and winter were the periods with the highest and lowest number of records, respectively (Fig. 13.3).

When we analyze the seasonal frequency of strandings in a comparative way for each degree of latitude, it is possible to observe that in the

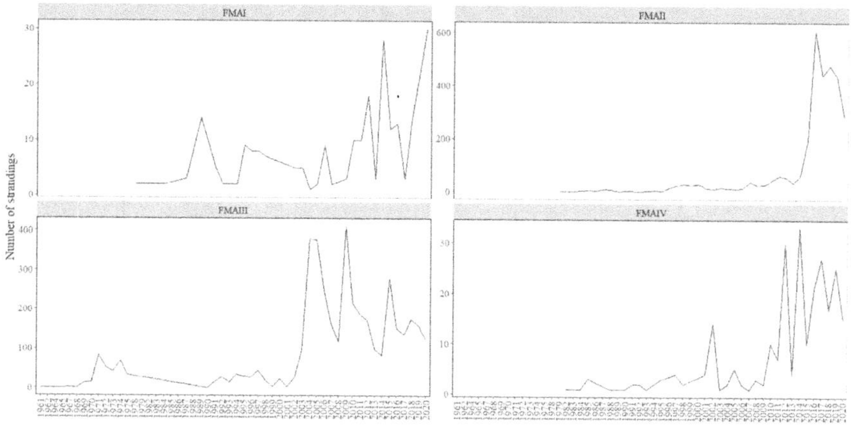

Fig. 13.2 ***Interannual strandings of franciscana,*** **Pontoporia blainvillei,** ***in the different franciscana management areas*** **(FMAs)** ***from 1961 to 2020.*** All stranding records with dates and locations were used ($n = 7,160$). FMA I = 160; FMA II = 2,985; FMA III = 3,541; FMA IV = 474.

north portion of FMA I (17°S) the highest values occurred in November, while in the center (18°S) and south (19°S) occurred in January and February, respectively (Fig. 13.4). Between FMA II and FMA IV, there is a trend in the number of strandings that advances over the months from north to south.

Despite the large variations in the registration of carcasses over time, it is possible to see that there is a tendency for an increase in the number of strandings in winter and spring, starting to decline with the arrival of summer (Fig. 13.5). The occurrence of a peak in early winter seems to be a trend for all decades, whereas the peak in spring occurred in the last two decades (2000–2020).

Inferring the factors that can be contributing to the seasonality of franciscana stranding is not a trivial task. Stranding is the result of at-sea mortality, drift patterns, and beach survey effort (Saavedra et al., 2017). Therefore, interpreting stranding records is difficult and must be analyzed carefully. Despite these constraints, some studies were able to link franciscana mortality and strandings. For instance, in the south of Rio Grande do Sul State (31o to 33oS), southern Brazil, the highest numbers of strandings in spring and summer are due to the high levels of franciscana by-caught in gillnet fisheries during the whitemouth croaker (*Migropogonia furnieri*) season (Secchi et al., 1997; Secchi et al., 2004; Prado et al., 2013, 2016, 2021). Although there are no studies relating franciscana by-catch and strandings

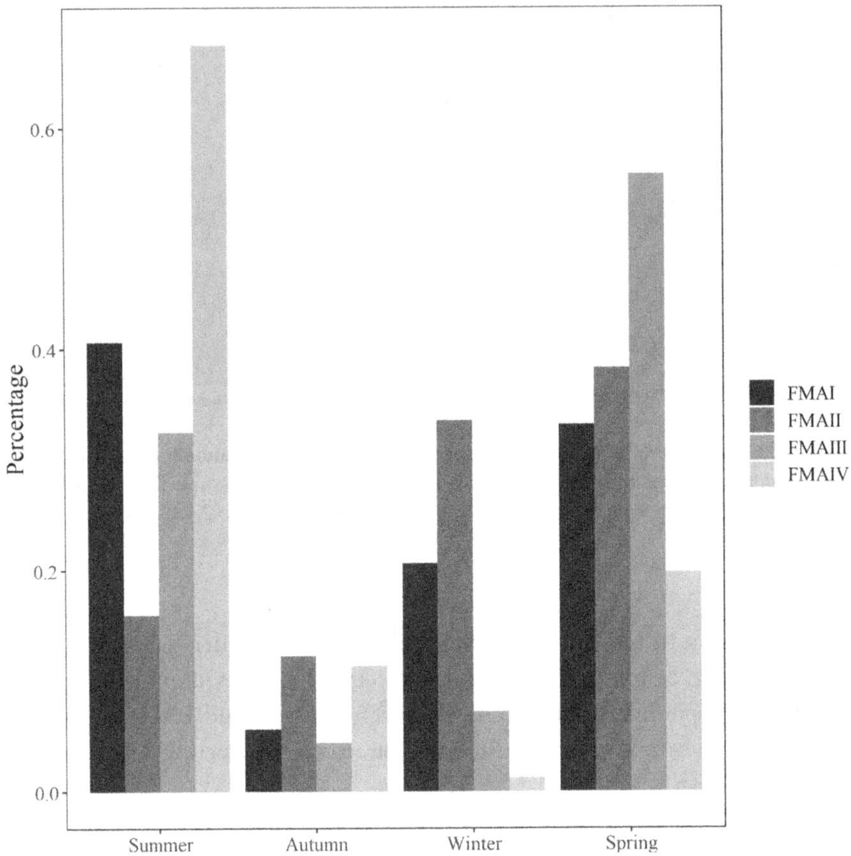

Fig. 13.3 ***Percentage of carcasses of franciscanas,* Pontoporia blainvillei, *recorded per season for each franciscana management area (FMA) from 1961 to 2020.*** All stranding records with dates and locations were used (n = 7,160). FMA I = 160; FMA II = 2,985; FMA III = 3,541; FMA IV = 474.

in Uruguay, the highest levels of franciscana mortality in gillnet and bottom trawl fisheries in this country in spring (Praderi, 1997; Franco-Trecu, et al. 2009; Franco-Trecu et al., 2019) seems to be contributing to the high numbers of franciscana strandings observed in FMA III in this season. For the remaining FMAs the franciscana by-caught estimates are punctual in space and time. Therefore, establishing a direct cause-effect relationship between fishery mortality and stranding patterns is not trivial. Future studies examining the carcasses along franciscana's distribution would be invaluable to identify its causes of mortality and help elucidate the potential drivers of franciscana stranding patterns.

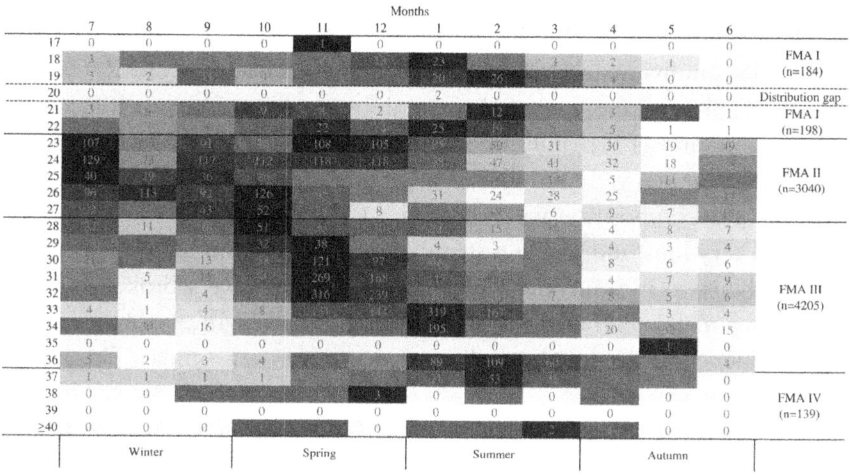

Fig. 13.4 ***Number of franciscana,*** **Pontoporia blainvillei*****, strandings grouped by month and latitude from 1961 to 2020, and respective seasons.*** Monthly values were compared within each latitude from white (lowest values) to black (highest values). All stranding records with date and location were used (n = 7,160). FMA I = 160; FMA II = 2,985; FMA III = 3,541; FMA IV = 474.

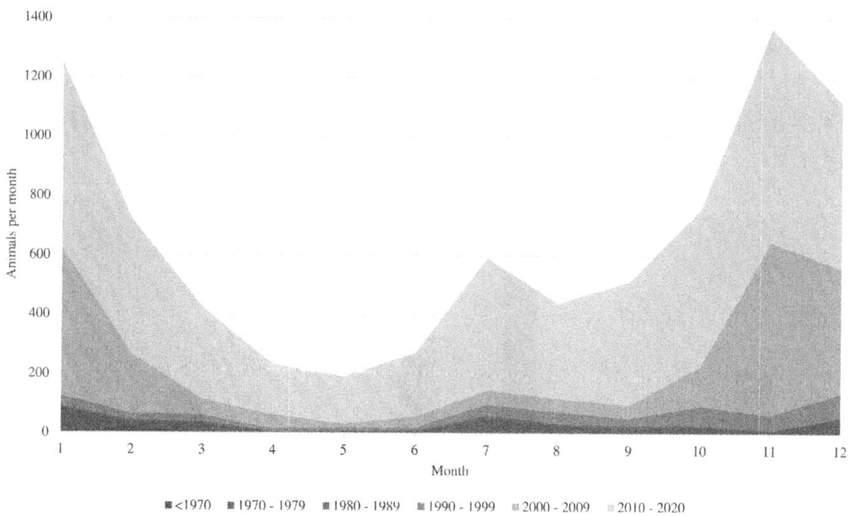

Fig. 13.5 ***Distribution of franciscana,*** **Pontoporia blainvillei*****, strandings per month stratified by decades from 1961 to 2020.*** All stranding records with dates were used from 1961 to 2020 (n = 7,160). FMA I = 160; FMA II = 2,985; FMA III = 3,541; FMA IV = 474. The number 1 corresponds to the month of January, and so on, until the number 12, which corresponds to December.

Characterization of the individuals

For 1,454 individuals (18%), TL and sex had been recorded (animals less than 70 cm were excluded). The absence of this information for the remaining animals is mainly due to the advanced stage of decomposition in which carcasses were found. For the 1,096 carcasses (15%) for which the degree of decomposition was recorded, 54.1% were code 4 or 5. The franciscanas very long and fragile rostrum also contributes, in many cases, to the impossibility to measure the TL, as many carcasses have a broken rostrum. Analyzing males and females by season within each FMA, we observed that the frequency of dead males and females, in general, is similar (Fig. 13.6).

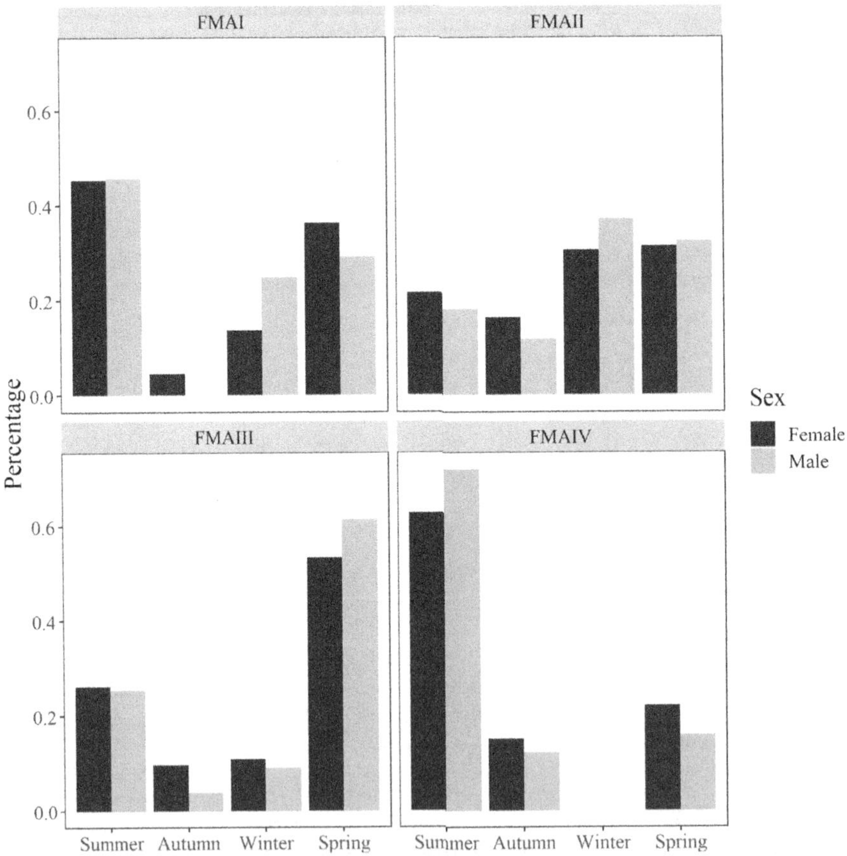

Fig. 13.6 ***Percentage of male and female franciscanas,*** **Pontoporia blainvillei,** ***found dead per season for each franciscana management area (FMA) (n = 1,607) from 1961 to 2020.*** FMA I: female = 22, male = 24; FMA II: female = 379, male = 454; FMA III: female = 178, male = 382; FMA IV: female = 86, male = 82. Records with TL less than 70 cm were excluded.

Table 13.3 Mean, standard deviation (SD), maximum (max), and minimum (min) of the total length (cm) of franciscana, *Pontoporia blainvillei,* males and females recorded in the different FMAs from 1961 to 2020. Records with a total length lower than 70 cm were excluded ($n = 400$).

			Total length			
FMAs	**Sex**	***N***	**Mean**	**SD**	**Minimum**	**Maximum**
FMA I	Female	22	109.8	21.3	74.0	138.2
	Male	24	109.6	12.6	85.0	136.0
FMA II	Female	379	116.4	21.6	70.0	165.0
	Male	454	108.4	15.2	70.0	173.3
FMA III	Female	176	129.6	24.8	72.2	170.0
	Male	231	120.3	18.6	70.0	167.0
FMA IV	Female	86	117.4	22.7	70.0	161.0
	Male	82	111.5	20.2	70.0	150.0

The average TL was higher in FMA III (129.6 cm for females; 120.3 cm for males), followed by FMA IV, II, and I (Table 13.3 and Fig. 13.7). This pattern agrees fairly well with previous studies, in which larger animals are observed in Rio Grande do Sul State, Brazil, Uruguay, and Argentina, followed by animals from FMA I, and smaller animals in São Paulo, Paraná, and Santa Catarina States, Brazil (Barbato et al., 2012; Barreto and Rosas, 2006; Kasuya and Brownell, 1979; Pinedo, 1991; Ramos and Di Beneditto, 2000). Differences between FMA I and FMA II may be biased related to the small sample size of FMA I (Table 13.3).

The pattern of females larger than males was observed in all FMAs (Table 13.3 and Fig. 13.7). Sexual dimorphism in relation to size is already well known for franciscanas, with females larger than males (Botta et al., 2010; Conversani et al., 2020; Kasuya and Brownell, 1979; Ramos et al., 2002).

Mortality is higher in females than in males, considering individuals with TC greater than 130 cm, with the exception of FMA III. For smaller individuals, we observed an opposite trend, with the exception of individuals smaller than 90 cm (Fig. 13.8).

Spatial variation in the mean length at sexual maturity is due to the marked interpopulation differences in body size (Secchi et al., 2021). However, despite these differences, we can observe high mortality of immature individuals in all FMAs, except for males in FMA III (Fig. 13.9). Studies that analyzed the age of dead animals, both stranded and rescued from incidental capture, indicate high mortality of young and immature individuals, at most two years old, throughout the species distribution (e.g., 74% in FMA I—Di Beneditto and Ramos, 2001; more than 80% in

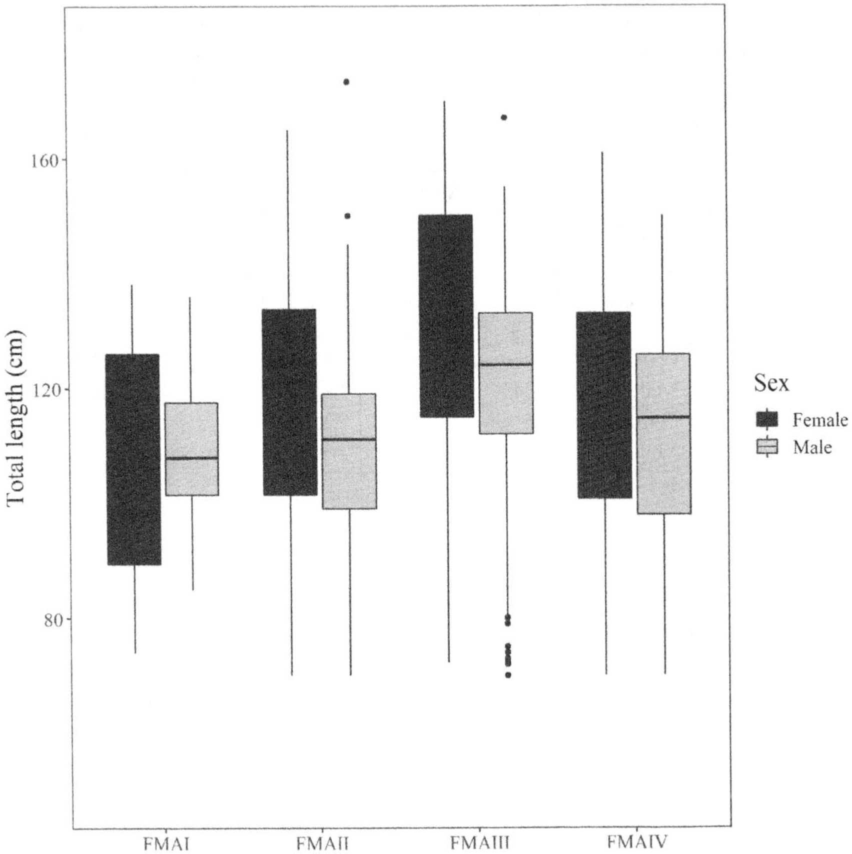

Fig. 13.7 ***Total length (cm) of male and female franciscanas,*** **Pontoporia blainvillei,** ***recorded in the different franciscana management areas*** **(FMAs)** ***from 1961 to 2020.*** The box represents the first and third quartiles, and the horizontal line indicates the median ($n = 1{,}607$) (FMA I: female = 22, male = 24; FMA II: female = 379, male = 454; FMA III: female = 178, male = 382; FMA IV: female = 86, male = 82). The dots represent outliers. Records with total length lower than 70 cm were excluded ($n = 400$).

FMA II—Conversani et al., 2020; 80% of the sample in FMA III—Botta et al., 2010 and 77.5% in Kasuya and Brownell, 1979; 92% in FMA IV—Denuncio et al., 2018).

The high mortality of juveniles is a common pattern of many long-lived marine vertebrate predators, mainly related to the insufficient development in dive and forage ability (e.g., Cox et al., 2019; Daunt et al., 2007; Orgeret et al., 2016). The period after the mother's care is then a critical phase of the life cycle of younger individuals. The reason for the higher mortality of immature individuals is not well understood in franciscanas and is

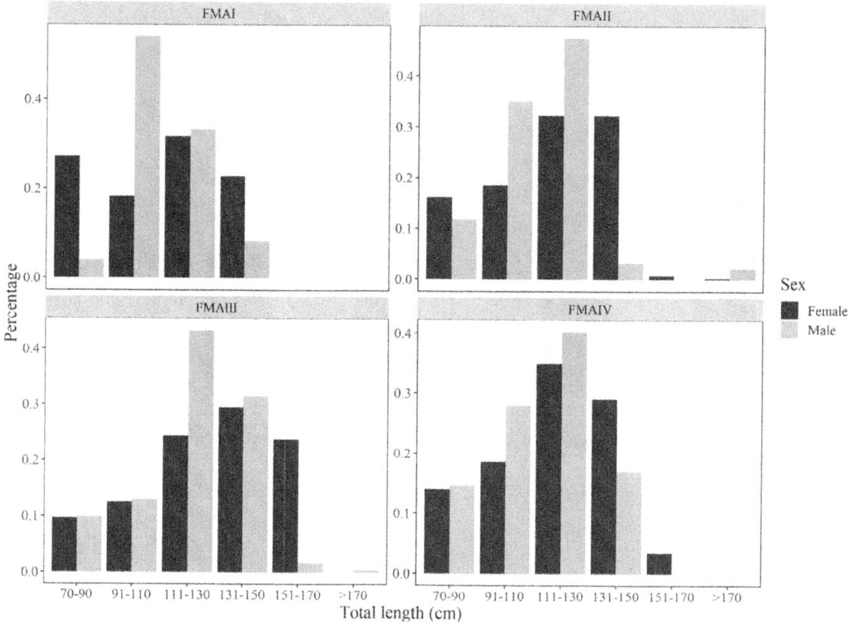

Fig. 13.8 ***Percentage of male and female franciscanas,*** **Pontoporia blainvillei,** ***found dead by total length (TL) (cm) category in the different franciscana management areas (FMAs) from 1961 to 2020 (n = 1,607; FMA I: female = 22, male = 24; FMA II: female = 379, male = 454; FMA III: female = 178, male = 382; FMA IV: female = 86, male = 82).*** Records with TL lower than 70 cm were excluded (n = 400).

probably a combination of factors. Frainer et al. (2015) associated the higher bycatch of young individuals with the peculiar development of the main biosonar structures. This hypothesis was supported by Mélcon et al. (2016), that observed lower frequency calls in the spectral domain of a newborn franciscana calf, which may be associated with the subdevelopment of structures related to sound production, suggesting that the capacity of younger individuals to detect the gillnets would be inferior to adults. Danilewicz et al. (2004) considered the possibility of an imbalanced age structure of the populations, with a low number of older individuals due to an extensive history of bycatch.

Live strandings

From 1991 to 2020, 159 strandings of live animals were recorded, 37% (n = 58) of them in FMA II, 35% (n = 39) in FMA IV, 28% (n = 46) in FMA III, and 0.01% in FMA I (n = 1).

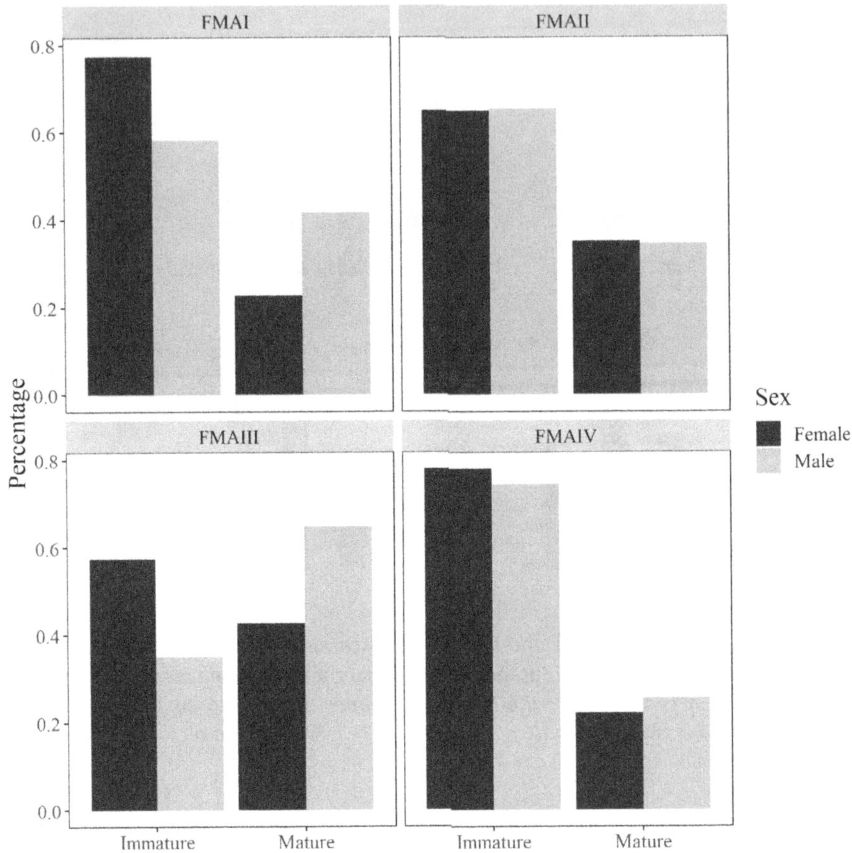

Fig. 13.9 ***Percentage of mature and immature franciscanas,*** **Pontoporia blainvillei,** ***both male and female, recorded in the different franciscana management areas (FMAs) from 1961 to 2020.*** For the definition of maturity was considered the mean length at sexual maturity, from information available in the literature ($n = 1607$; FMA I: female = 22, male = 24; FMA II: female = 379, male = 454; FMA III: female = 178, male = 382; FMA IV: female = 86, male = 82). Records with TL lower than 70 cm were excluded ($n = 400$).

The average TL by sex was similar between FMAs (Table 13.4), with the largest animal observed in FMA IV (155 cm) and the smallest in FMA II (56.5 cm). Of the total records in which sex and length were available, 54% were females and 46% were males. For 87 animals the body mass was recorded, with an average of 5 kg (SD = 4.4; maximum = 27 kg; minimum = 2.1 kg).

The live strandings occur mainly in spring for all FMAs which corresponds to the main birth period (Fig. 13.10) (Cordeiro, 2021; Danilewicz,

Table 13.4 Mean, standard deviation (SD), maximum (max), and minimum (min) of the total length (cm) of live stranded franciscanas, *Pontoporia blainvillei*, males and females recorded in the different franciscana management areas (FMAs) from 1961 to 2020 ($n = 159$).

FMA	Sex	*N*	Total length (cm)			
			Mean	SD	Maximum	Minimum
FMA I	Female	1	62.0	–	–	–
	Male	1	60.0	–	–	–
FMA II	Female	19	71.4	17.4	132.0	56.5
	Undetermined	1	66.0	–	–	–
	Male	14	71.0	18.6	113.0	55.0
FMA III	Female	6	71.4	11.0	91.0	65.0
	Undetermined	9	70.0	–	70.0	70.0
	Male	6	67.8	4.4	73.0	61.0
FMA IV	Female	32	72.6	19.6	155.0	58.5
	Male	18	74.0	15.9	119.0	63.5

SD, standard deviation.

2003; Denuncio et al., 2013). Of the total live records, 63% had less than 90 cm, indicating that they were possibly still offspring, with some degree of dependence on the mother for milk intake (Denuncio et al., 2013). The stranding of calves, in turn, can be related to the death of the mother, either by incidental capture in fishing nets or by any other factor.

Many animals were typically newborns, with the umbilical cord, or premature with poor body condition, and few or no gastric content. At necropsy, pulmonary alterations such as atelectasis in premature individuals and pneumonia were diagnosed. Skin lesions and lung congestion correlated to bycatch could also be seen.

Of the total number of live animals registered, four were immediately released, 46 survived less than one day, and only two were released after rehabilitation: one in São Paulo State in 2015 (Maranho and Farah, 2016) and another in Santa Catarina State, in 2017 (see Kolesnikovas, 2022). The first animal was 84 cm when rescued, and was released with 93 cm after 84 days in rehabilitation. The second one had 91 cm and was released after eight days in 2017. Both animals were eating by themselves when rescued. In São Paulo State, an animal rescued with less than 70 cm died after 82 days in captivity, and in Argentina, an adult female with 130 cm was maintained for 1,185 days in captivity and died after this period.

Until 2017, there were few institutions able to receive dolphins for rehabilitation in the area of occurrence of franciscanas. Since then, due to

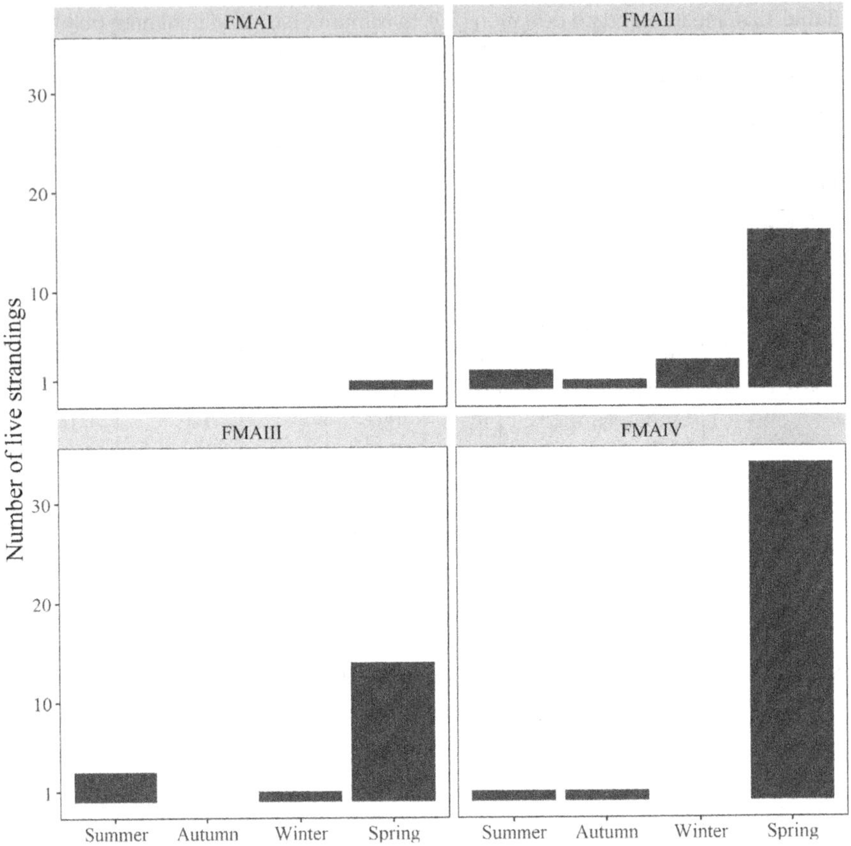

Fig. 13.10 ***Number of live franciscana,*** **Pontoporia blainvillei*****, strandings per season in the different franciscana management areas (FMAs) from 1991 to 2020 (n = 159).***

two great beach monitoring programs established in the south and southeast Brazil (PMP—BC; PMP—BS), six rehabilitation centers were built in this area, enabling better conditions for rehabilitation procedures in the north portion of FMA III, FMA II, and FMA I. Uruguay does not have a rehabilitation center for franciscana, but has recorded around ten animals from 2003 to 2021, most of them were released immediately and only three were maintained in makeshift structures for rehabilitation. On the north coast of Argentina, since 1991 live animals have been taken and rehabilitated by the Fundación Mundo Marino. Despite the establishment of new rehabilitation centers for cetaceans along the franciscana distribution, the rehabilitation of the species is still a great challenge, given its sensitivity to management (Kolesnikovas, 2022).

Final considerations

Strandings are an important tool to make inferences about the biology, ecology, and conservation of marine megafauna. The results presented in this chapter represent the first compilation of franciscana stranding throughout the entire distribution of the species.

Franciscana is a coastal species that occurs mainly up to the 25 m isobaths (Amaral et al., 2018), which can contribute greatly to the carcass stranding. Even so, Prado et al. (2013) estimated that only 10% of the franciscanas that die in the fishing activity strand on the beach in Rio Grande do Sul State, in Southern Brazil. There is a lack of estimates of the proportion of animals killed offshore that end up ashore for the entire franciscana distribution, unable to quantify the real magnitude of fishery mortality. Besides that, it is also difficult to identify marks of interaction with fishing gears in carcasses found on beaches due to the advanced state of decomposition (Conversani et al., 2020). However, the stranding data offer potential insights into the franciscana mortality patterns (Prado et al., 2016), as presented in this chapter.

Monitoring the beaches proved to be a very important strategy to assess the mortality of the franciscana, together with several other species, as demonstrated in the FMA II. Until August 2015, existing knowledge about the mortality of the species was incipient in this region, masking the real threat situation. The number of dead animals jumped from about 50 to more than 500 from one year to another. The fact that the coastline is very jagged, with several small beaches, always made it difficult to record strandings, and only with strong institutional support was this possible. The FMA III is where the longest-term record of strandings of the species exists, starting in the 1990s, and is the area with the highest population abundance (Sucunza et al., 2020). Industrial fishing is very intense in this region, being the main responsible for the mortality of the species. An average of 120 animals are stranded per year (1976–2013), which represents about 10% of animals caught incidentally in gillnets (Prado et al., 2013, 2016).

The number of carcasses registered in FMA I is a cause for great concern, as mentioned by considering the reduced abundance and isolation of the populations in this region. From 2013 to 2017, an average of five dead animals per year was recorded for Espírito Santo State. The *potential biological removal* was estimated in 0.7 individuals per year in this area (Sucunza et al., 2020).

Other impacts include those related to fishing activities, such as the depletion in fish stocks, specifically the species included in the franciscana diet, affecting nutritional resources necessary for the species' survival (Secchi et al., 2021). However, coastal development and the consequences of multiple anthropogenic activities expose franciscana dolphins to an array of impacts and might also be responsible for the morbidity and mortality of many individuals (see Domit et al., 2022, this book). Cumulative impacts might be affecting franciscanas' health condition, habitat quality, and survival, requiring a strong and coordinated management action, including national and international efforts, to guarantee the conservation of the species.

Acknowledgments

The authors are deeply grateful to all those who contributed to the collection of data in the field and in the laboratory, as well as to the institutions that contributed financially to the work. We would like to thank PETROBRAS for sponsoring the beach monitoring projects of the Santos (PMP-BS) and Campos (PMP-ES) basins.

References

Amaral, K.B., Danilewicz, D., Zerbini, A., Beneditto, Di, A.P., A., A., A., D., J., Secchi, E., Ferreira, E., Sucunza, F., Martins, M.B., de Oliveira Santos, M.C., Cremer, M., Denuncio, P., Ott, P., Moreno, I.B., 2018. eassessment of the franciscana *Pontoporia blainvillei* (Gervais & d'Orbigny, 1844) distribution and niche characteristics in Brazil. J. Exp. Mar. Biol. Ecol. 508, 1–12. doi:10.1016/j.jembe.2018.07.010.

Barbato, B.H.A., Secchi, E.R., Di Beneditto, A.P.M., Ramos, R.M.A., Bertozzi, C., Marigo, J., 2012. Geographical variation in franciscana (*Pontoporia blainvillei*) external morphology. J. Mar. Biol. Assoc. U.K. 92, 1645–1656. doi:10.1017/S0025315411000725.

Barreto, A.S., Rosas, F.C.W., 2006. Comparative growth analysis of two populations of *Pontoporia blainvillei* on the Brazilian coast. Mar. Mamm. Sci. 22, 644–653. doi:10.1111/j.1748-7692.2006.00040.x.

Bastida, R., Rodríguez, D., Secchi, E., da Silva, V., 2007. Mamíferos Marinos Sudamericanos. Vazquez Mazzini Editores, Buenos Aires.

Bertozzi, C.P., Zerbini, A.N., 2002. Incidental mortality of franciscana (*Pontoporia blainvillei*) in the artisanal fishery of Praia Grande, São Paulo State. Brazil. Lat. Am. J. Aq. Mamm. 1, 153–160.

Botta, S., Secchi, E.R., Muelbert, M.M., Danilewicz, D., Negri, M.F., Cappozzo, H.L., 2010. Age and growth of franciscana dolphins, *Pontoporia blainvillei* (Cetacea: Pontoporiidae) incidentally caught off southern Brazil and northern Argentina. J. Mar. Biol. Ass. UK 90, 1493–1500. doi:10.1017/s0025315410001141.

Bracarense, A.P.F.R.L., Domiciano., I.G., 2022. Pathologies and diseases. In: Simões-Lopes, P.C., Cremer, M.J. (Eds.), The Franciscana Dolphin. Elsevier, Cambridge.

Conversani, V., Silva, D., Barbosa, R.A., Hohn, A.A., Santos, M.C.O., 2020. Age and growth of franciscana, *Pontoporia blainvillei*, and Guiana, *Sotalia guianensis*, dolphins from southeastern Brazil. Mar. Mam Sci. 2020, 1–15.

Cordeiro, E., 2021. Biologia reprodutiva e padrões de crescimento de Pontoporia blainvillei (Mammalia, Cetacea), no litoral de Santa Catarina, Brasil. Master's Monograph. Universidade da Região de Joinville.

Cox, S.L., Authier, M., Orgeret, F., Weimerskirch, H., Guinet, C., 2019. High mortality rates in a juvenile free-ranging marine predator and links to dive and forage ability. Ecol. Evolution, 1–21. doi:10.1002/ece3.5905.

Cremer, M.J., Barreto, A.S., Castilho, P.V., Dick, J., Domit, C., Barbosa, C.B., et al. 2018. Franciscana Dolphin Mortality in Franciscana Management Area II (FMA II), Southern Brazil. SM WP03.

Crespo, E.A., Harris, G., González, R., 1998. Group size and distributional range of the franciscana, *Pontoporia blainvillei*. Mar. Mamm. Sci. 14, 845–849. doi:10.1111/j.1748-7692.1998.tb00768.x.

Crespo, E.A., 2002. Franciscana *Pontoporia blainvillei*. In: Perrin, W.P, Wursig, B., Thewissen, J.G.M. (Eds.), Encyclopedia of Marine Mammals. Academic Press, San Diego, CA, p. 484.

Crespo, E.A., Pedraza, S.N., Grandi, M.F., Dans, S.L., Graffo, G.V., 2010. Abundance and distribution of endangered Franciscana dolphins (*Pontoporia blainvillei*) in Argentine waters and conservation implications.. Mar. Mammal Sci. 26, 17–35. doi:10.1111/j.1748-7692.2009.00313.x.

Cunha, H.A., Medeiros, B.V., Barbosa, L.A., Cremer, M.J., Marigo, J., Lailson-Brito, J., Azevedo, A.F., Solé-Cava, A.M., 2014. Population structure of the endangered franciscana dolphin (*Pontoporia blainvillei*): reassessing management units. PLoS One 9, e85633. doi:10.1371/journal.pone.0085633.

Danilewicz, D.S., 2003. Reproduction of female franciscana (*Pontoporia blainvillei*) in Rio Grande do Sul, southern Brazil. Lat. Am. J. Aq. Mamm. 2, 67–78.

Danilewicz, D., Claver, J.A., Carrera, A.L.P., Secchi, E.R., Fontoura, N.F., 2004. Reproductive biology of male franciscanas (*Pontoporia blainvillei*) (Mammalia: cetacea) from Rio Grande do Sul, southern Brazil. Fish. Bull. 102, 581–592.

Danilewicz, D., Moreno, I.B., Ott, P.H., Tavares, M., Azevedo, A.F., Secchi, E.R., et al., 2010. Abundance estimate for a threatened population of franciscana dolphins in southern coastal Brazil: uncertainties and management implications. J. Mar. Biol. Assoc. UK 90, 1649–1657. doi:10.1017/s0025315409991482.

Danilewicz, D., Rosas, F., Bastida, R., Marigo, J., Muelbert, M., Rodriguez, D., et al., 2002. Special issue on the biology and conservation of franciscana. Lat. Am. J. Aq. Mamm. 1, 25–42. doi:10.31931/fmbc.v22i2.2019.25-25.

Danilewicz, D.S., Secchi, E.R., Ott, P.H., Moreno, I.B., Bassoi, M., Martins, M.B., 2009. Habitat use patterns of franciscana dolphin (*Pontoporia blainvillei*) off southern Brazil in relation to water depth. J. Mar. Biol. Assoc. UK 89, 943–949. doi:10.1017/s002531540900054x.

Daunt, F., Afanasyev, V., Adam, A., Croxall, J.P., Wanless, S., 2007. From cradle to early grave: Juvenile mortality in European shags *Phalacrocorax aristotelis* results from inadequate development of foraging proficiency. Biol. Lett. 3, 371–374.

Denuncio, P., Negri, M.F., Bastida, R., Rodriguez, D., 2018. Age and growth of Franciscana dolphins in northern Argentina. J. Mar. Biol. Assoc. UK 98, 119–1203. doi:10.1017/S0025315417000765.

Denuncio, P., Bastida, R., Danilewicz, D., Morón, S., Rodrigues-Heredia, S., Rodriguez, D., 2013. Calf chronology of the franciscana dolphin: birth, lactation and onset of feeding ecology in coastal waters of Argentina Aq. Mamm. 39, 73–80.

Denuncio, P., Bastida, R., Dassis, M., Giardino, G., Gerpe, M., Rodríguez, D., 2011. Plastic ingestion in Franciscana dolphins, *Pontoporia blainvillei* (Gervais and d'Orbigny, 1844), from Argentina. Mar. Pol. Bul. 62, 1836–1841.

Di Beneditto, A.P.M., Ramos, R.M.A., Lima, N.R.W., 1998. Fishing activity in northern Rio de Janeiro State (Brazil) and its relation with small cetaceans. Braz. Arch. Biol. Tech. 41, 296–302.

Di Beneditto, A.P.M., Ramos, R.M.A., 2001. Biology and conservation of the franciscana (*Pontoporia blainvillei*) in the north of Rio de Janeiro State. Brazil. J. Cetacean Res. Manage. 3, 185–192.

Di Tullio, J.C., Gandra, T.B.R., Zerbini, A., Secchi, E.R., 2016. Diversity and distribution patterns of cetaceans in the subtropical southwestern Atlantic outer continental shelf and slope. PLoS One 11 (5), e0155841. doi:10.1371/journal.pone.0155841.

Evans, K., Thresher, R., Warneke, R.M., Bradshaw, C.J.A., Pook, M., Thiele, D., Hindell, M.A., 2005. Periodic variability in cetacean strandings: links to large-scale climate events. Biol. Letters 1, 147–150.

Frainer, G., Huggenberger, S., Moreno, I.B., 2015. Postnatal development of franciscana's (*Pontoporia blainvillei*) biosonar relevant structures with potential implications for function, life history, and bycatch. Mar. Mamm. Sci. 31, 1193–1212. doi:10.1111/mms.12211.

Franco-Trecu, V., Costa, P., Abud, C., Dimitriadis, C., Laporta, P., Passadore, C., et al., 2009. Bycatch of franciscana *Pontoporia blainvillei* in Uruguayan artisanal gillnet fisheries: an evaluation after a twelve-year gap in data collection. Lat. Am. J. Aq. Mamm. 7, 11–22.

Franco-Trecu, V., Szephegyi, M.N., Doño, F., Forselledo, R., Reyes, F., Passadore, C., Crespo, E.A., Inchausti, P., 2019. Marine mammal bycatch by the industrial bottom trawl fishery at the Río de la Plata Estuary and the adjacent Atlantic Ocean. Lat. Am. J. Aquat. Res 47 (1), 89–101. doi:10.3856/vol47-issue1-fulltext-10.

Frizzera, F.C., Tosi, C., Pinheiro, H., Marcondes, M., 2012. Captura acidental de toninha (*Pontoporia blainvillei*) na costa norte do Espírito Santo. Brasil. Bol. Mus. Biol. Mello Leitão 29, 81–86.

GBIF.org, 2021. GBIF Occurrence Download. doi:10.15468/dl.24rcwj.

Geraci, J.R., Lounsbury, V.J., 2005. Marine Mammals Ashore—A Field Guide for Strandings, National Aquarium in Baltimore, Baltimore, pp. 1–371.

Guerrero, R., Piola, A.R., Fenco, H., et al., 2014. The salinity signature of the cross-shelf exchanges in the southwestern Atlantic Ocean: satellite observations. J. Geophys. Res. Oceans 119. doi:10.1002/2014JC010113.

Guinder, V.A., Tillmann, U., Krock, B., Delgado, A.L., Krohn, T., Cardona, J.E.G., Metfies, K., Abbate, C.L., Silva, R., Lara, R., 2018. Plankton multiproxy analyses in the Northern Patagonian Shelf, Argentina: Community structure, phycotoxins, and characterization of toxic Alexandrium strains. Front. Marine Sci. 5, 394.

Jepson, P.D., Arbelo, M., Deaville, R., Patterson, I.A.P., Castro, P., Baker, J.R., Degollada, E., Ross, H.M., Herráez, P., Pocknell, A.M., Rodríguez, F., Howie, F.E., Espinosa, A., Reid, R.J., Jaber, J.R., Martin, V., Cunningham, A.A., Fernández, A., 2003. Gas-bubble lesions in stranded cetaceans. Nature 425, 575–576.

Kasuya, T., Brownell, R.L., 1979. Age determination, reproduction, and growth of the franciscana dolphin, *Pontoporia blainvillei*. Sci. Rep. Whal. Res. Inst. 31, 45–67.

Kinas, P., 2002. The impact of incidental kills by gill nets on the franciscana dolphin (*Pontoporia blainvillei*) in Southern Brazil. Bull. Mar. Sci. 70 (2), 409–421.

Kirschvink, J.L., 1983. Biogenic Ferrimagnetism: A New Biomagnetism. In: Williamson, S.J., Romani, G.L., Kaufman, L., Modena, I. (Eds.), Plenum Publishing Corp., New York, NY, pp. 501–532.

Kolesnikovas, C.K.M., 2022. Rehabilitation of stranded specimens. In: Simões-Lopes, P.C., Cremer, M.J. (Eds.), The Franciscana Dolphin. Elsevier, Cambridge.

Lailson-Brito, J., Dorneles, P.R., Azevedo-Silva, C.E., Azevedo, A.F., Vidal, L.G., 2011. Organochlorine concentrations in franciscana dolphins, *Pontoporia blainvillei*, from Brazilian waters. Chemosphere 84, 882–887. doi:10.1016/j.chemosphere.2011.06.018.

Lucas, A.J., Guerrero, R.A., Mianzán, H.W., Acha, M., Lasta, C.A., 2005. Coastal oceanographic regimes of the Northern Argentine Continental Shelf (34-43 degrees S). Estuar. Coast. Shelf Sci. 65 (3), 405–420.

Maranho, A., Farah, R.F., 2016. First successful case of franciscana pontoporia blainvillei rehabilitation. XI Congreso de la Sociedad Latinoamericana De Especialistas Mamíferos Acuáticos, 2016 Valparaído, v. XI, p. 157.

Matano, R.P., Combes, V., Piola, A.R., Guerrero, R., Palma, E.D., Strub, P.T., James, C., Fenco, H., Chao, Y., Saraceno, M., 2014. The salinity signature of the cross-shelf exchanges in the southwestern Atlantic Ocean: numerical simulations. J. Geophys. Res. Oceans 119. doi:10.1002/2014JC010116.

Melcón, M.L., Failla, M., Iñíguez, M.A., 2016. Towards understanding the ontogeny of echolocation in franciscana dolphins (*Pontoporia blainvillei*). Mar. Mamm. Sci. 32, 1516–1521. doi:10.1111/mms.12336.

Méndez-Fernandez, P.M., Webster, L., Chouvelon, T., Bustamante, P., Ferreira, M., González, A.F., Fernandez, A.F., Moffat, C.F., Pierce, G.J., Read, F.L., Russell, M., Santos, M.B., Spitz, J., Vingada, J.V., Caurant, F., 2014. An assessment of contaminant concentrations in toothed whale species of the NW Iberian Peninsula: Part II. Trace element concentrations. Sci. Total Environ. 484, 206–217.

Monteiro, D.S., Estima, S.C., Gandra, T.B.R., Silva, A.P., Bugoni, L., Swimmer, Y., Seminoff, J.A., Secchi, E.R., 2016. Long-term spatial and temporal patterns of sea turtle strandings in southern Brazil. Mar. Biol. 163, 247.

Moller, O.O., Piola, R.A., Freitas, A.C., Campos, E.J.D., 2008. The effects of river discharge and seasonal winds on the shelf off southeastern South America. Cont. Shelf Res. 28, 1607–1624.

Negri, M.F., Denuncio, P., Panebianco, M.V., Cappozzo, H.L., 2012. Bycatch of franciscana dolphins *Pontoporia blainvillei* and the dynamic of artisanal fisheries in the species' southernmost area of distribution. Braz. J. Ocean 60, 149–158. doi:10.1590/s1679-87592012000200005.

Norris, K.S., 1961. Standardized methods for measuring and recording data on the smaller cetaceans. J. Mamm. 42, 471–476.

Orgeret, F., Weimerskirch, H., Bost, C.A., 2016. Early diving behaviour in juvenile penguins: improvement or selection processes. Biology Lett. 12, 20160490. doi:10.1098/rsbl.2016.0490.

Ortega, L., Martinez, A., 2007. Multiannual and seasonal variability of water masses and fronts over the Uruguayan shelf. J. Coast. Res. 23, 625–629.

Ott, P.H., Secchi, E.R., Moreno, I.B., Danilewicz, D., Crespo, E.A., Bordino, P., Ramos, R., Beneditto, Di, Bertozzi, A.P., Bastida, C., Zanelatto, R., Perez, J., Kinas, P.G., 2002. Report of the working group on fishery interactions. Lat. Am. J. Aq. Mamm. 1, 55–64.

Palma, E.D., Matano, R.P., Piola, A.R., 2008. A numerical study of the Southwestern Atlantic Shelf circulation: stratified ocean response to local and offshore forcing. J. Geoph. Res. 113, C11010.

Panebianco, M.V., Negri, M.F., Cappozzo, H.L., 2012. Reproductive aspects of male franciscana dolphins (*Pontoporia blainvillei*) off Argentina. Anim. Rep. Sci. 131, 41–48. doi:10.1016/j.anireprosci.2012.02.005.

Panebianco, M.V., Negri, M.F., Botte, S.E., Marcovecchio, J.E., Cappozzo, H.L., 2013. Essential and non-essential heavy metals in skin and muscle tissues of franciscana dolphins (*Pontoporia blainvillei*) from the southern Argentina coast. Chem. Ecol. 29 (6), 511–518. doi:10.1080/02757540.2013.810727.

Panebianco, M.V., Del Castillo, D.L., Denuncio, P.E., Negri, M.F., Bastida, R., Failla, M., Vitullo, A.D., Cappozzo, H.E., 2016. Reproductive biology of female franciscana dolphins (*Pontoporia blainvillei*) from Argentina. J. Mar. Biol. Assoc. UK 96, 831–840.

Peltier, H., Dabinb, W., Daniel, C.P., Vancanneyt, O., Doremush, G., Huon, M., Ridoux, V., 2012. The significance of stranding data as indicators of cetacean populations at sea: Modelling the drift of cetacean carcasses. Ecol. Indic. 18, 278–290. doi:10.1016/j.ecolind.2011.11.014.

Pinedo, M.C., Polacheck, T., 1999. Trends in franciscana (*Pontoporia blainvillei*) stranding rates in Rio de Grande do Sul, Southern Brazil (1979–1998). J. Cetacean Res. Manage. 1 (2), 179–189.

Pinedo, M.C., 1991. Development and Variation of the Franciscana *Pontoporia blainvillei* Ph.D. thesis. University of California, Santa Cruz.

Pinheiro, F.C.F., Pinheiro, H.T., Teixeira, J.B., Martins, A.S., Cremer, M.J., 2019. Opportunistic development and environmental disaster threat franciscana dolphins in the Southeast of Brazil. Trop. Cons. Sci. 12, 1940082919847886.

Pinheiro, L., Cremer, M.J., 2003. Etnoecologia e captura acidental de golfinhos (Cetacea: Pontoporiidae e Delphinidae) na Baía da Babitonga, Santa Catarina. Desenvolvimento e Meio Ambiente 8, 69–76.

Polizzi, P., Romero, M., Boudet, L., Das, K., Denuncio, P., Rodríguez, D., Gerpe, M.S., 2014. Metallothioneins pattern during ontogeny of coastal dolphin, *Pontoporia blainvillei*, from Argentina. Mar. Pollut. Bull. 80, 275–281. doi:10.1016/j.marpolbul.2013.10.037.

Praderi, R., Pinedo, M.C., Crespo, E.A., 1989. Conservation and management of *Pontoporia blainvillei* in Uruguay, Brazil and Argentina. In: Perrin, W.F., Brownell, R.L., Kaiyaand, Z., Jiankang, L. (Eds.), Biology and Conservation of the River Dolphins, Occasional Papers. IUCN SSC 3, Gland, pp. 52–56.

Praderi, R., 1997. Análisis comparativo de estadísticas de captura y mortalidad incidental de *Pontoporia blainvillei* en Uruguay durante 20 años. In: Pinedo, M.C., Barreto, A.S. (Eds.), Anais do 2 ° Encontro sobre Coordenação de Pesquisa e Manejo da Franciscana. Fundação Universidades do Rio Grande, Rio Grande, pp. 42–53.

Prado, J.H.F., Secchi, E.R., Kinas, P.G., 2013. Mark-recapture of the endangered franciscana dolphin (*Pontoporia blainvillei*) killed in gillnet fisheries to estimate past bycatch from time series of stranded carcasses in southern Brazil. Ecol. Indicators. 32, 35–41. doi:10.1016/j.ecolind.2013.03.005.

Prado, J.H.F., Mattos, P.H., Silva, K.G., Secchi, E.R., 2016. Long-term seasonal and interannual patterns of marine mammal strandings in subtropical western South Atlantic. PLoS One 11, e0146339. doi:10.1371/journal.pone.0146339.

Quirós, Y.B., Hartwick, M., Rotstein, D.S., Garner, M.M., Bogomolni, A., Greer, W., Niemeyer, M.E., Early, G., Wenzel, F., Moore, M., 2018. Discrimination between bycatch and other causes of cetacean and pinniped stranding. Diseases Aquatic Organ. 127, 83–95.

Ramos, R.M.A., Di Beneditto, A.P.M., Lima, N.R.W., 2000. Growth parameters of *Pontoporia blainvillei* and *Sotalia fluviatilis* (Cetacea) in northern Rio de Janeiro, Brazil. Aq. Mamm. 26 (1), 65–75.

Ramos, R.M.A., Di Beneditto, A.P., Siciliano, S., Santos, M., Zerbini, A., Bertozzi, C., Vicente, A., Zampirolli, E., Siqueira, R., De Vivo, M., Lima, N.R.W., 2002. Morphology of franciscana (*Pontoporia blainvillei*) off southeastern Brazil: sexual dimorphism,growth and geographic variation. Latin Am. J. Aquatic Mammals 1 (1), 129–144.

Reid, J.L., Nowlin, W.D.J., Patzert, W.C., 1977. On the characteristics and circulation of the Southwestern Atlantic Ocean. J. Phy. Oceanog. 7, 62–91.

Rosas, F.C.W., Monteiro-Filho, E.L.A., 2002. Reproductive parameters of *Pontoporia blainvillei* (Cetacea, Pontoporiidae), on the coast of São Paulo and Paraná States. Brazil. Mammalia 66, 231–245.

Rosas, F.C.W., Monteiro-Filho, E.L.A., Oliveira, M.R., 2002. Incidental catches of franciscana (*Pontoporia blainvillei*) on the southern coast of São Paulo State and the coast of Paraná State. Brazil. Lat. Am. J. Aq. Mamm. 1, 161–168.

Saavedra, C., Pierce, G.J., Gago, J., Jusufovski, D., Cabrero, A., Cerviño, S., López, A., Martínez-Cedeira, J.A., Begoña Santos, M., 2017. Factors driving patterns and trends in strandings of small cetaceans. Mar. Biol. 164, Article no. 165. doi:10.1007/s00227-017-3200-3, 164/165.

Seabra, C.B., Ribeiro, V.L., Bertozzi, C.P., Junior, I.R.B., Luiz, V.S., Valle, R.R., 2018. Ingestão de resíduos sólidos por toninhas (Pontoporia blainvillei) do litoral centro-sul de São Paulo, Brasil. 4° Congresso Latino Americano de Reabilitação de Fauna Marinha, p. 50–52.

Secchi, E.R., Zerbini, A.N., Bassoi, M., Dalla Rosa, L., Möller, L.M., Rocha-Campos, C.C., 1997. Mortality of franciscanas, *Pontoporia blainvillei*, in coastal gillnetting in southern Brazil: 1994–1995. Rep. Int. Whal. Commn. 47, 653–658.

Secchi, E.R., Ott, P.H., Crespo, E.A., Kinas, P.G., Pedraza, S.N., Bordino, P., 2001. A first estimate of franciscana (*Pontoporia blainvillei*) abundance off southern Brazil. J. Cet. Res. Manag. 3, 95–100.

Secchi, E.R., Danilewicz, D., Ott, P.H., 2003. Applying the phylogeographic concept to identify franciscana dolphin stocks: implications to meet management objectives. J. Cet. Res. Manag. 5, 61–68.

Secchi, E.R., Ott, P.H., Danilewicz, D.S., 2003. Effects of fishing by-catch and conservation status of the franciscana dolphin, Pontoporia blainvillei. In: Gales, N., Hindell, M., Kirkwood, R. (Eds.), Marine Mammals: Fisheries, Tourism and Management Issues. CSIRO Publishing, Collingwood, pp. 174–191.

Secchi, E.R., Kinas, P.G., Muelbert, M., 2004. Incidental catches of franciscana in coastal gillnet fisheries in the Franciscana Management Area III: period 1999-2000. Lat. Am. J. Aq. Mamm. 3, 61–68.

Secchi, E.R., Cremer, M.J., Danilewicz, D., Lailson-Brito, J., 2021. A synthesis of the ecology, human-related threats and conservation perspectives for the endangered franciscana dolphin. Front. Marine Sci. 8. doi:10.3389/fmars.2021.617956.

Silva, D.F., Barbosa, R.A., Conversani, V.R.M., Botta, S., Hohn, A.A., Santos, M.C.O., 2020. Reproductive parameters of franciscana dolphins (*Pontoporia blainvillei*) of Southeastern Brazil. Mar. Mamm. Sci. 36, 1291–1308. doi:10.1111/mms.12720.

SIMBA (Sistema Integrado de Monitoramento da Biota Aquática), 2021. Available at: https://simba.petrobras.com.br/simba. (Accessed April 18, 2021).

SIMMAM (Sistema de Apoio ao Monitoramento de Mamíferos Marinhos), 2021. Available at: http://simmam.acad.univali.br/site/. (Accessed October 30, 2021).

Sucunza, F., Danilewicz, D., Andriolo, A., Azevedo, A.F., Secchi, E.R., Zerbini, A.N., 2020. Distribution, habitat use, and abundance of the endangered franciscana in southeastern and southern Brazil. Mar. Mamm. Sci. 36, 421–435. doi:10.1111/mms.12650.

Trinchin, R., Ortega, L., Barreiro, M., 2019. Spatiotemporal characterization of summer coastal upwelling events in Uruguay, South America. Regional Studies Marine Sci. 31, 100787.

Vélez-Rubio, G.M., Estrades, A., Fallabrino, A., Tomás, J., 2013. Marine turtle threats in Uruguayan waters: insights from 12 years of stranding data. Marine Biol. 160 (11), 2797–2811.

Yochum, N., Starr, R.M., Wendt, D.E., 2011. Utilizing fishermen knowledge and expertise: keys to success for collaborative fisheries research. Fisheries 36 (12), 593–605.

Zerbini, A.N., Secchi, E., Crespo, E., Danilewicz, D., Reeves, R., 2017. Pontoporia blainvillei (errata version published in 2018). The IUCN Red List of Threatened Species 2017. doi: doi:10.2305/IUCN.UK.20173.RLTS.T17978A50371075.en. Downloaded on April 18, 2021.

CHAPTER 14

Rehabilitation of stranded specimens

Cristiane Kiyomi Miyagi Kolesnikovas
Associação R3 Animal, Florianópolis, Santa Catarina, Brazil

Introduction

Rehabilitation of wild animals is an important tool to understand their basic biology, physiology, and behavior. It also contributes to species conservation. For franciscanas, a single successful rehabilitation would contribute a lot, since most stranding cases occur due to bycatch (Rocha-Campos et al., 2010), putting species at risk of extinction.

Franciscana's rehabilitation still is a big challenge as most stranded animals are newborns that survive for only a few hours or days. Until now there are only two records of juvenile franciscanas that were successfully rehabilitated and released in Brazil (Maranho, 2016; Kolesnikovas, data not published): a juvenile that has been kept for 90 days in rehabilitation and another one that was released after 8 days in rehabilitation. One adult has been kept in captivity for three years in Mundo Marino, Argentina (Sergio Heredia, personal communication).

Recently, the International Union for Conservation of Nature has proposed the one plan approach (OPA) for endangered species conservation. This plan has joined many efforts of integrated and united plans under the same title (Byers, 2014). Traditionally, species conservation planning follows two parallels, but separate tracks (www.cpsg.org), the *ex-situ* and *in situ* strategies. OPA has put conservationists that monitor the population and develop conservation strategies together with the zoo's and aquarium's community to have plans that link wild and captive management.

Recent efforts to capture vaquitas (*Phocoena sinus*) and have them in captivity to protect the species from bycatch (Rojas-Bracho, 2019) showed that rehabilitating and keeping animals in captivity before a critical situation of conservation status is needed. In 2017, an international effort to protect the remaining vaquitas took place in Baja California–Mexico. The goal was to capture the animals and keep them in captivity to learn more about the species and, if successful, breed them and have a reintroduction

The Franciscana Dolphin
DOI: https://doi.org/10.1016/B978-0-323-90974-7.00011-2

plan. Prior to the initiation of the Vaquita CPR project and facilities were built, there were no records of capture, handling, or housing vaquitas. Two animals were captured, one juvenile female that showed signs of stress and was immediately released, and an adult female, that suffered cardiac arrest after four hours from capture (Rojas-Bracho, 2019).

In China, the effort to save the Yangtze finless porpoise *Neophocaena asiaeorientalis* ssp. *asiaeorientalis)*, a critically endangered species, started in 1990 with the relocation of some animals into a 21 km long natural *ex-situ* reserve, the Tian-E-Zhou Oxbow (Nabi et al., 2020). Currently, there are more than 60 animals in the reserve and the first porpoise born in captivity was released in nature in 2020 (Chinadaily, 2020). All years of work and hard effort were important to understand animals' physiology and behavior, as well as how to keep them safe in captivity for conservation purposes. In an effort to contribute to franciscana´s conservatio reasearchers from all over th world have created The Alliance for Franciscana-Dolphin Conservation, Research, Rescue and Rehabilitation. It first effort was to publish the Neonatal Care and Hand-rearing Protocol-Rescue and Rehabiltiation of Franciscana Dolphins (Meegan et al, 2022).

Since 2019 reasearchers from all over the world have been working on specific protocols for franciscana rehabilitation. The Alliance for Franciscana-dolphin has been created in order to promote its conservation, research, rescue and rehabiitation and in 2022 published a protocol for "Rescue and Rehabilitation of Franciscana Dolphins" (Meegan et al., 2022).

Our main goal is to rehabilitate animals and take them back to nature, learn the most about the species so that we will be able to contribute to franciscanas's conservation.

Strandings

Cetaceans can strand from natural and anthropogenic causes. Among natural causes, malnutrition, disease, trauma, predation, maternal separation, poor navigation (due to weather or other condition), and exposure to biotoxins can be cited (Simeone and Moore, 2018). Anthropogenic causes include entanglement, entrapment, entrainment (i.e., power plant intake), exposure to noise (i.e., seismic surveys), ingestion of debris, vessel strikes, gunshots, trauma because of direct human aggression, and chemical contamination (i.e., oil spill) (Simeone and Moore, 2018).

As written before, bycatch is the main cause of stranding and death of francicanas. Every year, hundreds or even thousands of animals are found at

Brazilian beaches because of bycatch (Rocha-Campos et al., 2010). When the rescue team arrives, in most cases, the animal is already dead.

Rescue and transport

When a live cetacean strand, the immediate release is an option if there is adequate logistic support, favorable beach and environmental conditions, no risks to wild populations or public safety, and if it is a juvenile that does not depend on the mother anymore or a juvenile/adult. The animal must also be healthy and able to function normally, and if the veterinarian considers that the animal has no major trauma and it is in good condition after clinical exams (Geraci and Lounsbury, 2005). If it is a calf that depends on the mother for feeding and the mother is not around or a sick/injured animal, the rescue must be considered. If the initial assessment indicates that the animal is suffering or it has injuries incompatible with life, euthanasia will be the choice. Among euthanasia indications, it can be cited: serious locomotors injuries; wounds in a large percentage of the body or that penetrates abdominal and/or thoracic cavity; extensive skin blistering or damage due to scavenger's predation; extensive lesion in eyes or blowhole; significant hemorrhage from the anus, genital opening, blowhole, tongue, eyelids, and eyes; neurological abnormalities; prolonged hypo or hyperthermia and persistent restrandings (Geraci and Lounsbury, 2005; Harms et al., 2018).

It is also important to consider if there is an available facility ready to receive a live dolphin. If there is a facility with trained professionals, adequate pool to the animal's size with saltwater, rescue can take place.

People at the stranding site must be oriented to keep animals comfortable and minimize stress. It is necessary to avoid animal handling, protect eyes, and blowhole from blowing sand, avoid water to go into the blowhole, minimize noise, dig holes in the sand to allow flippers and flukes to lie in a natural position, protect skin with wet towels, and provide shadow (Geraci and Lounsbury, 2005). It is also important to ask for a picture of the animal to have an idea of the animal's size and body condition and prepare all the material that will be necessary for the rescue. Always remember to instruct the rescue crew not to sniff or breathe the air exhaled from the blowhole, as wild animals can carry organisms pathogenic to man (Gage, 2001).

Depending on the size of the animal, stretchers, foam pads, mattresses, and towels will be needed. Wet towels can be placed on the animal's body to keep skin wet and avoid lesions, but do not put them on the blowhole and eyes. Zinc oxide or anhydrous lanolin can be applied on uncovered skin, such as melon, and around the blowhole to protect the skin (Gage, 1990).

Once the rescue team arrives, the veterinary will make an initial assessment: state of consciousness, reflexes response, presence of injuries, respiratory and cardiac rate. Respiratory rates in small cetaceans should be recorded for 5 min rather than 1 min. Normal respiratory rates are 5–25 breaths in 5 min, rates lower or above it would be considered abnormal. The character of the respiration is particularly important to evaluate, with complete blowhole opening and closing, and full inspiration and expiration with a thoracic excursion (Van Bonn, 2007). Normal heart rate for franciscanas is 99–139 beats/min (Baldassin et al., 2007). For the Atlantic bottlenose dolphins, the normal heart rate varies from 40 to 130 bpm. Small cetaceans have pronounced sinus arrhythmia, which corresponds to the respiratory cycle. The heart rate is higher immediately after the breath and slows until the next (Van Bonn, 2007). If the animal is not in good condition, immediate treatment must be done to stabilize it before transport.

Newborns size varies from 70 to 80 cm (Di Beneditto et al., 2010), but smaller ones have been rescued, that is, a female with 68 cm was rescued in Santa Catarina (Fig. 14.1), Brazil, in 2016. They can be rescued by only one person and transportation must be made on a foam pad or mattress. It

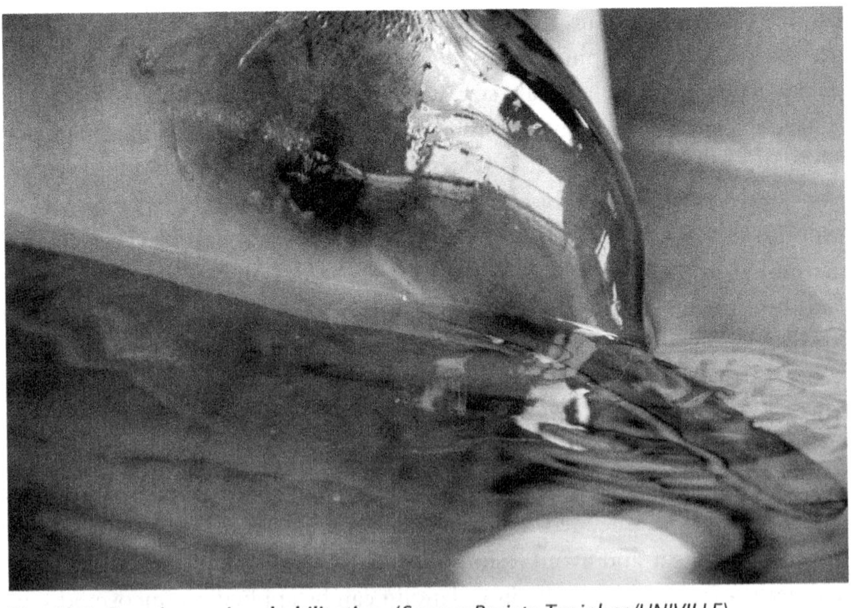

Fig. 14.1 ***Franciscana in rehabilitation.*** *(Source: Projeto Toninhas/UNIVILLE).*

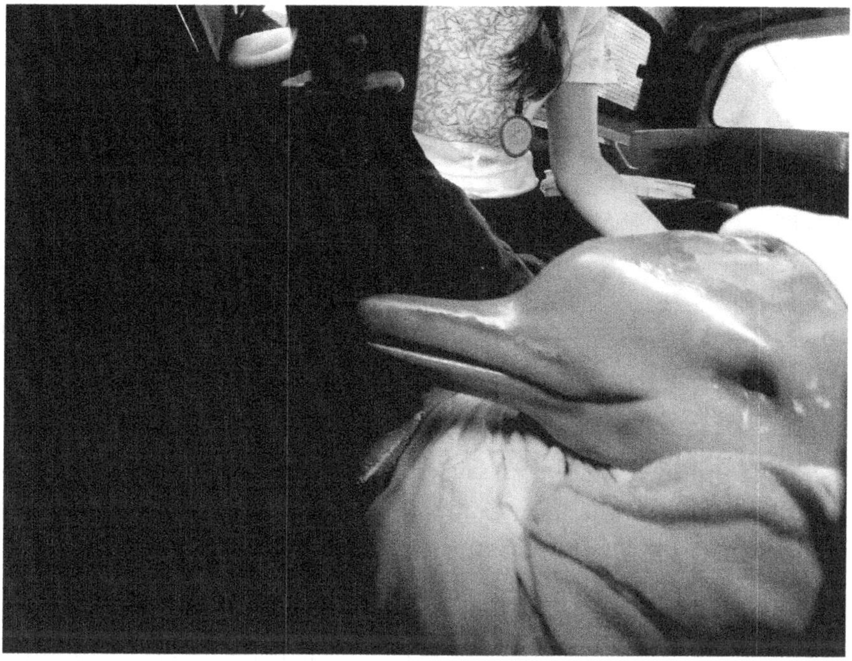

Fig. 14.2 ***Transport in truck bodywork with constant veterinary support.*** *(Source: Associação R3 Animal).*

has been noted that some newborns get calmer if on the lap once in the car (Sergio Heredia, personal communication) (Fig. 14.2).

If it is a juvenile or adult, a stretcher will be the best choice. Placing a dolphin on a stretcher is done by rolling up half the stretcher and pushing the rolled edge close to the side of the animal. The animal, then, can be pushed over the roll onto the flat side of the stretcher. Care must be taken with pectoral flippers that must be placed flat against the side of the body to avoid injuries (Gage, 1990). Once on the stretcher, the animal can be lifted by hand and transported to the vehicle. The juvenile/adult must be put on a foam pad or mattress in a truck's bodywork or a vehicle adapted to cetaceans' transport (Figs. 14.3 and 14.4). If it is available, the use of a water container is indicated to minimize stress. In this case, the eyes must be under the water and blowhole, outside.

It is important to plan and know the route in advance to avoid any unexpected events. Silence must be kept all over the time to minimize stress.

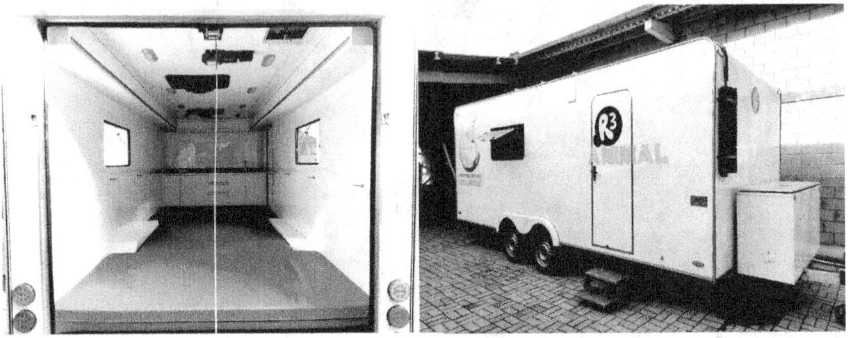

Fig. 14.3 ***Adapted vehicle for cetacean rescue.*** *(Source: Associaçao R3 Animal).*

Clinical evaluation and first aid

Because bycatch is the main cause of stranding and death, the veterinary staff must have in mind that treatment for stress, drowning and water aspiration may be necessary on-site before transportation.

Health evaluation follows some criteria as quality of respiration, heart rate, absence of substantial eye, blowhole or integument trauma, minimal flatulence,

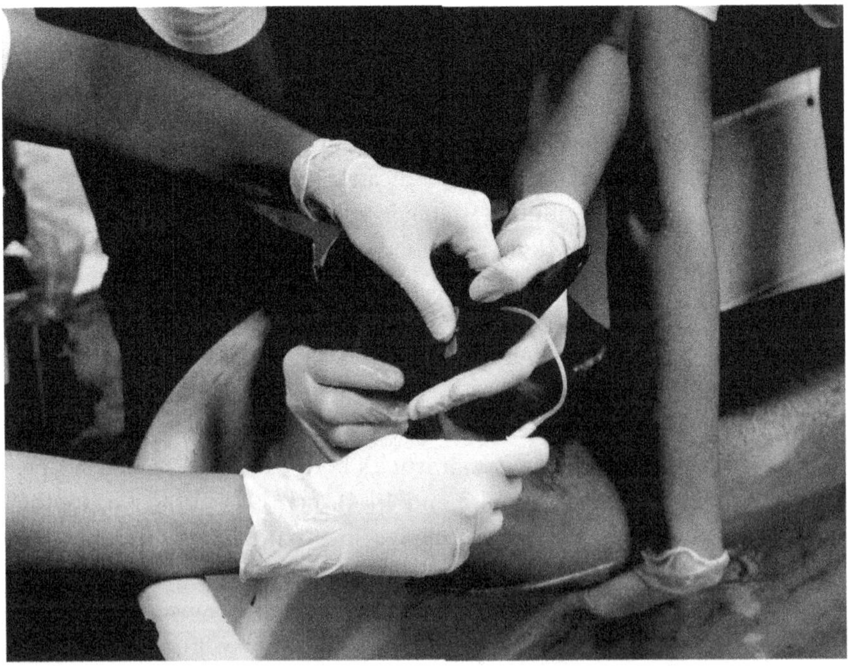

Fig. 14.4 ***Blood collection from periarterial venous rete.*** *(Source: Associação R3 Animal).*

belching or gaseous feces, minimal thrashing, lack of opaque colored or foamy discharge from the blowhole, palpebral reflexes and menace, sufficient muscle, and blubber mass. The animal must be alert and responsive, with no seizures/ convulsions, body arching, or rigidity (Marigo, 2007; Sharp et al., 2016).

Blood glucose levels must be monitored constantly. A stranded cetacean can be assumed to be dehydrated and hypoglycemic (Nollens et al., 2018) and fasting levels of blood glucose in marine mammals tend to be higher than in domestic mammals (>100 mg/dL) (Bossart et al., 2001). A study carried with 47 *Tursiops truncatus* calves showed that glucose levels in wild animals vary from 80 to 153 mg/dL; for juveniles (n = 101), the values varied from 66 to 141 mg/dL (Scwacke et al., 2009). Since there are no published data about glucose levels in franciscanas, it can be assumed that animals with levels below the ones for T. *truncatus* can also be considered hypoglycemic. Glucose administration can be made orally at a concentration of 5%–25%, 2 mL/Kg, diluted in 0.9% NaCl or Ringer Lactate (Bossart et al., 2004). The concentration of the solution will depend on the glucose level. If the animal is hypoglycemic, glucose levels must be checked twice a day (Gage and Walsh, 2018).

Hydration status is also important to be checked. A dehydrated animal will have high hematocrit (>55%) and high total serum protein (Walsh and Gearhart, 2001). The fluid requirement for marine mammals has been estimated to be 40 mL/Kg/day (Walsh and Bossard, 2001) and can also be given orally. Hypoglycemia often accompanies hypothermia. If a cetacean is hypothermic, weak, and rapidly declining, the probability of hypoglycemia is high (Nollens, 2018). In newborns, it is indicated to hydrate the animal every hour with small volumes to avoid regurgitation (Meegan et al., 2022).

The first clinical evaluation takes place as soon as the animal arrives at the rehabilitation facility. Collecting minimal parameters to avoid more stress to the animal includes body weight and temperature (36–38°C) and external exam. At any sign of stress, the procedure must be aborted until the animal acclimated to the situation. As soon as the animal is adapted, a complete clinical evaluation must be performed, including blood samples, body temperature, and total length. Complete blood count, serum biochemical parameters (glucose, electrolytes, BUN/Creatinine, minerals, iron, albumin, globulin, alkaline phosphatase, transaminases, bilirubin, fibrinogen, etc.) and sample collection for microbiological analysis must be performed to help the veterinary staff decide for the best treatment. Blood can be collected easily from the periarterial venous rete from either the fluke blade or from the caudal peduncle using a 21–23 gauge butterfly and a 3-cc syringe (Fig. 14.5). Complementary urine and stool analysis, ultrasound (Fig. 14.6), and X-ray are also recommended as soon the animal's condition is stable.

Fig. 14.5 *Ultrasound exam.*

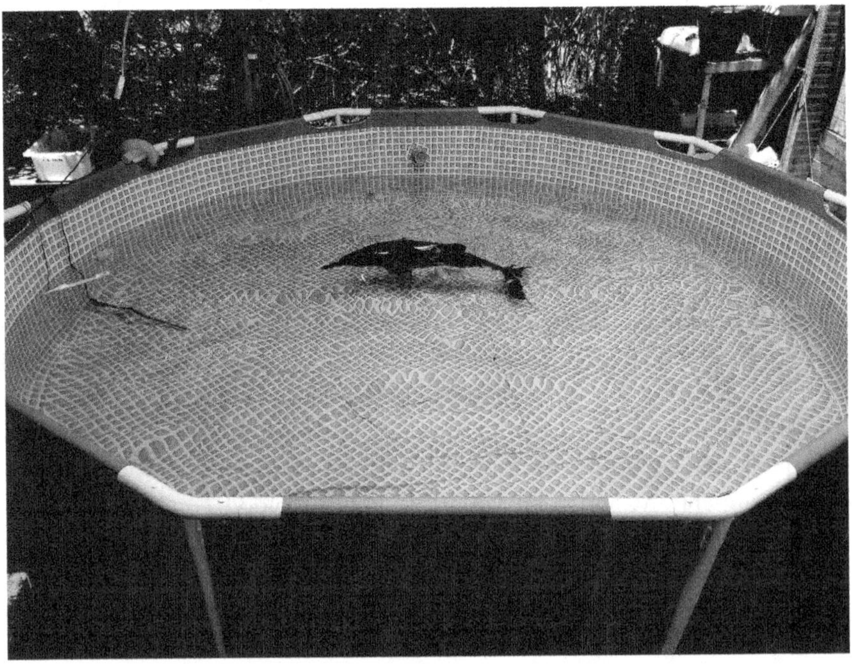

Fig. 14.6 *Neonate-juvenile rehabilitation swimming pool.*

Monitor weight daily as it provides important background for interpreting clinical signs, such as anorexia and decreased appetite (Nollen et al., 2018). Heart and respiratory rate must be taken every 30 min, and the animal must be observed 24 h/day. Once the animal is in better condition, weight can be measured weekly, respiratory and heart rate, three times a day. Regular patient monitoring should continue, even when clinical signs appear to resolve (Nollen et al., 2018).

Keep records from all data, treatments, and clinical observations.

Rehabilitation

Considering the small size of the newborns and juveniles, a child vinyl pool can be used to rehabilitate franciscanas. A 6000–10,000 L pool is ideal to start newborns' rehabilitation as it facilitates animal handling and provides enough space for swimming (Fig. 14.7).

Water salinity must be at 25–35 ppm and temperature at 23–28°C, depending on the animal's condition (Meegan et al., 2022). A severely debilitated animal must be kept in warmer temperatures to avoid energy loss trying to keep the body temperature. A 24 h water filtration system

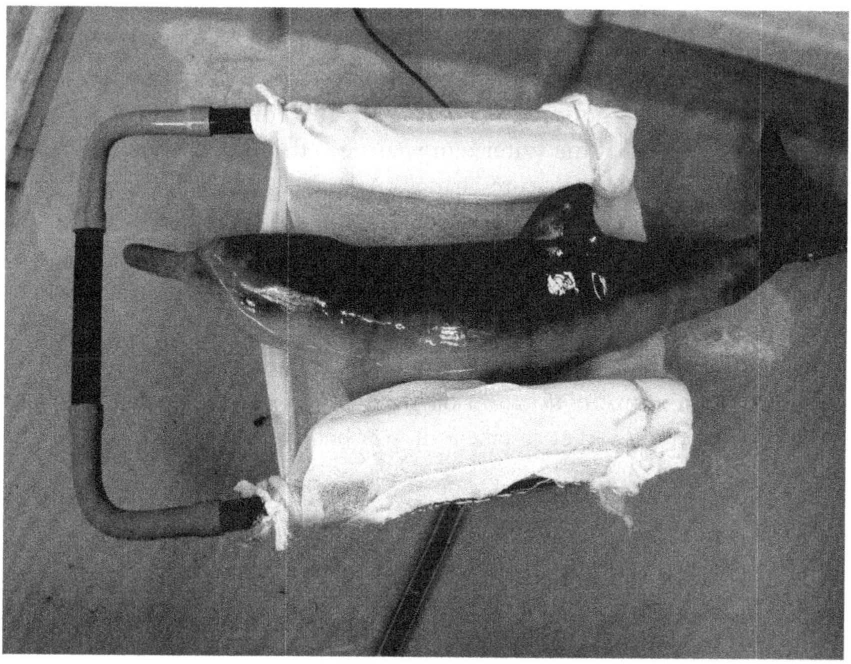

Fig. 14.7 ***Flotation device.*** *(Source: Associação R3 Animal).*

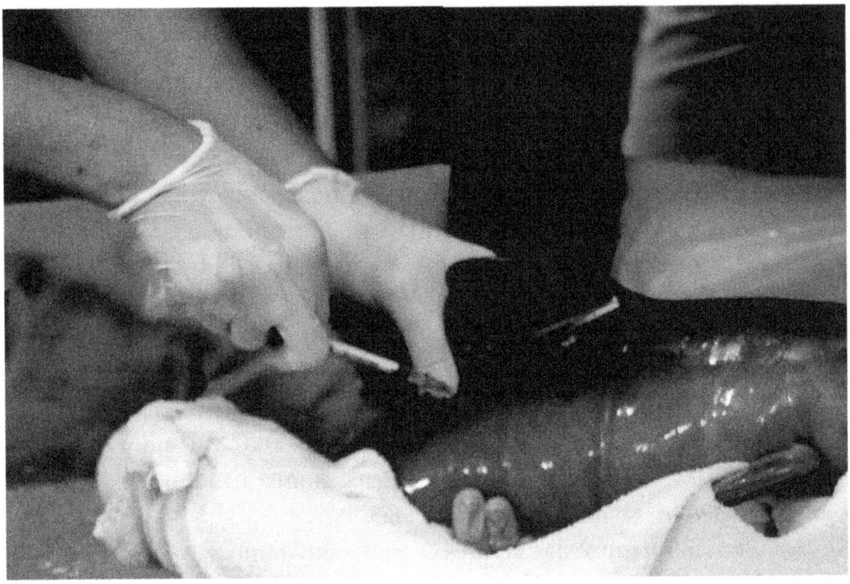

Fig. 14.8 ***Intramuscular injection.*** *(Source: Associação R3 Animal).*

is important to keep water quality and complementary water treatment as UV light is recommended. Daily salinity, temperature, pH (7.2–8.5), and ammonia (<2 ppm) must be checked. Total coliform count (<1000 NMP) must be performed weekly. A water change of 25% every other day is recommended to keep good water parameters. If a complete water change is necessary, be sure that the water temperature is the same to avoid any stress because of sudden temperature change (Arkush, 2001).

If the animal cannot float by itself, flotation devices can be used to help (Fig. 14.8).

As calves tend to have skin contact with their mothers, manual contact can be helpful to minimize stress, alternating period on the flotation device with free-swimming or animal handling. It will also prevent injuries from staying too long on the flotation device. If the choice is to have someone in the pool, do not forget to take a shower before entering the water, wear a face mask, and clean clothes.

Nutrition

As mentioned above, stranded animals tend to be dehydrated and hypoglycemic. A hydration/glucose solution must be offered every hour as the animal arrives. Calculate a volume of 2 mL/Kg given slowly via stomach tube. Appropriate tube length is calculated by measuring the distance from the end of the rostrum to the anterior insertion of the dorsal fin (Gage and Walsh, 2018). If there is any sign of vomiting or aspiration, reduce the volume until the animal is adapted to the tube feeding and extend the interval to one and a half hours. As there is improvement in the animal's glycemic and hydration, feeding with formula can be started. Start with a mix of 25% of the formula with hydration solution. As the animal shows no signs of regurgitation, vomiting, or impact, use 50%, in the third day 75% and then 100% of the formula. The formula must be offered at body temperature. If the animal accepts, bottle feeding will be the best choice for formula administration and sheep nipples may be adequate for franciscana's calves. During the initial adaptation to the nipple, lift the rostrum out of the water to avoid saltwater ingestion (Gage and Walsh, 2018). After 2 weeks with artificial milk formula, start adding fish to the recipe.

Total calories daily intake ranges from 150 to 200 Kcal/kg/day and daily weight gain for calves must be 0.125 kg/day (Marigo, 2004). If it is losing or not gaining weight, formula adaptation or volume input, if the animal supports it, must be made.

Many formulas have been successfully used with other cetacean species, such as *Turisops truncatus* and other species, but none with franciscanas. Formulas are listed in Table 14.1.

Table 14.1 Artificial milk formula (adapted from Sweeney et al., 2010).

Igredients	Dolphin quest	Busch entertainment corp
Herring—Fish Fillet(no skin, viscera (gut), spine or head)	750 g	1.135 g
Zoologic Milk Marix 30/55	450 g	200 mL
Zoologic Milk Matrix 33/40	300 g	345 g
Salmo Oil	100 mL	500 mL
Heavy whipping cream	100 mL	200 mL
Water/Pedialyte	1.600 mL	2.200 mL
Lecythin (Granules)	15 mL	8 g
Taurine	500 mg	500 mg
Lactobacillus (Acidophilus)	3 tabs	5 g (10 tabs 500 mg/tab)
$DiCaPO_4$	75 g (10 tabs)	
Multivitamin	¼ to ½ tab	
Dextrose		30 g
Salt		18g
Total volume (aprox.)	3.300 mL	4.000 mL
Total calories (aprox.)	2.21 kcal/mL	

For juveniles and adults, after balancing glucose and hydration status, start tube feeding. Force-feeding a dehydrated animal with whole fish can result in regurgitation or accumulation of dehydrated fish debris in the stomach, followed by vomiting or impact (Walsh and Gearhart, 2001). Franciscana's prey consists of around 80 species that had been registered in Brazil, Uruguay, and Argentina, consisting of teleost and squid of small size, generally around 10 cm. Calves start to ingest solid prey with 2–3 months and 75–80 cm length. In this period, shrimps are important components of the diet (Rocha-Campos et al., 2010).

Drugs and routes of administration

No drugs have been licensed for marine mammals, and veterinarians must be aware at all times that data on drug dosages for cetaceans have been used on a limited number of individuals from a narrow range of species. For drug dosage, published pharmacopeias from different institutions can be used as a basis (Simeone and Stoskopf, 2018).

The oral route (PO) is the first option for the administration of fluids and pharmaceuticals in cetaceans, as it can be hidden in the fish. If the animal is not eating by itself, a stomach tube can be used for pharmaceutical administration. As cetaceans have a thick bubbler layer, the needle for intramuscular injection might be longer than considered for other species. The injection is made off the midline, slightly anterior to, parallel to, or just posterior to the dorsal fin (Nollen et al., 2018). Before the injection, the site must be disinfected with chlorhexidine or alcohol and it is indicated to use an antibiotic ointment after the injection to protect against secondary infections. The subcutaneous route can be used for cetaceans by inserting the needle between the bubbler layer and the skeletal muscle layer. To administer subcutaneous fluids, use fluid bags with a standard intravenous infusion setup with tubing clamp opened. Use long-term intravenous (IV) therapy is rarely used or considered in cetaceans. For IV administration use fluke vessels if the volume is low and the medication is not harmful perivascularly (Nollen et al., 2018).

Release

The animal is ready for release if it is not under any kind of treatment, it has normal blood and biochemical parameters, it has normal clinical parameters, it is able to feed by itself, and has normal behavior and buoyancy.

The decision for release must be made in accordance with the veterinary and biology staff.

Tagging or marking released animals is highly recommended for the validation of veterinary procedures and treatments, but also to allow for the recovery of animals that are unable to adapt to the wild (Lander et al., 2018). For endangered populations in which rehabilitation is used to enhance the numbers of individuals, monitoring the ability of individuals to forage, survive, and ultimately reproduce the following release is essential for evaluating all efforts given to the rehabilitation program (Lander et al., 2018).

Satellite tags are used in cetaceans because they do not require filed-based monitoring and may operate 24 h/day and in all weather conditions. If satellite tags are not available, plastic rototags may be used (Lander et al., 2018). Rototags are attached near the edge of the dorsal fin and often migrate out of the tissue, leaving a notch (Lander et al., 2018). In franciscanas SPLASH time-depth recording, satellite-linked transmitters and SPOT5 location-only transmitters have been used in adult individuals, for calves, small livestock ear tags can be used (Wells et al., 2013) (Fig. 14.9).

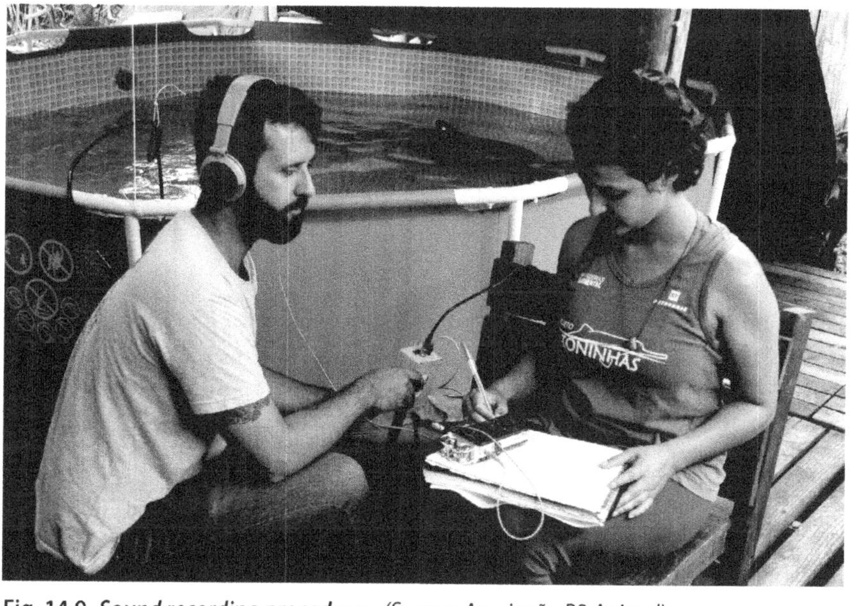

Fig. 14.9 ***Sound recording procedures.*** *(Source: Asociação R3 Animal).*

Zoonotic considerations

Marine mammals are also susceptible to a variety of zoonosis caused by bacteria, viruses, protozoans, fungi, and parasites. Special attention must be taken to multiresistant bacteria. Multiresistant bacteria have been detected worldwide in the microbiota of many wildlife species, including marine mammals. The indiscriminate use of antibiotics in human and veterinary medicine contributes to the development and global dissemination of antibiotic resistance including in wildlife species. Many marine species are migratory and, on such migrations, these animals are exposed to many different qualities of waters, with effluents from a diversity of communities. They may thus be exposed to not only a range of potential pathogens but also a wide variety of antibiotics (Tryland et al., 2018).

All the staff must be oriented about zoonotic diseases before working with rehabilitation.

NOOA Guidelines for rehabilitation of stranded animals (Gage, 2009) recommends restricting public access and direct contact with animals in rehabilitation due to public health hazards of non-trained individuals interacting with sick animals; train staff about how to prevent contracting zoonotic diseases; train staff and personnel working directly with stranded/sick animals how to recognize symptoms of zoonotic diseases; provide safety equipment such as protective clothing, eye protection, and face masks; staff with open wound shall not enter the pool of animals carrying potentially infectious diseases; people with disabilities, respiratory conditions, infectious diseases or infectious skin conditions must not enter pools; train staff basics of sanitation and properly handling contaminated equipment.

Final considerations

Wildlife rehabilitation has been done successfully for decades. Since the 1970s, dolphin rehabilitation and maintenance has been based on veterinary-based knowledge, not just as empiric trials (Ridgway, 1988). During all those years many research had taken place and lots of information, especially about *Tursiops truncatus,* have been published. Unfortunately, just a few franciscanas had been released after rehabilitation or kept in captivity to enable biologists and veterinarians enough data to have standardized rehabilitation protocols. A single opportunity to have a live franciscana in rehabilitation brings lots of knowledge. The opportunity has to be used not only to record veterinary information but also biological such as behavior using ethograms, sound records Specialists from many countries are working to have the best protocol for franciscana rehabilitation and help species conservation.

References

Arkush, K.D., 2001. Water quality. In: Dierauf, L., Gulland, F.M.D. (Eds.), CRC Handbook of Marine Mammal Medicine: Health, Disease, and Rehabilitation. CRC press, Boca Raton, FL, pp. 779–790.

Baldassin, P., Werneck, M., Barbosa, C.B., Gallo, H., Walsh, M., 2007. Veterinary treatmentof an injured wild Franciscana dolphin calf. LAJAM 6 (2), 185–187.

Bossart, G.D., Reiderson, L.A., Dierauf, D.A., Duffield, D.D., 2001 In: Dierauf, L., Gulland, F.M.D. (Eds). CRC handbook of marine mammal medicine: health, disease, and rehabilitation. CRC press, Boca Raton, FL, pp. 383-436.

Byers, O., 2014. The one plan approach. Connect Magazine, 18–20. https://www.cpsg.org/sites/cbsg.org/files/Pages%20from%20CONNECT%20Magazine%20March%20 2014%20-%20Final.pdf.

Chinadaily, 2020. First Yangtze porpoise born in captivity released into the wild. https://www.chinadaily.com.cn/a/202007/16/WS5f0faa0ba310834817259ca4.html. (Accessed July 2021).

Gage, L.J., 1990. Rescue And Rehabilitation of Cetaceans. In: Hhandbook of marine mammal medicine: health, disease, and rehabilitation. CRC Press, Boca Raton, USA, pp. 685–692.

Gage, L.J., 2001. Rescue and rehabilitation of cetaceans. In: Dierauf, L., Gulland, F.M.D. (Eds.), CRC Handbook of Marine Mammal Medicine: Health, Disease, and Rehabilitation. CRC press, Boca Raton, FL, pp. 685–692.

Gage, L., 2009. Marine mammal stranding response, rehabilitation ans release. NOAA, Silver Spring, Maryland.

Gage, L.G., Walsh., M., 2018. Hand-rearing and artificial milk formula. In: Gulland, F.M., Dierauf, L.A., Whitman, K.L. (Eds.), CRC Handbook of Marine Mammal Medicine. CRC Press, Boca Raton, FL, pp. 239–756.

Geraci, J.R., Lounsbury, V.J., 2005. Marine Mammals Ashore: A Field Guide for Strandings. National Aquarium in Baltimore, Baltimore, USA.

Harms, C., Greer, L.L., Whaley, J., Rowles, T.K., 2018. Euthanasia. In: Gulland, F.M., Dierauf, L.A., Whitman, K.L. (Eds.). CRC Handbook of Marine Mammal Medicine. CRC Press, Boca Raton, FL, pp. 675–694.

Lander, M.E., Westgate, A.J., Balmer, B.C., Reid, J.P., Murray, M.L.J., Laidre, K.L., 2018. Tagging and tracking. In: Gulland, F.M., Dierauf, L.A., Whitman, K.L. (Eds.), CRC Handbook of Marine Mammal Medicine. CRC Press, Boca Raton, Florida, pp. 767–798.

Maranho, A., Farah, R.F., 2016. First Succesful case of Fanciscana Pontoporia blainvillei Rehabilitation. XI Congreso de la Sociedad Latinoamericana de Especialistas Mamíferos Acuáticos. Valparaiso.

Marigo, J., 2007. Cetacea. In: Silva, J.C.R., Dias, J.L.C., Cubas, Z.S. (Eds.), Tratado de animais selvagens-medicina veterinária. Editora Roca, Rio de Janeiro, pp. 2512.

Meegan, J., Gomez, F., Barratclough, A., Smith, C., Sweeney, J., Ruopollo, V., et al., 2022. Rescue and rehabilitation of franciscana dolphin. Neonatal care and hand-rearing protocol. Alliance for Franciscana Dolphin Conservation Research, Rescue and Rehabilitation.

Nabi, G., Li, Y., McLaughlin, R.W., Mei, Z., Wang, K., Hao, Y., et al., 2020. Immune responses of the critically endangered yangtze finless porpoises (Neophocaena asiaeorientalis ssp.

asiaeorientalis) to escalating anthropogenic stressors in the wild and seminatural environments. Front. Physiol. 10, 1594.

Nollens, H.H., Venn-Watson, S., Gili, C., McBain, J.F., 2018. Cetacean Medicine. In: Gulland, F.M., Dierauf, L.A., Whitman, K.L. (Eds.), CRC Handbook of Marine Mammal Medicine. CRC Press, Boca Raton, FL, pp. 887–908.

Ridgway, S.H., 1988. The Dolphin Doctor. Fawcett, New York, NY.

Rocha-Campos, C.C., Danilewicz, D., Siciliano, S., 2010. Plano de Ação Nacional para a conservação do pequeno cetáceo toninha Pontoporia blainvillei. Instituto Chico Mendes de Conservação da Biodiversidade, Brasília.

Rojas-Bracho, L., Gulland, F.M.D., Smith, C.R., Taylor, B., Wells, R.S., Thomas, P.O., et al., 2019. A field effort to capture critically endangered vaquitas Phocoena sinus for protection from entanglement in illegal gillnets. Endanger. Species Res. 38, 11–27.

Schwacke, L.H., Hall, A.J., Townsend, F.I., Wells, R.S., Hansen, L.J., Hohn, A.A., et al., 2009. Hematologic and serum biochemical reference intervals for free-ranging common bottlenose dolphins (Tursiops truncatus) and variation in the distributions of clinicopathologic values related to geographic sampling site. Am. J. Vet. Res. 70 (8), 973–985.

Sharp, S.M., Harry, C.T., Hoppe, J.M., Moore, K.M., Niemeyer, M.E., Robinson, I., et al., 2016. A comparison of postrelease survival parameters between single and mass stranded delphinids from Cape Cod, Massachusetts, USA. Mar. Mamm. Sci. 32 (1), 161–180.

Simeone, C.A., Moore, K.M.T., 2018. Stranding response. In: Gulland, F.M., Dierauf, L.A., Whitman, K.L. (Eds.), CRC Handbook of Marine Mammal Medicine. CRC Press, Boca Raton, FL, pp. 3–18.

Simeone, C.A., Stokopf, M.K., 2018. Pharmaceuticals and formularies. In: Gulland, F.M., Dierauf, L.A., Whitman, K.L. (Eds.), CRC Handbook of Marine Mammal Medicine. CRC Press, Boca Raton, FL, pp. 607–674.

Sweeney, J.C., Stone, R., Campbell, M., McBain, J., Leger, J.S., Xitco, M., et al., 2010. Comparative survivability of Tursiops neonates from three US institutions for the decades 1990-1999 and 2000-2009. Aq. Mamm. 36 (3), 248–261.

Tryland, M., Larsen, A.K., Nymo, I., 2018. Bacterial infections and diseases. In: Gulland, F.M., Dierauf, L.A., Whitman, K.L. (Eds.), CRC Handbook of Marine Mammal Medicine. CRC Press, Boca Raton, FL, pp. 367–388.

Van Bonn, W., 2007. Clinical Concepts and Applications. Concepts Marine Mammal Medicine. Marvet, Sarasota, FL.

Walsh, M., Gearhart, S., 2001. Intensive care. In: Dierauf, L., Gulland, F.M.D. (Eds.), CRC Handbook of Marine Mammal Medicine: Health, Disease, and Rehabilitation. CRC press, Boca Raton, FL, pp. 689–702.

Wells, R.S., Bordino, P., Douglas, D.C., 2013. Patterns of social association in the franciscana, Pontoporia blainvillei. Mar. Mamm. Sci. 29 (4), 520–528.

CHAPTER 15

Alternative fishing methods, the potential use of "pingers," and other solutions to reduce the bycatch of franciscana dolphins (*Pontoporia blainvillei*)

Leonardo G. Berninsone[a], Sebastián Jiménez[b], Rodrigo Forselledo[b], Martín Laporta[c], Timothy B. Werner[d]

[a]Fundación AquaMarina, Centro de Estudios en Ciencias Marinas, Pinamar, Buenos Aires, Argentina
[b]Laboratorio de Recursos Pelágicos, Dirección Nacional de Recursos Acuáticos, Constituyente, Montevideo, Uruguay
[c]Unidad de Gestión Pesquera Atlántica, Dirección Nacional de Recursos Acuáticos, Rocha, Uruguay
[d]Ocean Associates Inc., Arlington, VA, United States

Bycatch is considered the main threat to the persistence of marine megafauna worldwide (Leaper and Calderan, 2018). Records show marine mammal bycatch occurs in all types of fishing gear such as gillnets, trawl nets, and surrounding nets among others (Read et al., 2006), and during both, industrial and artisanal fishing activities. Small cetaceans are most susceptible to become entangled in gillnets, especially those inhabiting coastal waters where small-scale artisanal fisheries are more common (Reeves et al., 2013). Entanglements often result in mortality which explains the critically endangered status of species such as the vaquita (*Phocoena sinus*) and the now likely extinct Yangtze river dolphin (*Lipotes vexillifer*).

Mortality due to bycatch in gillnets represents the biggest threat to marine mammals (Read et al., 2006), especially for Franciscanas (*Pontoporia blainvillei*) (Secchi et al., 2021) (Fig. 15.1). To prevent and reduce the bycatch of these small dolphins in artisanal fishing gear, a large amount of effort has been directed towards bycatch mitigation measures. Several strategies have been developed for other cetacean species such as acoustic deterrents (pingers and other alerting sound emitting devices), acoustically reflective nets, alternative fishing gears, escape or excluder devices, and time-area closures.

Several projects have been carried out during the past two decades in Uruguay, Brazil, and Argentina along with the local communities of

The Franciscana Dolphin
DOI: https://doi.org/10.1016/B978-0-323-90974-7.00016-1

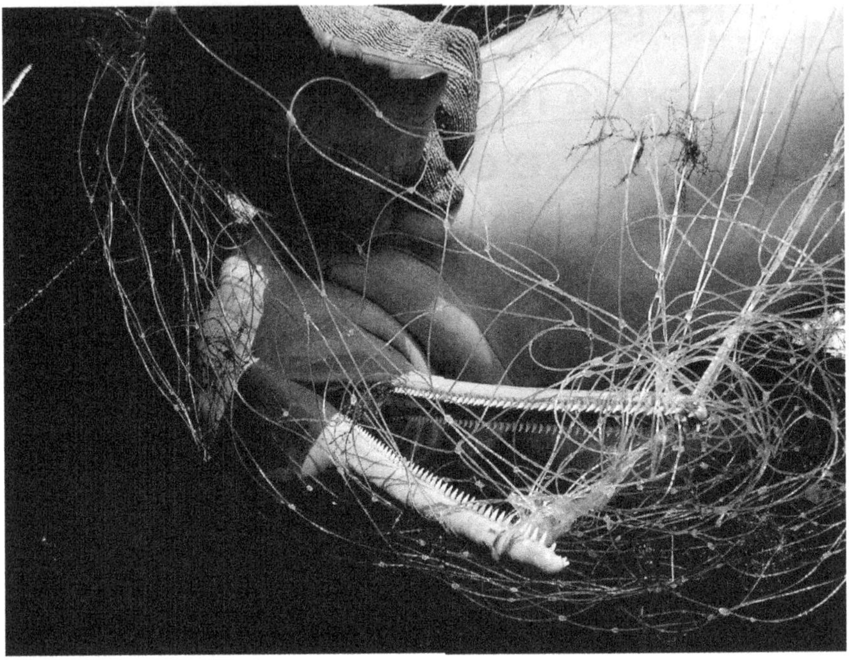

Fig. 15.1 ***Bycatch of a franciscana in a gillnet.*** *(Credit: AquaMarina).*

artisanal fishermen in order to find strategies to mitigate the bycatch of franciscanas in artisanal fishing gillnets. In this chapter, we review the different results obtained from these studies.

Acoustic deterrent devices (Pingers)

Pingers are acoustic deterrent devices that emits pulses of sounds in a certain frequency underwater. When these devices are attached to a net they sometimes generate a response in dolphins preventing the incidental bycatch. When pingers are deployed properly (sufficient spatial coverage on nets, correct spacing, and with timely replacement of batteries) they enable a range of responses of cetaceans to pingers which can result in bycatch reduction without changing the way fishermen operate when fishing (Leaper and Calderan, 2018).

The first experiment with pingers and franciscanas was undertaken in Cabo San Antonio, Argentina by Bordino et al. (2002) (Fig. 15.2). The pinger used was a Dukane NetMark 1000 pinger that emitted a broadband

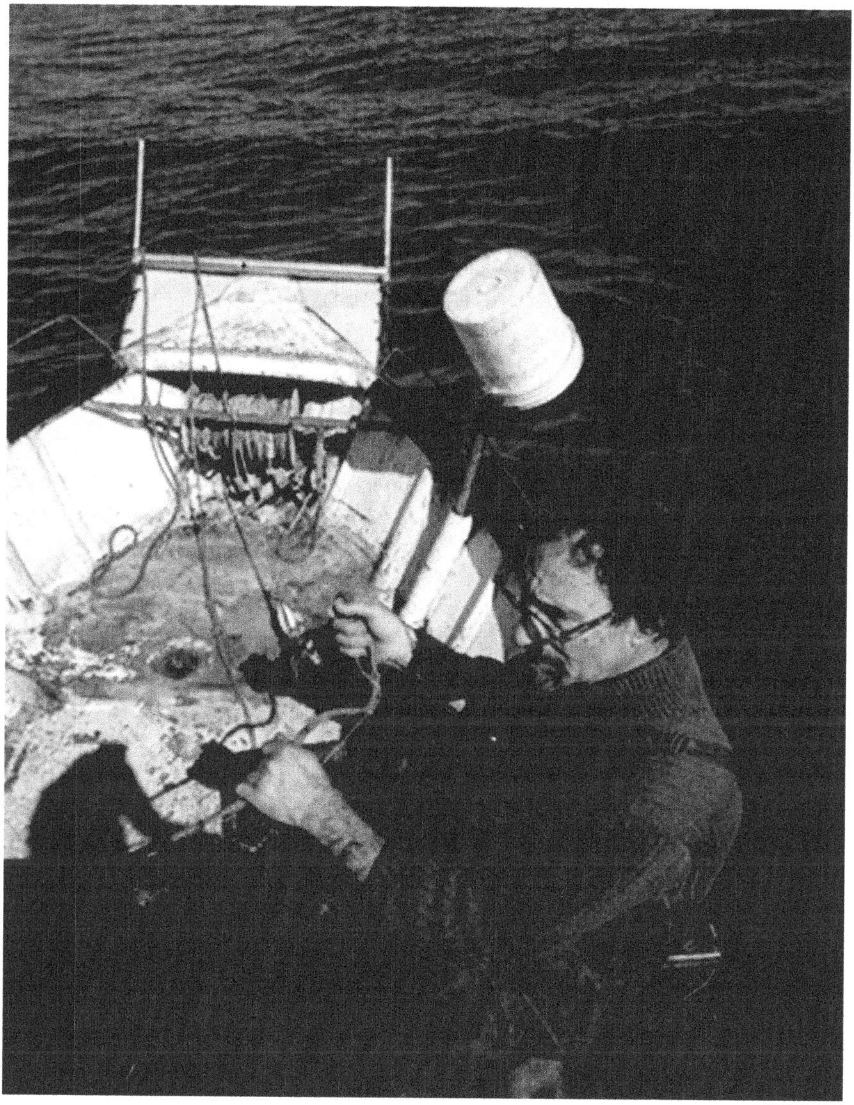

Fig. 15.2 ***Artisanal fisherman deploying a gillnet with a pinger in Cabo San Antonio (Argentina).*** *(Credit: AquaMarina).*

signal every four seconds centered at 10kHz, with a source level of 132 dB re 1 pPa @ 1 m. This project was done in cooperation with fishermen who took onboard observers in their daily fishing trips to record the incidental mortality in gillnets with active pingers and nonactive pingers. Results showed that the pinger was effective in reducing the bycatch of franciscana dolphin. However, interactions of South American sea lions (*Otaria flavescens*) with gillnets appeared to increase when pingers were active, presumable from the "dinner bell effect" (Barlow and Cameron, 2003; Bordino et al., 2002; Carretta and Barlow, 2011; Dawson et al., 2013), in which the animal becomes attracted to the net by associating the sound with the presence of prey caught in the net. When increasing the frequency of the pingers to 70kHz, the sound of the signal becomes less audible to pinnipeds, eliminating this unwanted behavior, and showing a similar reduction in franciscana dolphin bycatch (Bordino et al., 2004).

The effectiveness of pingers depends on many variables, including the experimental design, the fishery in which they are tested, the sound they create, the ambient noise level, the gear type, the fishing practices and the species. Therefore, implementation of pingers should be considered only for the area where pingers were tested by following the same conditions of the scientific evaluation (Dawson et al., 2013). Monitoring the use of pinger is crucial for ensuring the results of the implementation of a bycatch mitigation plan (FAO, 2021).

After several years of testing pingers (Future Oceans 70kHz Dolphin pingers) in Cabo San Antonio, the National Government of Argentina (Ministerio de Ambiente y Desarrollo Sustentable) are promoting the implementation of a bycatch mitigation plan co-designed with the local fishermen. This plan is being developed using the advice of local organizations and with the support of Provincial and Municipal Governments and is part of the National Action Plan for bycatch mitigation of marine mammals (https://cfp.gob.ar/wp-content/uploads/2017/09/LIBRO_PAN-MAM.pdf).

In Uruguay, the Franciscana dolphin is captured as bycatch mainly in artisanal fisheries (Praderi, 1997; Franco-Trecu et al., 2009; Szephegyi, 2012) (Fig. 15.3). Since 2018, DINARA has been developing a research program in collaboration with these fisheries for assessing the bycatch of Franciscana and testing, under commercial fishing conditions, the effectiveness of pingers in reducing its incidental mortality (Jiménez et al., 2021a, 2021b). The acoustic deterrent devices being tested are Banana Pingers (Porpoise & dolphin deterrent pinger; https://www.fishtekmarine.com/product/deterrent-pinger-50-120/), which emit sounds within the range of 50–120 kHz (SPL: 145 dB re 1 μPa at 1 m) (Fig. 15.4). These frequencies are considered beyond the audible range of pinnipeds, and therefore may protect against the "dinner bell" effect caused by pingers emitting at low frequencies, that is also associated with an increase in damage to target

Fig. 15.3 ***Bycatch of a franciscana dolphin in La Palona, Rocha, Uruguay.*** *Unidad de Gestión Pesquera Atlántica, Dirección Nacional de Recursos Acuáticos (DINARA).*

Fig. 15.4 ***Banana Pingers on an artisanal fishing boat of Punta Brava, Montevideo, Uruguay.*** Dirección Nacional de Recursos Acuáticos (DINARA).

catch (Bordino et al., 2002). All data were obtained by participating fishers and reported either in fishing logbooks or during frequent interviews with the fishermen (1–3 times per week). The protocol had nets without pingers (control treatment) and nets with pingers (experimental treatment). The results obtained up to March 2021 indicated that Banana Pingers substantially reduced the rate of by catch (Jiménez et al., 2021a).

Bycatch of franciscanas has also been recorded in the industrial coastal pair trawling fishery of Uruguay (Franco-Trecu et al., 2019). An experiment in collaboration with their skippers and the crew was carried out to evaluate the efficancy of pingers in reducing the bycatch produced by the industrial fleet. The pinger tested was the Dolphin Anti-Depredation Pinger from Fishtek Marine (https://www.fishtekmarine.com/product/dolphin-anti-depredation-pinger/) that emit sound signals of high intensity (175 dB re 1 μPa at 1 m; 40 Khz). The protocol considered a control treatment (trawl sets without pingers) and an experimental treatment of trawl sets with pingers (Jiménez et al., 2021b). Preliminary results suggest that the use of pingers reduces the bycatch of franciscana in this coastal pair trawl fishery. However, further trials are still required to confirm this reduction. In addition, this should be complemented with an analysis determining if these devices have any effect on fish catch.

Individual behavior of franciscanas reacting to pingers was studied by Bordino (2018) in Bahía San Blas, Argentina. In the experiment, the reaction and echolocation clicks of franciscanas were recorded using two different C-PODs when the animals were swimming toward an active pinger (Future Oceans 70kHz Dolphin Pinger). Results showed that franciscanas have an "aversive" behavior (Dawson et al., 2013), where the sound of the pinger produced a displacement of the dolphins away from the pinger location after an absence of echolocation clicks.

There has been a concern about whether the encounters of dolphins with pingers over time would lead either to a displacement of the cetaceans to another foraging area (Culik et al., 2001; Carlström et al., 2002) or to a reduction in the effectiveness of the pingers due to habituation (Amano et al., 2017).

Paitach et al. (2018) (Paitach et al., 2022) developed an experiment to test the deterrent effect of Banana pingers in Baia da Babitonga (Brazil) using five C-PODs at different distances from the source of sound. The data indicated that franciscanas avoided ensonified areas and no habituation effects were recorded during 65 days of testing. Habituation behavior for other cetaceans was also found not to occur in other studies (Koschinski and Culik, 1997; Palka et al., 2008; Carretta and Barlow, 2011).

Modified fishing gear

The reason why dolphins get entangled in gillnets is still unknown. However one hypothesis is that gillnets cannot be detected easily by dolphins through echolocation. In this sense, several studies evaluated the bycatch reduction by improving the acoustic reflectivity of the net (Dawson, 1994; Northridge et al., 2003; Trippel et al., 2003, 2009; Larsen et al., 2007).

Bordino et al. (2013) reported experimental trials that compared the bycatch rates of franciscanas in regular gillnets and two gillnets, one modified to have increased acoustic detectability. Acoustically reflective gillnets were made of a nylon enriched with BaSO4 that provides a better reflection of echolocation waves. This change in the chemical composition of the nets can also create a greater stiffness in the net twine. Another type of gillnet made to have greater stiffness than standard ones but with no BaSO4 was also tested in order to determine if the stiffness of the net would play a role in the reduction of the bycatch rates.

The study was carried out in Bahia Samborombón (Argentina) involving local artisanal fishermen and onboard observers for recording data during the trials. The results showed no significant differences of the bycatch rates in the three different types of gillnets. A similar experiment was developed in São Paulo (Brazil) with similar results, showing that bycatch of franciscanas was not reduced neither by acoustically reflective gillnets nor gillnets made stifffer (Fig. 15.5) (C. Bertozzi, personal communication).

These results suggest different hypotheses regarding the mechanism by which franciscanas are bycaught; (1) gillnets, including the reflective and stiffer nets, cannot be detected by the dolphins, (2) the dolphins might be able to detect some nets but are not echolocating when they head towards them, (3) the dolphins detect the nets but reaction time is insufficient or ineffective to avoid them, or (4) they detect the net but they do not perceive it as a threat (Kastelein et al., 2000). However, it is still not clear which one of these hypothesis can explain the inefficiency of reflective and stiffer nets in reducing the bycatch of franciscanas.

Alternative fishing gear

Small-scale artisanal fisheries involve traditions and fishing culture that can strongly influence how likely fishermen might change how they fish. However, some fishermen are opportunistic and willing to change their way of fishing if they have the chance to choose an alternative fishing gear

Fig. 15.5 ***Bycatch of a franciscana in an acoustically reflective gillnet.*** *(Credit: AquaMarina).*

that is more profitable by reducing operational costs or increasingfinancial returns.

Berninsone et al. (2020) compared target and non-target catch rates between gillnets and bottom longlines in Bahía Samborombón and Cabo San Antonio, Argentina. Target and non-target catch composition, fishing yield, catch size distribution, and quality of catch, as well as bycatch of dolphins, sea turtles, and interaction with sea lions were compared between the two fishing methods to assess the profitability of switching fishing gears.

Hauls of both gear types deployed simultaneously in the same locations showed similar fish catch composition and catch size with both gears but the reduced catch of juvenile fishes in longlines.

Bycatch of franciscana in bottom longlines was limited to only one dolphin in three consecutive years of trials, and no direct interaction between turtles and hooks were recorded whereas 85 franciscana were bycaught in gillnets during the same period. The study showed strong potential for the use of bottom longlines as a viable alternative fishing gear for Bahía Samborombón in terms of the sustainable use of living resources due to minimal capture of undersized fish, low discard and limited bycatch of non-target species. In general, gillnets and bottom longlines had a relatively similar efficiency/selectivity rate, however the complexity of socio-economic factors have so far prevented a transition from gillnets to handlines. This finding might be due to the complexity of technical reasons related to the effort required for operating with longlines and perhaps because even with some economic advantages such as improved quality in target catch, the current fish market line is not giving an added value to this better quality fish that would result in an immediate economic bonus for the fishermen. Fishing pots are another alternative gear that have undergone preliminary testing in Bahia Samborombón, Argentina. Results show that they were effective for reducing bycatch rates although there was a reduction in the catch size of one target species (Werner, personal communication). A new type of fishing pot is being co-designed with the local fishermen from Cabo San Antonio where the currents, sea bed and fishing practices are different than in Bahía Samborombón. Tests of these fishing pots will compare catch rates over the same duration of a single morning of gillnet fishing after which fishermen sell the fish to tourists at the beach.

Considerations for franciscana bycatch mitigation strategies

Bycatch mitigation and the ecosystem-based approach to fisheries management implies modifying behavior and decisions made by stakeholders (Squires et al 2021) by working along with local fishermen for finding the best solutions and co-designing management actions that account for the needs of fishing communities. Pingers appear to be the most viable approach to mitigate bycatch, at least in the fisheries discussed above, since they are effective in terms of bycatch mitigation and they do not require fishermen to make any major change in their traditional way of fishing.

Nevertheless, there are many aspects that need to be taken into account for pinger implementation including the expense of purchasing and maintaining the pingers.

From an ecological point of view, there are also considerations regarding the amount of sound introduced in the environment when using pingers and the potential effects these might have on habitat use. Information about movement patterns and spatial fishing effort is crucial in this matter, and may vary for franciscanas in estuarine areas versus open waters considering that they move in and out of the bay following tide patterns (Wells et al., 2022).

Adopting alternative fishing gear represents a cultural challenge for the fishermen that have to change the way they operate during their fishing practices. Testing of bottom longlines has proven to be an effective method for reducing bycatch rates; however, no fisherman had opted for exchanging gillnets for bottom longlines. Using bottom longlines represents a disadvantage for the fishermen since the amount of time required for operating the fishing gear is higher than when using gillnets. An alternative fishing gear that is viable should take into account the needs of the fishermen not only from an economic but also logistic, operational and cultural point of view. According to Tucker et al. (1997), fishing gear designed to reduce incidental mortality of non-target catch should be safe, simple, practicable to use, and does not increase operating costs.

Modifications on gillnets are primarily based on changing the detection power of the net. The hypothesis that gillnets cannot be detected by franciscanas still needs to be confirmed, and can be further examined by testing the efficiency of passive acoustic devices such as air-filled beads attached to the net, or by addditional behavioral studies using bioacoustics. Other novel modifications such as attaching LED lights to gillnets to increase their visual detectability (Bielli et al., 2020) may be worth examination, however if effective their application may be limited to clearer waters and not the murky waters adjacent to the La Plata estuary.

In order to be effective, a bycatch reduction technique must be shown to reduce bycatch and be viable for a fisherman to implement. In considering all the techniques evaluated to date in Argentina, Brazil, and Uruguay—with the exception of time-area closures which were not reviewed here - pingers so far provide the most viable method for bycatch mitigation of franciscanas. In contrast, reflective gillnets and stiffened gillnets were shown to be ineffective even though they could be easily adopted by fishermen and the uptake of gillnet alternatives such as bottom longlines and pots require further socio-economic studies and engagement with fishermen if there is any potential for them to be adopted (Fig. 15.6).

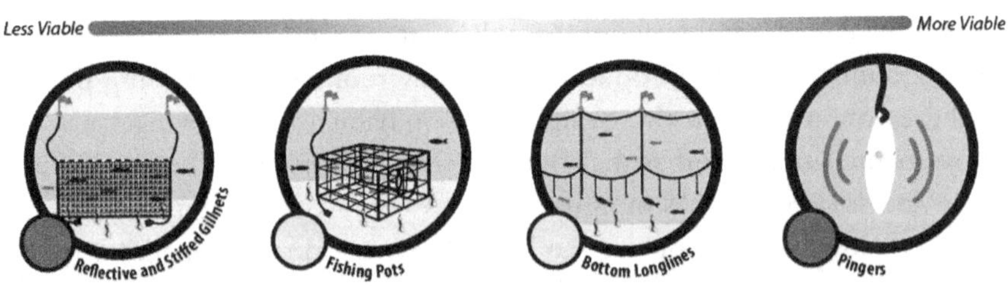

Fig. 15.6 ***Effectiveness of different bycatch mitigation methods.*** *(Credit: AquaMarina).*

An effective strategy for franciscana populations may involve a combination of approaches (e.g.: reduce the number of gillnets by replacing them with an alternative fishing gear while attaching pingers to the remaining gillnets) and should consider the local context in which they might be implemented. Management actions for conservation of franciscanas require improving on techniques for bycatch mitigation. The aim to stop bycatch of franciscanas is unreal, both from a theoretical point of view (no strategies are 100% effective) and from a practical point of view (it is impossible to control 100% of the fishing effort along the distribution). A more realistic approach would be to reduce the bycatch rates to sustainable levels for each population.

References

Amano, et al., 2017. Long-term effectiveness of pingers on a small population of finless porpoises in Japan. Endangered Species Res. 32, 35–40.

Barlow, J., Cameron, G.A, 2003. Field experiments show that acoustic pingers reduce marine mammal bycatch in the California drift gill net fishery. Marine Mammal Sci. 19 (2), 265–283.

Berninsone, L.G., Bordino, P., Gnecco, M., Foutel, M., Mackay, A.I., Werner, T.B., 2020. Switching gillnets to longlines: an alternative to mitigate the bycatch of Franciscana Dolphins (Pontoporia blainvillei) in Argentina. Front. Mar. Sci. 699.

Bielli, A., Alfaro-Shigueto, J., Doherty, P.D., Godley, B.J., Ortiz, C., Pasara, A., Mangel, J.C., 2020. An illuminating idea to reduce bycatch in the Peruvian small-scale gillnet fishery. Biological Cons. 241, 108277.

Bordino, P., 2018. Understanding the sonar behavior of franciscana dolphins in response to active pingers. Final Report: https://www.bycatch.org/sites/default/files/r_Franciscana%20CPODS_FINAL_140918.pdf (accessed December 1, 2021).

Bordino, P., Kraus, S., Albareda, D., et al., 2002. Reducing incidental mortality of Franciscana dolphin Pontoporia blainvillei with acoustic warning devices attached to fishing nets. Marine Mammal Sci. 18, 833–842.

Bordino, P., Kraus, S., Albareda, D., Baldwin, K., 2004. Acoustic Devices Help to Reduce Incidental Mortality of the Franciscana Dolphin (Pontoporia blainvillei) in Coastal Gillnets, Scientific Committee of the International Whaling Commission, Sorrento.

Bordino, P., Mackay, A.I., Werner, T.B., Northridge, S.P., Read, A.J., 2013. Franciscana bycatch is not reduced by acoustically reflective or physically stiffened gillnets. Endangered Species Res. 21 (1), 1–12.

Carlström, J., Berggren, P., Dinnétz, F., Börjesson, P., 2002. A field experiment using acoustic alarms (pingers) to reduce harbour porpoise by-catch in bottom-set gillnets. ICES Journal of Marine Sci. 59 (4), 816–824.

Carretta, J.V., Barlow, J., 2011. Long-term effectiveness, failure rate, and "dinner bell" properties of acoustic pingers in a gillnet fishery. Marine Technol. Soc. J. 45 (5), 7–19.

Cox, T.M., Read, A.J., Solow, A., Tregenza, N., 2001. Will Harbour porpoises (Phocoena phocoena) habituate to pingers? J. Cetacean Res. Manag. 3 (1), 81–86.

Cox, T.M., Read, A.J., Swanner, D., Urian, K., Waples, D., 2004. Behavioral responses of bottlenose dolphins, Tursiops truncatus to gillnets and acoustic alarms. Biological Cons. 115 (2), 203–212.

Culik, B.M., Koschinski, S., Tregenza, N., Ellis, G.M., 2001. Reactions of harbor porpoises Phocoena phocoena and herring Clupea harengus to acoustic alarms. Marine Ecol. Progress Series 211, 255–260.

Dawson, S., 1994. The potential for reducing entanglement of dolphins and porpoises with acoustic modifications to gillnets. Rep. Int. Whal. Commun. 15 (Spec Issue), 573–578.

Dawson, S.M., Northridge, S., Waples, D., Read, A.J, 2013. To ping or not to ping: the use of active acoustic devices in mitigating interactions between small cetaceans and gillnet fisheries. Endangered Species Res. 19, 201–221.

FAO, 2021. Fishing operations. Guidelines to prevent and reduce bycatch of marine mammals in capture fisheries. FAO Technical Guidelines for Responsible Fisheries No.1. Suppl 4. https://doi.org/10.4060/cb2887en.

Franco-Trecu, V., Costa, P., Abud, C., et al., 2009. By-catch of franciscana Pontoporia blainvillei in Uruguayan artisanal gillnet fisheries: an evaluation after a twelve-year gap in data collection. Latin American J. Aquatic Mammals 7, 11–22. http://dx.doi.org/10.5597/lajam00129.

Franco-Trecu, V., Szephegyi, M.N., Doño, F., et al., 2019. Marine mammal bycatch by the industrial bottom trawl fishery at the Río de la Plata Estuary and the adjacent Atlantic Ocean. Latin Am. J. Aquatic Res. 47, 89–101. https://doi.org/10.3856/vol47-issue1-fulltext-10.

Kastelein, R.A., Rippe, J.T., Vaughan, N., Schooneman, N.M., Verboom, W.C., de Haan, D., 2000. The effects of acoustic alarms on the behavior of harbour porpoises (Phocoena phocoena) in a floating pen. Marine Mammal Sci. 16 (1), 46–64.

Koschinski, Culik, 1997. Deterring harbor porpoises (Phocoena phocoena) from gillnets: observed reactions to passive reflectors and pingers. Rep. Int. Whaling Commun. 47, 659–668.

Jiménez, S., Laporta, M., Forselledo, R., et al., 2021a. Effectiveness of pingers to reduce the bycatch of Franciscana (Pontoporia blainvillei) in the Uruguayan artisanal gillnet fishery. Presented at the Scientific Committee of the International Whaling Commission. SC/68C/HIM/10.

Jiménez, S., Forselledo, R., Laporta, M., et al., 2021b. Assessing the effectiveness of dolphin antipredation pingers to reduce the bycatch of Franciscana (Pontoporia blainvillei) in demersal pair trawl fishing. Presented at the Scientific Committee of the International Whaling Commission. SC/68C/HIM/11.

Larsen, F., Eigaard, O.R., Tougaard, J., 2007. Reduction of harbor porpoise Phocoena phocoena bycatch by iron-oxide gillnets. Fish Res. 85, 270–278.

Leaper, R., Calderan, S., 2018. Review of methods used to reduce risks of cetacean bycatch and entanglements. UNEP/CMS Secretariat, Bonn, Germany. 76 pages. CMS Technical Series No. 38.

Northridge, S., Sanderson, D., Mackay, A., Hammond, P., 2003. Analysis and mitigation of cetacean bycatch in UK fisheries: final report to DEFRA Project MF0726. Sea Mammal Research Unit, St. Andrews.

Paitach, R.L., Amundin, M., Cremer, M.J., 2018. "A "Seal safe" acoustic deterrent may mitigate the most serious threat to franciscana dolphin," in XII Congreso de la SOLAMAC, Lima, Peru. Livro de Resumos do XII Congreso de la SOLAMAC, 258.

Paitach, R.L., Amundin, M., Königson, S., Cremer, M.J., 2022. Assessing effectiveness and side effects of likely "seal safe" pinger sounds to ward off endangered franciscana dolphins (Pontoporia blainvillei). Marine Mammal Science.

Palka, et al., 2008. Effect of pingers on harbour porpoise (Phocoena phocoena) bycatch in the US Northeast gillnet fishery. J. Cetacean Res. Manag. 10 (3), 217–226.

Praderi, R., 1997. Estado actual de la mortalidad de franciscana en las pesquerías artesanales de Uruguay. Convention on Migratory Species, pp. 13–15.

Read, A.J., Drinker, P., Northridge, S., 2006. Bycatch of marine mammals in US and global fisheries. Conservation Biol. 20 (1), 163–169.

Reeves, R.R., McClellan, K., Werner, T.B., 2013. Marine mammal bycatch in gillnet and other entangling net fisheries, 1990 to 2011. Endangered Species Res. 20 (1), 71–97.

Secchi, E.R., Cremer, M.J., Danilewicz, D., Lailson-Brito, J., 2021. A synthesis of the ecology, human-related threats and conservation perspectives for the endangered franciscana dolphin. Front. Marine Sci. 8, 617956.

Squires, D., Ballance, L.T., Dagorn, L., Dutton, P.H., Lent, R., 2021. Mitigating bycatch: novel insights to multidisciplinary approaches. Front. Marine Sci. 3.

Szephegyi, M.N., 2012. Captura incidental y uso de hábitat del delfín franciscana (Pontoporia blainvillei) en el Río de la Plata y la costa atlántica uruguaya a partir de información de las flotas pesqueras. PEDECIBA. Universidad de la República.

Trippel, E.A., Holy, N.L., Palka, D.L., Shepherd, T.D., Melvin, G.D., Terhune, J.M., 2003. Nylon barium sulphate gillnet reduces porpoise and seabird mortality. Mar. Mamm. Sci. 19, 240–243.

Trippel, E.A., Holy, N.L., Shepherd, T.D., 2009. Barium sulphate modified fishing gear as a mitigative measure for cetacean incidental mortalities. J. Cetacean. Res. Manag. 10, 235–246.

Tucker, A., Robins, J., McPhee, D., 1997. Adopting Turtle Exclusion Devices in Australia and the US: what are the differences in technology transfer, promotion and acceptance? Coastal Manag. 25, 405–421.

Wells, R.S., Cremer, M.J., Berninsone, L.G., Albareda, D., Wilkinson, K.A., Stamper, M.A., Bordino, P., 2022. Tagging, ranging patterns, and behavior of franciscana dolphins (Pontoporia blainvillei) off Argentina and Brazil: Considerations for conservation. Mar. Mamm. Sci. 38(2), 571–605.

CHAPTER 16

Gillnet fishing characteristics in Franciscana Management Areas

Isabel Gonçalves[a], Camila Carvalho de Carvalho[b], Liane Amaral Dias[c], Carolina Pacheco Bertozzi[d], Camila Domit[e], Juliana Di Tullio[a]

[a]KAOSA, Rio Grande, Rio Grande do Sul, Brazil
[b]Laboratório de Ecologia e Conservação da Megafauna Marinha, Universidade Federal do Rio Grande, Rio Grande, Rio Grande do Sul, Brazil
[c]Laboratório de Ecologia e Conservação da Megafauna Marinha, Instituto de Oceanografia, Universidade Federal do Rio Grande, Rio Grande, Rio Grande do Sul, Brazil
[d]Instituto de Biociências (Câmpus do Litoral Paulista), Departamento de Ciências Biológicas e Ambientais, Universidade Estadual Paulista - UNESP, São Vicente, São Paulo, Brasil
[e]Laboratório de Ecologia e Conservação, Universidade Federal do Paraná, Pontal do Sul, Paraná, Brazil

Introduction

Records of incidental catch (hereafter bycatch) in gillnet fisheries date over 50 years back and are the primary cause of high mortality rates of the franciscana dolphin (*Pontoporia blainvillei*) throughout this species distribution on the Brazilian coast (Siciliano, 1994; Secchi and Ott, 1997; Secchi et al., 2003; Di Benedito and Ramos, 2001; Bertozzi and Zerbini, 2002). Although there are restrictions for gillnet use to protect the franciscana and other endangered species, these measures have been ineffective given that bycatch rates remain high (Monteiro et al., 2019). In addition, other anthropogenic activities such as chemical and noise pollution also contribute to habitat degradation and threaten this species' conservation (Di Beneditto et al., 2010).

Over the last 20 years, researchers collaborated with decision-makers to mitigate the number of franciscana dolphins caught by the fleet of artisanal and industrial gillnet fishing. The contribution includes proposing and modifying existing fishing policies dedicated to the franciscana occurrence areas in Brazil, including gear restrictions, limits to net and mesh sizes, and fishing exclusion zones. The decision-makers considered some of the suggestions and elaborated the Interministerial Regulation of the Ministry of Fisheries and Aquaculture and Ministry of Environment No. 12 of August

The Franciscana Dolphin
DOI: https://doi.org/10.1016/B978-0-323-90974-7.00006-9

22, 2012 (INI MPA/MMA 12/2012) to regulate gillnet fisheries in the southeastern and southern regions that cover FMA I, II and III.

It is noteworthy that conservation strategies and regulations in a top-down approach and without socioeconomic and ethnoecological data from the dependent communities and fisheries might be the formula for conflictful and ineffective measures. One of the most evident deficiencies in this practice is establishing rules from quantitative data, putting aside the qualitative aspects (social, economic, and cultural dynamics) that tend to be decisive in the interface between science and politics (Maxim and Van der Sluijs, 2011).

Understanding the relevance of socioeconomic, sociopolitical, socio-environmental, and ethnological aspects of fishing communities related to gillnet fishing activities, recent projects were design to collect qualitative data about the mentioned topics. Between 2016 and 2021 these projects covered 38 cities in the southeastern and southern regions of Brazil (Fig. 16.1), including the states of Rio Grande do Sul (RS), Santa Catarina (SC), Paraná (PR), and São Paulo (SP). A total of 311 fishers working with gillnets were interviewed to (1) characterize the fishing activity of each location, (2) characterize fishers' perception of their profession and fishing

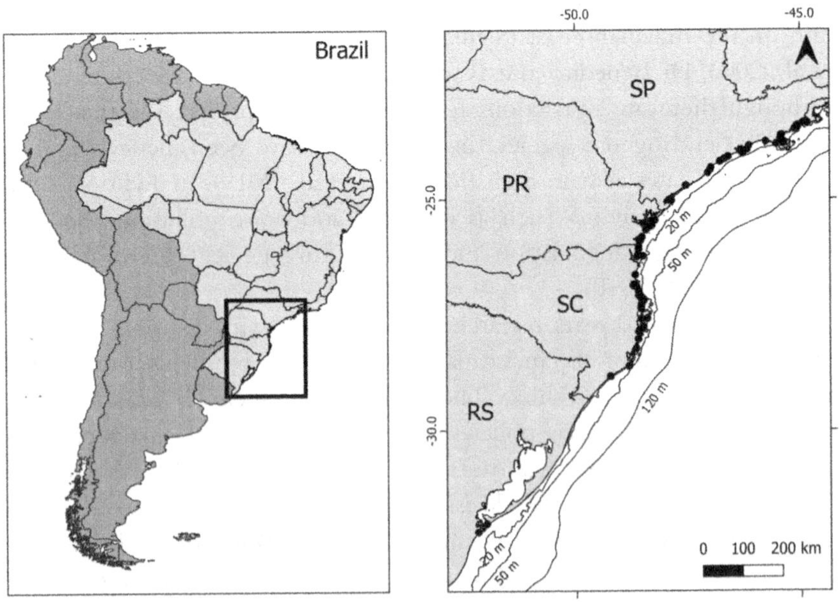

Fig. 16.1 ***Interview locations visited between 2016 and 2020.*** *RS*, Rio Grande do Sul; *SC*, Santa Catarina, *PR*, Paraná; *SP*, São Paulo.

activity, and (3) investigate bycatch of endangered species, focusing on the franciscana dolphin.

Gillnet fisheries definition

According to Brazilian Federal Law No. 11.959/2009, fishing is classified into two types considering the boat size and technology, income sharing, and fishing effort. Artisanal commercial fishing includes professional fishers, independently or in a family economy regime, by their means of production or a partnership contract, landed or using vessels with gross tonnage (GT) less than or equal to 20 GT.

Conversely, industrial commercial fishing includes individuals or legal entities, employs professional fishers or in a partnership regime, uses small, medium, or large vessels (of any GT), and has commercial purposes.

However, due to the great variety of vessel types and sizes in artisanal fishing, some authors proposed dividing gillnet fisheries in Brazil into small-scale (artisanal), medium-scale (semi-industrial), and large-scale (industrial) fishing (Boffo and Reis, 2003; CEPERG, 2011; Lucena and Reis, 1998; Reis, 1992, 1993; Vasconcellos et al., 2014). According to this definition, adopted hereafter, semi-industrial vessels comprise boats from 12 m onwards with GT less than or equal to 20 (Nogueira, 2016) and will be included in industrial fishery topics.

Production and monitoring of gillnet fishing

Artisanal gillnet fishing is distributed in several locations in the south and southeast of Brazil. A large part of the production and production units are concentrated in Santa Catarina (PMAP-BS, 2019a; PMAP-BS, 2019b; Fig. 16.2). Industrial gillnet fishing occurs in the states of Rio Grande do Sul, Santa Catarina and São Paulo. The activity of this sector is concentrated in the cities of Rio Grande and São José do Norte, in the south of Rio Grande do Sul; Itajaí, Navegantes, Porto Belo and Laguna in Santa Catarina; and Cananéia, Santos and Ubatuba in São Paulo. The state of Santa Catarina concentrates the largest number of vessels, fishing effort and production (FURG/SEMA, 2020; PMAP-BS, 2019a; PMAP-BS, 2019b; Fig. 16.2). There are no industrial gillnet fleet in Paraná, but vessels from other states fish in this region (PMAP-BS, 2019a, 2019b).

Fishing production data for São Paulo, Santa Catarina, and Paraná were obtained from reports of the Fishing Activity Monitoring Project (PMAP-BS), an initiative required during the environmental licensing process for

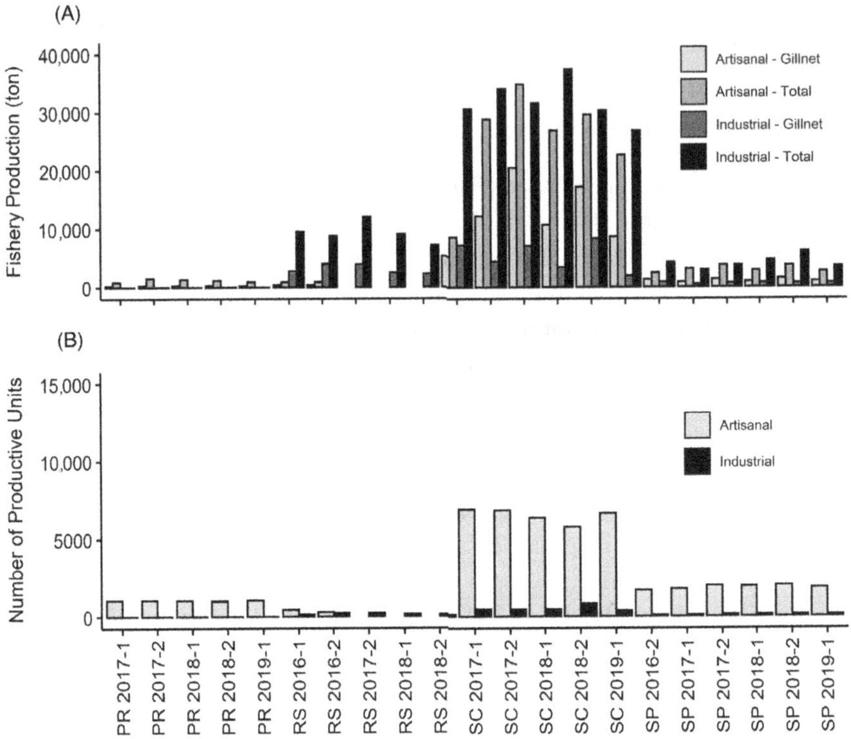

Fig. 16.2 ***Production data of artisanal and industrial fisheries.*** (A) Total artisanal and industrial fishery production and gillnet production; (B) Number of artisanal and industrial production units. *RS*, Rio Grande do Sul; *SC*, Santa Catarina; *PR*, Paraná; *SP*, São Paulo. The subsequent number represents the year, and numbers 1 and 2 represent the semester of the year. Data source: Fishing Activity Monitoring Project—PMAP-BS reports, FURG/MPA (2017–2019), and FURG/SEMA (2020).

the oil exploration company. In these reports, fishing is classified as artisanal and industrial.

Unfortunately, the state of Rio Grande do Sul does not monitor systematically the artisanal and industrial fishing landings; therefore, the production data presented here are from distinct projects of short duration and refer only to two municipalities, Rio Grande and São José do Norte (Fig. 16.1).

Gillnet fishing characterization based on interviews

Artisanal fishing

Concerning the vessels employed by the artisanal fishers, the material can be wood, fiber, or aluminum (Table 16.1; Fig. 16.3). Wooden vessels

Table 16.1 Characteristics of the artisanal, semi-industrial, and industrial vessels working with gillnets in the studied states.

State in Brazil	Vessel type	Length (m)	Material	Boathouse	Engine (HP)	Crew
Rio Grande do Sul (RS)	Bote	8–12	Wood	No	Stern (6–24)	1–3
	Bote	11–12	Wood	Yes	Center (62–115)	2–4
	Semi-industrial	12 (<20 GT)	Wood	Yes	Center (240–260)	3–7
	Industrial	19–24	Wood	Yes	Center (325–360)	8–10
Santa Catarina (SC)	Canoe	8–12	Wood	No	Center (24–90)	3–10
	Trawler	14	Wood	Yes	Center (220)	10
	Botes	8–13	Wood	Yes	Center (33–115)	4–6
	Botes	7–10	Wood	No	Center (18–22)	2
	Fiber boat	6–9	Fiber	No	Center (10–24)	2–3
	Semi-industrial	11–12	Wood	Yes	Center (200)	4–6
	Industrial	16–22	Wood	Yes	Center (200)	4–6
Paraná (PR)	Canoe	8–10	Fiber	No	Center (18)	2–3
	Bote	8–9	Wood	No	Center (18)	2–3
	Voadeira	5–13	Aluminum	No	Stern (15–60)	1–5
São Paulo (SP)	Boat	8–15	Wood	Yes	Center (18–150)	2–6
	Canoe	6–9	Wood or fiber	No	Center (11–60)	2–3
	Speedboat	5–13	Aluminum	No	Stern (15–60)	1–5
	Canoe	8	Wood	No	Rowboat	4
	Semi-industrial	12–15	Wood	Yes	Center (175–265)	4–6
	Industrial	15–19	Wood	Yes	Center (175–265)	4–6

Length (m) = minimum and maximum size. The "engine" column refers to the location (stern or center), of the motor and the horsepower (HP) reported by the fishers.

Fig. 16.3 ***Types of artisanal fishing vessels.*** (A) Boat of up to 12 m with the central engine in the south of Rio Grande do Sul. (B) Boat of up to 14 m with the central engine in Rio Grande do Sul. (C) Canoe or whaling boat of up to 12 m for 3–10 crew members in Santa Catarina. (D) Boats from 8 to 13 m with boathouses in Santa Catarina. (E) Fiber boat or canoe reported in all studied states. (F) Aluminum speedboats.

predominate in Rio Grande do Sul and up to the central portion of Santa Catarina. In the north of Santa Catarina, Paraná, and São Paulo, most of the current vessels have fiber and aluminum hulls due to the lightness and agility to overpass the surf zone, ease of repair, and high durability. Changes in the vessels' material were stimulated mostly by the difficult access to the wood of legal origin. Most artisanal vessels do not have a boathouse and operate in shallow waters close to the shore. Larger artisanal vessels usually have a boathouse and explore greater distances to shore and depths.

The vast coastal distribution of artisanal fishing under varying physical and oceanographic features, the diversity of target resources and fishing practices, and the number of people involved make this activity considerably diverse and complex in terms of monitoring and management actions. In addition, artisanal fishing adapts faster and more independently to new fishing nets and methods. Gillnets can be set on the surface or at the bottom, fixed or adrift (locally known as "caceio"), and actively from throws (seine fishing). Fixed nets with "poitas" for anchoring, can remain submerged for up to 24 h without the need for monitoring. Gillnets without this anchoring system require the fishers' monitoring, as both vessel and net drift and might move with the currents. The soaking time of drift nets is shorter than that of fixed nets, ranging from 1 to 10 h.

The fishers can cast gillnets actively by fishing throws (seine) to aggregate and surround the schools. The operation of seine fishing, as well as the corresponding name, varies according to the site. In this method, the vessel hangs the net to produce an enclosure (round) or a spiral movement (snail). Some regions (e.g. Ubatuba/SP, Matinhos/PR) allow a technique locally known as "bate-bate" at the end of net casting. This fishing technique consists of banging the paddles on the water or the boat to drive fish to the enclosure, where they entangle. Laguna city in Santa Catarina employs gillnet fishing throws to enclosure schools of mullets (*Mugil* sp.). The net used stands out for the height, reaching up to 60 m.

Most artisanal fisheries occur on a sun-to-sun basis (they depart and return on the same day) in daylight time, but night departures can occur occasionally. The usual fishing area locates near the municipalities, close to the shore, ranging from 10 to 20 m deep, but can reach up to 60 m, depending on the local shelf geomorphology (Table 16.2).

The period of the year when fishing is most profitable and productive also varies across the study area. In the south of Rio Grande do Sul, the best months for the artisanal bottom-set gillnet fishing are November and December, corresponding to late spring and early summer, when is the white mouth croaker (*Micropogonias furnieri*) season. In autumn and winter, the artisanal fishers prioritize surface-set gillnet fishing to catch Lebranche mullet (*Mugil liza*) and bluefish (*Pomatomus saltatrix*). Summer is the worst period because no target species with high economic value is abundant in the region.

In Santa Catarina and Paraná states, winter is the best season for artisanal fishing, mainly when the mullet schools are close to the coast. Spring is also

Table 16.2 Fishing strategies are applied by artisanal and industrial fisheries in different states.

Site	Target species	Gear	Depth (m)	Days	Time of day
ARTISANAL					
Rio Grande do Sul (RS)	Pelagic	Surface gillnet	5–25	Sun-to-sun	Day
Santa Catarina (SC)	Whitemouth croaker	Bottom gillnet	10–50	Sun-to-sun	Day and night
	Other demersal fish	Bottom gillnet	15–60	2–3	Day
	Lebranche mullet	Surface gillnet (ring seine)	10–50	Sun-to-sun	Day
	Other pelagic fish	Surface gillnet	10–60	Sun-to-sun 2–5	Day and night
Paraná (PR)	Demersal	Bottom gillnet	10–40	Sun-to-sun	Day and night
	Lebranche mullet	Seine	10–40	Sun-to-sun 2–4	Day
	Pelagic	Surface gillnet	10–40	Sun-to-sun	Day and night
São Paulo (SP)	Demersal	Bottom gillnet	4–50	Sun-to-sun 2–5	Day and night
	Lebranche mullet	Surface gillnet (seine)	4–50	Sun-to-sun	Day
	Pelagic	Surface gillnet	4–50	Sun-to-sun 2–4 or 5–10	Day and night
SEMI-INDUSTRIAL					
Rio Grande do Sul (RS)	Demersal	Bottom gillnet	5–25	Sun-to-sun (2–5)	Day
INDUSTRIAL					
Rio Grande do Sul (RS)	Demersal	Bottom gillnet	14–28	14–23	Day
Santa Catarina (SC)	Demersal	Bottom gillnet	10–100	15–30	Day
São Paulo (SP)	Demersal	Bottom gillnet	15–50	12–20	Day

The target species is detailed when it has a specific fishing method; otherwise, the species were divided into demersal and pelagic. The main target species of artisanal and industrial fisheries in the study site are (i) demersal: Brazilian codling (*Urophycis brasiliensis*), whitemouth croaker (*Micropogonias furnieri*), seatrout (*Cynoscion* sp.), acoupa weakfish (*Cynoscion acoupa*), smooth weakfish (*Cynoscion leiarchus*), stripped weakfish (*Cynoscion guatucupa*), southern king weakfish (*Macrodon atricauda*), kingcroaker (*Menticirrhus* sp.), shorthead drum (*Larimus breviceps*), Brazilian flounder (*Paralichthys orbignyanus*), snook (*Centropomus* sp.), white sea catfish (*Genidens barbus*); and (ii) pelagic: bluefish (*Pomatomus saltatrix*), Castin leatherjacket (*Oligoplites saliens*), Serra Spanish mackerel (*Scomberomorus brasiliensis*), and Lebranche mullet (*Mugil liza*). Gear refers to the fishing method, Depth = maximum depth of the fishing site, Days = days at sea (sun-to-sun = departure and return on the same day), Time of Day = time of day the fishing occurred.

favorable to harvest bluefish and white mouth croakers in Santa Catarina. Summer is the worst fishing season for both states due to the low abundance of fish and warmer water temperatures, which accelerate fish decomposition in nets. Despite the low productivity, summer provides good income to the fishers when selling directly to the consumer, usually vacationers.

In São Paulo, the best periods for artisanal fisheries extend from winter to summer. Summer is considered the best fishing season; although fishes move far from the coast due to the water temperature, the hake (*Cynoscion* sp.) moves closer and becomes the primary catch. In addition, the fishers obtain a better income from marketing directly to the consumer. Winter is a good fishing season due to the harvest of Lebranche mullet, Serra Spanish mackerel (*Scomberomorus brasiliensis*), and bluefish.

According to fishers from Santa Catarina, Paraná, and São Paulo, August is not a good month for fishing due to adverse weather conditions (intense winds and rough seas). During this period, the fishers work to repair and prepare their vessels and nets, waiting for the next fishing season.

Industrial fishing

Industrial vessels from all states have in common wooden hulls (Fig. 16.4) and greater fishing autonomy to operate in a broad area for many days at sea. The industrial vessels usually have three types of amateur radio, including a long-range communication system, besides radar, global positioning system (GPS), echosounder (for navigation but also detection of schools), autopilot, and tracking device required by law (PREPS-National Program for Satellite Tracking of Fishing Vessels).

The semi-industrial vessels (Fig. 16.4), although presenting considerable catch power (many get close to industrial levels), are exempt by the PREPS of using a tracker due to their size (<20 GT), despite their potential to catch endangered species.

The fleets of Santa Catarina and Rio Grande do Sul have higher catch and production power but differ in fishing structure, effort, and catch capacity, mainly in terms of political control and financial resources. Conversely, most of the larger vessels of the gillnet fleet in São Paulo have characteristics of semi-industrial fishing.

Regarding the industrial vessels' strategy, the net soaking time varies from 3 to 6 h according to the amount of fish captured. In general, the net is thrown in the early morning and collected in the late afternoon. The industrial vessels usually spend 14 to 24 days at sea, but the largest ones can stay for up to 30 days. In the south of Rio Grande do Sul, according

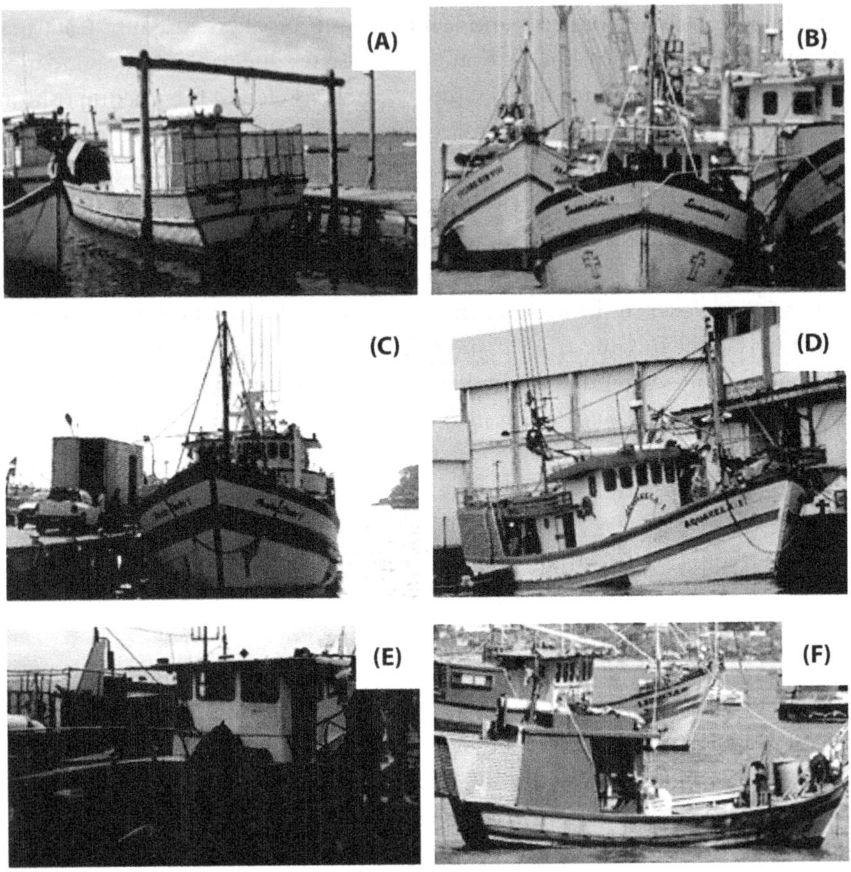

Fig. 16.4 ***Vessel types of the semi-industrial and industrial fisheries in different cities of the southeastern and southern regions in Brazil.*** (A) São José do Norte/RS (semi-industrial), (B) Rio Grande/RS (industrial), (C) Porto Belo/SC (industrial), (D) Itajaí/SC (industrial), (E) Cananéia/SP (semi-industrial), and (F) Ubatuba/SP (semi-industrial).

to shipowners, masters, and fishers, the semi-industrial vessels work with 4–5-mile (8–9 km) nets and have autonomy for up to 5 days at sea.

Industrial fishing catches white mouth croakers year-round, with only one break between May 15 to June 15, which corresponds to the closed fishing season (INI MPA/MMA 12/2012). Santa Catarina fleet targets white mouth croakers and operates in the south of Rio Grande do Sul in summer and the north during autumn and winter (Haimovici, 1997). Between autumn to early spring, the white mouth croaker is in greater depths (up to 100 m), while in summer, it occupies shallower areas (less than 50 m deep). Part of the fleet from Rio Grande do Sul alternates between the white mouth croaker and other bottom fish during the year.

Means of production and wages in gillnet fishing

The fishers in the study area working in both industrial and artisanal sectors have low education levels, therefore, little specialization, and work largely informally and without a labor contract (Gonçalves et al., 2018; Haimovici et al., 2014; Kalikoski and Vasconcellos, 2013). About 96% of workers in the fishing sector fall into four categories: (1) employed without a formal contract, (2) self-employed, (3) workers for their own consumption, and (4) unpaid workers (Garcia et al., 2018).

Artisanal fishers' average income per capita lies around 1.5–2 minimum wages in Paraná cities dominated by maritime fishing (minimum wage = 192 USD; PCSPA-PR, 2015). In São Paulo, the southern coast has a similar income compared to that of the northern coast of Paraná, with an income per capita above one minimum wage, while in other parts of São Paulo, it is around 1.5 minimum wage (PCSPA-PR, 2015; PCSPA-SP, 2015).

During the interviews, artisanal fishers declared fishing as their main economic activity throughout the study area, and most have the General Fishing Registry (RGP) required to engage in fishing activities. Wage results from the division of the total net catch. The income is usually divided into two equal parts to the proprietor and the crew. The crew divides their share into the number of workers, varying according to the vessel type (Table 16.2). The fish buyer can be an intermediary, the wet market, or directly the consumer, which guarantees a higher income to the fisher.

In terms of the operation mode of industrial fishing, the economic structure lies upon capital goods. There is a relation of dependency by the fishers toward the shipowners since the first do not own boats. In Rio Grande do Sul and Santa Catarina, shipowners usually do not work in the catch. Instead, they provide the means of production (boat, fuel, ice, gear, and ranch) and own the infrastructure (warehouses, berths), employing both fishers and warehouse workers.

Regarding wages, many fishers and masters who work on industrial vessels reported having a formal contract, especially in Itajaí (SC) and Cananéia (SP). However, in the south of Rio Grande do Sul, wages result from the division of the total net catch. Even though some masters have a formal contract and a fixed salary, it merely contributes to their retirement. Unlike artisanal fishing, landed fish goes to intermediaries in the industrial chain structure.

Environmental vulnerability

The reported conflicts by fishers evidence that a large part of the fishing community is subject to extreme social and environmental vulnerability, which result in disproportionate impacts and socioenvironmental risks experienced by these vulnerable populations (Acselrad et al., 2009). They are often victims of social and environmental issues which they are not responsible for (or are not the primary responsible), characterizing an injustice (Vasconcellos et al., 2007).

The socioenvironmental and socioeconomic vulnerability of fishing communities results from territorial dispute; lack of infrastructure services to store and sell their products (reducing the quality and value of fish); the informality and seasonality of fishing; the environment contamination; lack of public access to education and health care; the effects of climate change and overfishing; poverty; poor political representation; and finally, conflicting public policy (FAO, 2016). All the aspects impact the fishing communities studied, whether it is to a greater or lesser extent, associated or separately.

Communities under environmental injustice have a low resilience capacity due to scarce resources to adapt to significant changes in their environment, which amplifies the impact of new external pressures, such as, for instance, reducing their fishing area and resource availability.

Conflicts

For decades, the Brazilian State stimulated fishing through public policy that promoted tax incentives, oil subsidies, and programs that funded the acquisition of boats, nets, and engines to increase national fish production. Industrial fishing obtained more subsidies from the federal government through public policy to improve the fleet and increase the catching power (Azevedo and Pierri, 2014; Haimovici et al., 2014; MPA, 2011; Perez and Goméz, 2014). Instead, there was an attempt to regulate fishing through legislation that established limits for the construction of new vessels and release of licenses, fishing effort, length and height of nets, mesh size, and prohibited the fishing of endangered or overexploited species. The legislation also determined fishing quotes, closed seasons, and fishing exclusion zones. It is evident that the initiatives were conflicting, and there were failures in communication between the different sectors of the federal government. The analysis of the work entitled Fisheries Audit stated that there were systemic failures in the governance structure of Brazilian fisheries due to the fragility of public policy for the sector (Zamboni, 2020).

To understand the difficulties faced by artisanal and industrial fishing related to execution and profitability, it is crucial to understand the main conflicts reported and how these activities interact with all the actors and institutions related to fishing practices (Fig. 16.5).

Other fishing types

In general, gillnet fishers criticized trawling fishing, which they define as fisheries of low selectivity. According to the fishers, trawling catches a large amount of bycatch and small-sized species, which are usually discarded. Such operation mode precludes fish from achieving the proper size for catch and commercialization and reduces prey availability for gillnet-targeted species.

Ring seine fishing mode, especially in Rio Grande do Sul, is also considered by the gillnet fishers the responsible for the steep decline of fish over the last years. The vessels are smaller than 20 GT, allowing them to fish close to the coast and feed the market with a large amount of fresh and unmarked fish, which has a higher quality compared with the fish from gillnet fishing (Gonçalves et al., 2018). Ring seine fishing also catches founding

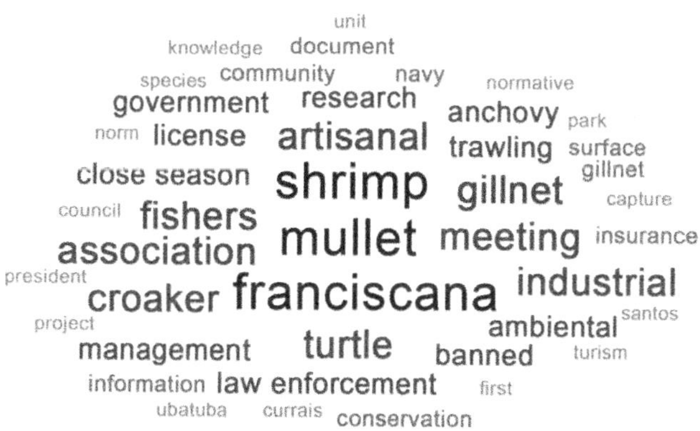

Fig. 16.5 ***Word cloud representing the words that appear more frequently in the conflicts session of the interviews realized in São Paulo, Santa Catarina, and Paraná.***

stocks of white mouth croakers, besides fishing in banned sites without any supervision or even regulation.

Overfishing

The decrease in fish is a topic commonly raised by the fishers throughout the study area. Most interviewees mentioned that the primary cause for the production drop is the growth of fishing effort, which includes an increase in net size and number of vessels, especially the industrial and semi-industrial types. The artisanal fishers complain it is an unfair competition given that they use smaller vessels, depend on favorable weather conditions, and the fish proximity to the shore. In addition, the higher catch potential of industrial vessels influences the fish price on the market, causing a decrease in value and, consequently, in the income of the artisanal fishers, according to them.

The industrial fishers in Rio Grande do Sul complain about the competition with semi-industrial vessels, which have a high catching power due to modifications in the stern (capacity goes from 8 to 25 tons), and for being smaller than 20 GT they can fish closer to the coast, making this competition even more uneven. Therefore, the interviewees declare the semi-industrial vessels are mainly responsible for catching endangered species.

Anthropogenic threats (pollution, coastal development, and climate change)

It is often difficult for the fishers to see their impacts on the marine environment in the face of many other anthropogenic threats. In many places, the fishers reported a decrease in fish and change in behavior caused by dredging activities, outfalls, water contamination by chemical contaminants, litter, activities to extract and transport natural gas and oil, and other coastal activities and infrastructure development. In addition, the fishers cited that climate change directly affects the marine environment, causing changes in the wind patterns, rougher sea, increase in water temperature, and spread of unwanted organisms.

Discard of elasmobranchs and other banned species

Under Brazilian law, catches of endangered species must return to the sea (dead or alive), but the landing is illegal and subject to a fine. Thus, according to interviewees, the industrial sector is pressured to discard tons (3–10 tons) of elasmobranchs bycaught. The discard causes outrage and displeasure among fishing actors, who do not understand why tons of food cannot serve as extra income for the fishers when other harvests are scarce (given the high commercial value of these species) or go to charities upon landing. They also believe that returning dead animals to the sea could harm the ecosystem; however, no alternatives for avoiding catching these species were proposed by the fishers during the studies.

Due to discards, the fishers, masters, and shipowners believe that these species populations (elasmobranchs) have already recovered and do not understand the ban. There is a lack in the understanding that other population parameters must be known, such as the size and maturation age of individuals, in addition to robust abundance estimates so that, over time, there is a review of the main bycatch species ban.

Scientific researchers and environmental managers

In general, the fishers nurture a relationship of distrust with the scientific researchers and managers from environmental agencies throughout the study area (Abreu et al., 2017). The suspicion relies on the fishers' belief that the sampling and use of research data aim to elaborate legislation that comes the opposite to their interests.

The conflict intensified with the implementation of INI MPA/MMA 12/2012, which, according to the fishers is based on their bycatch data.

Since then, there has been a strong resistance on the fishers' side to share the information obtained from catches of franciscana dolphins and other endangered species. They keep from reporting those species to the responsible bodies for fisheries statistics and hesitate to have onboard observers on their vessels.

The interviewees understand that working together, developing bottom-up approaches, is the only way to reveal the current state of fisheries stocks, the threats to endangered species and establish effective legislation that suits reality. Supplementing insufficient data will only happen with the involvement and protagonism of the fishers in the researches, the recognition of traditional knowledge, and the specificities of socioeconomic and socioenvironmental needs. It is necessary to establish trust-based relationships and programs of onboard observers that consider the fishers in the planning process, for the artisanal, semi-industrial and industrial fishing. A well-planned long-term program would contribute to a better understanding of fishing effort (temporal and spatial efforts), the targeted-species habits, and abundance, and mapping areas where there is the bycatch of endangered species.

Legislation

The conflicts tend to arise from implementing legislation aimed to rule the fishing activity and protect endangered species. When the fishing community notices that fishing areas and gear have to change, they usually resist such changes, react, and attempt to revoke the public policy. In many cases, they claim there was no effective comanagement process neither respectful of their cultural and traditional practices.

Even with the current conflicts reported below, publishing INI MPA/MMA 12/2012 was a remark in the regulation process of gillnet fishing. According to Occhialini et al. (2012), the INI MPA/MMA 12/2012 set relevant positive aspects, such as limits to the fishing effort of the industrial fleet and fishing exclusion zones in significant areas for endangered species conservation. However, according to the authors, the norm is hard to interpret and apply because it did not consider the regional specificities and diversity of the ecosystems and gillnet fisheries across the southeastern and southern coasts. Thus, the norm contributed to the breach in ongoing co-management processes within federal coastal and marine protected areas and other territorial laws/agreements.

Article 6 of the norm banned artisanal fishing vessels of the southeastern and southern regions within the first mile from the coastline, which, according to the fishers, precluded fishing in many localities, especially in São Paulo state, due to the diverse coastal geomorphology.

Moreover, the first mile overlaps with the INI MPA/MMA 10/2011 that limits the engine horsepower (maximum 18 HP), and the navy legislation restrains small fishing vessels past the first mile.

It seems that the fishers started to comply (not entirely) with the legislation only when law enforcement intensified, even though they do not understand the norm content thoroughly. In general, the interviewees related to artisanal fishing informed they did not participate in norming meetings and did not receive information on legislation demands.

The fishers solely knew they could not fish in certain areas or use some of their gear, but in most cases, they continued fishing, nonetheless. In Santa Catarina, the interviewees were unaware of this fishing restriction, claimed they did not hear about it, and thus declared they fish in the first mile.

In São Paulo, when law enforcement intensified by the year 2015, the fishers said they organized themselves to resist and react to several articles in the norm INI MPA/MMA 12/2012. At first, they held marine protection areas ("APAs Marinhas"- MPAs) responsible for creating the norm, which enhanced old conflicts. Ultimately, these conflicts required that social and institutional actors linked to fishing discussed and searched an effective co-management process. Once the Working Groups, established by the INI MPA/MMA 12/2012 to evaluate the norm impacts, were not put into practice, MPAs replaced the center of debates, which caused the modification of paragraph 4 of article 6 of the INI MPA/MMA 12/2012. Thus, a new Normative Instruction got published, the Normative Instruction No. 15/2020, which allows São Paulo fishing vessels (powered or not) to develop their activities in the first mile, as long as they are vessels of up to 10 m with nets of up to 1.000 m.

Law enforcement

The Environmental Police of each locality is involved in the most intense law enforcement actions, with the highest number of reported conflicts. In general, the fishers reported several conflicts with law enforcement officers due to a lack of dialogue and empathy toward the fishers' situation.

In the states where law enforcement is cited as stricter, like São Paulo and Paraná, the fishers reported they complain about some aggressive approaches by the officers. The fishers also declared they are treated differently by the policemen compared to the tourists or non-professional fishers, complying that this behavior demotivates many of them to initiate or continue in fishing activities.

The outrageous value of fines for fishing offenses was also a recurrent topic cited by the interviewees. In addition, the penalties often include confiscating the fishing gear and caught fish. In that way, the fishers cannot resume the fishing activity to earn their monthly income and pay their debts. The material to produce the fishing nets are expensive and require a high investment. Therefore, the fishers demand that their fishing material and fish are not confiscated and suggest a noticing system to avoid a penalty in the first offense. The fishers reported that the times they fish in the wrong places or use unallowed nets owe to the fact that they are unfamiliar with the legislation, or the legislation establishes fishing areas of difficult identification for the small-scale artisanal fishers, which do not have GPS on their vessels.

Bycatch

Even though this is a sensitive topic considering all the conflicts, most interviewees reported the bycatch of endangered species. The industrial and artisanal fishers reported catching several elasmobranchs and bycatch of other marine megafauna species, mainly marine turtles and dolphins, including franciscana.

Concerning the bycatch of franciscanas, the fishers observed they catch the species in varying fishing gear, such as bottom and surface gillnet, both in drifting ("caceio") and fixed nets, and that dolphins die when they entangle in the net. None of the fishers reported cutting the animals free while alive, but they believe that fishing as in "caceio" would enable them to do so.

In general, the fishers declared they catch franciscanas occasionally. Some of them mentioned their regions had higher abundances in the past compared to the current years, which they attribute to possible changes in dolphin behavior. None of the interviewees attributed the decrease of the franciscana to the bycatch rates. The fishers relate the drop in the number of franciscanas to other factors, such as pollution by effluents, chemical contaminants, and underwater noise.

In several opportunities, the fishers declared they disagree with the magnitude of the data on franciscana stranding reported by the different monitoring projects and programs. According to them, bycatch is a rare and sparse event in their communities. Nonetheless, they might not notice that, on a large scale, many localities combined can contribute to a high number of bycatches.

Concluding remarks

According to the perception of the interviewees, the fishing activity has faced several challenges: climate change, overfishing, a surfeit of vessels, issues to complete the crew and dispatch the vessel, disqualified labor (from deckhand to inexperienced masters), higher costs, lower fish price, and new fishing types that exceed the norms and shatter the fish stocks and law enforcement. In addition, nowadays, artisanal and industrial fishing need greater efforts to obtain a profitable production, resulting in the use of longer fishing nets and extended working hours. For example, the industrial fishing travels went from one week to 23–30 days.

The shipowners and fishers feel overwhelmed amidst several conflicts, regulations, restrictions, and law enforcement agencies. They claim to be the victims of a fining industry, unrealistic working laws, and criminal-like treatment instead of the fair treatment they deserve as employers and workers.

Conversely, there is an eagerness on their side to reverse the scenario. The fishers and shipowners understand the need for scientific research, management actions (e.g. closed seasons), strict law enforcement - to everyone - and penalties that fit their reality. However, it is important to highlight that even the fishing sector cares for its interests, they are making room for dialogue and agreement.

The public policy of a people-oriented government between 2003 and 2009 gave hopes of benefiting artisanal fishing. Through the speech on global sustainable development (UN's 2030 Agenda for Sustainable Development), there was an attempt to reconcile environmental conservation with economic growth and reduce poverty (Azevedo and Pierri, 2014). The closed season insurance (monthly amount for the fishers during mandatory cease of fishing activities by law) is an example of effective public policy to mitigate poverty and protect endangered species. Nonetheless, the public policy overly concerns and benefits industrial fishing, leading fish stocks to an overfishing status, which directly affects artisanal fishing, an aspect neglected by decision-makers.

It is vital to get a better understanding of the socioeconomic and socio-environmental realities of the fishing communities. It is relevant especially from the perspective of social and institutional actors who intend to get involved in fisheries through management. It is also relevant to know and determine the roles and responsibilities of the institutional agents that play a direct or indirect part in the sector.

In addition, the decision-making must rely on both quantitative and qualitative data. The qualitative aspects are often decisive in the interface between science and politics and contribute to a better understanding of the actions and reactions of the actors involved. Ultimately, the latter indicates whether there will be a collaboration by a large part of the sectors involved, whether the proposal has objectives in common and means to an agreement, or whether there are conflicting objectives to which the sides will follow in opposite directions (Howlett, 2011).

It is also significant to understand that public policy directed to the fishing sector is often generalist and establishes shared norms to diverse realities without attention to the local and regional particularities and needs. Thus, public policy can be detrimental to the fishers in some areas and intensify the environmental, social, and economic vulnerability of many fishing communities, especially the artisanal ones (Vasconcellos et al., 2007). For that reason, it is essential to understand that public policy impacts the national territory in diverse ways, which makes it meaningful to comprehend its limitations to optimize the expected outcomes (Resende, 2017).

It is essential to recognize institutional competencies to manage the fisheries sector keeping the endangered species conservation in mind, or even for the effective management of fisheries resources. It is necessary to go beyond it and search for an agreement through the co-management of resources. Thus, it is paramount to consider the traditional knowledge, traditional fishing communities' rights over their territory and fishing resources within, and environmental and economic vulnerabilities to achieve sustainable artisanal and industrial fishing and guarantee that the lifestyle and practices of these communities will thrive (Azevedo and Pierri, 2014).

The principle is taking the opposite direction than the one paved by conservationism, which comes from a scientific process engaging society instead of a social process engaging science. In addition, it requires a novel mindset, skills, and resources to navigate the spaces efficiently instead of persisting in producing the best evidence to bridge knowledge gaps (Toomey et al., 2016). This paradigm shift is vital and represents a challenge for the constitution of conservationist-biased public policy.

References

Abreu, J.S., Domit, C., Zappes, C.A., 2017. Is there dialogue between researchers and traditional community members? The importance of integration between traditional knowledge and scientific knowledge to coastal management. Ocean Coast. Manag. 141, 10–19.

Acselrad, H., Mello, C.C.A., Bezerra, G., das, N., 2009. O que é Justiça Ambiental? Editora Garamond, Rio de Janeiro.

Azevedo, N.T., Pierri, N., 2014. A política pesqueira no Brasil (2003-2011): a escolha pelo crescimento produtivo e o lugar da pesca artesanal. Desenv. e Meio Ambiente 32, 61–80.

Bertozzi, C.P., Zerbini, A.N., 2002. Incidental mortality of Franciscana (Pontoporia blainvillei) in the artisanal fishery of Praia Grande, São Paulo State, Brazil. LAJAM 1 (1), 153–160. doi: 10.5597/lajam00019.

Boffo, M.S., Reis, E.G., 2003. Atividade pesqueira da frota de média escala no extremo sul do Brasil. Atlântica (Rio Grande) 25 (2), 171–178.

CEPERG, 2011. Desembarque de Pescados no Rio Grande do Sul. Sistema controle de desembarque. https://www.icmbio.gov.br/cepsul/images/stories/biblioteca/download/estatistica/rs/est_2011.pdf (accessed July 20, 2020).

Di Beneditto, A.P., Ramos, R., 2001. Biologia e conservação de pequenos cetáceos no Norte do Estado do Rio de Janeiro. Fundação Estadual do Norte Fluminense, Campos dos Goytacazes, Brasil. 94.

Di Beneditto, A.P.M., Rocha-Campos, C.C., Danilewicz, D., Secchi, E.R., Moreno, I.B., Hassel, L.B., Tavares, M., Ott, P.H., Siciliano, S., Souza, S.P., Alves, V.C., 2010. Plano de Ação Nacional para a Conservação do Pequeno Cetáceo Toninha: Pontoporia blainvillei. Brasília: Instituto Chico Mendes de Conservação da Biodiversidade, 76 ICMBio. Série Espécies Ameaçadas.

FAO - Food and Agriculture Organization, 2016. The state of world fisheries and aquaculture 2016. Contributing to Food Security and Nutrition for All. Rome. p. 200.

FURG/SEMA, 2020. Boletim da pesca industrial marinha no Rio Grande do Sul–2019. Laboratório de Recursos Pesqueiros Demersais e Cefalópodes - Instituto de

Oceanografia - FURG. p. 28. https://demersais.furg.br/images/Boletim_da_pesca_industrial_marinha_no_Rio_Grande_do_Sul__2019_23_abr_2020.pdf. (accessed July 20, 2020).

Garcia, F., Abdallah, P. R., & Sachsida, A., 2018. Avaliação de efeitos do Programa Profrota Pesqueira sobre indicadores do mercado de trabalho (No. 2391). Texto para Discussão. Instituto de Pesquisa Econômica Aplicada (IPEA), Brasília.

Gonçalves, I.C., Di Tullio, J.C., Fruet, P.F., 2018. Avaliação da efetividade e a percepção dos pescadores artesanais sobre a área de proteção ao boto, Tursiops truncatus, no estuário da Lagoa dos Patos e áreas adjacentes: Relatório sobre o Meio Socioeconômico, Etnoecologia e Educação Ambiental. ONG KAOSA, Rio Grande.

Haimovici, M., 1997. Recursos pesqueiros demersais da Região Sul. FEMAR, Rio de Janeiro.

Haimovici, M., Castello, J.P., Abdallah, P.R., 2014. Desenvolvimento da pesca industrial sediada em Rio Grande: Uma visão histórica sob a ótica de atores privilegiados. In: Haimovici, M., Andriguetto, M., Sunye, P. (Eds.), A pesca marinha e estuarina no Brasil: Estudos de casos multidisciplinares. Editora da FURG, Rio Grande.

Howlett, M., 2011. Public managers as the missing variable in policy studies: An empirical investigation using Canadian data. Rev. Policy Res. 28 (3), 247–263.

Kalikoski, D.C., Vasconcellos, M., 2013. Estudo das condições técnicas, econômicas e ambientais da pesca de pequena escala no estuário da Lagos dos Patos, Brasil. Fisheries and Aquaculture Circular No. 1075. Rome. FAO, p. 200.

Lucena, F.M., Reis, E.G., 1998. Estrutura e estratégia de pesca da anchova Pomatomus saltatrix (Pisces, Pomatomidae) na costa do Rio Grande do Sul. Atlântica 20, 87–103.

Maxim, L., van der Sluijs, J.P., 2011. Quality in environmental science for policy: assessing uncertainty as a component of policy analysis. Environ. Sci. Policy 14 (4), 482–492.

Monteiro, D., Prado, J., Botta, S., Estima, S., Secchi, E., 2019. Temporal trends in mortality patterns of the endangered franciscana dolphin (Pontoporia blainvillei) in southern Brazil. In: World Marine Mammal Conference, 19, 2019. Book of Abstracts. Wmmc, Barcelona, pp. 295. Available from: https://www.wmmconference.org/wp-content/uploads/2020/02/WMMC-Book-of-Abstracts.pdf (accessed March 1, 2021).

MPA, 2011. Ministério da Pesca e Aquicultura. Boletim Estatístico da Pesca e Aquicultura, 2011. https://www.icmbio.gov.br/cepsul/images/stories/biblioteca/download/estatistica/est_2011_bol__bra.pdf (accessed July 20, 2020).

Nogueira, B.G., 2016. Aplicação do conhecimento de pescadores artesanais para entender a captura incidental de tartarugas marinhas no sul do Brasil. Dissertação (mestrado em Ecologia). Rio Grande do Sul, Universidade Federal do Rio Grande do Sul - UFRGS. p. 109.

Occhialini, D.S., Rodrigues, A.M.T., Kotas, J.E., 2012. Caracterização e análise comparativa da pesca industrial de emalhe-de-fundo costeiro, considerando a evolução física da frota a partir de 1995 e a produção pesqueira entre 2001 e 2010. Revista CEPSUL-Biodiversidade e Conservação Marinha 3 (1), 1–21.

PCSPA-PR, 2015. Caracterização socioeconômica da atividade pesqueira e aquícola nos municípios do litoral dos estados de São Paulo e do Paraná – PCSPA-BS. Relatório técnico final - Paraná. https://www.comunicabaciadesantos.com.br/sites/default/files/resultados_pr_pcspa_bs.pdf (accessed July 20, 2020).

PCSPA-SP, 2015. Caracterização socioeconômica da atividade pesqueira e aquícola nos municípios do litoral dos estados de São Paulo e do Paraná – PCSPA-BS. Relatório técnico final – São Paulo. https://www.comunicabaciadesantos.com.br/sites/default/files/resultados_sp_pcspa_bs.pdf (accessed July 20, 2020).

Pérez, M.S., Gómez, J.R.M., 2014. Políticas de desenvolvimento da pesca e aquicultura: Conflitos e resistências nos territórios dos pescadores e pescadoras artesanais da vila do Superagüi 26. Sociedade & Natureza, Paraná, pp. 35–47.

PMAP-BS, 2019a. Projeto de Monitoramento da Atividade Pesqueira na Bacia de Santos PMAP-BS – Relatório Técnico Semestral – Janeiro a Junho de 2019. https://www.

comunicabaciadesantos.com.br/sites/default/files/Consolidado_PMAP-BS_1SEM2019.pdf (accessed July 20, 2020).

PMAP-BS, 2019b. Projeto de Monitoramento da Atividade Pesqueira na Bacia de Santos PMAP-BS – Relatório Técnico Semestral – Julho a Dezembro de 2019. https://www.comunicabaciadesantos.com.br/sites/default/files/PMAP-_16_07_20.7z (accessed July 20, 2020).

Reis, E.G., 1992. An assessment of the exploitation of the white croaker Micropogonias furnieri (Pisces, Sciaenidae) by the artisanal and industrial fisheries in coastal waters of southern Brazil PhD Thesis. University of East Anglia, p. 253.

Reis, E.G., 1993. Classificação das atividades pesqueiras na costa do Rio Grande do Sul e qualidade das estatísticas de desembarque. Atlântica 15, 107–114.

Resende, G.M., 2017. Avaliação de políticas públicas no Brasil: uma análise da política nacional de desenvolvimento regional; vol. 3. Instituto de Pesquisa Econômica Aplicada.

Secchi, E.R., Danilewicz, D., Ott, P.H., 2003. Applying the phylogeographic concept to identify franciscana dolphin stocks: implications to meet management objectives. J. Cetacean Res. Manag. 5 (1), 61–68.

Secchi, E.R., Ott, P.H., 1997. The depth factor determining the distribution and abundance of franciscana, Pontoporia blainvillei, as demonstrated by a CPUE index approach, Paper DT18 presented to III Taller para la Coordinación de la Investigación y la Conservación de la Franciscana en el Atlántico Sudoccidental. Buenos Aires, Argentina.

Siciliano, S., 1994. Review of small cetaceans and fishery interactions in coastal waters of Brazil. Rep. Int. Whaling Comm. 15 (Special Issue), 241–250.

Toomey, A.H., Knight, A.T., Barlow, J., 2016. Navigating the space between research and implementation in conservation: research-implementation spaces. Conserv. Lett. 10 (5), 619–625.

Vasconcellos, M., Diegues, A.C., Sales, R.R., 2007. Limites e possibilidades na gestão da pesca artesanal costeira. In: Costa, A.L. (Ed.), (Org.) Nas redes da pesca artesanal. IBAMA, Brasília, pp. 2–70.

Vasconcellos, M., Haimovici, M., Ramos, K., 2014. Pesca de emalhe demersal no sul do Brasil: evolução, conflitos e (des) ordenamento. A pesca marinha e estuarina no Brasil: estudos de caso. Editora da FURG, Rio Grande, pp. 29-40.

Zamboni, A., 2020. Auditoria da pesca: Brasil 2020: uma avaliação integrada da governança, da situação dos estoques e das pescarias. /Ademilson Zamboni, Martin Dias, Lara Iwanicki. – 1. ed. – Brasília, DF: Oceana Brasil, 2020. https://brasil.oceana.org/sites/default/files/auditoria_da_pesca_-_brasil_2020.zip (accessed July 20, 2020).

CHAPTER 17

Ethnoecological approaches and fishermen perception

Camilah Antunes Zappes[a], Ana Paula Madeira Di Beneditto[b], Camila Domit[c]

[a]Departamento de Oceanografia, Centro de Ciências Humanas e Naturais, Universidade Federal do Espírito Santo, Vitória, Espírito Santo, Brazil
[b]Laboratório de Ciências Ambientais, Universidade Estadual do Norte Fluminense Darcy Ribeiro, Campos dos Goytacazes, Rio de Janeiro, Brazil
[c]Laboratório de Ecologia e Conservação, Universidade Federal do Paraná, Pontal do Sul, Paraná, Brazil

Artisanal fishermen and traditional knowledge

Artisanal fishing is an economic activity practiced as an exclusive activity by approximately 39 million fishermen worldwide (FAO, 2020). This traditional activity is responsible for approximately 55% of the world's food provision, with men and women working in the production chain to ensure food security and nutrition, in addition to positively participating in the eradication of poverty (FAO, 2018, 2020).

Artisanal fishermen work in environments close to the coast and make use of vessels with little technological apparatus and artifacts considered to be artisanal, capturing all kinds of aquatic species for subsistence or for small-scale commercial purposes (FAO, 2020) (Fig. 17.1). In this activity, fishermen can act independently or in a family economy regime and without formal employment (Begossi et al., 2012; Diegues, 2000). Due to the daily practice of artisanal fishing, artisanal fishermen maintain direct and constant contact with the marine environment, which enables them to acquire traditional knowledge about species (Begossi et al., 2012; Zappes et al., 2011).

Traditional knowledge in Brazil belongs to communities considered traditional (e.g., *caiçaras*, artisanal fishermen, *quilombolas*, and indigenous peoples). It is conceptualized as individual knowledge and know-how in relation to the natural and spiritual worlds, with a link between the environment and the supernatural and social organization in these populations (Begossi et al., 2012). To build this local knowledge, communities must develop systems for classifying everything that surrounds them based on empiricism (Diegues, 2000). This classification system belongs

The Franciscana Dolphin
DOI: https://doi.org/10.1016/B978-0-323-90974-7.00013-6

Fig. 17.1 (A), (B), (C), and (D)—Vessels used for artisanal fishing along the coastal area of the Southwest Atlantic Ocean; (E) and (F)—Artisanal fishermen fishing without vessels; (G)—Popular market for artisanal fish in Brazil.

to the cultural heritage of each people and expresses how the community adapts to the environment. For this reason, it must be interpreted specifically for each culture in which it is generated. Each people has its own classificatory form and its own relationship with the environment; in this respect, each community must be studied in a specific manner. In addition to using a popular taxonomic classification, local actors from different communities hold information on climate and the behavior, occurrence, and distribution of species (Alves et al., 2018, 2020; Zappes et al., 2014, 2016).

These traditional populations have developed techniques adapted to each region and time, and their survival depends on the knowledge and skills that members have about the places where they live. Their actions in a territory as well as the determination of which resources are needed are based on years of observation of how the environment reacts to human interference. From this observation, new information is added with each generation and passed on orally from the oldest members of the community to the youngest members (Begossi, 2001; Zappes et al., 2014).

Since the 1970s, proposals to incorporate traditional communities, their knowledge, and their cultural practices into the process of environmental conservation have been implemented in countries in the Southern Hemisphere (Diegues, 2000). Interest in local knowledge is about not only understanding the environment but also learning how to use and maintain it (Abreu et al., 2017; Begossi, 2001). Thus, traditional knowledge from artisanal fishermen about the franciscana dolphin and its interactions with fisheries is an important contribution to understanding this species, its habitat, and its relationships. This traditional knowledge added to scientific knowledge is the basis for building more efficient mechanisms for bringing science and society closer together. Gathering traditional information is crucial for the elaboration of plans and decision-making related to fisheries management and species conservation measures (Rosa et al., 2012; Zappes et al., 2016).

The ethnoecology of the franciscana dolphin

Ethnoscience is the study of the knowledge from human populations about natural phenomena; that is, it identifies the logic present in human knowledge about the natural world, taxonomies, and complete classifications (Diegues, 2000). This science compares the knowledge practiced by human communities and that described in the scientific literature, using linguistics as a methodological framework (Fisher and Young, 2007). Ethnoecology, as a branch of Ethnoscience, allows the integration between traditional knowledge and academic knowledge in the areas of ecology and biology, since these are sciences aimed at the study of the knowledge of human groups, their practices, and their beliefs in relation to ecosystems (Diegues, 2000).

The daily contact of fishing communities with small cetaceans provides a unique opportunity to monitor the impacts of artisanal fishing activities on these animals (Souza and Begossi, 2007; Zappes et al., 2014, 2016). Thus, ethnoecological studies related to the franciscana dolphin contribute to the understanding of different interpretations of information for this dolphin, from its biology to its interactions with human activities (Rosa et al., 2012; Zappes et al., 2016). To address the ethnoecology of fishermen regarding the Franciscana dolphin, studies have been carried out on the coast of the South Atlantic Ocean, more specifically on the Brazilian coast (e.g., Araujo, 2016, 2020; Castilhos, 2014; Freitas-Netto and Di Beneditto, 2008; Hassel, 2006; Pinheiro and Cremer, 2003; Przbylski and Monteiro-Filho, 2001; Rosa et al., 2012; Zappes et al., 2016). In these studies, information was collected mainly by conducting interviews with artisanal fishermen to understand their traditional knowledge of the franciscana dolphin (Fig. 17.2).

In Brazil, the identification of the franciscana dolphin by artisanal fishing communities is reported to the following franciscana management areas (suggested by Cunha et al., 2014): (1) FMA Ia [communities at Conceição da Barra (18° 35'S; 39° 43'W), Regência (19° 38'S; 39° 38'W), and Barra do Riacho (19°49'S; 40°16'W), state of Espírito Santo] (Araujo, 2016, 2020; Freitas Netto and Di Beneditto, 2008), (2) FMA Ib [community at Atafona (21°37'S; 41°01'W) (Araujo, 2020; Rosa et al., 2012) and the Lagos microregion (22°46'S; 22°36'S) (Hassel, 2006), state of Rio de Janeiro], 3) FMA IIa [São Sebastião, state of São Paulo (23°42′-23°45′S; 45°25′-45°53′W)] (Souza and Begossi, 2007), (4) FMA IIb [Cananéia-Iguape Estuary Lagoon Complex, state of São Paulo (24°40′-25°05′S; 47°25′-47°56′W) (Desvaux, 2013, Silva, 2007), and communities located on the coast of the state of Paraná (~25°12'S; 25°58'S) (Araujo, 2020; Gama, 2013; Przbylski and Monteiro-Filho, 2001; Zappes et al., 2016) and Babitonga Bay (26°16'S; 48°41'S), state of Santa Catarina (Pinheiro and Cremer, 2003)], and 5) FMA IIIa [communities located on the north coast of the state of Rio Grande do Sul (~29°19'S; 29°58'S)] (Castilhos, 2014).

Fig. 17.2 ***Collection of ethnoecological information on the franciscana dolphin from artisanal fishing communities off the coast of Brazil.*** *Source*: Grupo Ecologia Humana.

The popular names given to the Franciscana dolphin by Brazilian fishermen are shown in Table 17.1.

Descriptions given by artisanal fishermen in Brazil about the color (e.g., yellowish, brown, pink), body size (e.g., the total length between 1 and 1.5 m), and range of the franciscana dolphin (shallow coastal areas, up to 30 m deep) (Araujo, 2016, 2020; Gama, 2013; Rosa et al., 2012) coincides with information in the literature on the physical characteristics and areas of occurrence of this species (e.g., Amaral et al., 2018; Crespo, 2002; Culik, 2011; Cunha et al., 2014; Di Beneditto and Ramos, 2001; Jefferson et al., 2015; Rosas et al., 2002, Secchi et al., 2002, Trimble and Praderi, 2006). The factors that limit the occurrence of species populations involve variables such as depth, temperature, salinity, ichthyofauna distribution, and water turbidity (Amaral et al., 2018; Gomez and Cassini, 2014; Siciliano et al., 2002), and the areas of greatest use by P. blainvillei are also those with

Table 17.1 Popular names given to the franciscana dolphin by artisanal fishermen in the Brazilian coastal zone.

Site	Franciscana management areas (FMA)	Popular name	References
Regência, ES(19°98'S; 39°38'W)	FMA Ia	Manico or cachimbo	Freitas Netto and Di Beneditto (2008)
District of São Sebastião, SP (23°42′-23°45′S; 45°25′-45°53′W)	FMA IIa	Toninha	Souza and Begossi (2007)
Island of Peças, PR(25°27'S; 48°20'W)	FMA IIb	Toninha	Gama (2013)
Island of Superagui, PR(25°28'S; 48°13'W)	FMA IIb	Toninha	Gama (2013)
North coast of the state of Rio Grande do Sul, RS(~29°19'S; 29°58'W)	FMA IIIa	Toninha	Castilhos (2014)

the highest concentration of artisanal fishing, which directly influences the opportunities for fishermen to observe this dolphin (Rosa et al., 2012; Zappes et al., 2016).

According to reports by fishermen, the franciscana dolphin is mainly observed in the winter (from June to August) and summer (from December to February) due to the availability of prey, but sightings vary between FMAs and can also occur in other seasons (Hassel, 2006; Zappes et al., 2016). In FMA Ib, there are records of species sightings throughout the year close to the coastline, with greater occurrence in the spring (September to November) and summer (Di Beneditto and Ramos, 2001). In FMA IIb, artisanal fishermen have described seeing adult franciscana dolphins throughout the year, with a higher frequency of sightings of infants in the summer and winter (Zappes et al., 2016). The difference in seasonal occurrence noted between the distribution areas (and studies) is probably due to environmental factors, such as water temperature, the presence of water bodies, photoperiod, predation, and the availability of food resources for the species (Bertozzi, 2009).

Ethnoecological records have described group sizes for the franciscana dolphin ranging from 11 to 30 individuals in FMA IIa (Hassel, 2006) and 2 to 5 individuals in FMA IIb (Zappes et al., 2016). This information converges with literature based on field observations of the species, including variations in the average size of groups (e.g., Bordino et al., 1999; Cremer and Simões-Lopes, 2005; Santos et al., 2009; Secchi et al., 2001, Sucunza et al., 2018). The frequency of small groups of this small animal is one of the difficulties of sighting in the field by researchers; however, it does not seem to be a barrier to the construction of knowledge among fishermen, since they work at sea daily.

Among some fishing communities in Brazil, the franciscana dolphin is considered "a mammalian fish" since it lives in an aquatic environment, despite fishermen having identified breastfeeding behavior (Hassel, 2006). The observation of breastfeeding in this species was also mentioned following a study carried out on islands of the state of Paraná (Zappes et al., 2016), a fact that underscores the detailed knowledge and opportunities for recognition of individuals by fishermen. Ethnoecological studies on the behavior of the franciscana dolphin have described behavioral categories of feeding/fishing as the ones most frequently observed by fishermen; observations of socialization, displacement, rest, parental care, and bait theft have been recorded but were rare behavioral events (Devaux, 2013; Hassel, 2006; Zappes et al., 2016). These results are similar to the descriptions found in the literature based on field research to record species behaviors (Bordino et al., 2002; Cremer and Simões-Lopes, 2005; Cremer et al., 2006; Jefferson et al., 2015; Melcón et al., 2012). The feeding/fishing behavior is probably the most described by fishermen because they observe these animals in the areas used for artisanal fishing or in association with fishing activities. Fishermen have described the franciscana dolphin diet as being composed of fish from the families Clupeidae, Engraulidae, and Mugilidae; squids from the family Loliginidae; and shrimp from the family Penaeidae (Hassel, 2006; Zappes et al., 2016). This corroborates literature records based on analysis of the stomach contents of the franciscana dolphin (e.g., Bittar and Di Beneditto, 2009; Botta et al., 2015; Di Beneditto and Ramos, 2001).

Franciscana dolphin prey species are generally common in estuarine and coastal regions and are known to be targets for artisanal fishing (Bittar and Di Beneditto, 2009; Botta et al., 2015; Di Beneditto and Ramos, 2001). The overlap of the areas of use and areas of fish occurrence impacts the interactions between artisanal fishermen and

P. blainvillei individuals, which are very common in the Southwest Atlantic Ocean. Incidental catches of this species in coastal fisheries that use gillnets (or casting nets) are frequent and consequently are the main threats to this species' conservation throughout its range (e.g., Bertozzi and Zerbini, 2002; Crespo et al., 2010; Danilewicz et al., 2010, Di Beneditto, 2003; Franco-Trecu et al., 2009; Ott et al., 2002; Rosas et al., 2002; Santos et al., 2009; Secchi et al., 2002). Over time, this type of interaction has reduced the number of individuals in franciscana dolphin populations (FMAs) (Danielwicz et al., 2010; Secchi et al., 2002). Ethnoecology studies have indicated that this type of negative interaction is recognized by fishermen, who also called attention to the fact that dolphins damage casting nets when they are caught incidentally. Records of this perception by fishing communities have been recorded in FMA Ia (Araujo, 2016, 2020; Freitas Netto and Di Beneditto, 2008), FMA Ib (Araujo, 2020; Hassel, 2006; Rosas et al., 2012), FMA IIb (Araujo, 2020; Desvaux, 2013; Gama, 2013; Pinheiro and Cremer, 2003; Przbylski and Monteiro-Filho, 2001; Silva, 2007), and FMA IIIa (Castilhos, 2014).

Fishermen believe that the franciscana dolphin is caught because it gets entangled in the net when trying to feed on the fish attached to an artifact because it does not see the net, or because it feeds (or occurs) in the same area where artisanal fishing is practiced (Hassel, 2006; Gama, 2013; Araujo, 2016, 2020). According to reports, incidental capture of these animals occurs in casting nets that remain in the sea for a long period of time, for example, nets that are cast at night and collected the next morning (Araujo, 2016, 2020; Pinheiro and Cremer, 2003; Gama, 2013). The casting nets in which incidental catches occur have a mesh size ranging from 10 to 200 mm (between nonadjacent knots) and operate at depths of 10–30 m at a distance of up to 5 miles from the coastline (Castilhos, 2014; Desvaux, 2013; Gama, 2013; Pinheiro and Cremer, 2003; Przbylski and Monteiro-Filho, 2001). The characteristics reported by fishermen regarding the depth and distance from the coastline overlap with the preferred area of occurrence of P. blainvillei (Amaral et al., 2018; Bertozzi, 2009; Cunha et al., 2014; Danilewicz et al., 2010).

To reduce the incidental capture of the franciscana dolphin, fishermen have suggested stopping the use of this type of artifact, despite claiming, at the same time, that this would be financially unfeasible for the socioeconomic maintenance of artisanal fishing (Araujo, 2016, 2020; Gama, 2013). Other suggestions given by fishermen to mitigate the negative impacts of fishing practices on the franciscana dolphin were (1) to prohibit the use of trawls and casting nets with large meshes, (2) to decrease the number of vessels in operation, (3) to create defense periods for franciscana dolphin prey species that coincide with target species of the fisheries, and (4) to establish environmental protection areas. Castilhos (2014) suggested as a mitigating measure the establishment of a bathymetric limit for fishing operations be defined by a common agreement between stakeholders. Di Beneditto (2003) suggested that casting nets should be cast beyond 10 nautical miles away from the shoreline of FMA Ib due to the higher rates of incidental catches of this species occurring within that bathymetric interval. Different measures were suggested by different communities, possibly encompassing the local realities and even the ability of each group to reach deeper areas that are further from the coast or use a greater diversity of fishing gear.

Regarding the franciscana dolphin carcasses that are captured, the fishermen reported that they discard them at sea, use them as bait when fishing for elasmobranchs and, less frequently, use them for human consumption or oil extraction to waterproof the hull of vessels (Araujo, 2016; Castilhos, 2014; Devaux, 2013; Di Beneditto et al., 1998; Gama, 2013; Hassel, 2006; Przbylski and Monteiro-Filho, 2001; Rosa et al., 2012; Siciliano, 1994). In some situations, discarding carcasses at sea was linked to fishermen's fear of legislation in Brazil, which prohibits intentional capture and molestation of cetaceans (Brazil, 1987; Federal Law no. 7.643/87). It should be noted that the use of franciscana dolphin carcasses for human consumption has been reported, in most cases, as part of an ancestral culture but one that no longer occurs (Araújo, 2020; Gama, 2013; Rosa et al., 2012). There have been reports of food taboos regarding the consumption of franciscana dolphin carcasses, as their dark and greasy muscle is not appealing as food (Pinheiro and Cremer, 2003).

Artisanal fishermen have also reported that there are positive interactions with franciscana dolphins, with individuals assisting in fishing practices. This type of interaction has been described in FMA Ia (Araujo, 2016, 2020), FMA IIa (Hassel, 2006), and FMA IIb (Araujo, 2016, 2020; Gama, 2013; Przbylski and Monteiro-Filho, 2001; Zappes et al., 2020). According to the fishermen, the feeding behavior of the franciscana dolphin helps them to locate schools of fish targeted by fisheries (Zappes et al., 2016). The use of sightings for locating schools justifies the fishermen's interest in the behaviors of franciscana dolphins throughout the sampled areas. However, this practice also shows that fishermen recognize the areas where the franciscana dolphin occurs because they coincide with the areas used for artisanal fishing (Araújo, 2016, 2020; Gama, 2013; Hassel, 2006).

Ethnoecological studies have identified the preferred areas of distribution of the franciscana dolphin as being close to estuaries and river mouths, which are characterized by high water turbidity (Araujo, 2016, 2020; Freitas Netto and Di Beneditto, 2008; Gama, 2013; Hassel, 2006). This coincides with the assessment of the distribution pattern of this species previously presented in Siciliano et al. (2002), who attributed this pattern to the ability to obtain prey and reduce predation risks. Some ethnoecological studies have found evidence for the spatial overlap between fishing fields, the areas of occurrence of the franciscana dolphin, and areas of incidental capture by gathering data from traditional and scientific knowledge (FMA Ia, Ib, and IIb: Araújo, 2016, 2020; Gama, 2013). This overlap is reflected in incidental catches and poses an even greater challenge to fisheries and territorial management to ensure the sustainability of fisheries and conservation of the franciscana dolphin.

The gathering of data on the areas of occurrence and incidental capture of the franciscana dolphin based on traditional and scientific knowledge should be considered in fisheries management actions to help minimize the anthropogenic impact on this species (Rosa et al., 2012). In addition, only with the aggregation of knowledge is it possible to build proposals for integrated management, which ensure the social and economic development of artisanal fishing communities and the survival of species. Understanding the different interpretations of interactions involving fishermen and the franciscana dolphin can provide opportunities for the development of participatory and regionalized management actions along the Southwest Atlantic Ocean. Mitigation actions for bottom-up environmental impacts are more likely to achieve positive and long-term results, as all the actors involved feel valued and heard in the process (FAO, 2021). Fishermen working on the Brazilian coast within the range of the franciscana dolphin have shown a positive perception of this animal, with reports that suggest admiration and respect for the fact that the franciscana dolphin is a "beautiful" and "powerful" animal and one that helps in fishing (Hassel, 2006). This positive perception facilitates the addition of educational activities about the franciscana dolphin to the routine of fishing communities, which strengthens the need to understand traditional knowledge, the perception of natural elements, and the cultural relationships of artisanal fishermen to achieve the goal of conservation of dolphins such as P. blainvillei (Zappes et al., 2016).

Conservationist aspects

In the Southwest Atlantic Ocean, only Brazil, together with governmental conservation goals, has specific indications related to the need to obtain information from fishing communities about the franciscana dolphin. These indications are aimed at building actions to mitigate intentional capture in a participatory manner in Brazilian government documents [Plano de Ação Nacional para a Conservação do Pequeno Cetáceo—Toninha—Pontoporia blainvillei (PAN/ICMBIO, 2018) and the "Plano de Ação Nacional para a Conservação de Cetáceos Marinhos" (PAN/ICMBIO, 2019)] (Fig. 17.2). At the international level, joining the recommendations from Brazil, Uruguay, and Argentina, the Conservation Management Plan proposed together with the International Whaling Commission reinforces actions to assess and reduce interactions between fishing practices and mortality of Franciscana dolphins but does not recommend ethnobiology as a complementary approach. However, it is important to underscore that the "Guidelines to prevent and reduce bycatch of marine mammals in capture fisheries," published in 2021 by the FAO, highlight the importance of uniting knowledge (traditional and scientific) to build more efficient conservation strategies.

Ethnoecological studies on the franciscana dolphin in Brazil are still scarce, especially when compared to research efforts in other areas of knowledge about this species. In addition, there are no studies that use this approach in Uruguay or Argentina. This scarcity of information makes it difficult to develop strategies for the co-management of fisheries throughout the distribution range of this species with the participation and appreciation of stakeholders. Thus, it is important to encourage actions to mitigate negative interactions between fishing practices and the franciscana dolphin based on the integration of knowledge. It is also important to invest in more research efforts on the subject in addition to building standardized methodological approaches for collecting ethnoecological data to facilitate future comparisons between the species range areas, as proposed in Box 17.1 and Table 17.2.

BOX 17.1 Proposals for actions to collect and integrate knowledge aimed at building franciscana dolphin comanagement and conservation processes.

(1) Carry out ethnoecological studies to understand traditional knowledge and the language used by local actors, and thus promote the integration of interdisciplinary researchers and government together with communities. (2) Obtain detailed information about the routines of fisheries, for example, the characteristics of their vessels and gear, their fishing areas, and the periods of their fishing practices. These data must be obtained together with the traditional knowledge from artisanal fishermen in addition to studies about artisanal fisheries being carried out by researchers and governmental fishing departments. (3) Regulate and supervise fishing activities involving casting nets, described as the method that most impact the franciscana dolphin. Collaboration among governmental fishing departments, universities, and environmental institutes can be leveraged to collect information for annual fishing statistics and identify the main overfished species. These data can be collected together or by fishermen and will make it possible to establish fishing regulations specific for each region. (4) Encourage mitigation actions through the application of technical measures, including spatial closures, the use of acoustic deterrents or alerting devices, modifications to fishing gear, changes in fishing operations, and other strategies related to casting nets. These technical interventions might be an option to minimize the incidental capture of franciscana dolphins. The development and implementation of these alternatives must occur in collaboration with artisanal fishermen, enterprises, government, and research institutions. (5) Determine whether traditional fishery management activities are conducted by local communities in areas of occurrence of the franciscana dolphin in Brazil, Argentina, and Uruguay and seek alternatives for fishing techniques together with fishermen to decrease impacts on this species. (6) Promote educational partnerships and the exchange of information about the Franciscana dolphin between Brazil, Argentina, and Uruguay. As a first step, the national fishing departments should contact the fishermen unions of their respective countries to promote international partnerships. To ensure this network, it will be necessary to promote annual meetings between stakeholders. (7) Propose communication and capacity-building actions related to biology, ecology and the interactions between fishing practices and the franciscana dolphin, uniting traditional knowledge and scientific knowledge. These educational programs must be performed in partnership with educational and research institutes, governmental fishing departments, and fishermen unions in order to disseminate information on this species. To facilitate the presence of artisanal fishermen, these activities can occur in fish markets, local public schools, and at fishermen union buildings.

Table 17.2 Proposals for the standardization of surveys of ethnoecological data related to the franciscana dolphin to facilitate comparisons between areas of study.

Proposal for data collection	Method/technique	Why?
Strengthening partnerships between stakeholders	Form partnerships with fishermen unions to exchange information on their fishing routine, such as meteorology, fishermen's health, and fishing legislation. These actions must be carried out in a language that is easy for fishermen to understand. A technique for approaching and building trust is participant observation, used in the study by Zappes et al. (2016)	Bringing managers and researchers closer to the community, encouraging dialogue between the actors involved, uniting traditional and technical/scientific knowledge on the biology and ecology of the franciscana dolphin
Interviews	Use of questionnaires with questions related to the biology and ecology of franciscana dolphins, in addition to information on artisanal fishing. For the questionnaires, it is important to know a little about the community, their place of residence and fishing locations so it is possible to address topics and points of interest with fishermen. As presented by Araujo (2020) and Gama (2013)	Traditional knowledge holds a wealth of details on the behavior, life model and ecological relationships of marine species, and this is an important approach to access more detailed data on the franciscana dolphin, the ecosystem under study, and interactions with artisanal fisheries
Fishermen identifying the species	Use of boards containing photographs of franciscana dolphins along with photos of other dolphins. This is an important technique that contributes to the confirmation of whether the information obtained refers to the focus of the study. It is important to include clear photos that highlight the main anatomical and morphological characteristics of the species. Examples in studies by Araujo (2020) and Gama (2013)	To ensure the identification of P. blainvillei by the fishermen, as from this recognition, it will be possible to connect the biological, ecological and fishing interaction information provided by the local actors, contributing to future proposals for measures aimed at conservation in accordance with the relationship of the fishermen and the environment that surrounds them

(Continued)

Table 17.2 (Cont'd)

Proposal for data collection	Method/technique	Why?
Mapping the areas used by the Franciscana dolphin	Elaboration of maps based on traditional knowledge and with the fishing communities, cross-referencing information about the preferred areas of the franciscana dolphin that are used for fishing and for other activities. Example in the studies by Araujo (2020) and Gama (2013)	Fisheries planning and management measures require territorial planning and thus, the identification of areas of occurrence and risk of accidental capture of the franciscana dolphin. Knowing the main areas of economic and cultural interest for fishing is a basic factor to guide strategies for mitigating negative interactions between dolphins and fishermen. In addition, the maps reveal areas of occurrence of the species and even catching and stranding of carcasses

References

Abreu, J.S., Domit, C., Zappes, C.A., 2017. Is there dialogue between researchers and traditional community members? The importance of integration between traditional knowledge and scientific knowledge to coastal management. Ocean Coast. Manag. 141, 10–19. doi:10.1016/j.ocecoaman.2017.03.003.

Alves, L.D., Bulhões, E.M.R., Di Beneditto, A.P.M., 2018. Ethnoclimatology of artisanal fishermen: interference in coastal fishing in southeastern Brazil. Mar. Policy 96, 69–76. doi:10.1016/j.marpol.2018.07.003.

Alves, L.D., Di Beneditto, A.P.M., Ghisolfi, R.D., Quaresma, V.S., Zappes, C.A., 2020. Comparisons between ethnooceanographic predictions by fishermen and official weather forecast in Brazil. Ocean Coast. Manag. 198, 105347. doi:10.1016/j.ocecoaman.2020.105347.

Amaral, K.B., Danilewicz, D., Zerbini, A., Di Beneditto, A.P., Andriolo, A., Alvares, D.J., Secchi, E., Ferreira, E., Sucunza, F., Borges-Martins, M., Santos, M.C.O., Cremer, M., Denuncio, P., Ott, P.H., Moreno, I.B., 2018. Reassessment of the franciscana Pontoporia blainvillei (Gervais & d'Orbigny, 1844) distribution and niche characteristics in Brazil. J. Exp. Mar. Biol. Ecol. 508, 1–12. doi:10.1016/j.jembe.2018.07.010.

Araujo, S.C., 2016. A toninha (Pontoporia blainvillei Gervais D'Orbigny, 1844) e a pesca artesanal no norte do Espírito Santo: Etnoecologia e aspectos de conservação. Monograph (Faculdade de Oceanografia). Universidade Federal do Espírito Santo, p. 74.

Araujo, S.A., 2020. Caracterização da pesca e conhecimento de pescadores artesanais sobre populações de toninha (Pontoporia blainvillei) no sudeste e sul do Brasil. Dissertation (Programa de Pós-Graduação em Oceanografia Ambiental). Universidade Federal do Espírito Santo, p. 121.

Begossi, A., Burger, J., Norgaard, R., Ostrom, E., Policansky, D., Goldstein, B., 2001. Cooperative and territorial resources: Brazilian artisanal fisheries. The Commons Revisited: an American Perspective. Island Press, Washington, DC, pp. 109–130.

Begossi, A., Salyvonchyk, S., Nora, V., Lopes, P.F., Silvano, R.A.M., 2012. The Paraty artisanal fishery (southeastern Brazilian coast): ethnoecology and management of a social-ecological system (SES). J. Ethnobiol. Ethnomedicine 8 (22), 2–18.

Bertozzi, C.P., 2009. Interação com a pesca: Implicações na conservação da Toninha, Pontoporia blainvillei (Cetacea, Pontoporiidae) no litoral do estado de São Paulo, SP PhD thesis, Programa de Pós-Graduação em Ciências, Oceanografia Biológica. Universidade de São Paulo, p. 189.

Bertozzi, C.P., Zerbini, A.N., 2002. Incidental mortality of franciscana, Pontoporia blainvillei, in the artisanal fishery of Praia Grande, São Paulo State, Brazil. Lat. Am. J. Aquat. Mamm. 1 (1), 153–160. doi:10.5597/LAJAM00019.

Bittar, V.T., Di Beneditto, A.P.M., 2009. Diet and potential feeding overlap between Trichiurus lepturus (Osteichthyes: Perciformes) and Pontoporia blainvillei (Mammalia: Cetacea) in Northern Rio de Janeiro, Brazil. Zoologia (Curitiba): Int. J. Zool. 26 (2), 374–378. doi:10.1590/S1984-46702009000200023.

Bordino, P., Siciliano, S., Bastida, R., Cremer, M., 2002. Report of the working group on distribution and behavior. LAJAM 1 (1), 21–23. http://doi.org/10.5597/lajam00004.

Bordino, P., Thompson, G., Iniguez, M., 1999. Ecology and behaviour of the franciscana dolphin Pontoporia blainvillei in Bahia Anegada. Argentina. J. Cetacean Res. Manag 1 (2), 213–222.

Botta, S., Albuquerque, C., Hohn, A.A., Da Silva, V.M.F., Santos, M., Meirelles, C., Barbosa, L., Di Beneditto, A.P.M., Ramos, R.M.A., Bertozzi, C., Cremer, M.J., Franco-Trecu, V., Miekeley, N., Secchi, E.R., 2015. Ba/Ca ratios in teeth reveal habitat use patterns of dolphins. Mar. Ecol. Prog. Ser. 521, 249–263. doi:10.3354/meps11158.

Brasil. Law n° 7.643, de 18 de dezembro de 1987. Proíbe a pesca de cetáceo nas águas jurisdicionais brasileiras, e dá outras providências. http://www.planalto.gov.br/ccivil_03/Leis/L7643.htm. (Accessed February 2, 2021).

Castilhos, G.G., 2014. Estudo etnobiológico sobre a captura acidental de Pontoporia blainvillei (Gervais; D'Orbigny, 1844), toninha, em comunidades pesqueiras no litoral norte do Rio Grande do Sul. Monograph (Faculdade de Ciências Biológicas). Universidade Federal do Rio Grande do Sul, Brasil. p. 50.

Cremer, M.J., Hardt, F.A.S., Tonello Júnior, A.J., 2006. Evidence of epimeletic behavior involving a Pontoporia blainvillei calf (Cetacea, Pontoporiidae). Biotemas 19 (2), 83–86. https://doi.org/10.5007/%25x.

Cremer, M.J., Simões-Lopes, P.C., 2005. The occurrence of Pontoporia blainvillei (Gervais & d'Orbigny) (Cetacea, Pontoporiidae) in an estuarine area in southern Brazil. Rev. Bras. Zool. 22 (3), 717–723. doi:10.1590/S0101-81752005000300032.

Crespo, E.A., 2002. Franciscana, Pontoporia blainvillei. In: Perrin, W.F., Würsig, B., Thewissen, J.G.M. (Eds.), Encyclopedia of Marine Mammals. Academic Press, San Diego, CA, pp. 482–485.

Crespo, E.A., Pedraza, S.N., Grandi, M.F., Dans, S.L., Garaffo, G.V., 2010. Abundance and distribution of endangered franciscana dolphins in Argentine waters and conservation implications. Mar. Mammal Sci. 26 (1), 17–35. doi:10.1111/j.1748-7692.2009.00313.x.

Culik, B.M., 2011. Odontocetes—The toothed whales. Distribution, behaviour, migration and threats. Convention on Migratory Species/United Nations Environment Programme, Bonn.

Cunha, H.A., Medeiros, B.V., Barbosa, L.A., Cremer, M.J., Marigo, J., Lailson-Brito, J., Azevedo, A.F., Solé-Cava, A.M., 2014. Population structure of the endangered franciscana (Pontoporia blainvillei): reassessing management units. Plos One 9 (1), e8563. doi:10.1371/journal.pone.0085633.

Danilewicz, D., Moreno, I.B., Ott, P.H., Tavares, M., Azevedo, A.F., Secchi, E.R., Andriolo, A., 2010. Abundance estimate for a threatened population of franciscanas in southern coastal Brazil: uncertainties and management implications. J. Mar. Biol. Assoc. UK 90 (8), 1649–1657. doi:10.1017/S0025315409991482 (special issue).

Desvaux, J.A.S., 2013. Captura acidental da toninha, Pontoporia blainvillei (Cetacea: Pontoporiidae) e do boto-cinza, Sotalia guianensis (Cetacea: Delphinidae) em redes de pesca no Complexo Estuarino Lagunar de Cananéia, litoral sul do estado de São Paulo. Dissertation (Programa de Pós-Graduação em Ciências Biológicas – Zoologia). Universidade Federal do Paraná, Brasil, p. 132.

Di Beneditto, A.P.M., 2003. Interactions between gillnet fisheries and small cetaceans in Northern Rio de Janeiro, Brazil: 2001–2002. Lat. Am. J. Aquat. Mamm. 2 (2), 79–86. doi:10.5597/LAJAM00035.

Di Beneditto, A.P., Ramos, R.M., 2001. Biology and conservation of the franciscana (Pontoporia blainvillei) in the north of Rio de Janeiro State. Brazil. J. Cetacean Res. Manage. 3 (2), 185–192.

Di Beneditto, A.P., Ramos, R.M.A., Lima, N.R.W., 1998. Fishing Activity in Northern Rio de Janeiro State (Brazil) and its Relation with Small Cetaceans. Braz. Arch. Biol. Technol. 41 (3), 296–302. doi:10.1590/S1516-89131998000300004.

Diegues, A.C., 2000. Etnoconservação da natureza: enfoques alternativos. In: Diegues, A.C. (Ed.). Etnoconservação: novos rumos para a proteção da natureza nos trópicos, 2. NUPAUB LTDA, São Paulo, pp. 1–46.

FAO - Food and Agriculture Organization, 2018. The State of World Fisheries and Aquaculture (Food and Agriculture Organization). Roma. http://www.fao.org/3/i9540en/i9540en.pdf (Accessed March 14, 2020).

FAO - Food and Agriculture Organization. 2020. The state of world fisheries and aquaculture. Sustainability in action, Rome. doi: 10.4060/ca9229en (Accessed February 17, 2021).

FAO - Food and Agriculture Organization. 2021. Fishing operations. Guidelines to prevent and reduce bycatch of marine mammals in capture fisheries. FAO Technical Guidelines for Responsible Fisheries, 1, Suppl. 4. Rome. doi: 10.4060/cb2887en (Accessed February 17, 2021).

Fisher, A., Young, J.C., 2007. Understanding mental constructs of biodiversity: Implications for biodiversity management and conservation. Biol. Conserv. 136 (2), 271–282. doi:10.1016/j.biocon.2006.11.024.

Franco-Trecu, V., Costa, P., Abud, C., Dimitriadis, C., Laporta, P., Passadore, C., Szephegyi, M., 2009. By-catch of franciscana Pontoporia blainvillei in uruguayan artisanal gillnet fisheries: an evaluation after a twelve-year gap in data collection. Lat. Am. J. Aquat. Mamm. 7 (1-2), 11–22. doi:10.5597/LAJAM00129.

Freitas Netto, R., Di Beneditto, A.P.M., 2008. Interactions between fisheries and cetaceans in Espírito Santo State coast, southestern Brazil. Rev. Bras. Zoociências 10 (1), 55–63.

Gama, R.M., 2013. Conservação da toninha (Pontoporia blainvillei Gervais & D´Orbigny, 1844) no sul do brasil com base no conhecimento ecológico local. Monograph (Ciências Biológicas). Universidade Estadual do Norte Fluminense, Brasil, p. 79.

Gomez, J.J., Cassini, M.H., 2015. Environmental predictors of habitat suitability and biogeographical range of Franciscana dolphins (Pontoporia blainvillei). Glob. Ecol. Conserv. 3, 90–99. doi:10.1016/j.gecco.2014.11.007.

Hassel, L.B., 2006. Conhecimentos e Práticas de Comunidades Pesqueiras sobre a Conservação de Mamíferos Marinhos na Costa Leste do Estado do Rio de Janeiro, Brasil: Estudo de Caso das Comunidades de Barra de São João e Armação dos Búzios. Dissertation (Programa de Pós-Graduação em Ensino de Biociências e Saúde). Instituto Oswaldo Cruz – Fundação Oswaldo Cruz, Brasil. p. 82.

Jefferson, T.A., Webber, M.A., Pitman, R.L., 2015. Marine mammals of the World: a comprehensive guide to their identification. Academic Press, San Diego, CA.

Melcón, M.L., Failla, M., Iñíguez, M., 2012. Echolocation behavior of franciscana dolphins (Pontoporia blainvillei) in the wild. J. Acoust. Soc. Am. 131 (6), EL448–EL453. doi:10.1121/1.4710837.

Ott, P.H., Secchi, E.R., Moreno, I.B., Danilewicz, D., Crespo, E.A., Bordino, P., Ramos, R., Di Beneditto, A.P.M., Bertozzi, C., Bastida, R., Zanelatto, R., Perez, J., Kinas, P.G., 2002. Report of the working group on fishery interactions. LAJAM 1 (1), 55–64. doi:10.5597/lajam00008 (Special Issue).

PAN/ICMBio, 2018. Plano de ação nacional para a conservação da Toninha: Pontoporia blainvillei. Brasília: Instituto Chico Mendes de Conservação da Biodiversidade (ICMBio). https://www.gov.br/icmbio/pt-br/assuntos/biodiversidade/pan/pan-toninha. (Accessed March 1, 2021).

PAN/ICMBio, 2019. Plano de ação nacional para a conservação de Cetáceos Marinhos. Instituto Chico Mendes de Conservação da Biodiversidade (ICMBio, Brasília. https://www.gov.br/icmbio/pt-br/assuntos/biodiversidade/pan/pan-cetaceos-marinhos. (Accessed March 1, 2021).

Pinheiro, L., Cremer, M., 2003. Etnoecologia e captura acidental de golfinhos (Cetacea: Pontoporiidae e Delphinidae) na Baía de Babitonga, Santa Catarina. Desenvolvimento e Meio Ambiente 8, 69–75. doi:10.5380/dma.v8i0.22053.

Przbylski, C.B., Monteiro-Filho, E.L.A., 2001. Interação entre pescadores e mamíferos marinhos no litoral do Estado do Paraná - Brasil. Biotemas 14 (2), 141–156. doi:10.5007/%25x.

Rosa, G.A., Zappes, C.A, Di Beneditto, A.P.M., 2012. Etnoecologia de pequenos cetáceos: interações entre a pesca artesanal e golfinhos no norte do estado do Rio de Janeiro. Brasil. Biotemas 25 (3), 293–304. doi:10.5007/2175-7925.2012v25n3p293.

Rosas, F.C.W., Monteiro-Filho, E.L.A., Oliveira, M.R., 2002. Incidental catches of franciscana (Pontoporia blainvillei) on the southern coast of São Paulo state and the coast of Paraná state, Brazil. Lat. Am. J. Aquat. Mamm. 1 (1), 161–167 (special issue).

Santos, M.C.O., Oshima, J.E.F., Silva, E., 2009. Sightings of franciscana dolphins (Pontoporia blainvillei): the discovery of a population in the Paranaguá Estuarine Complex, Southern Brazil. Braz. J. Oceanogr. 57 (1), 57–63. doi:10.1590/S1679-87592009000100006.

Secchi, E.R., Ott, P.H., Crespo, E.A., Kinas, P.G., Pedraza, S.N., Bordino, P., 2001. A first estimate of franciscana (Pontoporia blainvillei) abundance off Southern Brazil. J. Cetacean Res. Manag. 3 (1), 95–100.

Secchi, E.R., Ott, P.H., Danilewicz, D., 2002. Report of the fourth workshop for the coordinated research and conservation of the franciscana dolphin (Pontoporia blainvillei) in the Western South Atlantic. Lat. Am. J. Aquat. Mamm. 1 (1), 11–20. doi:10.5597/lajam00003 (Special Issue).

Siciliano, S., 1994. Review of small cetaceans and fishery interactions in coastal waters of Brazil. Rep. Int. Whaling Comm. 15, 241–250 (Special Issue).

Siciliano, S., Di Beneditto, A.P.M., Ramos, R., 2002. A toninha, Pontoporia blainvillei (Gervais & d´Orbigny, 1844) (Mammalia, Cetacea, Pontoporiidae), nos estados do Rio de Janeiro e Espírito Santo, costa sudeste do Brasil: Caracterizações dos hábitats e fatores de isolamento das populações. Bolm. Mus. Nac., Zool. 476, 1–15.

Silva, F.O., 2007. Conhecimento tradicional e etnoconservação de cetáceos em comunidades caiçaras do município de Cananéia, litoral sul de São Paulo. Dissertation (Programa de Pós-Graduação em Ecologia Aplicada). Escola Superior de Agricultura Luiz de Queiroz - Universidade de São Paulo, Brasil. p. 110.

Souza, S.P., Begossi, A., 2007. Whales, dolphins or fishes? The ethnotaxonomy of cetaceans in São Sebastião, Brazil. J. Ethnobiol. Ethnomed. 3 (9). doi:10.1186/1746-4269-3-9.

Sucunza, F., Danilewicz, D., Cremer, M., Andriolo, A., Zerbini, A.N., 2018. Refining estimates of availability bias to improve assessments of the conservation status of an endangered dolphin. PLoS One 13 (3), e0194213. doi:10.1371/journal.pone.0194213.

Trimble, M., Praderi, R., 2006. What is the colour of the franciscana (Pontoporia blainvillei)? A review and a proposed assessment method. LAJAM 5 (1), 55–63. doi:10.5597/lajam00092.

Zappes, C.A., Andriolo, A., Simões-Lopes, P.C., Di Beneditto, A.P.M., 2011. Human-dolphin (Tursiops truncatus Montagu, 1821) cooperative fishery' and its influence on cast net fishing activities in Barra de Imbé/Tramandaí, Southern Brazil. Ocean Coast. Manag. 54, 427–432. doi:10.1016/j.ocecoaman.2011.02.003.

Zappes, C.A, Danielski, M.L., Di Beneditto, A.P.M., 2020. Etnoecologia comportamental da baleia-franca-austral Eubalaena australis (Desmoulins, 1822) por meio do conhecimento ecológico local de pescadores artesanais na Área de Proteção Ambiental da Baleia Franca, Sul do Brasil. Bol. Mus. Para. Emilio Goeldi. Cienc. Nat. 15 (3), 601–616. doi:10.46357/bcnaturais.v15i3.183.

Zappes, C.A., Gama, R.M., Domit, C., Gatts, C.E.N., Di Beneditto, A.P.M., 2016. Artisanal fishing and the franciscana (Pontoporia blainvillei) in Southern Brazil: ethnoecology from the fishing practice. J. Mar. Biolog. Assoc. U.K. 96, 1–11. doi:10.1017/S0025315416001788.

Zappes, C.A., Gatts, C.N., Lodi, L.F., Simões-Lopes, P.C., Laporta, P., Andriolo, A., Di Beneditto, A.P.M., 2014. Comparison of local knowledge about the bottlenose dolphin (Tursiops truncatus Montagu, 1821) in the Southwest Atlantic Ocean: New research needed to develop conservation management strategies. Ocean Coast. Manag. 98, 120–129. doi:10.1016/j.ocecoaman.2014.06.014.

CHAPTER 18

Marine protected areas: Their potential for the conservation of the franciscana

Paulo Henrique Ott[a], Juliana Di Tullio[b], Jonatas Henrique Prado[c], Daniel Danilewicz[d], Cecilia Passadore[e,f], Mauricio Failla[g], Miguel Iñíguez Bessega[g], Diego Rodríguez[h]

[a]Universidade Estadual do Rio Grande do Sul (Uergs), Osório, Rio Grande do Sul, Brazil
[b]Kaosa, Rio Grande, Rio Grande do Sul, Brazil
[c]Cooperativa para Conservação e Proteção dos Recursos Naturais, Florianópolis, Santa Catarina, Brazil
[d]Grupo de Estudos de Mamíferos Aquáticos do Rio Grande do Sul, Torres, Rio Grande do Sul, Brazil
[e]Vida Silvestre Uruguay, Montevideo, Uruguay
[f]Dirección Nacional de Recursos Acuáticos (DINARA), Ministerio de Ganadería Agricultura y Pesca (MGAP), Montevideo, Uruguay
[g]Fundación Cethus, Olivos, Buenos Aires, Argentina
[h]Instituto de Investigaciones Marinas y Costeras, Facultad de Ciencias Exactas y Naturales, Universidad Nacional de Mar del Plata, Mar del Plata, Argentina

Introduction

The conservation of franciscana (*Pontoporia blainvillei*) is a challenging and urgent undertaking. The species is restricted to the coastal waters of southeastern and southern Brazil, Uruguay, and northern Argentina (Crespo et al., 1998; Amaral et al., 2018). Multiple human activities are developed in this coastal area, resulting in numerous modifications in its ecosystems, including pollution and overexploitation of marine resources. Nevertheless, the main threat to the conservation of the species is the high level of incidental mortality in coastal gillnets across its distribution (Ott et al., 2002; Zerbini et al., 2017; Secchi et al., 2021). Currently, there is no estimate of total annual bycatch of the species throughout its range, but previous data gathered from different regions indicates that this number is in the range of, at least, 2,000–3,000 individuals (e.g., Ott et al., 2002; Secchi et al., 2003a; Franco-Trecu et al., 2009, 2019; Negri et al., 2012; Prado et al., 2013). Although there is no estimate of the total population size of franciscana for its entire range, the rates of annual mortality in gillnets, in some regions, may exceed 5% or more of the estimated local population, as reported for the Rio Grande do Sul state, in southern Brazil (Ott et al., 2002; Sucunza et al., 2020a). In Argentina, bycatches are somewhat

The Franciscana Dolphin
DOI: https://doi.org/10.1016/B978-0-323-90974-7.00005-7

lower; however, they still exceed 3–5% of its population per year (Crespo et al, 2020). Incidental captures of franciscanas have also been recorded in trawling fisheries in southern Brazil and Uruguay (Montealegre-Quijano and Ferreira, 2010; Franco-Trecu et al., 2019), even though not in the same order of magnitude as in gillnets. Additional threats to the species include different types of habitat degradation, such as pollution, coastal modification, and prey depletion (Secchi et al., 2003a, 2021; Denuncio et al., 2011; Di Beneditto and Ramos, 2014), although these threats are more difficult to assess.

The mortality of franciscana in gillnet is typically incidental and occurs during commercial fishing operations directed toward several fish species (e.g., Di Beneditto et al., 1998; Kinas and Secchi, 1998; Ott et al., 2002; Franco-Trecu et al., 2009; Prado et al., 2021). Incidental mortality in fisheries is a well-known threat for many cetacean species around the world but no easy solution has been found (Read et al., 2006; Brownell et al., 2019). At first glance, the fact that the capture is incidental could be seen as an opportunity for the implementation of management actions once the fishers do not deliberately target the species. Nevertheless, as most of the captures occur in gillnets, which are basically size-selective (Alves et al., 2012), the catches of the species are somewhat unpredictable, especially regarding time. Moreover, the existence of high fishing effort overlapping with the main distribution areas of franciscana results in high levels of incidental captures throughout the year (Prado et al., 2021).

These high bycatch levels of the species combined with its low intrinsic reproductive rate (e.g., Danilewicz et al., 2002) resulted in unsustainable mortality over time. In fact, based on population simulation analysis, the species is suspected to be decreasing and is currently classified as Vulnerable (VU) in the Red List of Threatened Species by the International Union for Conservation of Nature (IUCN) (Zerbini et al., 2017). The species is also officially considered Critically Endangered (CR) in Brazil (MMA, 2014) and Vulnerable (VU) in Argentina (Denuncio et al., 2019). In Uruguay, even though a Red List of mammals has yet to be created, franciscana is considered threatened and is listed as a Priority Species for Conservation by the National System of Protected Areas (SNAP), consequently requiring special attention when planning conservation strategies in the country (Soutullo et al., 2013).

Therefore, considering that the main threat to the conservation of franciscana is the incidental catch in coastal gillnets, the reduction of fishing effort along the species distribution is considered the most effective way to reduce the mortality of the species and guarantee its long-term persistence

(Ott et al., 2015; Brownell et al., 2019; Prado et al., 2021, Secchi et al., 2021). However, besides the traditional fishing regulations (e.g., number of boats, net size, time-area closures or quotas) and other measures to reduce bycatch, such as gear modifications (Bordino et al., 2002, 2013) and switching gillnets to longlines (Bernisone et al., 2020), the establishment and strengthening of marine protected areas (MPAs) is also a very promising way to protect the species from an ecosystem-based perspective. Indeed, the International Whaling Commission's (IWC) Conservation Management Plan (CMP) for franciscana classifies the creation and implementation of MPAs for the species as one of the priority actions (IWC, 2016).

In this chapter, an unprecedented review of the MPAs current status aiming specifically to assess their potential contribution to the conservation of franciscana is presented. First, we briefly present the MPA as a biodiversity conservation management tool, especially regarding marine mammals. Then, we provide an overview of the status of the MPAs across the franciscana distribution range and, finally, present some future perspectives and recommendations.

MPAs as a management tool

One of the most commonly used definitions of protected area (PA) is that provided by the IUCN, which define it as "clearly defined geographical space, recognized, dedicated and managed, through legal or other effective means, to achieve the long-term conservation of nature with associated ecosystem services and cultural values" (Day et al., 2019). Therefore, MPAs are defined as sites that fulfill these criteria and that are designed to fully or partially protect any portion of the marine ecosystem (Thomas et al., 2014; Day et al., 2019).

The names and categories used for MPAs as well as for terrestrial PAs vary greatly across countries. There is also a wide variety of different approaches to protect the biodiversity within these areas. Nevertheless, to allow a more comprehensive system worldwide, the IUCN has developed an international scheme composed of six management categories (one with a subdivision), according to their management objectives and degree of human intervention (Day et al., 2019). These categories range from "strict nature reserves" (Category Ia), where extractive use (of living or dead material) and public visitation are not permitted, to multiple-use reserves, where there is a focus on the sustainable extraction of natural resources (Categories V and VI) (Table 18.1).

Table 18.1 Summary of IUCN protected area categories (modified from Day et al., 2019).

IUCN category	Definition/primary objective	No-take	Extractive use and habitation
Ia	Strict nature reserves. To conserve outstanding ecosystems, species, and/or geodiversity features	Yes	Not allowed
Ib	Wilderness areas. To protect the long-term ecological integrity of natural areas	Yes	In some circumstances, used by indigenous people to conserve their traditional, spiritual, and cultural values
II	Large natural areas set aside to protect large-scale ecological processes. To promote ecosystem protection through education and recreation (e.g., National Parks)	Yes	Not allowed
III	Conservation of natural features and their associated biodiversity and habitats. They are generally quite small areas and often have high visitor value (e.g., Natural Monuments)	Yes	Not allowed
IV	Conservation of particular species or habitats through active management	No	Sustainable use is permitted, as long as it is compatible with the MPA's objectives. Industrial fishing, industrial-scale aquaculture and habitation are not allowed
V	Sustainable use of natural resources through traditional management practices	No	Local communities living within and sustainably using the seascape are allowed, including local fishing practices or small-scale aquaculture
VI	Sustainable use of natural resources (managed-resource protected area)	No	Sustainable use is permitted, as long as it is considered to be mutually beneficial to conservation. Industrial fishing, industrial-scale aquaculture and habitation are not permitted

In addition to MPAs, many countries or international bodies have established specific marine areas where some human activities are regulated. These areas may not have nature conservation as a primary goal but can also contribute to the objectives of an MPA network or indirectly to the conservation of some species. Many of these areas in coastal regions have a primary fishing goal and may be only temporary. Therefore, according to the IUCN's guidelines, they should not be classified as MPAs (Day et al., 2019).

Historically, the MPAs were established after the terrestrial PAs, resulting in a broad scenario of lower spatial coverage and usually lower management experience. This means that prior to the establishment of the MPAs, many socioeconomic activities had previously been developed for many years, mainly in coastal areas. In Argentina, for instance, the first terrestrial national park was created in 1903 (Nahuel Huapi National Park) and the first PA including coastal areas was Tierra del Fuego National Park in 1960. Similarly, in Brazil, the first terrestrial PA, following the "Yellowstone Model," was established in 1937 (Itatiaia National Park), whereas the first MPA (Atol das Rocas Biological Reserve) was created in 1979. In Uruguay, despite some previous initiatives, the National System of Protected Areas (SNAP) was only established in 2000 and formally regulated in 2005. Within this system, in 2008, the Protected Landscape Quebrada de Los Cuervos was the first terrestrial area created in the country. One year later, the Cabo Polonio National Park, which included a portion of the marine surface, was also incorporated into the SNAP.

Nevertheless, the implementation of MPAs is as complex and challenging as for terrestrial reserves. MPAs are generally multiple-use areas aiming to protect marine biodiversity along with the goods and services they provide to society (Di Franco et al., 2016). Therefore, the MPAs usually involve many different stakeholders with very distinct perspectives and expectations. Moreover, as for the terrestrial ecosystems, to maximize the benefits for biodiversity the delineation of MPAs should be based on clear objectives and detailed spatial planning (Magris et al., 2013). The spatial planning process should be used to identify, for instance, biologically and ecologically sensitive areas in time and space, potential connectivity with other MPAs or other types of management areas, as well as existing and potential human uses in the region (Ehler and Douvere, 2009; Magris et al., 2013).

According to the World Database on Protected Areas (WDPA), currently, there are more than 18,500 MPAs around the world. The number and spatial extent of MPAs have increased rapidly over the last decades.

While in 2000, the MPAs had covered an area of approximately 2 million km^2, representing 0.7% of the global ocean, by 2021 they covered almost 30 million km^2 (more precisely 27,841,368 km^2 or 7.68%) (UNEP-WCMC and IUCN, 2021). In Argentina, for instance, six MPAs, including inter-jurisdictional coastal marine parks, have been created in the last decades: Patagonia Austral in 2009, Makenke in 2012, Isla Pingüino in 2012, Namuncurá-Banco Burdwood I in 2013, Namuncurá-Banco Burdwood II in 2018, and Yaganes in 2018 (SAyDS, 2018), covering 117,600.2 km^2. A new National Marine Park in the Complejo Islote de Lobos is also in the process of being created. Nevertheless, some of the recent large MPAs created in South America were established with the intent of fulfilling primarily international commitments (e.g., Aichi Target 11) rather than specific conservation targets (Magris and Pressey, 2018). As an illustration, the Brazilian government created four very large MPAs in 2018, increasing the coverage of MPAs in the country from 1.5% to 25% (Giglio et al., 2018). These MPAs, however, were established in the open ocean where many human activities (e.g., fishing and mining) are already unlikely or rare, giving the false impression of a great advance in the protection of marine biodiversity (Magris and Pressey, 2018).

The effectiveness of MPAs is also closely related to some important attributes, such as the existence of a management plan stating clear management objectives and mechanisms consistent with the goals of the site (Reeves, 2000; Di Franco et al., 2016). However, many MPAs in South American countries, including Brazil, Uruguay, and Argentina, besides being outdated, either lack management plans or fail to present clear and feasible objectives, making it difficult to translate them into action (Mills et al., 2020). Among other key attributes that have been identified in order to increase the performance of MPAs are high enforcement and the existence of a management council that allows the engagement of different stakeholders, usually including fishers (Di Franco et al., 2016; Mills et al., 2020). The size and years of protection are also important attributes related to the success of MPAs in conserving marine biodiversity (Claudet et al., 2008; Edgar et al., 2014). However, the geographic extension of PAs is a component that, by itself, does not guarantee the conservation of biodiversity and ecosystems, and must be complemented by other actions such as improvement of representativeness and governance, assessment of the management effectiveness, integration of MPAs into broader landscapes and identification of other effective conservation measures (Alvarez Malvido et al., 2021).

MPAs as a tool to protect marine mammals

Historically, the MPAs were created to recover depleted fish stocks or protect other exploited marine resources. Therefore, these areas are defined usually based on the knowledge of spawning fish and nursery areas, or zones of high fishing pressure. Among the reported and expected outcomes of MPAs are an increase in species diversity, fish biomass, and larger sizes of individuals, as well as exportation of eggs, larvae, juveniles, and adults (i.e., spillover) to adjacent areas (Worm et al., 2006). Besides, the implementation of MPAs can increase the economic income of local communities due to the development of eco-tourism activities (Worm et al., 2006), as well as provide effective tools for managing and ensuring the conservation of marine top predators such as marine mammals and their associated habitats (e.g., Hughes et al., 2003; Hooker and Gerber, 2004; Worm et al., 2006; Notarbartolo di Sciara et al., 2016). Yet, when it comes to designing and implementing effective MPAs for coastal long-living and mobile species, such as the franciscana, it is quite challenging.

The optimal design of MPA intended to conserve a marine mammal population would encompass its entire distribution year-round. However, considering that most species have a very large home range and some exhibit migratory behavior, this area-based approach is generally quite unrealistic (Reeves, 2000; Notarbartolo di Sciara et al., 2016). In practice, most MPAs encompass only a portion of the range of the marine mammal population and it may not be large enough alone to guarantee the long-term viability of these populations. Meanwhile, species that form small populations and occupy restricted ranges are more prone to be managed through the establishment of MPAs (Notarbartolo di Sciara et al., 2016), especially when these areas encompass the critical habitat of a population (Hooker et al., 2011). Therefore, the identification and protection of critical areas of a species assume even a greater role when delineating MPAs for marine mammals. Although the definition of critical areas can vary among different legal frameworks, the concept usually involves recognizing essential areas used for breeding, feeding, and calving. Accordingly, these sites are not necessarily related to high-density areas (Hoyt, 2011). In order to protect critical habitats for marine mammals in multiple-use MPAs, establishing management zones in which some activities (e.g., fishing, tourism, and vessel traffic) can be strictly regulated, has proven to be an important strategy (Lusseau and Higham, 2004; Hoyt, 2011; Gormley et al., 2012; Di Tullio et al., 2015). Creating networks of MPAs has also been considered particularly appropriate to protect marine mammals (Reeves, 2000; Hoyt, 2018).

Globally, the first areas created for marine mammal conservation were established in the early 1970s. Even though, currently, there are hundreds of MPAs around the world encompassing important marine mammal communities and the areas they inhabit, only a small number has been created explicitly to protect small cetaceans (Reeves, 2000; Hoyt, 2011). In the western South Atlantic, among the pioneer MPAs created for the purpose of conserving small cetaceans is the Anhatomirim Environmental Protection Area located in Santa Catarina, southern Brazil. This multiple-use MPA (Category V) was established in 1992 and includes both terrestrial and marine ecosystems. The boundaries of its marine zone were designed explicitly to protect the resident population of the Guiana dolphin (*Sotalia guianensis*) in the southernmost distribution range of the species. However, despite a large amount of time spent by dolphins within the MPA, the percentage of the overall home range of the individuals within the limits of the PA was considered moderate and the creation of a buffer zone in its southern limit was recommended (Flores and Bazzalo, 2004). These results point to the importance of a monitoring process as well as an adaptive approach to MPAs so that their objectives and effectiveness could be constantly reviewed and adjusted as needed (Hoyt, 2018).

Primary source and analyses of MPAs across the franciscana distribution

In order to review the MPAs that can contribute to franciscana conservation, we considered all MPAs that include at least a marine or estuarine area across the species´ distribution. So, first, we mapped the distribution range of franciscana based on previous literature and the most recent records from aerial surveys and incidental captures of the species (e.g., Amaral et al., 2018; GEMARS, unp. data; Denuncio et al., 2019; Prado et al., 2021). In this sense, the locality of Itaúnas (18°25′S), in the state of Espírito Santo, southeastern Brazil, was considered the northernmost limit of the distribution of the species (Siciliano et al., 2002), whereas Puerto Lobos (42°00′S) in Golfo San Matías, Chubut, Argentina, was defined as the southernmost limit (Denuncio et al., 2019). Although a sighting of the species has been reported inside Golfo Nuevo (42°35′S, 64°48′W) by Crespo et al. (1998), we considered it to be an extralimital record. Moreover, since the distribution of franciscana is not continuous along its range, the existing MPAs within the two hiatuses proposed in the southeastern Brazilian coast

(Amaral et al., 2018) were not included. The exceptions were two MPAs in the region of Tamoios, Rio de Janeiro state (Annex 18.I), due to some recent records of the species.

Currently, five franciscana management areas (FMAs; see Secchi et al. 2003b) along the species range are recognized by the IWC (FMA-Ia, FMA-Ib, FMA-II, FMA-III, and FMA-IV) (IWC, 2016). Within these units, the franciscana seems to have a distinct pattern of distribution over the continental shelf (see Amaral et al., 2018). In this sense, although the external limit of 50 m depth seems to be feasible for some of the FMAs (e.g., FMA-III encompassing Rio Grande do Sul, southern Brazil, and Uruguay; see Danilewicz et al., 2009), it is not a fixed rule. In fact, franciscana is known to be more coastal in the northern portion of its distribution, mainly from São Paulo to Espírito Santo (both Brazilian states) as indicated by aerial surveys and incidental captures of the species (Amaral et al., 2018; GEMARS, unp. data). Therefore, the external limit of FMA-Ia was mapped at the 25 m isobaths, for FMA-Ib and FMA-II at 30 m, and for FMA-III and FMA-IV at 50 m; though a few areas deeper than 50 m were also included in the Golfo San Matías and northern Buenos Aires Province, Argentina, based on a recent review of records of the species (Denuncio et al., 2019). In addition to the marine ecosystem, PAs in estuarine zones with confirmed records of the species, such as in the Brazilian estuaries of Paranaguá and Cananéia (Santos et al., 2009), and in the La Plata River estuary on the border between Uruguay and Argentina (e.g., Franco-Trecu et al., 2009; Wells et al., 2021) were also included.

After mapping the franciscana range, we evaluated how much of it was covered by MPAs. For MPAs covering both terrestrial and marine ecosystems, only the size of the marine or estuarine area was considered. MPAs composed entirely of islands, with no water coverage, were not included. We also considered only officially created MPAs that fulfill the IUCN criteria and clearly have as a primary focus the long-term conservation of biodiversity (Day et al., 2019). A special case, however, was the Franciscana Provincial Natural Monument, in Argentina, which in spite of not following the strict IUCN scheme, has great potential to protect the species. Areas managed primarily for fishery goals were not included in the analyses. However, since these fishery management areas could also be important for the conservation of franciscana, as an example, a specific legislation in Brazil (INI 12/2012) is briefly presented in a separate section (see Box 18.1).

BOX 18.1 Additional fishery management areas: The case of INI-12 in Brazil

In addition to the MPAs, many countries have areas with temporary or permanent fishing closures or gear restrictions. Although these areas are not considered to be MPAs according to the IUCN criteria (Day et al., 2019), they can potentially contribute to the conservation of marine mammals. Currently, there are many fishery regulations along the coastal waters of Brazil, Uruguay, and Argentina (see Chapter 19), including gillnet fishery management areas. Although some of these areas cover a large proportion of the distribution of franciscana, many of them need to have their management tools revised and improved to be effective in the conservation of the species.

In 2012, for instance, the Brazilian Ministries of Environment and of Fisheries and Aquiculture published a norm (herein called INI-12) regulating gillnet fisheries in southern and southeastern Brazil (Brasil, 2012), which included, at least partially, four FMAs (FMA-Ia, FMA-Ib, FMA-II, and the Brazilian section of FMA-III). Among others, the items of the norm that were expected to benefit franciscana and other non-target species, as well as the recovering of the fishing stocks, include a halt in new gillnet fishing licenses, reduction of fishing effort in terms of net dimensions and a mosaic of no-fishing areas. For example, powered boats cannot use gillnets within the 5 nm between the southern limit of Brazil and the Albardão lighthouse (~85 km stretch of the coast), and within 1 nm from the Albardão lighthouse to the north limit of Espírito Santo state (>2,000 km). Moreover, no industrial fishing boat (20 GT or over) can operate within 4 nm from the Albardão lighthouse to the northern limit of Paraná (~1,000 km) or within 3 nm from the southern limit of São Paulo to the north limit of Espírito Santo (~1,200 km) (see also Secchi et al., 2021). In addition, there are some areas where gillnet fishery cannot operate (e.g., in the adjacency of Patos Lagoon estuary). Although the lack of enforcement and compliance hinders the evaluation of the effectiveness of the INI-12 in diminishing franciscana mortality, a study conducted on the coast of Rio Grande do Sul demonstrated that even with full fishery compliance, the INI-12 is not enough to reduce franciscana mortality to sustainable levels in southern Brazil (Prado et al., 2021).

Limiting or halting fishing efforts can work as intended to reduce mortality but only if the areas are established in the right place and if the threat factor is removed rather than simply shifted to another part of the population's range (Rojas-Bracho and Reeves, 2013). For instance, the southern portion of the state of Rio Grande do Sul (FMA-III) presents a relatively wide continental shelf and a smooth slope resulting in a distribution of franciscana farther from the offshore limit of the INI-12 exclusion zone, where the industrial gillnet fishery can operate legally with a net length of up to 16 km per vessel. Therefore, according to Prado et al. (2021), the establishment of no-fishing zones that cover the entire hotspot areas of franciscana mortality coupled with a reduction, by at least one order of magnitude, of the current total fishing effort are needed to bring down the mortality of the species to sustainable levels along the Rio Grande do Sul coast. Although there are no other similar studies in Brazil, information on gillnet fishing operations and the occurrence of franciscana in other parts of its distribution reveals that the INI-12 potential (with full compliance) to reduce the mortality of the species may vary greatly among the FMAs. In this sense, the INI-12 appears to be more effective in reducing the overlap between franciscanas and industrial gillnets in some areas of southeastern Brazil (FMA-Ia, FMA-Ib, and FMA-II), where the franciscana has a more coastal distribution (e.g., FMAIa—up to 4 nm near its southern limit) (Sucunza et al., 2020b). However, it should be noticed that an important artisanal gillnet fleet operates in nearshore coastal waters along all the Brazilian FMAs. Therefore, while INI-12 may be effective in reducing the overlap between franciscanas and industrial gillnets in some areas (e.g., southern region of FMA-Ia and FMA-Ib), it has little effect in reducing the overlap of the species with small-scale fisheries, which is also responsible for high bycatch rates of franciscana (Bertozzi and Zerbini, 2002; Ott et al., 2002).

In summary, the INI-12 represents an important legal framework to regulate the gillnet fishery and reduce the bycatch of endangered species, including franciscana. However, the norm was not specifically designed to reduce franciscana mortality to sustainable levels. To achieve this goal, two main criteria should be considered: (1) no-fishing zones must be created based on the information from the species mortality hotspot areas; (2) the fishing effort of both artisanal and industrial fleets outside the no-fishing zones must be defined based on population viability analyses of the species within each FMA.

Our primary source of MPAs data was the WDPA, available at the Protected Planet website (https://www.protectedplanet.net/), managed by the United Nations (UN) Environment World Conservation Monitoring Centre (UNEP-WCMC) and the International Union for Conservation of Nature (IUCN). Additional information was obtained at the official databases of national PAs within each country (Brazil—Instituto Chico Mendes de Conservação da Biodiversidade—ICMBio; Uruguay—Sistema Nacional de Áreas Protegidas—SNAP; and Argentina—Sistema Federal de Áreas Protegidas—SiFAP). For the latter, an integrated system of coastal and marine protected areas (AMP Argentina, 2020) was also consulted. Information about subnational MPAs were also obtained from official databases (e.g., Instituto Estadual do Ambiente—INEA for Rio de Janeiro, Fundação Florestal for São Paulo, and Organismo Provincial para el Desarrollo Sostenible—OPDS for Buenos Aires). The public data available for each MPA usually contain basic information, such as name, location, year of creation, size, national (or subnational) category, and the status of the management plan. The shapefile of most MPAs was also directly available. In the case of the Franciscana Provincial Natural Monument, in the Buenos Aires Province, once the species is widely distributed in the region, the boundaries of the area were defined based on the limits of the marine area of the province (i.e., up to 12 nm from the coast). Some MPAs contain a buffer zone that extends beyond the limits of the PA. However, this information is not consistently available, and these external areas were not included in the analyses. Additional information about the MPAs was obtained from the published management plans or directly from their managers through personal communications.

For a better understanding of the representativeness of franciscana´s MPAs, in addition to the distribution of the PAs across the three countries, two other spatial analyses were performed. First, the MPAs were analyzed in biogeographic terms based on the Marine Ecoregions of the World (MEOW) (Spalding et al., 2007). Then, an analysis of MPAs within each currently recognized FMA (sensu IWC, 2016) was conducted. Next, for a more comprehensive analysis of the MPAs' potential contribution to the conservation of the species, the total area covered by these PAs was compared to the size of each MEOW and FMA, based on the known distribution of the species. In these analyzes, all overlapping MPAs were removed to avoid double counting. The FMAs were also evaluated in terms of the percentage of coverage of MPAs by 10 m depth intervals (0–10; 10–20; 20–30; 30–40; 40–50; >50 m). Considering that the external limits of the distribution of franciscana vary among the FMAs, we identified 20 strata of 10 m depth intervals across the species distribution. In this scenario, FMA-Ia, FMA-Ib, and FMA-II contain three depth intervals (0–10; 10–20; 20–30 m), whereas FMA-III and FMA-IV present five (0–10; 10–20; 20–30; 30–40; 40–50 m) and six strata (0–10; 10–20; 20–30; 30–40; 40–50; >50 m), respectively. The geospatial analyses of the MPAs were performed using the ArcGIS 10.8 (ESRI) software. In terms of governance, the MPAs were also analyzed regarding the existence of established management councils and published management plans. The inclusion of the franciscana in the management plans and the existence of specific measures proposed directly for the protection of the species were evaluated as well. Finally, a putative analysis of the number of individuals and percentage of the population protected by MPAs in each FMA was conducted. Detailed information about these population analyses is presented in the corresponding section.

Existing MPAs along the franciscana distribution

Currently, there are 44 MPAs along the distribution of franciscana, including 25 in Brazil, six in Uruguay, and 13 in Argentina (Fig. 18.1; Table 18.2,

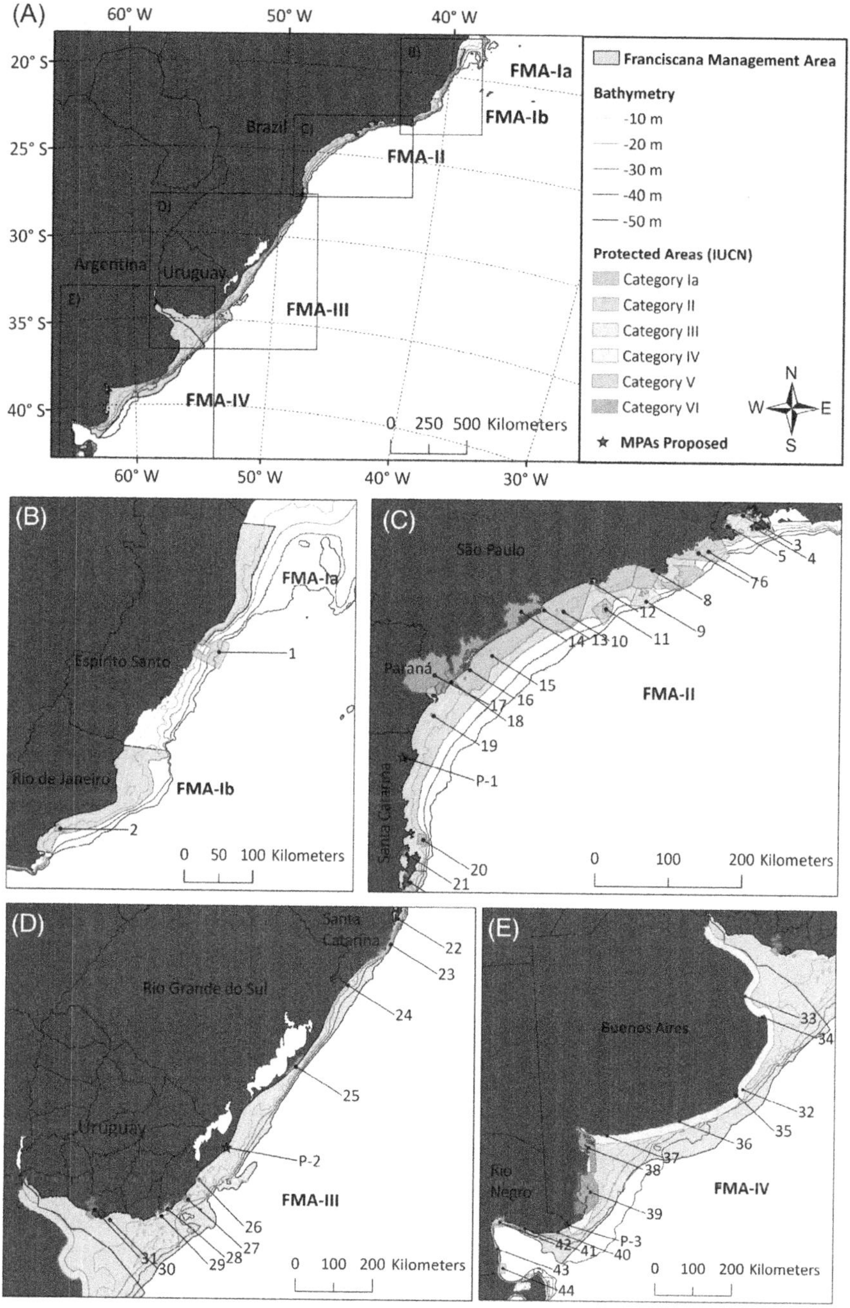

Fig. 18.1 Franciscana's distribution range (A). Marine protected areas (MPAs) that overlap with the distribution of franciscana across the five currently recognized franciscanas management areas (i.e., FMA-Ia and FMA-Ib [B], FMA-II [C], FMA-III [D], and FMA-IV [E]; IWC, 2016). The MPAs are labeled by colors representing the International Union for Conservation of Nature (IUCN) categories and indicated by numbers (1–44) which correspond to the names mentioned in Annex 18.I. New MPAs officially proposed and mentioned in Box 18.2 are indicated as follows: P-1- Babitonga Bay; P-2- Albardão; and P-3- Río Negro Estuary.

Table 18.2 Number of marine protected areas (MPAs) per country across the franciscana distribution according to the management categories of the International Union for Conservation of Nature (IUCN).

Country/ category	Ia	Ib	II	III	IV	V	VI	Total
Brazil	4	0	5	3	1	10	2	25
Uruguay	0	0	2[a]	0	2	1	1	6
Argentina	0	0	0	1[a]	5	2	5	13
Total	4	0	7	4	8	13	8	44

[a]MPAs that, while officially classified in categories of no-take areas (e.g., Natural Monument and National Park), perform as sustainable use as they currently allow fishing activities.
See Table 18.1 for a description of each category.

details in Annex I). None of the MPAs are shared among the countries. The country which provides more protection for the franciscana in terms of the total area of MPAs, without considering the overlaps, but regardless of their IUCN category, is Argentina (29,121.1 km^2), followed by Brazil (12,847.6 km^2) and Uruguay (1,010.1 km^2). In terms of representativeness, the MPAs cover, respectively, 27.5%, 17.6%, and 2.1% of the distribution range of the species in each country based on the mapping of the FMAs (described above, see Fig. 18.1). Argentina´s highest MPA coverage is largely due to the Franciscana Provincial Natural Monument, which alone covers 26,389.6 km^2 of the marine and estuarine waters of the Buenos Aires Province.

The first MPA established along the distribution of franciscana was Punta Bermeja, in the Province of Río Negro, Argentina, in 1971. This MPA was originally designed, mainly, to protect the coastal and marine ecosystems of an important reproductive colony of South American sea lions (*Otaria flavescens*). The number of MPAs started to increase significantly in Argentina and Brazil in the 1980s, whereas in Uruguay the PAs with marine portions were incorporated into the SNAP only about three decades later (Fig. 18.2). Apart from the Franciscana Natural Monument, created in 2018 in the Buenos Aires Province (Argentina), none of these areas were specifically designed to protect the species.

Nowadays, there are many types of designations for the MPAs across the three countries (e.g., national park, wildlife refuge) which correspond to six of the seven categories of the IUCN system (Day et al., 2019). The only IUCN category not existing along the distribution of franciscana is the Wilderness Area (Category Ib). In addition, the IUCN category of the Protected Nature Area of Punta Bermeja, Argentina, has not been officially

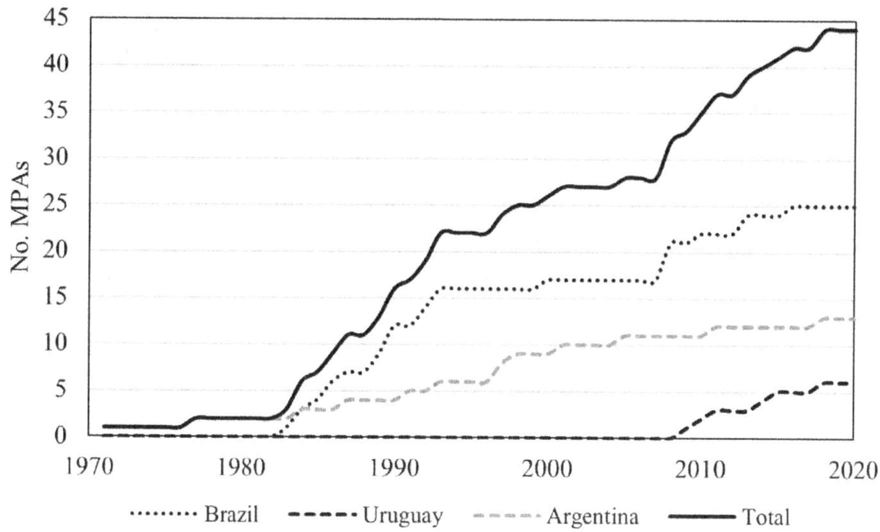

Fig. 18.2 ***Evolution of marine protected areas (MPAs) across the distribution of franciscana in terms of a cumulative number of sites in the three countries (Brazil, Uruguay, and Argentina) over the years.***

reported but seems to be broadly equivalent to IUCN Category IV. The most common management category along the distribution of franciscana was the Environmental Protection Area (V), mostly because of its large numbers in Brazil. The categories IV and VI are also largely represented, mainly due to their representativeness in Argentina (Table 18.2).

Officially, 15 out of the 44 MPAs can be classified as full protection, or no-take MPAs (Categories I–III), whereas the other 29 are categorized as sustainable use (Categories IV–VI), allowing some extractive activities, including fishing. Nevertheless, although most of the MPAs adopt the IUCN names and criteria, some of them do not completely follow its system. In Argentina, for instance, the category called Natural Monument ("Monumento Natural," in Spanish) has been used to establish the maximum levels of protection for species in accordance with its national legislation. The first marine mammal protected under this legal instrument was the southern right whale (*Eubalaena australis* - National Law No. 23,094, published in 1984), and more recently the franciscana (Provincial Law/Buenos Aires No. 14,992, published in 2018). However, differently from the IUCN guidelines (see Table 18.1), which considered Natural Monuments (Category III) as no-take PAs designed to preserve rare natural sites with unique or beautiful scenery (Day et al., 2019), this designation in Argentina has been applied to protect species and therefore it seems to be equivalent to the called Habitat/Species Management Areas (Category IV). In this sense, it should be noted that artisanal fishery is allowed within the Franciscana Provincial Natural Monument. Similarly, in Uruguay, there are also two parks (Cabo Polonio and Isla de Flores) that, although based on the country´s original designation are classified into the IUCN Category II, due to the lack of management plans or the small fraction of their no-take zones, they contain fishing activities and can be viewed as sustainable use MPAs. Therefore, these three MPAs were considered as sustainable use in the following analyses. In addition, differently from the IUCN guidelines, in some of the Brazilian MPAs classified as category V-IUCN and named as Environmental Protected Areas, both artisanal and industrial fishing (i.e., boats over 20 of gross tonnage) are allowed, at least in part of the area. Therefore, there is normally a large fishing effort within these MPAs, much greater than recommended for sites that should be known for promoting a balanced interaction of nature and culture and, at the same time encouraging the conservation of aquatic biodiversity while acting as models of sustainability for wider application (see Day et al., 2019).

The size of the MPAs ranged from 1.4 km^2 of the Wildlife Refuge of Ilha dos Lobos, Brazil, to 26,389.6 km^2 of the Franciscana Natural Monument, in Argentina. In Brazil and Uruguay, the largest MPAs are, respectively, the Environmental Protected Area of Litoral Centro (4,549.7 km^2) along the São Paulo state coast, and the Humedales de Santa Lucia Protected Area of Managed Resources (292.6 km^2) in the Rio de la Plata. Nevertheless, the majority (63.6%, i.e., 28 out of 44) of the MPAs cover a marine area smaller than 100 km^2, with the most frequent size interval ranging from 1 to 50 km^2 (Fig. 18.3). The largest no-take MPA is the Wildlife Refuge of Arquipélago de Alcatrazes (266.1 km^2), in southeastern Brazil. Clearly, the sustainable use MPAs are significantly larger (mean = 1,450.8 km^2, SD = 4,709.6 km^2) than the no-take areas (mean = 55.9 km^2, SD = 81.8), even when the Franciscana Natural Monument is not included in the analysis (mean = 646.3 km^2, SD = 1,232.8). Moreover, considering the spatial coverage of the 44 MPAs, the no-take areas represent just 6.7% of the total.

From the 44 MPAs, 26 include some associated continental areas, whereas 18 are fully marine/estuarine, although some of them could also include islands. Moreover, many MPAs are totally or mostly located in estuarine systems where franciscanas have been recorded (see, for instance, MPAs labeled as 14 in Cananéia, 17 and 18 in Paranaguá along the Brazilian coast, 30, 31, 33, and 34 in the La Plata River estuary, and 38 and 39 in the

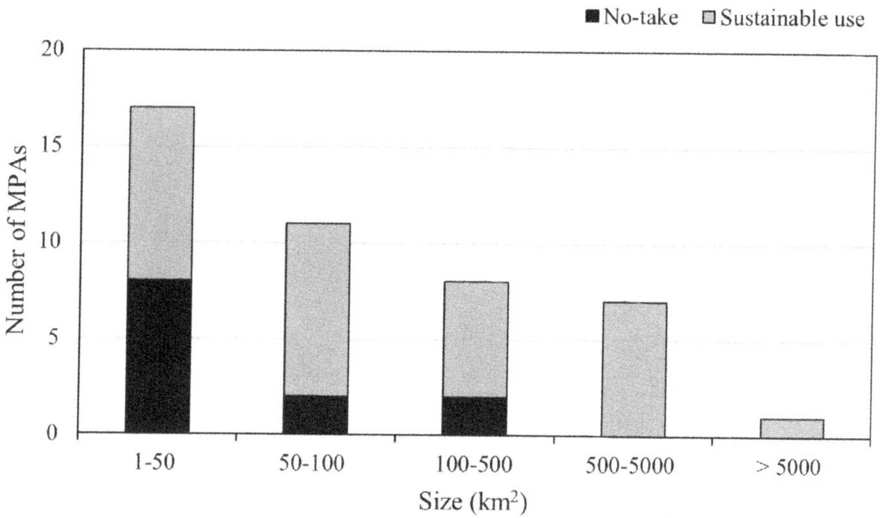

Fig. 18.3 ***Size intervals of the marine protected areas (MPAs) (no-take* versus *sustainable use) across the distribution of franciscana.***

southern Buenos Aires Province - Fig. 18.1 and Annex 18.I). Most of the MPAs are located in very shallow coastal waters. In this sense, although the franciscana can be distributed up to 50 m in some areas (e.g., southern Brazil—Danilewicz et al., 2009; Prado et al., 2021), around 60% of the marine area covered by the MPAs are located up to 10 m depth (Fig. 18.4).

In addition to the franciscana dolphin, the 44 MPAs also cover the usual distribution range of several marine mammal species, including, among others, the Guiana dolphin (*Sotalia guianensis*), the Lahille´s bottlenose dolphin (*Tursiops truncatus gephyreus*), the killer whales (*Orcinus orca*), the common dolphin (*Delphinus delphis*), the Burmeister's porpoise (*Phocoena spinipinnis*), the southern right whale (*Eubalaena australis*), the South American fur seal (*Arctocephalus australis*), and the South American sea lion (*Otaria flavescens*). Therefore, these species can also be benefited from these MPAs, although a more detailed analysis is needed to evaluate their effectiveness more reliably.

Management of MPAs along the franciscana distribution

There are some differences among the countries regarding MPAs governance type. In Brazil, most MPAs are under national administration, but there are also some subnational sites along the coasts of São Paulo and Rio

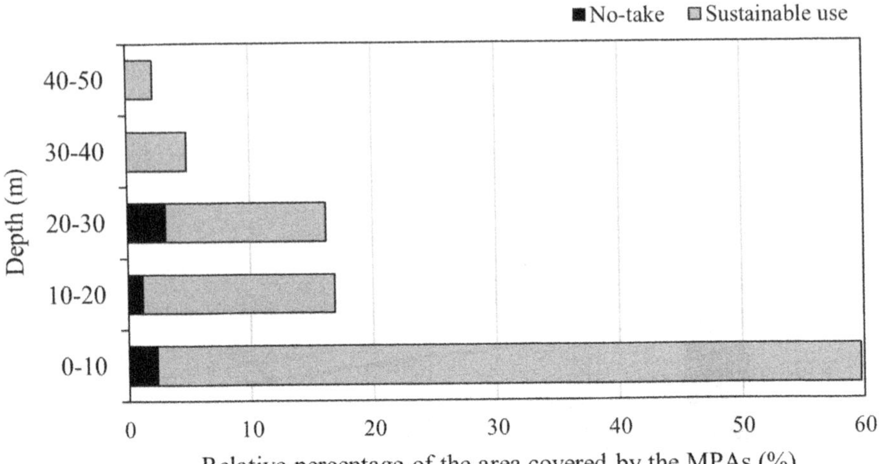

Fig. 18.4 ***Relative percentages of the area covered by the marine protected areas (MPAs) (no-take versus sustainable use) in different water depth intervals across the distribution of franciscana.***

de Janeiro managed by provinces or municipalities. In Uruguay, all MPAs with assigned administrators include the Ministry of the Environment along with other coadministrators such as the Ministry of National Defense or local governments (i.e., Intendencia de Rocha, Canelones, San José, and Montevideo). In Argentina, the MPAs are under the responsibility of the provinces and the provinces from Buenos Aires and Rio Negro administer the 13 MPAs along the franciscana distribution; Buenos Aires is responsible for eight of them while Rio Negro, the remaining five. Although none of them are under national administration, Complejo Islote Lobos, located in the Province of Río Negro, is projected to be a Marine National Park.

In Brazilian waters, 15 out of 25 MPAs that overlap with the distribution of franciscana has published management plans. Two other MPAs are waiting for their documents' official approval. Fourteen of these plans mention the franciscana as an occurring species in MPAs as well as in adjacent zones, but only seven considered it a focal conservation object (Fig. 18.5; see also Annex 18.I). The main actions proposed to protect the franciscana within these MPAs are establishing fishing regulations, protecting critical areas, supporting research, and monitoring the species. The majority (88%) of Brazilian MPAs have a management council (Fig. 18.5), which can be consultative or deliberative, depending on their main category (i.e., no-take or sustainable use, respectively).

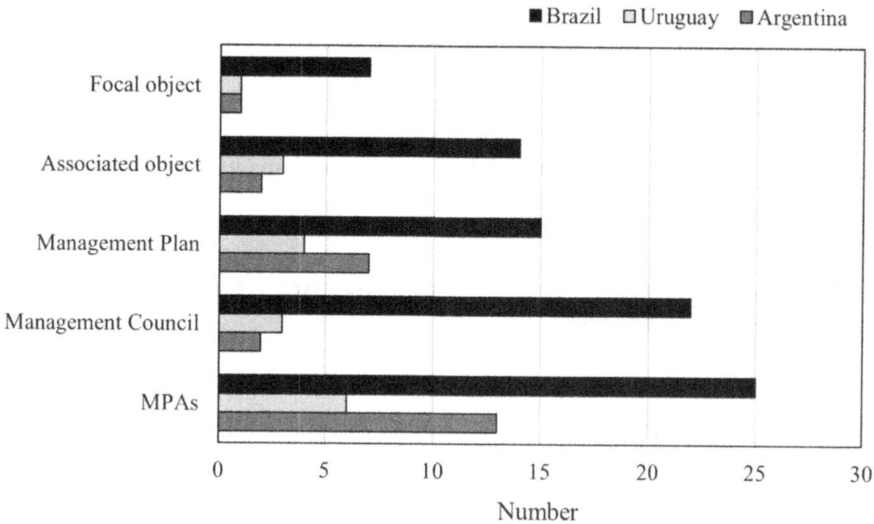

Fig. 18.5 ***Main management characteristics of the marine protected areas (MPAs) in the three countries across the distribution of franciscana.***

In Uruguay, by 2021, half of the MPAs (Cerro Verde and La Coronilla´s islands; Rocha lagoon; and Cabo Polonio) have already had management plans approved, while the management plans of two other areas (wetlands of Santa Lucía and Garzón lagoon) are still being prepared. Although some MPAs have approved management plans, their implementation—specifically within the marine and estuarine realm—is still under development. All these plans (and drafts) include the franciscana as a focal conservation object or as an associated object (Fig. 18.5); however, so far, none of the Uruguayan MPAs have any specific measures directed to monitoring or protecting the franciscana. Nevertheless, the same fishing regulations applied to the rest of the coastal zone of the country are applied in these areas (see Chapter 19).

In Argentina, from the 13 MPAs that incorporate the franciscana distribution in their waters, seven have management plans and only two of them have included the species (Fig. 18.5), the Multiple Use Natural Reserve of Bahía San Blas, and the Faunal Reserve of Punta Bermeja. The conservation of endangered species, such as the franciscana, is among the specific objectives mentioned in their management plans. In Punta Bermeja, there is a specific recommendation prohibiting the use of gillnets as well. Moreover, as previously mentioned, in 2018, the Buenos Aires Province declared the franciscana a Natural Monument under Law 14,992. Based on this regulation, the coastal waters of the province, up to 12 nm from the coast, have provided the highest level of protection for any species and it is equivalent to Category IV of IUCN. Currently, the only MPAs that have a management council or similar figures are Bahía de San Antonio and Complejo Islote Lobos, within the Rio Negro Province. In the Province of Buenos Aires, current legislation does not require an administrative council for MPAs.

MPAs of franciscana and the marine ecoregions of the world (MEOWs)

The distribution of franciscana in the southwestern Atlantic Ocean (SWAO) encompasses six out of the 232 Marine Ecoregions of the World (MEOW; Spalding et al., 2007), included in three different provinces: Tropical Southwestern Atlantic (1 MEOW), warm temperate Southwestern Atlantic (4 MEOWs) and Magellanic (1 MEOW). Although there is no direct correspondence between MEOWs and the FMAs, from an ecological perspective and marine spatial planning, it is interesting to examine the biogeographic representativeness of the MPAs that potentially contribute to the conservation of the franciscana.

The great majority of MPAs are located in two warm temperate MEOWs: Southeastern Brazil (21 MPAs) and Uruguay–Buenos Aires Shelf (10 MPAs) (Fig. 18.6), where MPAs cover 46% and 19% of the area of each MEOW, respectively (Table 18.3). These two MEOWs incorporate more than 60% of the estimated area of the franciscana distribution and contain almost 80% of the area covered by all MPAs (Table 18.3). In fact, the Uruguay–Buenos Aires Shelf alone contains more than 50% of the total area represented by all MPAs. On the other hand, Rio Grande is the MEOW with the lowest coverage of protection in terms of MPAs (less than 1%), even though it constitutes one of the largest areas of the franciscana distribution (around 34,000 km^2) (Table 18.3). It is also worth mentioning that Southeastern Brazil and Rio Grande are the only MEOWs containing no-take MPAs for franciscanas (Fig. 18.6). From a broader conservation perspective, these results are also helpfull for assessing the ecological representativeness of MPAs in the SWAO. In fact, considering that umbrella species can be useful for directing management strategies that benefit the ecosystems as a whole (Ward et al., 2020), the franciscana dolphin might be a good candidate to promote the increasing of representativeness of MPAs across the different MEOWs.

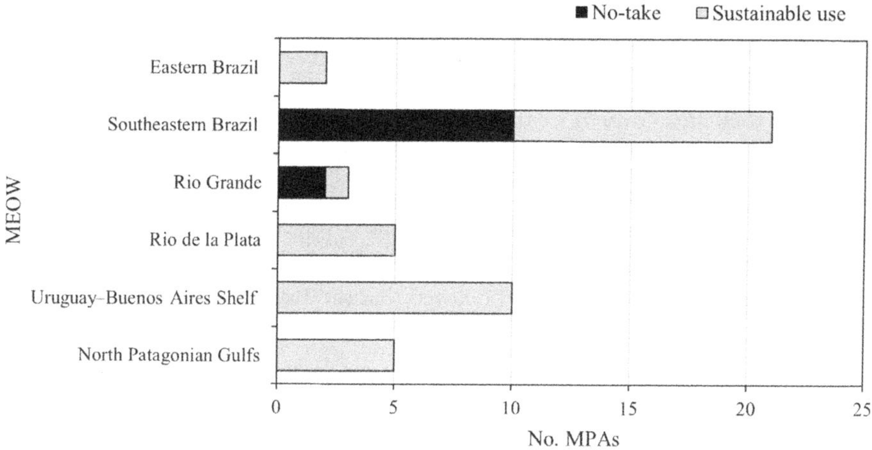

Fig. 18.6 ***Number and main category (no-take versus sustainable use) of marine protected areas (MPAs) that encompass the distribution of franciscana across the distinct Marine Ecoregions of the World (MEOWs).***

Table 18.3 Biogeographic representativeness of marine protected areas (MPAs) that potentially contribute to franciscana conservation across the Marine Ecoregions of the World (MEOW). Note that the sum of MPAs is different from the total (i.e., 44) since some MPAs are shared between MEOWs.

MEOW	Area used by franciscana (km^2)	No. MPAs	Area of MPAs (km^2)	Area covered by MPAs (%)
Eastern Brazil	11,950.6	2	146.1	1.2
Southeastern Brazil	26,987.4	21	12,366.7	45.8
Rio Grande	34,072.4	3	334.4	1.0
Rio de la Plata	31,163.8	5	6,960.7	22.3
Uruguay–Buenos Aires Shelf	111,642.2	10	21,613.5	19.4
North Patagonian Gulfs	10,499.6	5	1,556.0	14.8

MPAs across the franciscana management areas (FMAs)

Considering the five FMAs currently recognized by the IWC (i.e., FMA-Ia; FMA-Ib; FMA-II, FMA-III, FMA-IV) (IWC, 2016), the distribution of the MPAs is clearly uneven. The FMA covered by the largest number and area of MPAs is FMA-II (19 MPAs) while the smallest, FMA-Ia and FMA-Ib, each one with a single MPA. The number of sustainable uses MPAs (i.e., IUCN categories IV to VI) clearly surpasses the no-take areas (i.e., categories I to III). Moreover, when considering the Franciscana Provincial Natural Monument in Argentina as sustainable use, the FMAs II and III are the only ones that contain no-take MPAs (Fig. 18.7).

The total area covered by the 44 MPAs represents around 19% of the area occupied by franciscana across the three countries. Nevertheless, considering the size of each FMA, the area covered in each sector is also clearly uneven. Whereas FMA-II has almost half of its area covered by MPAs, FMA-Ia and FMA-Ib have only around 1% of their areas legally protected (Table 18.4). In terms of management, it should be noted that the two FMAs with the highest coverage by MPAs (FMA-II and FMA-IV) are markedly influenced by the existence of large sustainable use MPAs, particularly the three large environmental PAs in São Paulo (Litoral Norte, Sul, and Centro) and the Franciscana Natural Monument in Argentina. Moreover, despite FMA-III being considered one of the regions with the highest fishing effort and the highest annual bycatch of the species (e.g., Ott et al., 2002; Prado et al., 2021), its coverage by MPAs is also extremely low (2.7%) (Table 18.4 and Fig. 18.8). Therefore, the protection of species throughout its range is highly dependent on the fisheries management tools defined in the management plan of existing MPAs. Moreover, in those FMAs with little coverage of MPAs, the establishment and improvement of fisheries management measures beyond MPAs are considered crucial for the protection of the species (see Box 18.1).

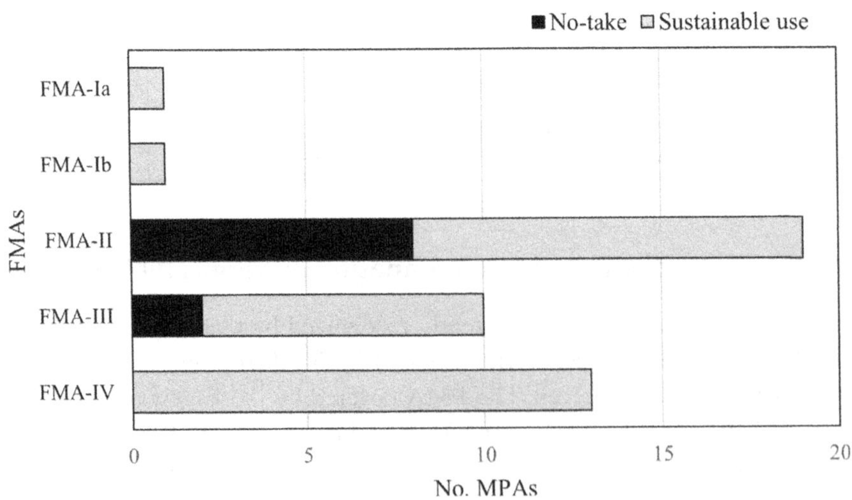

Fig. 18.7 ***Number and main category (no-take versus sustainable use) of marine protected areas (MPAs) across the distinct franciscana management areas (FMAs).***

Table 18.4 Main characteristics of the marine protected areas (MPAs) in each Franciscana Management Area (FMA). Details on the isobath used to define the external limit of each FMA and their estimated total area, number of MPAs, range size of MPAs, total area and percentage of the area covered by MPAs per FMA are shown.

FMA	Ia	Ib	II	III	IV	Total
External limit (isobath; m)	25	30	30	50	50[a]	–
Total area of FMA (km^2)	4,756.0	7,194.6	25,046.4	83,290.2	106,057.7	226,344.8
Number of MPAs	1	1	19	10	13	44
Range size of MPAs (km^2)	53.6	92.5	1.7–4,410.1	1.4-1,204.4	7.1-26,389.6	1.4– 26,389.6
Mean size of MPAs (km^2) [SD]	53.6	92.5	607.3 [1,343.7]	225.5 [360.2]	2,548.2 [7,240.1]	1,069.6 [4,047.1]
Total area of MPAs (km^2)	53.6	92.5	11,456.8	2,254.7	29,121.1	42,978.7
MPAs area coverage (%)	1.1	1.3	45.7	2.7	27.5	19.0

[a]In FMA-IV (Argentina), there are some records of franciscana beyond the 50 m depth isobath.

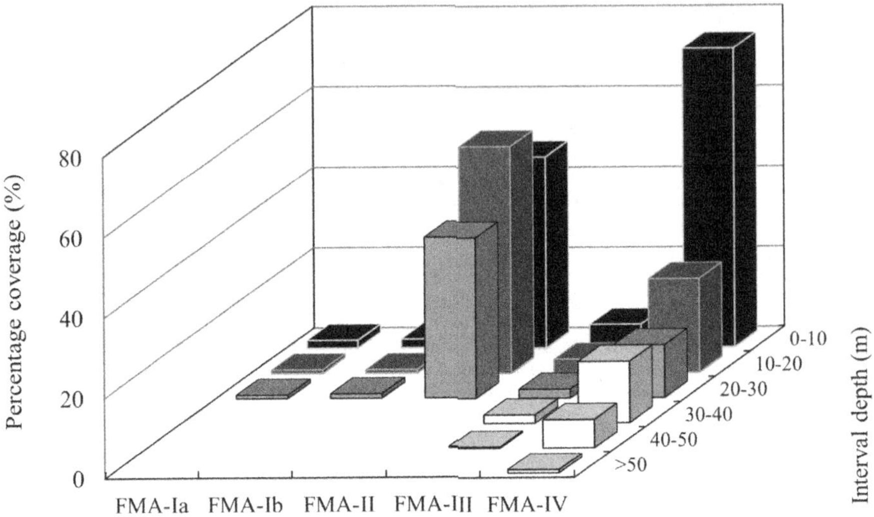

Fig. 18.8 ***Percentage of the area covered by marine protected areas (MPAs) across distinct water depths within the Franciscana Management Areas (FMAs).***

As previously mentioned, most MPAs are located in very shallow coastal waters (up to 10 m) (Fig. 18.4). However, when the FMAs are analyzed in terms of the percentage of coverage of MPAs by 10 m depth intervals, some additional results emerge (Fig. 18.8). Among the 20 strata analyzed, seven have more than 10% of their area covered by MPAs. All these strata with higher coverage are located within FMA-II and FMA-IV. It is interesting to note that FMA-II has not only a high percentage of MPA coverage but also a relatively homogeneous distribution of the MPAs across the different water depths. The two strata with the largest MPA coverage are FMA-IV/0-10m (74.3%) and FMA-II/10-20m (56.4%). In contrast, eight out of the 20 strata have only 2% or less of their total area protected. These strata are mainly located within FMA-Ia, FMA-Ib, and FMA-III (Fig. 18.8). Clearly, these are the main gaps in the marine protected network for franciscana. However, these data should ideally be compared to the areas of high density and threats to the species.

Potential population size of franciscana protected within the MPAs

The primary benefits expected from MPAs regarding the conservation of franciscana are the protection of its habitats and populations. In order to evaluate the number of individuals potentially protected within the MPAs, we tentatively combined the density of the species with the size of the largest no-take and sustainable use MPAs of each FMA. For these analyses, distinct densities of franciscana were used for each region, based on the most current and accurate information gathered from aerial surveys (see references in Table 18.5; Zerbini et al., 2020). The number of franciscanas encompassed within all MPAs combined (without overlapping) was also compared to the estimated population size for each FMA (Table 18.5), although this information is incomplete for some areas (e.g., FMA-III) and the distribution of species is known to be nonuniform (see Zerbini et al., 2020). In addition to the largest MPAs, we also analyzed the number of franciscanas hypothetically protected in an area equivalent to 100 km^2 that represent the upper size limit of the majority of MPAs (see Fig. 18.3). In this case, we putatively use the mean density of 0.31 franciscanas/km^2, obtained from the combined data of all FMAs.

As expected, the results from these analyses revealed a great disparity in the potential to protect franciscanas within the two main types of MPAs (i.e., no-take and sustainable use), as well as among FMAs (Table 18.5). The largest no-take MPA (Wildlife Refuge of Arquipélago de Alcatrazes) theoretically encompasses 223 franciscanas, which would be equivalent to 3.3% of the estimated population of the species in FMA-II. On the other hand, the largest sustainable use MPA (Franciscana Provincial Natural Monument) has the potential to protect almost 10,000 franciscanas, which corresponds to around 70% of the estimated population of FMA-IV. A high percentage of the estimated population of FMA-II (55.4%) is also distributed inside the MPAs. However, this legal protection has clearly not been enough to guarantee the conservation of the species in most of these areas, due to a lack of specific management instruments and enforcement to reduce bycatch rates. Therefore, many of these areas merely create an illusion of protecting the franciscana.

Another concerning result from these analyses is the very low percentage of the estimated populations covered by MPAs in FMA-Ia and FMA-Ib. These isolated FMAs have only about 2% of their populations protected by MPAs. Moreover, all these MPAs are sustainable use (Table 18.5). For FMA-III, it has to be acknowledged that the estimated population size currently available only refers to the Brazilian portion of the stock, once the abundance of franciscana in Uruguay has not been estimated yet. Therefore, the calculation of the percentage of franciscanas potentially protected by MPAs in FMA-III is feasible just if we consider exclusively the MPAs within this partial area. In this scenario, constituted by 1,244.7 km^2 of PAs, only about 4% of the franciscanas inhabiting the Brazilian portion of FMA-III are legally protected by some type of MPA.

Table 18.5 Potential population size of franciscana protected within the marine protected areas (MPAs) across the distinct Franciscana Management Areas (FMAs).

	Franciscana Management Area (FMA)				
	Ia	**Ib**	**II**	**III**	**IV**
Density of franciscana (individuals/km^2)	0.23[a]	0.32[b]	0.33[c]	0.31[d]	0.377[e]
Estimated population size	595[a]	1,280[b]	6,827[c]	9,651[d,f]	14,175[e,g]
Largest no-take MPA (km^2)	NE	NE	674.72	29.98	NE
No. of franciscanas within the largest no-take MPA [% of the population]	NA	NA	222.7 [3.3]	9.3 [NA]	NA
Largest sustainable use MPA (km^2)	53.5	92.5	4,410.1	1,204.4	26,389.6
No. of franciscanas within the largest sustainable use MPA [% of the population]	12.3 [2.1]	29.2 [2.3]	1,455.3 [21.3]	373.4 [NA]	9,948.9 [70.2][g]
Total area covered by all MPAs (km^2)	53.6	92.5	11,456.8	2,254.7	29,121.1
No. of franciscanas potentially protected within all MPAs [% of the population]	12.3 [2.1]	29.2 [2.3]	3,780.8 [55.4]	699.0 [NA]	10,978.7 [77.5][g]

[a]Sucunza et al. (2020b); [b]Danilewicz et al. (2020); [c]Sucunza et al. (2020c); [d]Sucunza et al. (2020a); [e]Crespo et al. (2010). [f]Estimates not including Uruguayan waters (i.e., covered a portion of FMA-III); [g]Estimates only for the coastal stratum up to 30 m isobath (i.e., covered a portion of FMA-IV).
NE, nonexisting; NA, not applicable.

Moreover, as most of the MPAs across the distribution of franciscana are smaller than 100 km^2 (Fig. 18.3), these areas potentially result in legal protection of only about 30 individuals. As a theoretical exercise, it is interesting to note that an MPA with 1,600 km^2 would be needed to protect around 500 franciscanas, which is the genetically effective population size (N_e) estimated to avoid inbreeding (Traill et al., 2010). However, as noted by Traill et al. (2010), this number is at least an order of magnitude smaller than the census population size necessary to effectively prevent the extinction of most species (i.e., the Minimum Viable Population size - MVP). Nevertheless, an area of that size (i.e., 1,600 km^2) could be used as initial guidance for the establishment of new MPAs, and also for the creation of a more effective network system of MPAs and no-fishing zones across the entire distribution of franciscana. It must be mentioned that tagging studies revealed that the home range size of franciscanas in two coastal areas of Argentina range between 51–312 km^2 (mean = 132.9 km^2; Bahía Samborombon) and 49–1,013 km^2 (mean = 463.5 km^2; Bahía San Blas) (Wells et al., 2021). Therefore, the most frequent size of MPAs of franciscanas (i.e., up to 100 km^2) is smaller than the area usually used by an individual.

Other potential benefits of MPAs for franciscana

The establishment of MPAs could certainly have additional benefits for the conservation of the franciscana. PAs represent spaces where contact with nature is ideal for promoting conservation and sustainable development strategies. In this context, and as the franciscana is one of the most difficult cetaceans to be watched in the wild, these are the areas that present opportunities to promote its conservation needs to the inhabitants and visitors of MPAs: a large part of people value contact with coastal environments. In MPAs, people are likely to be more conscious of a species they are completely unfamiliar with, that they have only seen dead on a net or stranded at the beach and hardly seen in the wild. Unfortunately, the vast majority of current MPAs where franciscanas are present have missed this great opportunity. Interpretation centers and even leaflets and posters providing information about the franciscana, its threats and, at the same time encouraging people to take an interest in this dolphin are scarce and usually outside of the current MPAs (e.g., Río Negro Estuary in Argentina, Babitonga Bay in Brazil). The franciscana conservation needs are practically unknown to the general public.

MPAs tend to often homogenize their visitors, lumping together tourists with recreationists. Although this may seem like a mere technicality, it is important to identify and characterize the target audience of each educational-communication-recreational-tourism campaign. So, for example, special strategies should be generated to raise awareness among MPAs' neighbors who visit an MPA near their home. These neighbors could vary from artisanal fishermen to local lodging service providers who are completely unaware of the existence of the franciscana and its local conservation needs. When dealing with them, messages and strategies should be concrete and respectful of their realities and may or may not differ significantly from those designed for visitors from other regions, realities, and expectations when visiting the same MPA. Given the limited economic and human resources allocated to the conservation of the franciscana, it would be important to address these differences as well as the sociocultural particularities of each MPA.

Future perspectives and recommendations

The recognition that MPAs are an important management tool for the conservation of franciscanas has been highlighted in several official documents, such as the CMP for franciscana, established in 2016 by the IWC (see Box 18.2). The recently reviewed National Action Plan (PAN) for the conservation of the species in Brazil (Directive ICMBio 655/2019) also includes, among its main proposed actions, the creation or extension of the Brazilian MPAs, paying special attention to the following areas (from north to south): Espiríto Santo (FMA-Ia); the marine zone adjacent to the Restinga de Jurubatiba National Park (Rio de Janeiro - FMA-Ib); Babitonga Bay (Santa Catarina – FMA-II) and Albardão (Rio Grande do Sul – FMA-III) (see Box 18.2). PAN has also recommended the creation of areas of special interest for the conservation of the species within the existing large sustainable use MPAs in the state of São Paulo (FMA-II). It is noteworthy that the data gathered from several aerial surveys ($n = 7$) conducted from 2017 to 2021, revealed that at least 50% of the estimated population of FMA-Ia (595 individuals) are distributed in an adjacent area close to the mouth of the Doce River (19°39′S; 39°49′W) (Fig. 18.9) (GEMARS, unp. data). This area of ca. 310 km^2 extends 25 km south and 5 km north of the river mouth and 10 km offshore. It is also worth noting that the area near to the mouth of Doce River concentrated most of the strandings records of the species in the region, including animals with clear indications of interaction with fisheries (Mayorga et al., 2020). Therefore, it represents one of the most critical areas for the conservation of this isolated and genetically distinct population.

BOX 18.2 Priority MPAs officially proposed for the conservation of the franciscana

The creation of new MPAs has been identified as an important tool for the conservation of the franciscana for a long time. In 2016, the IWC established a CMP for franciscana, which among its priority actions highlighted the establishment of two new areas in Brazil (Babitonga Bay and Albardão) and one in Argentina (Río Negro Estuary) (IWC, 2016) (see Figure 18.1). Here, we briefly describe those areas that had their processes submitted to government environmental agencies, but unfortunately, to this day, have not yet been approved nor implemented.

Babitonga Bay, Brazil - Santa Catarina

FMA II - 26°16′S; 48°42′W

Babitonga Bay is an estuarine area in southern Brazil, with the largest and most important representative area of mangrove in the threshold zone of its southern distribution in the Western Atlantic Ocean. The bay offers conditions for the spawning, recruitment, and growth of several organisms, including migratory and endangered species. Among the threatened species, are the Atlantic goliath grouper (*Epinephelus itajara*), the green turtle (*Chelonia mydas*), and a resident population of franciscana, with unique genetic characteristics (Wells et al., 2021). The bay is also largely used by Guiana dolphins (*Sotalia guianensis*). In 2005, the federal environmental agency of the Brazilian government, the Chico Mendes Institute for Biodiversity Conservation—ICMBio, initiated a process for the creation of a conservation unit, including all its water sheet, as well as the adjacent coastline. The first proposal suggested the creation of a Fauna Reserve (equivalent to IUCN Category IV), a sustainable use protected area, accordingly to the Federal Law nº 9.985/2000. A technical report was produced, and many public consultations were held to support the creation process, following the rites established in the law. However, the proposal was met with strong resistance from some socio-economic sectors operating in the region, mainly related to harbor activities. The creation process was then judicialized, which prevented it from being completed, and the conservation unit was not created. Even so, the understanding remains that the region is of great importance for the conservation of biodiversity, including the positioning of the Federal Public Ministry, through two Public Civil Actions. Currently, the process at the federal level remains stalled, but some initiatives have proposed a change of category, suggesting the creation of an Environmental Protection Area (IUCN Category V), a category that offers a greater flexibility. In view of the growing anthropogenic pressure on the region, it is understood that the creation of a protected area is the only possibility for the conservation of local biodiversity in the medium and long-term, including the survival of the unique franciscana population.

Albardão, Brazil - Rio Grande do Sul

FMA III - 33°12′S; 52°42′W

The region of Albardão comprises the marine coastal zone of the state of Rio Grande do Sul, located close to the Albardão lighthouse. Its high biological productivity sustains a great biodiversity of fish, sea birds, sea turtles and marine mammals

(Mattos and Ferreira, 2018). However, due to the high fishing effort observed in this region, many endangered vertebrate species are incidentally caught by fisheries, including the franciscana. Therefore, since 2003, the Albardão region has been indicated as a priority area for conservation of biodiversity. In 2008, the ICMBio officially started carrying out studies for the creation of an MPA in the Albardão region. In 2017, the agency resumed the technical studies and hired a consultant to support its creation. Although the final design has not been defined yet, the original proposal suggests the creation of a National Park (Category II – IUCN), including part of a coastal lagoon (Mangueira), the dunes and a singular stretch of the beach (called "Concheiros"), representing an important paleontological site, as well as the marine area almost up to 100 m isobath. Based on records of franciscana bycatch obtained from the industrial gillnet fleet of Rio Grande via logbooks throughout almost 10 years, Prado et al. (2021) identified the Mostardas and Albardão regions as hotspot areas of franciscana mortality in southern Brazil. Therefore, the establishment of an MPA in the Albardão region represents an important conservation measure to protect the franciscana. This MPA will also contribute to the recovery of commercially valuable and depleted fishery stocks, as well as ensure the conservation of many other threatened species.

Río Negro Estuary, Argentina - Rio Negro

FMA IV - 41°01′S; 62°48′W

Since 2007, the creation of a protected area (PA) in the Rio Negro Estuary has been proposed, based on the high environmental value of this region due to the presence of the largest parrot colony on the planet (Burrowing Parrot, *Cyanoliseus patagonus*) and one of the most outstanding breeding and feeding areas reported for franciscana in its southernmost range (see Masello et al., 2007). Since then, four bills have been presented to the Legislature of the Province of Río Negro, designed to create a multiple-use PA. In spite of receiving a first approval in all cases (in this region double approval is required to make a new law effective), each of the bills has lost parliamentary status and the PA has not yet been created. Today, there is no new parliamentary bill in force, despite the fact that several NGOs, along with regional and international social and scientific actors are promoting its creation. The proposal area includes the mouth of the estuary of Rio Negro, as well as a coastal and marine zone, extending from the sandstone cliff where the parrot colony is located to the 20 m isobath. Its implementation would be an indispensable tool for managing the conservation of the franciscana in its southern limit of distribution (Failla et al., 2012).

In Argentina, three new PAs have also been officially proposed. The first one, Río Negro Estuary (Masello et al., 2007), has been recognized as a priority for franciscana by the IWC CMP (IWC, 2016) and while it has been extensively discussed by the provincial legislature since 2007 it is yet to be created (see Box 18.2). The other two areas recognized as relevant for the biodiversity of the Argentinean Sea are the Río de la Plata Maritime Front

Fig. 18.9 ***Aerial view of the mouth of Doce River in Espírito Santo, southeastern Brazil (FMA-Ia).*** The establishment of a marine protected area in this region is considered one of the priority actions for the conservation of the species *(Photo: Paulo H. Ott Uergs/ GEMARS)*.

Ecosystem and Rincón, which together add up to a total area of 69,700 km^2 (22,000 km^2 and 47,700 km^2, respectively) (MAyDS, 2016). All three recognize the franciscana as a species that needs to be protected.

Considering the status of the species and the many unprotected areas of the FMAs, it is clear how important is to establish these new proposed MPAs. This is even more evident for those FMAs that have a very small fraction of their areas (<2%) covered by MPAs (i.e., FMA-Ia, FMA-Ib, and FMA-III). As each FMA corresponds to a distinct population (Secchi et al., 2003b; Cunha et al., 2014), it is mandatory that each one has good coverage of PAs. Furthermore, other measures should be urgently adopted within the current MPAs, including the establishment of specific measures to monitor and protect the species and the enforcement of fishery regulations, among others. It is worth that, currently, 40% of the MPAs do not have a management plan and only 20% of them include the franciscana as a focal object of conservation. Therefore, in order to effectively improve the conservation of the species within these areas, there is an urgent need for all MPAs to have a management plan that clearly incorporates the species in its objectives.

The lack of specific goals for the franciscana in many management plans has also been reflected in other important initiatives besides the direct protection of the species. Today, educational-recreational-tourist actions and campaigns including the franciscana are scarce and practically nonexistent. The current and future MPA management plans need to incorporate concrete strategies focused on the educational, valuation, and tourist-recreational use of this dolphin throughout its range: we are missing the opportunity to raise awareness among visitors as well as to generate potential sources of employment linked to tourism and recreation. These initiatives do not need to be restricted to the MPAs of franciscanas, but can also be explored by terrestrial or insular PAs along the distribution of the species. It should be noted that many of these areas have a buffer zone that encompasses the marine zone. Therefore, such initiatives could be incorporated into their management plans.

The types of existing MPAS are another important aspect that needs to be taken into consideration. Currently, the no-take MPAs represent a very small fraction (6.7%) of the PAs across the distribution of franciscana. Evidently, their coverage and connectivity need to increase substantially in order to provide clear conservation benefits for the species. It could be especially important for areas where the species shows a resident pattern (e.g., Babitonga Bay, in Santa Catarina) or a clear patchy distribution (e.g., mouth of Doce River, in Espírito Santo), although the potential socioeconomic impacts of creating a no-take MPA should also be addressed. Moreover, it has been pointed out that MPAs alone do not reduce fishing pressure, but simply move it to a different area (e.g., Agardy et al., 2011). Therefore, effective conservation of the species would only be achieved if large MPAs were created in a network system and additional fishing regulations were established within them, as well as in adjacent areas. In this context, if large sustainable use MPAs have specific fishing regulations or zoning plans and are well enforced they may have a great potential to protect not only the franciscana, but also other endangered species, including some of the depleted fish stocks. This is a challenging but promising task since currently 19% of the area occupied by franciscana is covered by MPAs. Clearly, these MPAs provide an important legal framework to protect the species, but their effectiveness greatly depends on additional steps, such as the establishment of management plans that have the protection and monitoring of the species as clear objectives. These MPAs could also be used to test and implement mitigation measures in fisheries so as to reduce the incidental mortality of the species. In fact, the conservation of the franciscana seems to be closely linked to the adoption of more sustainable fishery practices by the governments of the three countries, both inside and outside the MPAs. In this sense, a step change toward an integrated ecosystem approach to the management of fisheries is clearly necessary. That would not only increase the chances of long-term survival of franciscana and other endangered species but would also help sustain the fishing activities of some communities that are socioeconomically vulnerable (Secchi et al., 2021).

Finally, since the distribution of the franciscana includes three different countries, it is also very important to increase integration efforts between countries. It is necessary to explore consensual management instruments, especially in border areas such as the transitional zone between Brazil and Uruguay in FMA-III, and the La Plata River estuary between Uruguay and Argentina (FMA-III and FMA-IV). In this context, in order to enhance the complementarity of conservation objectives in existing MPAs, it is particularly important to take advantage of the binational management agreements that have been already established (e.g., the Treaty of the Rio de la Plata and its Maritime Front; http://ctmfm.org/sitio/documentos-basicos/tratado-rio-de-la-plata-frente-maritimo/). Likewise, within each country, it is necessary to integrate MPAs and implement corridors that ensure connectivity between them and favor the long-term viability of populations, not just of the franciscana, but also of other components of the marine megafauna.

Acknowledgments

We would like to thank all the managers of the MPAs who helped us to compile valuable information about the current status of these areas. We would also like to specially acknowledge Eliana Zuazquita, from Fundación Cethus, for assistance with information on MPAs in Argentina, as well as Marta Cremer, from the Universidade da Região de Joinville (UNIVILLE), and Ronaldo Cataldo Costa, from the Centro Nacional de Pesquisa e Conservação da Biodiversidade Marinha do Sudeste e Sul (CEPSUL/ICMBio), for providing updated information on the ongoing process on the Babitonga and Albardão proposals, in Brazil. Important support was provided by the Brazilian Biodiversity Fund (Funbio), through a conduct adjustment agreement between Petrorio and the Brazilian Ministry for the Environment, to carry out the aerial surveys to estimate the abundance of franciscanas in Brazilian waters. Thanks also to the entire GEMARS team for conducting these aerial surveys, especially Federico Sucunza. These data were essential to identify some of the critical areas presented for the conservation of the species.

[Annex 18.I]

Annex 18.I—Summary of the marine protected areas (MPAs) across the franciscana management areas (FMAs); St./Prov., State/Province[a]; ID, Code

Annex 18.I Summary of the marine protected areas (MPAs) across the Franciscana Management Areas (FMAs).

Country	St./ Prov.	FMA	ID	MPA English name	WDPA #	Size (km^2)	Level	Year	Type	IUCN	MC	MP	Species
Brazil	ES	Ia	1	Costa das Algas—EPA	555576473	53.55	N	2010	SU	V	Yes	No	–
Brazil	RJ	Ib	2	Arquipélago de Santana—EPA	555682425	92.54	M	1989	SU	V	No	No	–
Brazil	RJ	II	3	Tamoios—ES	116085	86.19	N	1990	NT	I[a]	Yes	Yes	No
Brazil	RJ	II	4	Aventureiro—SDR	555599930	18.17	R	1990	SU	VI	No	No	–
Brazil	RJ	II	5	Baía de Paraty—EPA	555682195	67.15	M	1984	SU	V	Yes	No	–
Brazil	SP	II	6	Litoral Norte—Marine EPA	555576431	2,530.70	R	2008	SU	V	Yes	Yes[d]	FO
Brazil	SP	II	7	Tupinambás—ES	18739	25.48	N	1987	NT	I[a]	Yes	Yes	FO
Brazil	SP	II	8	São Sebastião—AREI	555576359	6.29	R	2008	SU	IV	Yes	Yes[d]	AO
Brazil	SP	II	9	Arquipélago de Alcatrazes—WR	555636626	266.09	N	2016	NT	III	Yes	Yes	AO
Brazil	SP	II	10	Litoral Centro—Marine EPA	555576417	4,410.06	R	2008	SU	V	Yes	Yes	FO
Brazil	SP	II	11	Laje de Santos—Marine SP	168273	51.39	R	1993	NT	II	Yes	Yes	AO
Brazil	SP	II	12	Xixová-Japuí—SP	198374	3.67	R	1993	NT	II	Yes	Yes	AO
Brazil	SP	II	13	Ilhas do Abrigo e Guararitama—WR	478593	4.81	R	2013	NT	III	Yes	No	–
Brazil	SP	II	14	Cananéia-Iguapé-Peruíbe—EPA	19484	80.88	N	1984	SU	V	Yes	Yes	FO
Brazil	SP	II	15	Litoral Sul—Marine EPA	555576387	3,677.26	R	2008	SU	V	Yes	Yes	AO
Brazil	SP	II	16	Tupiniquins—ES	18740	17.28	N	1986	NT	I[a]	Yes	Yes	FO
Brazil	PR	II	17	Guaraqueçaba—EPA	7906	81.57	N	1985	SU	V	Yes	Yes	–
Brazil	PR	II	18	Superagui—NP	19274	1.74	N	1989	NT	II	Yes	Yes	FO
Brazil	PR	II	19	Ilhas dos Currais—Marine NP	555576325	13.60	N	2013	NT	II	No	No	–
Brazil	SC	II	20	Arvoredo—Marine BR	31751	167.48	N	1990	NT	I[a]	Yes	Yes	No
Brazil	SC	II	21	Anhatomirim—EPA	115868	26.53	N	1992	SU	V	Yes	Yes	AO
Brazil	SC	III	22	Pirajubaé—Marine ER	67715	8.88	N	1992	SU	VI	Yes	No	–
Brazil	SC	III	23	Baleia Franca—EPA	351716	1,204.39	N	2000	SU	V	Yes	Yes	FO
Brazil	RS	III	24	Ilha dos Lobos—WR	10843	1.42	N	1983	NT	III	Yes	No	–
Brazil	RS	III	25	Lagoa do Peixe—NP	19448	29.98	N	1986	NT	II	Yes	Yes	AO
Uruguay	RO	III	26	Cerro Verde—H/SMA	478063	72.68	N	2011	SU	IV	Yes	Yes	AO
Uruguay	RO	III	27	Cabo Polonio—NP	478062	211.51	N	2009	NT[c]	II	Yes	Yes	AO

Uruguay	RO	III	28	Laguna de Rocha—PL	478064	102.73	N	2010	SU	V	Yes	Yes	FO
Uruguay	MA/RO	III	29	Laguna Garzón—H/SMA	555624150	273.32	N	2014	SU	IV	No	No[e]	AO
Uruguay	MO/CA	III	30	Isla de Flores—NP	555542451	57.14	N	2018	NT[c]	II	No	No	–
Uruguay	SJ/MO/CA	III	31	Humedales del Santa Lucia—PAMR	555542447	292.63	N	2015	SU	VI	No	No[e]	AO
Argentina	BA	IV	32	Franciscana—Provincial NM	NA	26,389.59	R	2018	NA[c]	III	NA	NA	FO
Argentina	BA	IV	33	Bahía de Samborombón—Integral NR	145514	232.20	R	1997	SU	IV	No	Yes	No
Argentina	BA	IV	34	Rincón de Ajo—NR	32516	84.10	R	1997	SU	IV	No	Yes	No
Argentina	BA	IV	35	Restinga del Faro—Provincial NR	555558364	7.10	R	2011	SU	IV	No	NR	–
Argentina	BA	IV	36	Arroyo Zabala—MUNR	555558341	21.10	R	2001	SU	VI	No	NR	–
Argentina	BA	IV	37	Pehuén-Co Monte Hermoso—NR	555558360	25.60	R	2005	SU	IV	No	NR	–
Argentina	BA	IV	38	Bahías Blanca, Falsa y Verde—MUNR	555558342	1,985.30	R	1991	SU	VI	No	NR	–
Argentina	BA	IV	39	Bahía San Blas—MUNR	555558343	3,562.20	R	1987	SU	VI	No	Yes	FO
Argentina	RN	IV	40	Punta Bermeja—PNA	16880	12.40	R	1971	SU	IV	No	Yes	AO
Argentina	RN	IV	41	Caleta de los Loros—MUNR	16882	41.90	R	1984	SU	VI	No	Yes	No
Argentina	RN	IV	42	Bahía San Antonio—PL	97481	555.71	R	1993	SU	V	Yes	Yes	No
Argentina	RN	IV	43	Complejo Islote Lobos—SNR	30844	52.00	R	1977	SU	VI	Yes	Yes	No
Argentina	RN	IV	44	Puerto Lobos—PL	555558344	157.29	R	1998/2007	SU	V	No	No	–

[a]Abbreviations for states and provinces: ES, Espírito Santo; RJ, Rio de Janeiro; SP, São Paulo; PR, Paraná; SC, Santa Catarina; RS, Rio Grande do Sul; RO, Rocha; MA, Maldonado; MO, Montevideo; CA, Canelones; SJ, San Jose; BA, Buenos Aires; RN, Rio Negro. [b]MPA English Name: EPA, Environmental Protection Area; ES, Ecological Station; SDR, Sustainable Development Reserve; AREI, Area of Relevant Ecological Interest; WR, Wildlife Refuge; SP, State Park; BR, Biological Reserve; ER, Extractive Reserve; H/SMA, Habitat/Species Management Area; PL, Protected Landscape; PAMR, Protected Area of Managed Resources; NM, Natural Monument; NR, Nature Reserve; MUNR, Multiple Use Nature Reserve; PNA, Protected Nature Area; SNR, Scientific Nature Reserve. [c]MPAs that, due to allowing gillnet fishing, were considered as "sustainable use" despite the original IUCN classification; [d]MPAs that have a management plan, but their publication is still pending; [e]MPAs that have a preliminary draft of the management plan.

according to Fig. 18.1; MPA English Name[b]; WDPA, Code in the World Database on Protected Areas; Size, size of the MPA within the species' range (see text); Level: N, national; R, regional/provincial; M , municipal; Type: SU, sustainable use; NT , no-take; NA , not applicable; IUCN, category according to the International Union for Conservation of Nature; MC, management council; MP, management plan; NR, not reported; Species, Inclusion of the franciscana in the management plan as a focal (FO) or Associated (AO—i.e., secondary) conservation object.

References

Agardy, T., di Sciara, G.N., Christie, P., 2011. Mind the gap: addressing the shortcomings of marine protected areas through large scale marine spatial planning. Mar. Policy 35, 226–232. https://doi.org/10.1016/j.marpol.2010.10.006.

Álvarez Malvido, M., Lázaro, C., De Lamo, X., Juffe-Bignoli, D., Cao, R., Bueno, P., Sofrony, C., Maretti, C., Guerra, F. (Eds.), 2021. Informe Planeta Protegido 2020: Latinoamérica y el Caribe. RedParques, UNEP-WCMC, CMAP-UICN, WWF, CONANP y Proyecto IAPA. Ciudad de México, México; Cambridge UK; Gland, Switzerland; Bogotá, Colombia (in Spanish).

Alves, P., Arfelli, C., Tomás, A., 2012. Selectivity of bottom gillnet of southeastern Brazil. Bol. Inst. Pesca 38 (4), 275–284.

Amaral, K.B., Danilewicz, D., Zerbini, A., Di Beneditto, A.P., Andriolo, A., Alvares, D.J., Secchi, E., Ferreira, E., Sucunza, F., Martins, M.B., de Oliveira Santos, M.C., Cremer, M., Denuncio, P., Ott, P.H, Moreno, I.B., 2018. Reassessment of the franciscana *Pontoporia blainvillei* (Gervais & d'Orbigny, 1844) distribution and niche characteristics in Brazil. J. Exp. Mar. Biol. Ecol. 508, 1–12. https://doi.org/10.1016/j.jembe.2018.07.010.

AMP Argentina, 2020. Áreas Costeras y Marinas Protegidas de la Argentina. Compiled by WCS Argentina and collaborators (in Spanish). https://beta.ampargentina.org/areas . (accessed June 17, 2021).

Berninsone, L.G., Bordino, P., Gnecco, M., Foutel, M., Machay, A.I., Werner, T.B., 2020. Alternative to mitigate the bycatch of franciscana dolphin (*Pontoporia blainvillei*) in Argentina. Front. Mar. Sci. 7, 699. https://doi.org/10.3389/fmars.2020.00699.

Bertozzi, C., Zerbini, A.N., 2002. Incidental mortality of franciscana, *Pontoporia blainvillei*, in the artisanal fishery of Praia Grande, São Paulo State, Brazil. Lat. Am. J. Aquat. Mamm. 1, 153–160.

Bordino, P., Kraus, S., Albareda, D., Fazio, A., Palmerio, A., Mendez, M., Botta, S., 2002. Reducing incidental mortality of Franciscana dolphin, *Pontoporia blainvillei*, with acoustic warning devices attached to fishing nets. Mar. Mamm. Sci. 18, 833–842.

Bordino, P., Mackay, A.I., Werner, T.B., Northridge, S.P., Read, A.J., 2013. Franciscana bycatch is not reduced by acoustically reflective or physically stiffened gillnets. Endanger. Species Res. 21, 1–12.

Brasil, 2012. Instrução Normativa Interministerial INI 12/2012. 22 de Agosto de 2012. MPA/MMA. Diário Oficial da União, 24 de Agosto de 2012, 1, 39-40 (in Portuguese).

Brownell, R.L., Reeves, R.R., Read, A.J., Smith, B.D., Thomas, P.O., Ralls, K., Amano, M., Berggren, P., Chit, A.M., Collins, T., Currey, R., Dolar, M.L., Genov, T., Hobbs, R.C., Kreb, D., Marsh, H., Zhigang, M., Perrin, W., Phay, S., Rojas-Bracho, L., Ryan, G.E., Shelden, K.E.W., Slooten, E., Taylor, B.L., Vidal, O., Ding, W., Whitty, T.S., Wang, J.Y., 2019. Bycatch in gillnet fisheries threatens critically endangered small cetaceans and other aquatic megafauna. Endanger. Species Res. 40, 285–296.

Claudet, J., Osenberg, C., Benedetti-Cecchi, L., Domenici, P., García-Charton, J., Perez-Ruzafa, A., Badalamenti, F., Bayle-Sempere, J., Brito, A., Bulleri, F., Culioli, J.M., Dimech, M., Falcón, J., Guala, I., Milazzo, M., Sanchez-Meca, J., Somerfield, P., Stobart, B., Vandeperre, F., Planes, S., 2008. Marine reserves: size and age do matter. Ecol. Lett. 11, 481–489. https://doi.org/10.1111/j.1461-0248.2008.01166.x.

Crespo, E.A., Coscarella, M.A., Arias, M., Sueyro, N., 2020. Abundance estimation of franciscana dolphins by means of aerial surveys in Buenos Aires Province, Argentina. Paper SC/69B/ASI/03 presented to the IWC Scientific Committee, May 2020, p. 9.

Crespo, E.A., Harris, G., González, R., 1998. Group size and distributional range of the franciscana, *Pontoporia blainvillei*. Mar. Mamm. Sci. 14, 845–849.

Crespo, E.A., Pedraza, S.N., Grandi, M.F., Dans, S.L., Garaffo, G.V., 2010. Abundance and distribution of endangered franciscana dolphins in Argentine waters and conservation implication. Mar. Mamm. Sci. 26, 17–35.

Cunha, H.A., Medeiros, B.V., Barbosa, L.A., Cremer, M.J., Marigo, J., Lailson-Brito, J., Azevedo, A.F., Solé-Cava, A.M., 2014. Population structure of the endangered franciscana dolphin (*Pontoporia blainvillei*): reassessing management units. PLoS One 9 (1), e85633. https://doi.org/10.1371/journal.pone.0085633.

Danilewicz, D., Rosas, F., Bastida, R., Marigo, J., Muelbert, M., Rodriguez, D., Lailson-Brito Jr., J., Ruoppolo, V., Ramos, R., Bassoi, M., Ott, P.H., Caon, G., Rocha, A.M., Catão-Dias, J.L., Secchi, E.R., 2002. Report of the working group on biology and ecology. Lat. Am. J. Aquat. Mamm. 1, 25–42.

Danilewicz, D., Secchi, E.R., Ott, P.H., Moreno, I.B., Bassoi, M., Martins, M.B., 2009. Habitat use patterns of franciscana dolphins (*Pontoporia blainvillei*) off southern Brazil in relation to water depth. J. Mar. Biolog. Assoc. U.K. 89, 943–949.

Danilewicz, D., Sucunza, F., Ott, P.H., Ferreira, E., Perez, M.S., Berchieri, N., Alvares, D., Andriolo, A., Secchi, E.R., Flores, P.A.C., Farro, A.P., Martins, A., Zerbini, A.N., 2020. Abundance and distribution of franciscanas (*Pontoporia blainvillei*) in northern Rio de Janeiro (FMA Ib), Brazil, Paper SC/69B/ASI/07 presented to the IWC Scientific Committee, May 2020. p. 15.

Day, J., Dudley, N., Hockings, M., Holmes, G., Laffoley, D., Stolton, S., Wells, S., Wenzel, L, 2019. Guidelines for Applying the IUCN Protected Area Management Categories to Marine Protected Areas, 2nd ed. IUCN, Gland.

Denuncio, P., Bastida, R., Dassis, M., Giardino, G., Gerpe, M., Rodríguez, D., 2011. Plastic ingestion in Franciscana dolphins, *Pontoporia blainvillei* (Gervais & d'Orbigny, 1844), from Argentina. Mar. Pollut. Bull. 62 (8), 1836–1841.

Denuncio, P.E., Paso Viola, N., Cáceres-Saez, I., Cappozzo, H.L., Rodríguez, D., Mandiola, A., 2019. *Pontoporia blainvillei*. In: SAyDS–SAREM (Eds.) Categorización 2019 de los mamíferos de Argentina según su riesgo de extinción. Lista Roja de los mamíferos de Argentina http://cma.sarem.org.ar (in Spanish).

Di Beneditto, A., Ramos, R., 2014. Marine debris ingestion by coastal dolphins: what drives differences between sympatric species? Mar. Pollut. Bull. 83 (1), 298–301.

Di Beneditto, A.P.M., Ramos, R.M.A., Lima, N.R.W., 1998. Fishing activity in northern Rio de Janeiro State (Brazil) and its relation with small cetaceans. Braz. Arch. Biol. Technol. 41 (3), 296–302.

Di Franco, A., Thiriet, P., Di Carlo, G., Dimitriadis, C., Francour, P., Gutiérrez, N.L., de Grissac, A.J., Koutsoubas, D., Milazzo, M., del Mar Otero, M., Piante, C., 2016. Five key attributes can increase marine protected areas performance for small-scale fisheries management. Sci. Rep. 6 (1), 1–9.

Di Tullio, J.C., Fruet, P.F., Secchi, E.R., 2015. Identifying critical areas to reduce bycatch of coastal common bottlenose dolphins *Tursiops truncatus* in artisanal fisheries of the subtropical western South Atlantic. Endanger. Species Res. 29, 35–50. http://doi.org/10.3354/esr00698.

Edgar, G.J., Stuart-Smith, R.D., Willis, T.J., Kininmonth, S., Baker, S.C., Banks, S., Barrett, N.S., Becerro, M.A., Bernard, A.T.F., Berkhout, J., Buxton, C.D., Campbell, S.J., Cooper, A.T., Davey, M., Edgar, S.C., Försterra, G., Galván, D.E., Irigoyen, A.J., Kushner, D.J., Moura, R., Parnell, P.E., Shears, N.T., Soler, G., Strain, E.M.A., Thomson, R.J., 2014. Global conservation outcomes depend on marine protected areas with five key features. Nature 506, 216–220. https://doi.org/10.1038/nature13022.

Ehler, C., Douvere, F., 2009. Marine spatial planning: a step-by-step approach toward ecosystem-based management. Intergovernmental Oceanographic Commission and Man and the Biosphere Programme. IOC Manual and Guides No. 53, ICAM Dossier No. 6. UNESCO, Paris.

Failla, M., Seijas, V., Espósito, R., Iñíguez, M., 2012. Franciscana dolphins, *Pontoporia blainvillei*, of the Río Negro Estuary, Patagonia, Argentina. Mar. Biodivers. Rec. 5, E102. http://dx.doi.org/10.1017/S1755267212000875.

Flores, P.A.C., Bazzalo, M., 2004. Home ranges and movements patterns of the marine tucuxi *Sotalia fluviatilis* in Baía Norte, southern Brazil. Lat. Am. J. Aquat. Mamm. 3 (1), 37–52.

Franco-Trecu, V., Costa, P., Abud, C., Dimitriadis, C., Laporta, P., Passadore, C., Szephegyi, M., 2009. By-catch of Franciscana *Pontoporia blainvillei* in Uruguayan artisanal gillnet fisheries: an evaluation after a twelve-year gap in data collection. Lat. Am. J. Aquat. Mamm. 7, 11–22. http://dx.doi.org/10.5597/lajam00129.

Franco-Trecu, V., Szephegyi, M.N., Doño, F., Forselledo, R., Reyes, F., Passadore, C., Crespo, E.A., Inchausti, P., 2019. Marine mammal bycatch by the industrial bottom trawl fishery at the Río de la Plata Estuary and the adjacent Atlantic Ocean. Lat. Am. J. Aquat. Res. 47 (1), 89–101. https://doi.org/10.3856/vol47-issue1-fulltext-10.

Giglio, V.J., Pinheiro, H.T., Bender, M.G., Bonaldo, R.M., Costa-Lotufo, L.V., Ferreira, C.E.L., Floeter, S.R., Freire, A., Gaspari, J.L., 2018. Large and remote marine protected areas in the South Atlantic Ocean are flawed and raise concerns: comments on Soares and Lucas (2018). Mar. Policy 96, 13–17. https://doi.org/10.1016/j.marpol.2018.07.017.

Gormley, A.M., Slooten, E., Dawson, S., Barker, R.J., Rayment, W., du Fresne, S., Bräger, S., 2012. First evidence that marine protected areas can work for marine mammals. J. Appl. Ecol. 49, 474–480. https://doi.org/10.1111/j.1365-2664.2012.02121.x.

Hooker, S.K., Cañadas, A., Hyrenbach, K.D., Corrigan, C., Polovina, J.J., Reeves, R.R., 2011. Making protected area networks effective for marine top predators. Endanger. Species Res. 13, 203–218.

Hooker, S.K., Gerber, L.R., 2004. Marine reserves as a tool for ecosystem-based management: the potential importance of megafauna. Bioscience 54, 27–39.

Hoyt, E., 2011. Marine Protected Areas for Whales, Dolphins and Porpoises: a World Handbook for Cetacean Habitat Conservation and Planning. Earthscan/Routledge, London and New York.

Hoyt, E., 2018. Marine protected areas. In: Würsig, B., Thewissen, J.G.M., Kovacs, K.M. (Eds.), Encyclopedia of marine mammals. Academic Press, San Diego, CA, pp. 569–580.

Hughes, T.P., Baird, A.H., Bellwood, D.R., Card, M., Connolly, S.R., Folke, C., Grosberg, R., Hoegh-Guldberg, O., Jackson, J., Kleypas, J., 2003. Climate change, human impacts, and the resilience of coral reefs. Science 301, 929–933.

IWC, 2016. A conservation management plan for franciscana (*Pontoporia blainvillei*), Paper IWC/66/CC11 presented by Argentina, Brazil and Uruguay to the IWC Scientific Committee, p. 22.

Kinas, P.G., Secchi, E.R., 1998. Modelling truncated data to estimate incidental kills of franciscana, *Pontoporia blainvillei*, by gillnets. Rep. int. Whal. Commn. 48, 533–536.

Lusseau., D., Higham, J., 2004. Managing the impacts of dolphin-based tourism through the definition of critical habitats: the case of bottlenose dolphins (*Tursiops* spp.) in Doubtful Sound. New Zealand. Tour. Manag. 25, 657–667.

Magris, R.A., Pressey, R.L., 2018. Marine protected areas: Just for show? Science 360 (6390), 723–724.

Magris, R.A., Mills, M., Fuentes, M.M.P.B., Pressey, R.L., 2013. Analysis of progress towards a comprehensive system of marine protected areas in Brazil. Nat. Conserv. 11, 81–87.

Masello, J.F., Aramayo, S.A., Borella, F., Bosso, A., Favier Dubois, C.M., Failla, M., Harris, G., Lew, J., Lini, R., Masera, R.F., Quillfeldt, P., Vinci, C., 2007. Propuesta para la creación de un área natural protegida: Estuario del río Negro y zonas de influencia. Informe técnico Legislatura de la Provincia de Río Negro-Argentina, p. 15. (in Spanish).

Mattos, P.H., Ferreira, W.L.S., 2018. The propositional models for fisheries and environmental management in the Albardão region, south of Rio Grande do Sul State. Desenvolv. Meio Ambiente 44 [Edição especial: X Encontro Nacional de Gerenciamento Costeiro], 183-198. https://doi.org/10.5380/dma.v44i0.54973

MAyDS, 2016. El Sistema Nacional de Áreas Marinas Protegidas. Bases para su puesta en funcionamento. Ministerio de Ambiente y Desarrollo Sustentable de la Nación, Buenos Aires. Argentina. https://www.argentina.gob.ar/sites/default/files/ambiente-sistema-nacional-areas-marinas-protegidas.pdf (in Spanish). (Accessed October 30, 2021).

Mayorga, L.F.S.P., Vanstreels, R.E.T., Bhering, R.C.C., Mamede, N., Costa, L.M.B., Pinheiro, F.C.F., Reis, L.W.D., Trazzi, A., Meirelles, W.L.C., Ribeiro, A.M., Siciliano, S., 2020. Strandings of cetaceans on the Espírito Santo coast, southeast Brazil, 1975–2015. ZooKeys 948, 129–152. doi: 10.3897/zookeys.948.50468.

Mills, M, Magris, R.A., Fuentes, M.M.P.B., Bonaldo, R., Herbst, D.F., Lima, M.C.S., Kerber, I.K.G., Gerhardinger, L.C., Moura, R.L.de, Domit, C., Teixeira, J.B., Pinheiro, H.T., Vianna, G., Freitas, R.R.de, 2020. Opportunities to close the gap between science and practice for Marine Protected Areas in Brazil. Perspect. Ecol. Conserv. 18, 161–168 10.1016/j.pecon.2020.05.002.

MMA, 2014. Lista Nacional Oficial de Espécies da Fauna Ameaçadas de Extinção – Mamíferos, Aves, Répteis, Anfíbios e Invertebrados Terrestres. Portaria MMA no. 444, de 17 de dezembro de 2014, Brasil (in Portuguese).

Montealegre-Quijano, S., Ferreira, C.N., 2010. Incidental catch of a franciscana *Pontoporia blainvillei* (Cetacea, Pontoporiidae) in a pair trawl of southern Brazil. Pan-Am. J. Aquat. Sci. 5 (3), 465–468.

Negri, M.F., Denuncio, P., Panebianco, M.V., Cappozzo, H.L., 2012. Bycatch of franciscana dolphins *Pontoporia blainvillei* and the dynamic of artisanal fisheries in the species' southernmost area of distribution. Braz. J. Oceanogr. 60, 149–158. https://doi.org/10.1590/s1679-87592012000200005.

Notarbartolo di Sciara, G., Hoyt, E., Reeves, R., Ardron, J., Marsh, H., Vongraven, D., Barr, B., 2016. Place-based approaches to marine mammal conservation. Aquat. Conserv. 26, 85–100.

Ott, P.H., Secchi, E.R., Moreno, I.B., Danilewicz, D., Crespo, E.A., Bordino, P., Ramos, R., Di Beneditto, A.P., Bertozzi, C., Bastida, R., Zanelatto, R., Perez, J.E., Kinas, P.G., 2002. Report of the working group on fishery interactions. Lat. Am. J. Aquat. Mamm. 1, 55–64.

Ott, P.H., Siciliano, S., Danilewicz, D., 2015. Toninha: mais uma vítima da trágica legislação pesqueira no Brasil. Ciênc. Hoje 55, 46–47 (in Portuguese).

Prado, J.H.F., Secchi, E.R., Kinas, P.G., 2013. Mark-recapture of the endangered franciscana dolphin (*Pontoporia blainvillei*) killed in gillnet fisheries to estimate past bycatch from time series of stranded carcasses in southern Brazil. Ecol. Indic. 32, 35–41.

Prado, J.H.F., Kinas, P.G., Pennino, M.G., Seyboth, E., Silveira, F.R.G., Ferreira, E.C., Secchi, E.R., 2021. Definition of no-fishing zones and fishing effort limits to reduce franciscana bycatch to sustainable levels in southern Brazil. Animal Cons. https://doi.org/10.1111/acv.12679.

Read, A.J., Drinker, P., Northridge, S.P., 2006. Bycatch of marine mammals in U.S. and global fisheries. Conserv. Biol. 20, 163–169. https://doi.org/10.1111/j.1523-1739.2006.00338.x.

Reeves, R.R., 2000. The value of sanctuaries, parks, and reserves (protected areas) as tools for conserving marine mammals. Report to the Marine Mammal Commission, Contract No. T74465385.

Rojas-Bracho, L., Reeves, R.R., 2013. Vaquitas and gillnets: Mexico's ultimate cetacean conservation challenge. Endang. Species Res. 21, 77–87. https://doi.org/10.3354/esr00501.

Santos, M.C.O., Oshima, J.E.F., da Silva, E., 2009. Sightings of franciscana dolphins (*Pontoporia blainvillei*): the discovery of a population in the Paranaguá estuarine complex, Southern Brazil. Braz. J. Oceanogr. 57, 57–63.

SAyDS, 2018. Informe Nacional Ambiente y Áreas Protegidas de la Argentina: 2008-2018. Secretaría de Ambiente y Desarrollo Sustentable, Buenos Aires, Argentina. https://www.argentina.gob.ar/sites/default/files/informe_ambiente_y_ap_final_0.pdf (in Spanish). (accessed October 30, 2021).

Secchi, E.R., Ott, P.H., Danilewicz, D., 2003a. Effects of fishing bycatch and the conservation status of the franciscana dolphin, *Pontoporia blainvillei*. In: Gales, N., Hindell, M., Kirkwood, R. (Eds.), Marine Mammals: Fisheries, Tourism and Management Issues. CSIRO Publishing, Collingwood, pp. 174–191.

Secchi, E.R., Danilewicz, D., Ott, P.H., 2003b. Applying the phylogeographic concept to identify franciscana dolphin stocks: implications to meet management objectives. J. Cetacean Res. Manag. 5, 61–68.

Secchi, E.R., Cremer, M.J., Danilewicz, D., Lailson-Brito, J., 2021. A synthesis of the ecology, human-related threats and conservation perspectives for the endangered franciscana dolphin. Front. Mar. Sci. 8, 617956. https://doi.org/10.3389/fmars.2021.617956.

Siciliano, S., Di Beneditto, A.P.M., Ramos, R.M.A., 2002. A toninha, *Pontoporia blainvillei* (Gervais & d'Orbigny, 1844) (Mammalia, Cetacea, Pontoporiidae), nos estados do Rio de Janeiro e Espírito Santo, costa sudeste do Brasil: Caracterização dos habitats e fatores de isolamento das populações. Bol. Mus. Nac. 476, 1–16 (in Portuguese).

Soutullo, A., Clavijo, C., Martínez-Lanfranco, J.A. (Eds.), 2013. Especies prioritarias para la conservación en Uruguay. Vertebrados, moluscos continentales y plantas vasculares. SNAP/DINAMA/MVOTMA y DICYT/MEC, Montevideo, Uruguay (in Spanish).

Spalding, M.D., Fox, H.E., Allen, G.R., Davidson, N., Ferdaña, Z.A., Finlayson, M., Halpern, B.S., Jorge, M.A., Lombana, A.L., Lourie, S.A., Martin, K.D., McManus, E., Molnar, J., Recchia, C.A., Robertson, J., 2007. Marine ecoregions of the world: a bioregionalization of coastal and shelf areas. Bioscience 57 (7), 573–583.

Sucunza, F., Danilewicz, D., Cremer, M., Ferreira, E., Denuncio, P., 2020a. Abundance of the endangered franciscana in southern Brazil. Paper SC/69B/ASI/06 presented to the IWC Scientific Committee, May 2020, p. 16.

Sucunza, F., Danilewicz, D., Ott, P.H., Neves, M., Bercheri, N., Farro, A.P., Martins, A., Zerbini, A.N., 2020b. Population size and IUCN Red Listing of the isolated northern population of the franciscana (*Pontoporia blainvillei*). Paper SC/69B/ASI/05 presented to the IWC Scientific Committee, May 2020, p. 25.

Sucunza, F., Danilewicz, D., Andriolo, A., Azevedo, A.F., Secchi, E.R., Zerbini, A.N., 2020c. Distribution, habitat use, and abundance of the endangered franciscana in southeastern and southern Brazil. Mar. Mamm. Sci. 36 (2), 421–435. https://doi.org/10.1111/mms.12650.

Thomas, H.L., Macsharry, B., Morgan, L., Kingston, N., Moffitt, R., Stanwell-Smith, D., Wood, L., 2014. Evaluating official marine protected area coverage for Aichi Target 11: appraising the data and methods that define our progress. Aquat. Conserv 24 (S2), 8–23.

Traill, L.W., Brook, B.W., Frankham, R.R., Bradshaw, C.J.A., 2010. Pragmatic population viability targets in a rapidly changing world. Biol. Conserv. 143 (1), 28–34.

UNEP-WCMC, IUCN, 2021. Protected Planet Report 2020. UNEP-WCMC and IUCN, Cambridge UK, Gland, Switzerland. https://livereport.protectedplanet.net/ (accessed October 30, 2021).

Ward, M., Rhodes, J.R., Watson, J.E.M., Lefevre, J., Atkinson, S., Possingham, H.P., 2020. Use of surrogate species to cost-effectively prioritize conservation actions. Conserv. Biol. 34 (3), 600–610. doi:10.1111/cobi.13430. 31691376.

Wells, R.S., Cremer, M.J., Berninsone, L.G., Albareda, D., Wilkinson, K.A., Stamper, M.A., Paitach, R.L., Bordino, P., 2021. Tagging, ranging patterns, and behavior of franciscana dolphins (*Pontoporia blainvillei*) off Argentina and Brazil: Considerations for conservation. Mar. Mamm. Sci, 1–35. https://doi.org/10.1111/mms.12879.

Worm, B., Barbier, E.B., Beaumont, N., Duffy, J.E., Folke, C., Halpern, B.S., Jackson, J.B.C., Lotze, H.K., Micheli, F., Palumbi, S.R., Sala, E., Selkoe, K.A, Stachowicz, J.J., Watson, R., 2006. Impacts of biodiversity loss on ocean ecosystem services. Science 314, 787–790. http://dx.doi.org/10.1126/science.1132294.

Zerbini, A.N., Secchi, E., Crespo, E., Danilewicz, D., Reeves, R., 2017. *Pontoporia blainvillei* (errata version published in 2018). The IUCN Red List of Threatened Species 2017: e.T17978A123792204. doi:10.2305/IUCN.UK.2017-3.RLTS.T17978A50371075.en. Downloaded on March 22, 2021.

Zerbini, A.N., Andriolo, A., Cremer, M., Crespo, E., Danilewicz, D., Dmit, C., Sucunza, F., 2020. Abundance estimates of franciscana dolphins (*Pontoporia blainvillei*): a review and future recommendations, Paper SC/68B/ASI/08 presented to the IWC Scientific Committee, May 2020. p. 35.

CHAPTER 19

International and regional agreements, national legislation, and public policies

Miguel Iñíguez Bessega[a], Juliana Di Tullio[b], Mauricio Failla[a], Sue Fisher[c], Sebástian Jiménez[d], Cecilia Passadore[e,f], Juan Pablo Torres-Florez[g], Vanessa Tossenberger[h]

[a]Fundación Cethus, Olivos, Buenos Aires, Argentina
[b]Kaosa, Rio Grande, Rio Grande do Sul, Brazil
[c]Animal Welfare Institute, Washington, DC, United States
[d]Laboratorio de Recursos Pelágicos, Dirección Nacional de Recursos Acuáticos, Constituyente, Montevideo, Uruguay
[e]Vida Silvestre Uruguay, Montevideo, Uruguay
[f]Dirección Nacional de Recursos Acuáticos (DINARA), Ministerio de Ganadería Agricultura y Pesca (MGAP), Montevideo, Uruguay
[g]Centro Nacional de Pesquisa e Conservação de Mamíferos Aquáticos, Instituto Chico Mendes de Conservação da Biodiversidade, Santos, São Paulo, Brazil
[h]Whale and Dolphin Conservation (WDCS), Chippenham, United Kingdom

Introduction

The franciscana dolphin is a small odontocete inhabiting coastal waters of the Southwestern Atlantic Ocean from Itaúnas (18°25'S), Espírito Santo State, Brazil, to Golfo San Matías (42°10'S), Chubut, Argentina (Crespo et al., 1998; Crespo, 2002; Siciliano et al., 2002) (Fig. 19.1). Two gaps in distribution are found near the northern range of the species, one between the states of Espírito Santo and Rio de Janeiro and another between central Rio de Janeiro and southern Rio de Janeiro (Siciliano et al., 2002, 2015; Cunha et al., 2020a).

Based on information available at that time, Secchi et al. (2003a) divided the species distribution range into four 'Franciscana Management Areas' (FMAs I–IV), with FMA I and II located in Brazil, FMA III shared between Brazil and Uruguay, and FMA IV located in Argentina. Subsequent studies confirmed the subdivisions proposed by Secchi et al. (2003a), but suggested the existence of genetically differentiated populations within the FMAs. Cunha et al. (2014) found four genetically distinct populations in Brazil: FMA Ia (Espírito Santo), FMA Ib (northern Rio de Janeiro), FMA IIa (southern Rio de Janeiro to northern São Paulo), and FMA IIb (central

DOI: https://doi.org/10.1016/B978-0-323-90974-7.00020-3

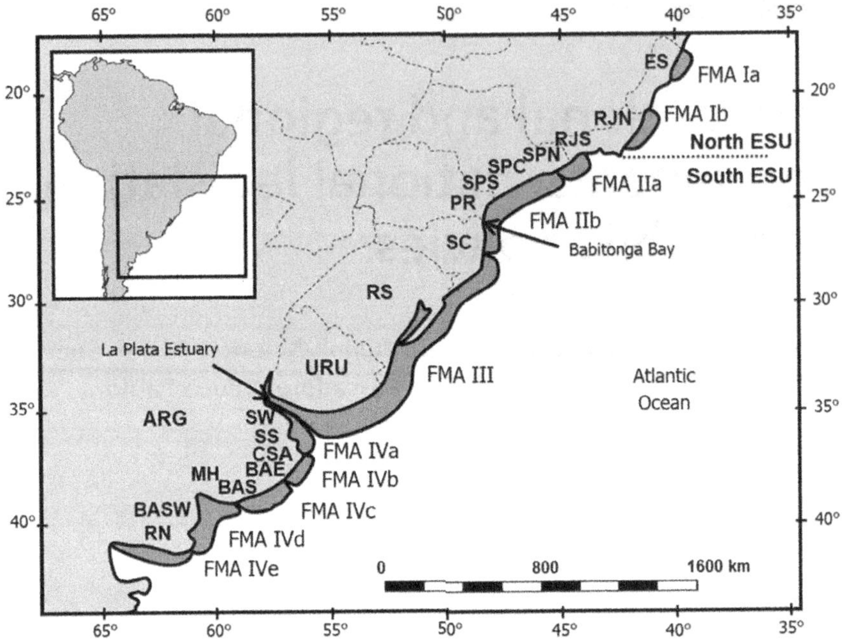

Fig. 19.1 ***Map showing the 11 proposed management units for franciscana, including the 10 labeled FMA subdivisions and Babitonga Bay (Cunha et al., 2020b; Cunha et al., 2020c).***

São Paulo to Santa Catarina). Costa Urrutia et al. (2012) distinguished two populations in FMA III: FMA IIIa (oceanic population) and FMAIIIb (estuarine population). Finally, Méndez et al. (2008, 2010) and Gariboldi et al. (2015, 2016) found 5 genetically distinct populations in FMA IV: FMA IVa (Samborombón West/Samborombón South), FMA IVb (Cabo San Antonio/Buenos Aires East), FMA IVc (Necochea/Claromecó/Buenos Aires Southwest), FMA IVd (Monte Hermoso), and FMA IVe (Río Negro).

The franciscana is considered the most threatened small cetacean in South America (Secchi, 2010a). Incidental mortality in gillnet fisheries has been a major conservation concern for the franciscana (Secchi, 2010b), but the species is also exposed to other threats of potential concern throughout its range, such as coastal development, marine debris, chemical and noise pollution, disease and vessel traffic (e.g., Di Beneditto and Ramos, 2014; Denuncio et al., 2011).

This chapter aims to give an overview of the legislative measures in place as well as public policies to protect the species.

Legal framework

International agreements

The international community regulates relations between states through policies that reflect common values, including nature protection. As a result of this consensus, several international and regional legal regimes have been agreed to address a variety of issues related to cetacean management and conservation and the threats they face.

Table 19.1 shows Argentina, Brazil, and Uruguay's membership of the United Nations (UN), International Whaling Commission (IWC), Convention on the Conservation of Migratory Species (CMS), Convention on International Trade in Endangered Species of Wild Fauna and Flora (CITES), Convention on Biological Diversity (CBD), Food and Agriculture Organization of the United Nations (FAO), and International Union for the Conservation Nature (IUCN).

UN Convention on the Law of the Sea (UNCLOS)

The United Nations Convention on the Law of the Sea (UNCLOS), which was established in 1982 and came into force in 1994, has been essential for the protection of marine living resources. The convention defines the rights and responsibilities of nations with respect to their use of the world's seas and oceans, giving coastal States rights and responsibilities for the management and use of fishery resources within areas of their national jurisdiction.

Table 19.1 Membership of Argentina, Brazil, and Uruguay to the UN, IWC, CMS, CITES, CBD, FAO, and IUCN (showing the year that the countries have adhered).

	Argentina	Brazil	Uruguay
United Nations (UN)	1945	1945	1945
International Whaling Commission (IWC)	1960	1974	2007
Convention on the Conservation of Migratory Species (CMS)	1992	2015	1990
Convention on International Trade in Endangered Species of Wild Fauna and Flora (CITES)	1981	1975	1975
Convention on Biological Diversity (CBD)	1995	1994	1994
Food and Agriculture Organization (FAO)	1951	1945	1945
International Union for the Conservation of Nature (IUCN)	Not applicable	Not applicable	Not applicable

Part XII of UNCLOS establishes provisions for the protection and preservation of marine ecosystems. These provisions also apply to fishing industries on a global scale. All States are obliged to protect the marine environment and to take all measures necessary to prevent, reduce and control pollution of the marine environment from any source (Articles 192 and 194). Provisions relating to the protection and preservation of the marine environment emphasize the importance of cooperation between states and require states to undertake monitoring of the activities they permit or engage in, in order to determine whether these activities are likely to pollute the marine environment (Article 204).

Parties to UNCLOS are required to establish measures for the conservation and management of living marine resources in their EEZs to ensure that marine living resources are not endangered by overexploitation. These measures must also address, interalia, the effects of the harvesting of target species on species associated with, or dependent on, the harvested species (Article 61(2) and (4)). Article 65 applies specifically to marine mammals and reaffirms the right of coastal States, and the competence of international organizations, to adopt stricter management measures than required in the convention. For cetaceans specifically, Article 65 requires cooperation through the appropriate international organizations. The conservation and management provisions in Article 65 are extended by Article 120 to the high seas. All States are therefore obliged to take measures to conserve the marine mammals of the high seas and, in doing so, should cooperate with each other, including by establishing regional or subregional coordination, as appropriate, to achieve this objective.

To protect marine mammals, among other species, from bycatch by large-scale driftnets, the United Nations General Assembly adopted Resolution 44/2255, which recommends that all UN members agree moratoria on all pelagic fishing with large-scale driftnets on the high seas by June 30, 1992.

UN Convention on Biological Diversity (CBD)

The Convention on Biological Diversity (CBD) was signed during the United Nations Conference on Environment and Development (the Rio "Earth Summit") on June 5, 1992, in Rio de Janeiro, Brazil, and entered into force on December 23, 1993 (www.CBD.int). It was conceived to provide an international framework for the conservation, sustainable development, and use of biodiversity and was supported by Agenda 21. The Convention applies to all terrestrial and marine biodiversity and outlines measures for conserving biodiversity as obligations of all parties. General measures for conserving biodiversity and ensuring sustainable development include a requirement to develop national policies, strategies, and programs that should inter alia reflect the principles espoused in the Convention (Article 6(a)). The Convention also urges Parties to integrate biodiversity conservation policies and strategies with cross-sectoral plans, programmes and policies (Article 6(b)).

Measures outlined for the in situ conservation of biodiversity encompass key issues including protected areas, ecosystems, and habitats. With respect to protected areas and ecosystems, the Convention (Art. 8) establishes, inter alia, the following obligations on all contracting parties: (a) establish a system of protected areas or areas where special measures need to be taken to conserve biological diversity; (b) develop, where necessary, guidelines for the selection, establishment and management of protected areas or areas where special measures need to be taken to conserve biological diversity; (c) regulate or manage biological resources important for the conservation of biological diversity whether within or outside protected areas, with a view to ensuring their conservation and sustainable use; (d) promote the protection of ecosystems, natural habitats and the maintenance of viable populations of species in natural surroundings; (e) promote environmentally sound and sustainable development in areas adjacent to protected areas with a view to furthering the protection of these areas; (f) rehabilitate and restore degraded ecosystems and promote the recovery of threatened species, inter alia, through the development and implementation of plans or other management strategies, and (k) Develop or maintain necessary legislation and/or other regulatory provisions for the protection of threatened species and populations.

The Ministerial Statement on the Implementation of the Convention on Biological Diversity was issued during the second meeting of the Conference of the Parties (COP) to the CBD, held in Jakarta in November 1995, as a result of the COP identifying marine and coastal biodiversity as a high priority issue. The statement essentially reaffirms the importance of the conservation and sustainable use of coastal and marine biodiversity and urges Parties to initiate the immediate development and implementation of actions concerning this issue. It specifically links conservation, the use of biodiversity and fishing activities, and establishes a new global consensus on the importance of marine and coastal biodiversity known as the Jakarta Mandate on Coastal and Marine Biodiversity.

UN Food and Agriculture Organization (FAO)

The FAO is a specialized agency of the UN established in 1945 with the goal of achieving food security for all and ensuring that people have regular access to enough high-quality food to lead active, healthy lives (www.fao.org). The FAO Code of Conduct for Responsible Fisheries was developed in 1995. It requires the protection of endangered species (including marine mammals) by adopting appropriate measures based on scientific evidence. In 2011, the FAO International Guidelines on Bycatch Management and Discard Reduction were adopted, identifying endangered, threatened and protected species (including marine mammals) as a bycatch problem. They recommend that the Member States assess their fisheries and identify bycatch of any protected and endangered species, including where bycatch species may overlap with fishing operations through the use of seabed maps and/or species ranges and distributions. Furthermore, in order to reduce bycatch, Member States should identify and establish areas where the use of all or some fishing gear should be limited or prohibited based on the best available scientific information. Most recently, at the thirty-fourth session of the FAO Committee on Fisheries (COFI) in 2021, Guidelines to Prevent and Reduce Bycatch of Marine Mammals in Capture Fisheries were adopted as part of the Technical Guidelines for Responsible Fisheries. These were developed to assist the international community in taking the necessary practical steps to develop and implement policy and technical interventions to reduce the bycatch of marine mammals in fisheries.

UN Convention on the Conservation of Migratory Species (CMS)

CMS is a global environmental treaty of the UN, also known as the Bonn Convention. It aims to conserve and manage migratory species as well as their habitats throughout their range (www.CMS.int).

"Migratory species" are defined under CMS as "the entire population or any geographically separate part of the population of any species or lower taxon of wild animals, a significant proportion of whose members cyclically and predictably cross one or more national jurisdictional boundaries" (Art I (1)(a)). CMS provides strict protection for endangered species, encourages the conclusion of multilateral agreements for species with unfavorable conservation status, and promotes cooperative research and conservation activities. CMS differentiates between two groups of species listed in the convention's appendices. Appendix I lists endangered species to be placed under strict protection by the Parties. Appendix II lists species with an unfavorable status that require international agreements for their conservation and management and those which would benefit significantly from international cooperation.

Franciscana is included in Appendix I (CMS, 2020) which requires parties to prohibit taking (subject to exceptions for science and to accommodate the needs of traditional subsistence users). In addition, "measures considered appropriate to benefit the species" may be recommended by the COP.

UN Convention on International Trade in Endangered Species of Wild Fauna and Flora (CITES)

CITES regulates international trade in endangered species of animals and plants (www.CITES.org). It entered into force in 1975 and was amended in 1979 and 1983. It lists species in three appendices: Appendix I includes all species threatened with extinction that are or may be affected by trade. International trade in specimens (which includes parts and derivatives) of these species must be subject to particularly strict regulation in order not to endanger further their survival and must only be authorized in exceptional circumstances. Appendix II includes all species that although not necessarily now threatened with extinction may become so unless trade is subject to strict regulation in order to avoid utilization incompatible with their survival. It also includes species not affected by trade but which must be subject to regulation in order to allow effective control of trade in species that need protection. Appendix III includes species that any Party declares are subject to regulation within its jurisdiction for the purpose of preventing or restricting its exploitation and that need global cooperation to control their trade. When a Party proposes an amendment to the appendices for a marine species, CITES Article XV 2(b) requires the Secretariat to "consult inter-governmental bodies having a function in relation to those species, especially with a view to obtaining scientific data these bodies may be able to provide and to ensuring co-ordination with any conservation measures enforced by such bodies. The Secretariat shall communicate the views expressed and data provided by these bodies and its own findings and recommendations to the Parties as soon as possible." CITES recognizes the IWC as the primary international organization responsible for the conservation and management of whales and dolphins. This cooperation began in 1973 at the first meeting of the Conference of Parties to CITES and was reinforced in a series of resolutions that were consolidated at COP 12 in Santiago de Chile (Conf. 11.4 (Rev. COP12).

Franciscana is listed in Appendix II of CITES. According to Article IV, export of live specimens, parts, and derivatives is permitted for commercial purposes subject to the exporting state first advising that the export will not be detrimental to the survival of that species, confirming that the specimen was not obtained in contravention of the laws of that state, and issuing an export permit. In the case of a live specimen, the exporting state must also be satisfied "that any living specimen will be so prepared and shipped as to minimize the risk of injury, damage to health or cruel treatment."

International Whaling Commission (IWC)

The mandate of the IWC, which was originally established in 1946 to conserve whales and regulate commercial whaling, is considered to have evolved (www.IWC.int). The IWC has directed its Scientific Committee (SC) (through its standing small cetacean subcommittee which was established in 1979) to review the distribution, structure, and status of small cetacean stocks and advise on ways to eliminate or mitigate threats (IWC, 2014). The SC regularly makes recommendations related to affected species, several of which have been taken up by the Commission in resolutions that call on range states to take, or desist from, specific action. For example to: "reduce the takes" (Resolution 1990:4 on Japan's Dall's porpoise hunt); "take appropriate conservation measures" (Resolution 1992:11 on belugas and narwhals); "share catch statistics and data on incidental takes" (1992: Appendix 9. Resolution on Small Cetaceans); "give high priority to reducing by-catches" (Resolution 1993-11 on Harbor Porpoise in the North Atlantic and the Baltic Sea); "provide information on directed and incidental catches" (Resolution 1997-8 Resolution on Small Cetaceans); "undertake genetic and telemetry studies to identify stocks and improve catch reporting" and "take all appropriate measures to prevent, minimize and mitigate by-catch of small cetaceans in fisheries operations" (Resolution 2001-13 Resolution on Small Cetaceans).

Franciscana has been a priority for the IWC. In this regard, the SC conducted a first review of the species in 2004 (IWC, 2005), and a task team for FMA I was established in 2015. A year later, the IWC endorsed a Conservation Management Plan (CMP) for franciscana—IWC/66/CC11 (IWC, 2016). This proposal was submitted by the governments of Argentina, Brazil, and Uruguay following recommendations of the VIII Workshop for the Research and Conservation of Franciscana (Franciscana Consortium, 2016). The overall objective of the CMP is to protect franciscana habitat and minimize anthropogenic threats, in particular by-catch.

It identified seven high-priority actions ranging from public awareness and capacity building through research to mitigation. The CMP is a living document that has a periodic review, at least every four years. In 2018, the IWC agreed to conduct a new review of the species by 2022 (IWC, 2019). This information will contribute to updating the actions identified as a priority in the franciscana CMP.

Argentina, Brazil, and Uruguay are parties of the IWC and they are also members of the Buenos Aires Group, a political block of Latin American countries within the IWC.

International Union for the Conservation Nature (IUCN)

IUCN is a membership Union composed of both government and civil society organizations dedicated to provide solutions to enable human progress, economic development, and nature conservation to take place together (www.iucn.org). The organization publishes the IUCN Red List of Threatened Species, which assesses the conservation status of species.

IUCN supports scientific research, manages field projects globally, and brings governments, nongovernment organizations, United Nations agencies, companies, and local communities together to develop and implement policy. IUCN is the world's oldest and largest global environmental network—a democratic membership union with more than 1000 government and NGO member organizations, and almost 11,000 volunteer scientists in more than 160 countries. IUCN's stated vision is "a just world that values and conserves nature." Its mission is to "Influence, encourage and assist societies to conserve the integrity and diversity of nature and ensure that any use of natural resources is equitable and ecologically sustainable."

The franciscana was last assessed by IUCN in 2017. The species is currently classified on the Red List as "Vulnerable" (Zerbini et al., 2017).

Regional Agreements

Treaty of the Río de la Plata and its Maritime Front

On November 19, 1973, Argentina and Uruguay signed the Treaty of the Río de la Plata and its Maritime Front. One of its main objectives is establishing general rules for the two countries' fishing activities both in the Río de la Plata and in the Maritime Front. Chapter XVI of the treaty which establishes procedures for fishing in the latter zone (including the definition of a common fishing zone, fixing the volumes of catches, etc.) highlights that none of the provisions are applicable to the capture of aquatic mammals. The *"Comisión Técnica Mixta del Frente Marítimo"* (CTMFM) is a binational organization that, among other responsibilities, regulates fisheries resources in the area of the Treaty of the Río de la Plata and its Maritime Front. So far there are no specific regulations with the objective of reducing the incidental capture of franciscana in fisheries managed within the framework of the CTMFM. However, there are two temporal closure areas for coastal trawl fisheries that could protect the species from bycatch to some extent. Resolution 1/04 of the CTMFM prohibits the use of bottom trawling upstream of the line established by the geographical points between Punta Piedras (Argentina) and Punta Yeguas (Uruguay) from October 1, March 31 of each year. This affects a large portion of the inner zone of the Río de la Plata. In addition in the last few years, from November 1 to March 31, the CTMFM has prohibited the use of bottom trawling in the area delimited to the north by the 36° S parallel and its intersection with the outer limit of the Río de la Plata, to the south by the 37° S parallel, to the west by the outer limit of the Argentine territorial sea and to the east by the 56° W meridian (CTMFM resolutions 13/17, 11/18, 12/19, 15/20).

National agreements and public policies

Argentina

Franciscana has been listed as *Vulnerable* on the Argentinean Red List of Mammals since 2019 (Denuncio et al., 2019).

Marine mammals are protected by the Law N° 22,421 on Fauna Conservation.

Argentina prohibits the hunting of cetaceans throughout the national territory, including the territorial sea, the exclusive economic zone and its internal waters under Law N° 25,577.

It has also adopted a National Action Plan to Reduce the Interaction of Marine Mammals with fisheries (Marine Mammals PAN) in 2015 (Consejo Federal Pesquero, 2016). This Plan was prepared jointly by the former Secretariat of Environment and Sustainable Development and the Undersecretariat of Fisheries and Aquaculture with the cooperation of national and provincial government agencies, scientific and academic institutions, and civil society. The Marine Mammals PAN aims to contribute to the ecosystem-based management of fisheries in waters under Argentine jurisdiction, assessing the interactions between them and marine mammals in order to reduce negative impacts on both. This PAN identified franciscana as one of five priority species to protect and establishes several actions to mitigate bycatch, including (1) placing gillnets farther offshore in order to protect females and young individuals, (2) using acoustic deterrent devices (pingers), and (3) using reflective and stiff nets (Bordino et al., 2002, 2004, 2013). The pingers have shown to be effective in reducing the bycatch of the species in an experimental setting (Bordino et al., 2002, 2004). However, it is still unclear whether the implementation of pingers in a real-scale fishery will have similar results and whether they will lead to habituation and habitat exclusion. Strategies need to be developed and evaluated for timely bycatch mitigation in gillnet fisheries. An assessment of the effectiveness of switching from gillnets to alternative fishing gear and the implementation of pingers in a real-scale fishery is ongoing in Buenos Aires province.

Other legal measures have been adopted more recently: in 2013 Balneario El Condor, municipality of Viedma (Río Negro province), introduced a ban on the use of gillnets from November 15 to March 15 (Ordinance N° 7326—Consejo Deliberante de Viedma) and it has been implemented every year since. Although the Ordinance was enacted to regulate nautical and fishing activity in the area, it indirectly benefits franciscana, since it is applied during the reproductive season of the species at the Río Negro Estuary. This region corresponds to the southern range of the species (Failla et al., 2012) and also to the FMA IVe (Gariboldi et al., 2015; Cunha et al., 2020a).

In 2018, the Buenos Aires province declared the franciscana to be a Natural Monument under Law 14,992. This is the highest level of protection for any species.

Since 2020 the Municipality of Viedma, province of Río Negro, has been working on the promulgation of an ordinance to protect emblematic species including the franciscana.

The IWC CMP for Franciscana (IWC, 2016) and subsequently Hevia et al. (in press) and Failla (pers. comm.) identified a system of nine Protected Areas (PA) (Bahía Samborombón, Rincón de Ajo, Mar Chiquita, Restinga del Faro, Arroyo Zabala, Pehuén-Co Monte Hermoso, Bahía Blanca—Falsa y Verde, Islote de la Gaviota Cangrejera y Bahía San Blas) for the Buenos Aires province, four (Punta Bermeja, Caleta de los Loros, Bahía San Antonio and Complejo Islote Lobos) for the Río Negro province and one (Península Valdés) for the Chubut province which partially overlap with the habitat of the franciscana, although these PAs had not been established specifically to protect the species. In Buenos Aires, management plans for the Bahia Samborombón, Pehuen Có-Monte Hermoso, Bahía

Blanca—Bahía Falsa—Bahía Verde, and Bahía San Blas Protected Areas have already been developed, but not implemented. Legal action taken to allow artisanal fishery operations prevents implementation of these plans in Bahía Blanca, Bahía Falsa and Bahía Verde, and Bahía San Blas. Moreover, management plans for the other areas (such as Arroyo Zabala) are yet to be developed. Finally, management plans have also been developed for all areas in Río Negro province, but these have not been fully implemented either. Península Valdés has a management plan in place which is being implemented.

Brazil

The first Brazilian legislation directed at small cetacean protection was approved in 1986 (Directive SUDEPE N11/1986) and all cetacean species were protected under Law 7.643 in 1987. Since the 1990s, the number of laws and regulations linked directly or indirectly to cetacean conservation has gradually increased, especially after the Brazilian Government signed the Convention on Biological Diversity (CBD) in 1993, ratified in 1994 and endorsed by the President in 1998 (Decree 2.519/1998). Many legal instruments were implemented to fulfill the goals of this convention, such as the creation of the National System of Protected Areas (SNUC; Law 9.985/2000), the National Biodiversity Policy (Decree 4.339/2002), and the National Program for Biodiversity (Decree 4.703/2003).

The franciscana has been considered an endangered species in Brazil since the publication of the National Red List of Threatened Species in 1989 (Directive IBAMA 1.522/1989). According to the last national assessment, which followed the IUCN Red List Criteria, it is currently classified as "Critically Endangered" (Ministério do Meio Ambiente, 2014; Ministério do Meio Ambiente, 2018). The species was also included in several State lists of Brazilian wildlife threatened with extinction, including Rio Grande do Sul (Critically Endangered—"CR"), Santa Catarina (Vulnerable—"VU"), Paraná (Endangered—"EN"), São Paulo (Endangered—"EN"), Rio de Janeiro (Vulnerable—"VU") and Espírito Santo (Endangered—"EN").

The first National Action Plan (PAN) for the conservation of aquatic mammals in Brazil was published in 1997 (IBAMA 1997) and included the franciscana and many other species. Subsequently, due to concerns about its conservation status and the considerable amount of information available about the species and its threats, an exclusive PAN for the franciscana was published in 2010 (ICMBio, 2010). The goal of the PAN was to avoid a decline in population numbers of the franciscana where it occurs in Brazil, from Espírito Santo (18°20') to Rio Grande do Sul (33°45'S). This first Brazilian PAN for the species foresaw the implementation of actions within 5 years and was valid until 2015.

A second cycle of the PAN for the conservation of the franciscana was approved by Ordinance 655 of November 4, 2019. The main goal of this second cycle is to prevent the population decline of the franciscana in all management areas, in particular by reducing incidental catches and protecting its habitat. In order to achieve this main objective, actions were established under eight specific objectives:

- **i.** Reduction of species mortality to sustainable levels through the establishment of efficient mechanisms to reduce incidental catches in gillnet fishing.
- **ii.** Creation and strengthening of local and regional initiatives for shared fisheries management as instruments for the protection of the franciscana.
- **iii.** Reduction and mitigation of habitat degradation and research into cumulative effects that affect the species.
- **iv.** Promotion of conservation and maintenance of critical areas for the species.
- **v.** Public awareness about franciscana, its importance and conservation status.
- **vi.** Integration of PAN Toninha (franciscana) with related public policies in different national and international spheres.
- **vii.** Systematization and availability of research data on franciscana for knowledge management.
- **viii.** Strengthening and articulation of research in the Franciscana Management Areas (FMAs).

The Centro Nacional de Pesquisas e Conservação de Mamíferos Aquáticos (CMA/ICMBio) is the coordinator of this PAN, which organizes annual monitoring in order to review and adjust actions. The PAN will be valid until September 2024.

Management of fisheries to reduce bycatch: In 2006, the Brazilian Institute of Environment and Natural Resources (IBAMA) created a working group (WG) on Incidental Capture in Fisheries (Directive IBAMA 83/2006) to provide strategies for monitoring and reducing incidental catches and evaluating appropriate mitigation measures, especially for endangered species. The establishment of this WG, composed of government representatives and scientists, initiated the discussion in Brazil about regulations to achieve a sustainable gillnet fishery. In the following years, industrial and artisanal gillnet fisheries representatives were also included in a co-management approach. This WG was renewed in 2011, composed only of government representatives from the Ministry of the Environment and Ministry of Fisheries and Aquaculture (Directive MPA 74/2011). The regulations resulting from these processes are summarized below:

- IN IBAMA 166/2007: Regulation that, among other measures, prohibits the use of surface and bottom gillnets at a depth less than twice their height and gillnets higher than 15 and 20 m, respectively.
- INI MPA/MMA 10/2011: Joint Inter-ministerial Regulation of the Ministry of Fisheries and Aquaculture and Ministry of Environment that establishes rules for issuing permits for fisheries, including definitions of all fishing modalities, target species, bycatch species, and permitted areas of operation.
- INI MPA/MMA 12/2012: Joint Inter-ministerial Regulation of the Ministry of Fisheries and Aquaculture and Ministry of Environment that regulates gillnet fisheries in south and southeast Brazil, where the Franciscana occurs (see also IWC/66/CC11).

The main mitigation measures of these regulations were the limitation of gillnet lengths and heights, a ban on the use of surface and bottom gillnets at a depth less than twice the nets' height, and the prohibition on motorized gillnet vessels operating within one nautical mile of the coastline. Furthermore, the regulations prevent the issuance of permits for new gillnet fishing boats. Although these regulations were based on robust scientific data (acquired since 2006) and negotiated with the gillnet fishing industry, the rules were poorly communicated, particularly to artisanal fishers. Moreover, compliance and monitoring programs are lacking and enforcement is weak or absent. Therefore, the effectiveness of the regulation remains unknown throughout the area.

One of the priority actions indicated during the franciscana PAN, in 2018, was the creation of regional WGs in the Brazilian states where the franciscana occurs. The aim of these regional WG was to discuss the INI MPA/MMA 12/2012 and IN IBAMA 166 with local fishermen, stakeholders, researchers, and governmental institutions in order to propose specific management and conservation actions to reduce bycatch. At this time, the only regional WG established is in the state of São Paulo.

At the end of 2019, following calls by fishing associations, worker unions, management councils of some protected areas, and also the WGs of São Paulo state, the Aquaculture and Fisheries Secretariat (formerly Ministry of Fisheries and Aquaculture) opened a public consultation seeking proposals and suggestions for the revision of the INI MPA/MMA 12/2012 by May 2020. Based on responses the Ministry of Agriculture, Livestock and Supply amended one of the regulations in the INI MPA/MMA 12/2012 for the state of São Paulo. This regulation now allows gill-net fishing within one nautical mile of the coastline, as long as it is carried out by boats up to 10 m in length and fishing nets shorter than 1000 m (INI MPA/MMA 15/2020).

Management of other threats than fisheries: Joint regulation of IBAMA and the Chico Mendes Institute for Biodiversity Conservation (ICMBio) (02/2011) ratified in 2011 established permanent and seasonal restrictions on seismic and oil and gas exploration activities in priority areas for the conservation of the Franciscana and other aquatic mammals in Brazilian coastal waters.

Uruguay

The franciscana is considered a threatened species for Uruguay (Soutullo et al., 2013).

Many national laws and decrees have been established to regulate the management and protection of wildlife in Uruguay, including the franciscana. According to the Fauna Law N° 9481 established in 1935, the State is responsible for regulating the exploitation and conservation of wild zoological species (mammals, birds, etc.). In 1996, the law N° 16,736 (Art. 269) declared the domain and jurisdiction of the State over "the living resources existing in the territorial sea, exclusive economic zone and Uruguayan continental shelf, as well as in the adjacent areas of current or eventual national jurisdiction, in accordance with international laws and treaties."

The protection of marine mammals specifically in Uruguay is regulated by Decree N° 238/998 as follows: Art.1 prohibits the persecution, hunting, fishing, or appropriation of pinnipeds and cetaceans in islands, coasts, and waters of national jurisdiction; Art. 2 prohibits the capture, acts of aggression or intentional harassment that lead to the death of said marine mammals well as any other alteration, damage or contamination of the areas that are used for reproduction, breeding or habitat; Art. 3 establishes that marine mammals caught in fishing operations (using longlines, nets, gillnets, and other gear) must be returned to the sea immediately after the fishing gear has been hauled, causing the least harm possible. Uruguay's territorial sea and exclusive economic zone were declared a "Sanctuary of Whales and Dolphins" in 2013 (Law N° 19,128), whereby (as stated by previous decrees) the following actions are prohibited: pursuit, hunting, fishing, removal, and processing; capture and transport of live whales and dolphins (except for scientific and sanitary purposes); and the intentional captivity, injury or harassment that leads to death.

Currently, Law N° 19,175 from 2013 on Hydrobiological Resources, implemented by Decree N° 115 (from 2018), establishes that the National Directorate of Aquatic Resources (DINARA), in coordination with the National Directorate of the Environment (DINAMA), are responsible for ensuring the conservation, management, sustainable development and responsible use of marine mammals (Art. 104). DINARA's main goal is to promote the sustainable use of fishery resources. Its responsibilities include establishing the minimum landing sizes and weights of species that can be caught, determining fishing gear and methods, and establishing periods, species, and closure areas, as well as reserve areas, refuges, or nurseries. DINARA has produced National Plans of Actions (NPOAs) for reducing the incidental mortality of seabirds (Domingo et al., 2007; Jiménez et al., 2015) and for the conservation of Chondrychtians (Domingo et al., 2008; Forselledo and Domingo, 2015) in Uruguayan fisheries. However, no management plans have been established yet to protect the franciscana from fisheries interaction in Uruguay. Bycatch constitutes the main threat for the species; therefore, decreasing its incidental mortality is a priority for the country. Since 2018, DINARA has developed a research program for testing the effectiveness of mitigation measures in reducing the incidental mortality of franciscana in the two main fisheries affecting the species; the artisanal gillnet fishery (Jiménez et al., 2021a) and the industrial pair trawl fishery (Jiménez et al., 2021b). These studies have assessed the effectiveness of acoustic alarms or "pingers" under commercial fishing conditions. The tested acoustic deterrent devices used in gillnets emit relatively low-intensity sounds (145 dB re 1 μPa at 1 m) at high frequency (50–120 Khz), compared to the devices employed in the industrial trawl fishery (175 dB re 1 μPa at 1 m; 40 Khz). In both fisheries, the use of pingers has shown promising results in reducing the bycatch of franciscana (Jiménez et al., 2021a, 2021b).

Some decrees and resolutions could also have a direct impact on reducing bycatch. The coastal pair trawling fishery is not allowed to operate within five or seven nautical miles of the coastline, depending on the area (Law N° 19,175; decree 115/2018). This decree covers an extensive coastal area of the habitat used by the franciscana in Uruguay.

In the Río de la Plata, gillnets are prohibited within 300 m of the mouth of any of its tributaries (MGAP's Resolution 988/008). In addition, artisanal vessels cannot fish within 300 m of the coast between November 1 and the last day of February (DINARA's Resolution 90/2014). None of these fishing area closures were established with the direct aim of protecting franciscana. However, they might have helped to prevent bycatch of the species. This hypothesis needs to be evaluated (Passadore et al., 2020).

In 2000, the National System of Protected Natural Areas (SNAP) was created to implement national environmental protection policies and plans (Law N° 17,234). To guide policies for the conservation of biodiversity and sustainable development in Uruguay, a List of Priority Species for Conservation was established with information on their distribution and condition (Soutullo et al., 2013). Eight small cetaceans are listed, including the franciscana dolphin. The franciscana was included in the list of priority species for conservation to the SNAP in Uruguay due to its threatened status and its particular characteristics that require special attention when planning conservation and sustainable development strategies in the country (Soutullo et al., 2013).

Role of public campaigns and education

The small and elusive franciscana is barely known to the public in Argentina, Brazil, and Uruguay. Consequently, few people recognize it as the most endangered small South American cetacean, and even fewer know how to actively collaborate in its conservation. Given the challenge of justifying action to preserve something that is effectively invisible, increasing its public status may be as important for its conservation as reducing incidental mortality in fishing operations.

Since the 1990s, several institutions have been taking action to inform and educate the public about the franciscana and the threats it faces (e.g. AquaMarina, Cetáceos Uruguay, Fundación Cethus, Fundación Vida Silvestre Argentina, Gemars, GEMM-Lagos, Projeto Toninhas/Univille, Whale and Dolphin Conservation, World Wildlife Fund, among others). They have applied a variety of strategies ranging from press releases to school textbooks, public art, and brochures. Some programs have reached hundreds of academic establishments and even included environmental workshops on summer beaches. Some educational materials have also been produced, generally disseminated at the local level. However, most of these actions have been developed in the short term by local NGOs; none of the governments have included information about this cetacean, or its threats, in their official educational curricula.

Although there are proposals to support initiatives to raise awareness of the species, these efforts should be undertaken by governments (e.g., Dia da Toninha; Brazil) and given the importance they deserve. Educating the public about the conservation needs of the most endangered dolphins in these countries is still not considered a priority by regional and federal governments. Yet, the franciscana has great potential as a classroom topic to illustrate biological concepts such as top predators, keystone and umbrella species, sustainable development and species extinction, as well as social and economic issues; today many Latin American textbooks illustrate these concepts with distant examples like large African mammals.

The promotion and development of a well-structured, properly regulated, and implemented coastal franciscana whale watching activity would be a recreational and tourism strategy to make this almost unknown species visible (Failla et al., 2014). Its nonconsumptive use, minimizing potential negative effects on coastal environments, could generate additional tourist resources. It would also help the inhabitants of these coasts, who have probably never seen more than a dead entangled specimen, to enjoy land-based whale watching and become aware of the conservation needs of this and other marine species. However, conditions for whale watching franciscana are restricted to a few areas within its distribution (e.g., Babitonga Bay in Brazil, Río Negro Estuary, and Anegada Bay in Argentina).

Although the franciscana is considered one of the most difficult cetaceans to observe in the wild, because it moves slowly, performs few aerial displays, and lives in small groups in turbid waters where the color of its body camouflages perfectly with the environment, responsible whale-watching of this species by tourists is a potentially valuable, opportunity for sustainable nonconsumptive use. Several whales and dolphins in the Southwest Atlantic are the focus of boat-based whale watching, but the development of this activity from the coast (perhaps one of the activities that generate the least impact on the animals) (Hoyt and Iñíguez, 2008) is in its early days. In Patagonia, Argentina, opportunities for land-based sightings of coastal franciscanas are being developed in the Río Negro Estuary, Río Negro province, based on previous feasibility studies and following best practices (Failla et al., 2014). At the moment, the activity is encouraged by NGOs and the local government through road signs and leaflets providing information on locations and techniques for land-based whale watching. These activities are free as, to date, no local companies are involved.

Currently, in Babitonga Bay, Brazil, the feasibility of responsible boat-based whale watching is being evaluated. Based on numerous previous studies on the behavior and habitat use of this species within the bay, the first naturalist scripts and training are being prepared so that local tour operators can develop the activity in a responsible manner (M. Cremer, pers. comm.).

Today it is urgent to implement large-scale awareness-raising campaigns and formal educational programs throughout the region, as well as to evaluate the feasibility of tourism-recreational activities, related to the franciscana and its conservation needs. Sustainable medium and long-term educational and tourism projects that take into account local distinctions and the conservation needs of the franciscana throughout its range are beyond the scope of NGOs; the survival of this dolphin depends on integrative local and federal government policies that recognize its ecological and economic value.

Future challenges

The Living Planet Index for marine populations shows a decline of 49% between 1970 and 2012, based on trends in 5829 populations of 1234 mammal, bird, reptile, and fish species (Tanzer et al., 2015). Cetaceans are particularly vulnerable to human activities such as fisheries, vessel traffic, seismic exploration, oil and gas extraction, and discharges of harmful substances at sea (Avila et al., 2018), as well as to climate disruption (Simmonds and Elliot, 2009; Whitehead et al., 2008). Out of a total of 90 species of cetaceans, 22 are listed as *Critically Endangered*, 22 as *Endangered*, and 16 as *Vulnerable* in the IUCN's Red List of Endangered Species (IUCN, 2021). Franciscana is listed as *Vulnerable* by the IUCN (Zerbini et al., 2017), and the Red List of Mammals of Argentina (Denuncio et al., 2019), and as *Critically Endangered* by the National Red List of Threatened Species of Brazil (Directive MMA 444/2014).

To date, there have been eight workshops for the Coordination of the Research and Conservation of Franciscana, which have gathered all the current scientific information on the conservation status of the species throughout its range. The first was carried out at the beginning of the 1990s and the most recent in 2015. The report of this last workshop (Franciscana Consortium, 2016) was the baseline for the development of the IWC's CMP for Franciscana (IWC, 2016). This species is perhaps the cetacean for which most information is available for Argentina, Brazil, and Uruguay, although this does not mean that it is well-used: it has become clear that, despite the scientific knowledge that exists about franciscana, this information has only begun to reach policy- and decision-makers in the last decade. As this chapter makes clear, all three countries have sufficient legislation or national plans to ensure the conservation of the franciscana; there is also an international framework that provides clear and concrete conservation guidelines, at least on paper, which, if implemented and enforced, would protect the species. Notwithstanding this legal framework, annual mortality of franciscana in fishing operations continues to range from several hundred to a few thousand (e.g., Franco-Trecu et al. 2009, 2019; Ott et al., 2002; Secchi et al., 2003b) as has been observed for almost 80 years (Van Erp, 1969).

The greatest challenge for all three range countries is to identify effective mitigation measures to reduce franciscana mortality in each of their fisheries, create appropriate strategies to implement them (within the existing legal framework) and provide sufficient resources to enforce them. Each of the countries must take their own measures and allocate real resources to ensure the long term survival of the species, such as through the NPOAs of Argentina and Brazil, but it is also essential that the three countries increase their collaboration to protect these dolphins, while there is still time to do so. For example, while the particularities of their fisheries differ, national experts should collaborate to identify practical commonalities and best practices, including in technical and operational mitigation methods to reduce bycatch and methods to ensure compliance. This cooperation could be achieved within the framework of IWC CMP for Franciscana.

As shown in chapter 18 of this book and briefly mentioned above, there are several MPAs that feature the franciscana within their territory, but none have been created and established with the specific purpose of protecting it. Moreover, recently developed MPA management plans do not include the species. There is an urgent need to establish protection measures for franciscana in all identified FMAs, particularly those with a more restricted distribution or under greater threat, such as the Río Negro Estuary in Argentina (Failla et al., 2012) and Babitonga Bay in Brazil (Cremer and Simões-Lopes, 2005). Both areas have been identified by the IWC CMP for franciscana as areas of importance for the species in order to be declared MPAs.

Although efforts are being made to devise effective measures for the conservation of the franciscana, public policies are less likely to be effective if they are imposed on the artisanal fishing communities concerned without consultation and taking into account their specific social and economic vulnerabilities. Efforts need to be made to elaborate conservation public policies in a more participatory way and consider local differences, adjusting these measures according to the social and economic needs of these fishing communities. For more details on this issue see Chapter 16 of this book.

Another weakness of conservation efforts is the ignorance of the general public. The best way to make a lasting impact on public policy is to change public opinion; when people's beliefs change, politicians and policies change to reflect those evolving norms. Conversely, if the public does not know what a franciscana is, let alone the threats it faces, they will not seek, or support, conservation measures to protect it. This leads to situations where ignorance has mainly compromised the animals involved as happened in Santa Teresita, Buenos Aires province in February 2016, when a group of tourists took a franciscana calf out of the water to take pictures with it, resulting in its death (Diario La Nación, 2016). While awareness-raising efforts by academic researchers and civil society are valuable and important, they are more effective when coordinated, especially nationally and regionally. As established in the IWC's CMP for franciscana, it is imperative that a regional awareness-raising program is developed jointly by the three countries in the corresponding languages.

In conclusion, the following statement by Sir David Attenborough sums up where we are and the great opportunity we have: "we are at a unique stage in our history. Never before have we had such an awareness of what we are doing to the planet and never before have we had the power to do something about that. Surely we have a responsibility to care for our blue planet. The future of humanity, and indeed all life on Earth, now depends on us."

References

Avila, I.C., Kaschner, K., Dormann, C.F., 2018. Current global risks to marine mammals: talking stock of the threats. Biol. Conserv 221, 44–58.

Bordino, P., Kraus, S., Albareda, D., Fazio, A., Palmerio, A., Mendez, M., Botta, S., 2002. Reducing incidental mortality of Franciscana dolphin, Pontoporia blainvillei, with acoustic warning devices attached to fishing nets. Mar. Mammal Sc. 18, 833–842.

Bordino, P., Kraus, S., Albareda, D., Baldwin, K., 2004. Acoustic devices help to reduce incidental mortality of the Franciscana dolphin (Pontoporia blainvillei) in coastal gillnets, SC56SM12 Document presented at the 56th IWC Meeting. Sorrento.

Bordino, P., Mackay, A.I., Werner, T.B., Northridge, S.P., Read, A.J., 2013. Franciscana bycatch is not reduced by acoustically reflective or physically stiffened gillnets. Endangered Species Res. 21, 1–12.

CMS, 2020. Appendices I and II of the convention of migratory species of wild animals (CMS), 13th Conference of the Parties. Effective May 22, 2020, p. 16.

Consejo Federal Pesquero, 2016. Plan de Acción Nacional para Reducir la Interacción de Mamíferos Marinos con pesquerías en la República Argentina. Buenos Aires, p. 168 1ra Ed.

Costa Urrutia, P., Abud, C., Secchi, E.R., Lessa, E.P, 2012. Population genetic structure and social kin associations of Franciscana Dolphin, Pontoporia blainvillei. J. Heredity 103, 92–102.

Cremer, M.J., Simões-Lopes, P.C., 2005. The occurrence of Pontoporia blainvillei (Gervais & d'Orbigny) (Cetacea, Pontoporiidae) in an estuarine area in southern Brazil. Rev. Bras. Zool. 22, 717–723.

Crespo, E.A., Harris, G., González, R., 1998. Group size and distributional range of the Franciscana, Pontoporia blainvillei. Mar. Mammal Sci. 14 (4), 845–849.

Crespo, E.A., 2002. Franciscana Dolphin, Pontoporia blainvillei. In: Perrin, W.F., Würsig, B., Thewissen, J.G.M. (Eds.), Encyclopedia of Marine Mammals. Academic Press, pp. 482–485.

Cunha, H.A., Bisi, T.L., Bertozzi, C.P., Dias, C.P., Santos-Neto, E.B., Manhães, B.M.R., Oliviera-Ferreira, N., Montanini, G.N., Ikeda, J., Carvalho, R.R., Azevedo, A.F., Lailson-Brito, J., 2020a. Population differentiation of endangered Franciscanas in Southeastern Brazil: new genetic, contaminant and stable isotope data support subdivision of FMAII, Paper SC/68b/SDDNA5 presented to the 68B International Whaling Commission's Scientific Committee. p. 11.

Cunha, H.A., Dias, C.P., Alvarenga, L.C., Wells, R.S., Cremer, M.J., 2020b. Microscale population structure and kinship analyses suggest philopatry of both sexes in franciscanas (Pontoporia blainvillei). Paper SC/68b/SDDNA4 presented to the International Whaling. Commission's Scientific Committee 23.

Cunha, HA, Gariboldi, MC, Mendez, M, Secchi, ER, Oliveira, LR, Ott, P, Torres-Florez, JP, Farro, APC, 2020c. Review on franciscana stock structure and Franciscana Management Areas (FMAs). Paper SC/68b/SDDNA7 presented to the International Whaling. Commission's Scientific Committee 9.

Cunha, H.A., Medeiros, B.V., Barbosa, L., Cremer, M., Marigo, J., Lailson Brito Jr, J., Azevedo, A., Sole-Cava, A.M., 2014. Population structure of the endangered franciscana dolphin (Pontoporia blainvillei): reassessing management units. PLoS One 9 (1), e85633.

Denuncio, P., Bastida, R., Dassis, M., Giardino, G., Gerpe, M., Rodríguez, D., 2011. Plastic ingestion in Franciscana dolphins, Pontoporia blainvillei (Gervais and d'Orbigny, 1844), from Argentina. Mar. Pollut. Bull. 62 (8), 1836–1841.

Denuncio, P. E., Paso Viola, N., Cáceres-Saez, I., Cappozzo, H.L., Rodríguez, D., Mandiola, A. 2019. Pontoporia blainvillei. In: SAyDS–SAREM (eds.) Categorización 2019 de los mamíferos de Argentina según su riesgo de extinción. Lista Roja de los mamíferos de Argentina. Digital version: http://cma.sarem.org.ar. Accessed March 29, 2022.

Diario La Nación. 2016. https://www.lanacion.com.ar/sociedad/santa-teresita-sacaron-un-delfin-del-mar-para-tomarse-una-foto-y-lo-mataron-por-accidente-nid1871616/. Accessed March 29, 2022.

Di Beneditto, A., Ramos, R., 2014. Marine debris ingestion by coastal dolphins: What drives differences between sympatric species? Mar. Pollut. Bull. 83 (1), 298–301.

Domingo, A., Jiménez, S., Passadore, C., 2007. Plan de Acción Nacional para Reducir la Captura Incidental de Aves Marinas en las Pesquerías Uruguayas. DINARA, Montevideo v. 1, p. 76, ISBN: 9974-563-35-3.

Domingo, A., Forselledo, R., Miller, P., Passadore, C., 2008. Plan de Acción Nacional para la Conservación de Condrictios en las Pesquerías Uruguayas. DINARA, Montevideo v. 1, p. 88, ISBN: 9974-563-44-5.

Failla, M., Seijas, V.A., Espósito, R., Iníguez, M., 2012. Franciscana dolphins, Pontoporia blainvillei, of the Río Negro Estuary, Patagonia, Argentina. Marine Biodiversity Records 5, 1–5.

Failla, M., Gasparrou, C., Melcón, M., Reyes, V., Seijas, V., Iñíguez Bessega, M., 2014. Potencialidad del avistaje costero responsable de delfines en el Estuario del Rio Negro,

Patagonia, Argentina Compiladores. In: Navarro y S. Ferrari, V. (Ed.), Turismo y recursos naturales: los recursos naturales como base del desarrollo turístico local, importancia de la conservación y la gestión para la sustentabilidad. Universidad Nacional de la Patagonia Austral. E-Book, Río Gallegos, Argentina, pp. 242–255 1ra ed2013.

Forselledo, R., Domingo, A. 2015. Plan de Acción Nacional para la Conservación de Condrictios en las Pesquerías Uruguayas, 2015. In: Domingo A, Forselledo R, Jiménez S (Eds.) Revisión de Planes de Acción Nacional para la Conservación de Aves Marinas y Condrictios en las Pesquerías Uruguayas. Dirección Nacional de Recursos Acuáticos, Montevideo. pp. 81–152 .

Franciscana Consortium, 2016. Report of the VIII Workshop for the Research and Conservation of the Franciscana (Pontoporia blainvillei), Presented at the 66b Scientific Committee of the International Whaling Commission. SC/66b/SM/05. Bled, Slovenia, 26.

Franco-Trecu, V., Costa, P., Abud, C., Dimitriadis, C., Laporta, P., Passadore, C., Szephegyi, M., 2009. By-catch of Franciscana Pontoporia blainvillei in Uruguayan artisanal gillnet fisheries: an evaluation after a twelve-year gap in data collection. Lat. J. of Aq. Mammals. 7, 11–22. http://dx.doi.org/10.5597/lajam00129.

Franco-Trecu, V., Szephegyi, M.N., Doño, F., Forselledo, R., Reyes, F., Passadore, C., Crespo, E.A., Inchausti, P., 2019. Marine mammal bycatch by the industrial bottom trawl fishery at the Río de la Plata Estuary and the adjacent Atlantic Ocean. Lat. Am. J. Aquat. Res. 47, 89–101.

Gariboldi, M.C., Túnez, J.I., Dejean, C.B., Failla, M., Vitullo, A.D., Negri, M.F., Cappozzo, H.L, 2015. Population genetics of Franciscana Dolphins (Pontoporia blainvillei): Introducing a new population from the southern edge of their distribution. PLoS One 10, e0132854.

Gariboldi, M.C., Túnez, J.I., Failla, M., Hevia, M., Panebianco, M.V., Paso Viola, M.N., Vitullo, A.D., Cappozzo, H.L., 2016. Patterns of population structure at microsatellite and mitochondrial DNA markers in the Franciscana dolphin (Pontoporia blainvillei). Ecol. Evol. 6, 8764–8776.

Hevia, M., Iñíguez Bessega, M.A., Reyes Reyes, M.V. and Zuazquita, E. (In Press). A review of marine protected areas in Argentina and their overlap with current cetacean distribution. Report for OceanCare, Switzerland. p. 69.

Hoyt, E., Iñíguez, M. 2008. El Estado del Avistamiento de Cetáceos en América Latina. WDCS, Chippenham, UK; IFAW, East Falmouth, EE.UU.; y Global Ocean, London, p. 60.

IBAMA, 1997. Mamíferos Aquáticos do Brasil: Plano de Ação. Brasília: IBAMA, p. 80.

ICMBio 2010. Plano de ação nacional para a conservação do pequeno cetáceo Toninha: Pontoporia blainvillei/Ana Paula Madeira Di Beneditto … [et al.]; organizadores Claudia Cavalcante Rocha Campos, Daniel Schiavon Danilewicz, Salvatore Siciliano. – Brasília: Instituto Chico Mendes de Conservação da Biodiversidade, ICMBio, 2010. p. 76.

IWC, 2005. Report of the sub-committee on small cetaceans. Report of the Scientific Committee. Annex L, 307–317.

IWC, 2014, Chair's Report of the 65th Meeting. Portoroz, Slovenia, 382p September 2014.

IWC, 2016. A conservation management plan for franciscana (Pontoporia blainvillei), Presented by Argentina, Brazil and Uruguay. IWC/66/CC11, p. 22.

IWC, 2019. Report of the sub-committee on small cetaceans. Report of the Scientific Committee. Appendix 2. Proposed process for the review of the status of the Franciscana (Pontoporia blainvillei) by the IWC Scientific Committee. pp. 339-340.

IUCN. 2021. The IUCN red list of threatened species. Version 2020-3. www.iucnredlist.org (accessed March 22, 2021)

Jiménez, S, Pin, O., Domingo, A. 2015. Plan de Acción Nacional para Reducir la Captura Incidental de Aves Marinas en las Pesquerías Uruguayas, 2015. In: Domingo

A., Forselledo R., Jiménez S. (Eds.) Revisión de Planes de Acción Nacional para la Conservación de Aves Marinas y Condrictios en las Pesquerías Uruguayas. Dirección Nacional de Recursos Acuáticos, Montevideo. pp. 11–79.

Jiménez, S., Laporta, M., Forselledo, R., Fabiano, G., Páez, E., Pereyra, I., Silveira, S., Domingo, A., 2021. Effectiveness of pingers to reduce the bycatch of Franciscana (Pontoporia blainvillei) in the Uruguayan artisanal gillnet fishery, Presented at the 68C Scientific Committee of the International Whaling Commission. SC/68C/HIM/10, p. 10.

Jiménez, S., Forselledo, R., Laporta, M., Páez, E., Domingo, A., 2021. Assessing the effectiveness of dolphin antipredation pingers to reduce the bycatch of Franciscana (Pontoporia blainvillei) in demersal pair trawl fishing, Presented at the 68C Scientific Committee of the International Whaling Commission. SC/68C/HIM/11, p. 3.

Méndez, M., Rosenbaum, H.C., Bordino, P., 2008. Conservation genetics of the Franciscana dolphin in Northern Argentina: population structure, by-catch impacts, and management implications. Conservation Genetics 9, 419–435.

Méndez, M., Rosenbaum, H.C., Subramaniam, A., Yackulic, C., Bordino, P., 2010. Isolation by environmental distance in mobile marine species: molecular ecology of Franciscana dolphins at their southern range. Molecular Ecol. 19, 2212–2228.

Ministério do Meio Ambiente – Portaria MMA n 444, de 17 de dezembro de 2014. 2014. https://www.icmbio.gov.br/cma/images/stories/Legislacao/Portarias/p_mma_444_2014_lista_espécies_ameçadas_extinção.pdf. Accessed March 29, 2022.

Ministério do Meio ambiente, 2018. Livro Vermelho da Fauna Brasileira Ameaçada de Extinção:Volume I /– 1. ed. – Brasília, DF: ICMBio/MMA, 2018. p. 492: il., gráfs., tabs.

Ott, P.H., Secchi, E.R., Moreno, I.B., Danilewicz, D., Crespo, E.A., Bordino, P., Ramos, R., Di Beneditto, A.P., Bertozzi, C., Bastida, R., Zanelatto, R., Perez, J., Kinas, P.G., 2002. Report of the working group on fishery interactions. Lat. Am. J. Aquat. Mamm. 1 (1), 55–64.

Passadore, C., Jiménez, S., Laporta, M., Forselledo, R., Ríos, M., Domingo, A., Fabiano, G., Lezama, C., Páez, E., Pereyra, I., Pereira, A., Ponce de León, A., Silveira, S., Szephegyi, M., 2020. Management and conservation of Franciscana (Pontoporia blainvillei) in Uruguay, Presented at the 68B Annual Meeting of the International Whaling Commission Scientific Committee. SC/68B/CMP/02, p. 10.

Secchi, E.R., Danilewicz, D., Ott, P.H., 2003a. Applying the phylogeographic concept to identify Franciscana dolphin stocks: implications to meet management objectives. J. Cetacean Res. Manage. 5 (1), 61–68.

Secchi, E.R., Ott, P.H., Danilewicz, D.S., 2003. Effects of fishing bycatch and conservation status of the Franciscana dolphin, Pontoporia blainvillei. In: Gales, N., Hindell, M., Kirkwood, R. (Eds.), Marine Mammals: Fisheries, Tourism and Management Issues. CSIRO Publishing, Collingwood, pp. 174–191.

Secchi, E.R., 2010. Life History and Ecology of Franciscana, Pontoporia blainvillei (Cetacea, Pontoporiidae) Biology,. In: Ruiz-García, M., Shostell, J. (Eds.), Evolution and Conservation of River Dolphins within South America and Asia. Nova Science, New York, NY, pp. 301–322.

Secchi, E.R., 2010. Review on the threats and conservation status of Franciscana, Pontoporia blainvillei (Cetacea, Pontoporiidae). In: Ruiz-García, M., Shostell, J. (Eds.), Biology, Evolution and Conservation of River Dolphins within South America and Asia. Nova Science, New York, NY, pp. 323–339.

Siciliano, S., Di Beneditto, A.P., Ramos, R.M.A., 2002. A toninha, Pontoporia blainvillei (Gervais and d'Orbigny, 1844) (Mammalia, Cetacea, Pontoporiidae), nos estados do Rio de Janeiro e Espírito Santo, costa sudeste do Brasil: caracterizaçoes dos habitats e fatores de isolamenta das populaçoes. Boletim do Museu Nacional, Zoologica 476, 1–15.

Siciliano, S., Moura, J.F., Secco, H.K.C., 2015. Considerações sobre a distribuição da toninha (Pontoporia blainvillei, Gervais & d'Orbigny, 1844) na costa centro-norte do estado do Rio de Janeiro, Brasil. Pg. 112-117. In: Ott, P.H., Domit, C., Siciliano, S., Flores,

P.A.C. (Eds.), Memórias do VII Workshop para a coordenação de pesquisa e conservação de Pontoporia blainvillei (Gervais & d'Orbigny, 1844), 22-24 de outubro de 2010. Florianópolis, Porto Alegre, Brasil, p. 163.

Simmonds, M.P., Eliott, W.J, 2009. Climate change and cetaceans: concerns and recent developments. J. Marine Biol. Assoc. United Kingdom 89 (1), 203–210.

Soutullo, A., Clavijo, C., Martínez-Lanfranco, J.A, 2013. Especies prioritarias para la conservación en Uruguay. Vertebrados, moluscos continentales y plantas vasculares. SNAP/DINAMA/MVOTMA y DICYT/MEC, Montevideo, p. 222.

Tanzer, J., Phua, C., Lawrence, A., Gonzales, A., Roxburgh, T., Gamblin, P., 2015. Living blue planet report: species, habitats and human well-being. WWF and Zoological Society of London. https://www.worldwildlife.org/publications/living-blue-planet-report-2015. (Accessed 29 March 2022).

Van Erp, I., 1969. In quest of the La Plata dolphin. Pacific Discovery 22, 18–24.

Whitehead, H., McGill, B., Worm, B., 2008. Diversity of deep-water cetaceans in relation to temperature: implications for ocean warming. Ecology Letters 11, 1198–1207.

Zerbini, A.N., Secchi, E., Crespo, E., Danilewicz, D. & Reeves, R. 2017. Pontoporia blainvillei (errata version published in 2018). The IUCN Red List of Threatened Species 2017: e.T17978A123792204. doi:10.2305/IUCN.UK.2017-3.RLTS.T17978A50371075.en. Downloaded March 22, 2021.

Index

Page numbers followed by "*f*" and "*t*" indicate, figures and tables respectively.

Made in the USA
Monee, IL
13 September 2026

62977351R20273